PLACE NAMES OF ALBERTA
VOLUME I

This book has been published with the help of a grant from
the Alberta Foundation for the Literary Arts.

Place Names of Alberta

Mountains, Mountain Parks and Foothills

Edited and Introduced by
Aphrodite Karamitsanis

Alberta Culture and Multiculturalism
and
Friends of Geographical Names of Alberta Society
and
University of Calgary Press

© 1991 Alberta Culture and Multiculturalism. All rights reserved.

ISBN 0-919813-91-7 (set)
ISBN 0-919813-73-9 (v.1)

University of Calgary Press
2500 University Drive N.W.
Calgary, Alberta, Canada T2N 1N4

Canadian Cataloguing in Publication Data

Main entry under title:

Place names of Alberta

Contents: v. 1. Mountains, mountain parks and foothills.

ISBN 0-919813-73-9

1. Names, Geographical - Alberta. 2. Alberta
- History, Local. I. Karamitsanis, Aphrodite, 1961-
FC3656.P62 1990   917.123′003   C90-091236-7
F1075.4.P62 1990

Cover photo: Totem Tower

Cover design and layout by Ouchi Design
Typography by Foy and Flanigan
Printed and bound in Canada by The Jasper Printing Group Ltd.

∞ This book is printed on acid-free paper.

For my father
Vlasios Panayiotis Karamitsanis

TABLE OF CONTENTS

ACKNOWLEDGEMENTS

The creation of a volume such as this is always an assisted project. Grateful acknowledgement is extended to Maryalice Stewart, Randy Freeman, Wendy Lefsrud, Tracey Harrison, Tom Monto, Sean Moir, Stephanie Muzyka, and Tom Jopling for their work and helpful comments. Ted Hart, from the Whyte Museum of the Canadian Rockies was especially helpful in pointing out deficiencies and augmenting many entries with his unique expertise. Penelope White, who devoted much of her free time to the painstaking search for and selection of photographs, deserves special mention for her volunteer efforts. This study also owes a great deal to the pioneering work of Eric Holmgren and Marie Dorsey and to the advice and support of Omar Broughton and Heather-Belle Dowling. A special thank you is extended to Lynn Pong for her tireless patience in typing and retyping the manuscript. The University of Calgary Press and its very capable staff, especially the Director, Linda Cameron, John King and Sharon Boyle, who provided the smooth transition from manuscript to publication. For their support, I thank Frits Pannekoek, Les Hurt, Carl Betke, my family, and all the "Friends."

INTRODUCTION

Romeo Montague was not, of course, thinking of the toponymy of Alberta when he asked his famous question, "What's in a name?" Alberta's place names have become so much a part of the landscape that we seldom stop to ask this question. Yet, toponymy, the study of place names, reveals a great deal about the fascinating history and unique cultural heritage of any populated region. Toponymic research is primarily concerned with the origins and meanings of place names. A geographical name unlocks a valuable store of information, not only as a key component of reliable maps and charts, but also as a reflection of culture and heritage and a clue to groups or individuals that have had an impact in an area.

Mutual sharing of information remains the focus of Alberta's Geographical Names Programme, and the purpose of this publication is twofold. Over the course of several years, much archival and field research has been conducted in this area of the province in order to provide accurate data for public use. Here, that information is made accessible in a major collection. So comprehensive is the range of information for so many names, however, that the refinement process is continual. Beyond presenting all the fascinating stories, therefore this book also invites those who may know more about any individual name to assist in the toponymic process by informing the Geographical Names Programme. Individuals with precious and sometimes guarded information are a vital resource to whom we owe great respect and from whom cooperation and help would be very much appreciated. Information may easily be entered by contacting the Geographical Names Programme, Alberta Culture and Multiculturalism, 8820 – 112 Street, Edmonton, Alberta, Canada T6G 2P8.

The formal process of geographical naming is a lengthy one, based on specific naming principles and procedures. It should be noted that the first function of geographical names is to ensure accurate identification of places. Recognition of the importance of rendering a correct spelling, use, and translation of a geographical name, as well as providing a generic definition of the type of feature it describes, led to the creation of the Geographic Board of Canada through Order-in-Council in 1897. Dedicated to the standardization of principles and procedures to be followed in naming geographical features, the Board has changed its centralized approach very little over the years. A name adjustment in 1948 to the Canadian Board on Geographical Names left provincial representation minimal. In 1961, however, the responsibility for geographical naming was left up to individual provinces and territories, though the necessity for the coordination of activities at a national level remained. Creation of the Canadian Permanent Committee on Geographical Names provided the opportunity for continued liaison between federal officials and the naming authorities of each province and territory, but provincial agencies like Alberta's Geographical Names Programme now research and maintain information on proposed names, name changes, and names previously established. Alberta's approach has drawn increasing attention to the cultural dimensions of the province's geographical names.

The accumulated Alberta research has shown that the connection between geographical naming and the development of an area is very close: as individuals develop, survey, and map a region, the need to give places specific names increases. Indeed, geographical naming both affects and reflects the culture of a place. Surveyors, chartmakers, cartographers, and businessmen all require familiar and common points of reference in order to continue their work, but in the choice of names they identify their landscape with their particular culture.

The specific area that this volume addresses, an area which contains the mountains, mountain parks and foothills of Alberta, has received names during exploration patterns unique to this kind of terrain. The study area is outlined on page xxv. The map located on this page is designed to aid the reader in locating in a general way some of the features, relative to nearby populated places, as they are described within the location information in the text. Many of the individuals and survey parties were commemorated in the names of various geographical features, and many of these names have survived over the decades. It is not surprising then to find the names of mountain climbers at the peaks of mountains; nor is it strange to have a name attached to an area because of a specific incident that occurred there, as at the Kicking Horse Pass (refer to page 131). The impact of the Hudson's Bay and North West Companies has been noted in names such as Simpson Pass (page 225), Mount Fraser (page 93), and McGillivray Ridge (page 159). This volume is a collection of names paying tribute to those individuals who contributed so much of their time and effort to a variety of enterprises in the area.

The history of Alberta is multifaceted, culturally and regionally. The mountain parks and foothills region of the province have a historically important and interesting heritage. The naming legacy left by the early fur traders such as Peter Fidler, the Palliser Expedition under John Palliser, the Dominion Land Surveyors who were charged with charting these vast expanses of land, and mountain climbers who recognized the majesty of the peaks and challenged them, reflects not just the men and their reasons for being in the West, but also the values of the societies from whence they emerged. Sir Sandford Fleming, a railway surveyor himself, noted the importance of retaining native heritage through toponymy. Through journals, diaries, and historical records left by these early antecedents, people of today and tomorrow have access to knowledge of the past and may come to appreciate the true significance to be found in the experiences of early Albertans.

The first white man to set foot in Alberta arrived on 11 September 1754. Anthony Henday, a fur trader associated with the Hudson's Bay Company, was sent west to make friends with the natives and report on their daily activities.

Henday journeyed from Hudson Bay to what has since become known as Rocky Mountain House and was the first white man to see the main chain of the Rocky Mountains. David Thompson carried out his own surveys and explorations beginning in 1787. He became one of Canada's greatest surveyors and also had an active interest in the native peoples he met and stayed with on his journeys. Thompson gained and passed down valuable insights into Indian culture and customs of that time. During 1787, James Gaddy, of the Hudson's Bay Company, settled and developed some of the south-west portion of Alberta. Peter Fidler followed in 1792. He, like Thompson, was a close observer of the natives and was a trained surveyor whose notes and journals are valuable reference materials. The North West Company, represented by John MacDonald, James Hughes, and twenty others, established a post north of Fort Saskatchewan in 1795. In 1799, the North West Company established Rocky Mountain Post, and the Hudson's Bay Company answered by establishing Acton House nearby. Sir George Simpson, Governor of the Hudson's Bay Company, traversed parts of western Canada with his Indian guide, Piché (since corrupted to Peechee), in 1841. In 1845, Father Pierre de Smet pursued his missionary calling through the Rockies from Oregon, leaving his mark and subsequently relating his impressions.

One of the first major exploration expeditions to the West occurred between 1857 and 1860 by order of the British Imperial Commission. The purpose of this expedition was to ascertain settlement potential, to explore the mountains for a possible traverse south of the Athabasca Pass, and to further the east-west connection of the Dominion of Canada. This mammoth task was headed by Captain John Palliser, the son of an Irish nobleman.

The Palliser Expedition, sponsored by the Colonial Office, was scheduled to last two years, but because of the vastness of the territory to be covered and the requirements set forward by both the British Government and the Royal Geographical Society, its duration was increased to three years — from 1857 to 1860. The territory to be covered was situated between parallels 49 degrees and 53 degrees north latitude and 100 degrees and 115 degrees west longitude. The expedition successfully documented the existence of twelve Rocky Mountain passes south of the Athabasca Pass.

Palliser was accompanied on this trip by Dr. James Hector, geologist and medical doctor, Mr. Eugène Bourgeau, botanist, Lieutenant Thomas Blakiston, astronomer, and Mr. John W. Sullivan, secretary. These men were expert in their given fields, but were adventurers at heart. Throughout the expedition, Hector and Bourgeau, in particular, would take side trips trying to cover as much territory as they could in as short a time as possible. On one of these side trips, Hector was kicked in the stomach by a horse while camped by a river that the group named Kicking Horse River. The river flows from the Kicking Horse Pass, where the Canadian Pacific Railway and Trans Canada Highway run today.

While Hector was exploring the area up the Bow River, Palliser and Sullivan went to look for the pass that he had discussed with James Sinclair ten years earlier. Palliser believed that the entrance to this pass was to be found via a tributary of the Bow River. He named the tributary the Kananaskis (page 129), after a native who had been struck with an axe and made a complete recovery. The trip through the Kananaskis Pass took them more than 1,740 metres above sea level, 510 metres above the old Hudson's Bay Company fort which was called "Peigan Post" but which was abandoned in 1834, and 645 metres above the Kootenay River. Palliser had no real proof that this was Sinclair's pass, but he knew he had crossed the mountains with horses and that the Columbia lay a short distance off through fallen timber, ravines, rocks, and gullies. Lack of time prevented further exploration.

In August of 1858, while Palliser was identifying the border between Canada and the United States for the Boundary Commission, Hector, Bourgeau, and Blakiston set out to ford the Bow River leading to the Kananaskis Pass. They traversed the territory of modern-day Banff en route to the region south of a craggy knob called the Devil's Head (page 69).

This was also the area known to the natives as "the Mountain-where-the-water-falls" or, as we know it, Cascade Mountain (page 42). Further on toward the pass that Sir George Simpson had discovered, Simpson Pass (page 225), they discovered the mountain named by Hector for Bourgeau. Through the valley of the Bow River, the group observed what Hector called Castle Mountain (page 43), since the rock formations gave it the appearance of towers and bastions of a castle.

Hector and his followers continued on through a wide valley then called "Little Vermilion," now unnamed, which contained Altrude Creek (page 5). The valley took them about 162 metres above the Bow River and about 1500 metres above sea level. The creek flows east into the Bow and Saskatchewan rivers. To the south-east, Hector named a large snowcapped mountain for John Ball (page 13), Palliser's friend and former Under-Secretary of State for the Colonies.

In September of 1858, Hector was determined to make his way to the North Saskatchewan River by travelling up the Bow River to its source. This trip took his party past the Waputik Mountains and Hector Lake (page 112) (named later by Dawson). They passed Bow Lake (page 28), which was fed from the Wapta Icefield. In the distance to the east was a huge mountain, which Hector called Mount Murchison (page 174), after his patron and the head of the Royal Geological Society. As Hector and his party progressed, they named mountains after people they had known on the trips, such as Henry John Moberly (1835-1931), a native of Penetanguishene, Upper Canada, a clerk in the Hudson's Bay Company and Chief Factor at Jasper House from 1855-1861, and W.J. Christie, a Chief Factor at Edmonton. Their trip up the valley of the Pipestone Creek led them to a mountain resembling a large tooth, which they called Molar Mountain (page 168).

The trip was long and arduous for all concerned, but the importance of their discoveries more than compensated for the hardships. On 22 October 1859, Palliser wrote the new Secretary of State of the Colonies indicating that the expedition was completed.

With Canadian Confederation and the 1870 acquisition of the West came increased Canadian interest in development, beginning with a transcontinental railway. The government launched a full-scale search for a southern route through the "Great Mountain Barrier," or what we call the Rocky Mountains. The Canadian Pacific Survey (C.P.S.) commenced in the summer of 1871, under the supervision of the government's chief engineer, Sandford Fleming. Fleming held that position until he retired in 1880, when the government turned the Canadian Pacific Railway (C.P.R.) project over to a private syndicate.

Sandford Fleming (1827-1915) had emigrated to Canada in 1845 after studying science and engineering in Scotland. He participated in and supervised several railway projects before assuming responsibility for the C.P.R. in 1871. Fleming quickly proved himself to be Canada's pre-eminent railway surveyor and construction engineer and later attained worldwide recognition as a distinguished inventor and scientist. He developed a universal system of standard time, was in charge of laying the Pacific Cable to Australia in 1902, and designed Canada's first postage stamp, the three penny beaver. In 1897, he was created a Knight Commander of St. Michael and St. George (K.C.M.G.).

The C.P.S. was an enormous undertaking and brought to the fore a wealth of information regarding vast tracts of uncharted territory throughout the Dominion of Canada. Of particular interest here is the information derived from those explorations carried out in the mountain regions of Alberta during the first six years of the survey. As far as can be determined, Fleming took no interest in the naming of geographical features. Fleming neither named anything after himself, nor has his name been affixed in a commemorative sense to any feature, at least officially, in Alberta. There are many features in British Columbia named for Sir Sandford Fleming, namely, Sandford Island (1965), Sir Sandford Glacier (1912), Mount Sir Sandford (1912), Sir Sandford Range (1912), Sir Sandford Pass (1965), Fleming Peak (1910), Fleming Bay (1879), and Fleming Island (1934); however, there is not one

feature that commemorates Sir Sandford Fleming in the study area covered in this volume. He stood out as an exception among the many explorers, surveyors, and alpine enthusiasts who readily and indiscriminately "donated" their names, or those of their fellow countrymen, in their search for some form of immortality. From Fleming's "Ocean to Ocean" trek in 1872 and steady C.P.S. work, however, came maps, reports, and diaries that have since provided vital information for generations of alpine enthusiasts, explorers, and surveyors about the territory beyond the realm of contemporary Canadian civilization.

Having dealt with a multitude of administrative details, gathering information, and dispensing supplies, Fleming was able to dispatch his first surveying parties in June 1871. By mid-summer, almost 800 men were in the field, and this number would rise to 2,000 the following year, including engineers, surveyors, levellers, assistants, axemen, boatmen, and packers. Unfortunately, a good number of these men were incompetent political appointees and party hacks, for whom work had to be created. This was one of the greatest problems Fleming faced on a recurring basis. After the first year in the field, he reported that it was almost impossible to obtain "the class of men required," resulting in considerable delays and difficulty in completing the work. Although this remained a major concern for Fleming, land to the extent of 74,000 km was reconnoitred from 1871 to 1876. A large percentage of this total was in the mountain regions of Alberta, as the search for the best route through the mountains continued. During this same period, an army of surveyors, engineers, and assistants charted, foot by foot, a total of 19,300 km, including the route Fleming had tentatively selected as early as 1871.

Fleming's unusual resistance to naming features throughout Alberta's mountains and foothill regions remains unexplained. Whether Fleming was under orders from the federal government not to assume the authority to "lend" his name to features, or whether he actually held certain beliefs and principles on the subject, is a purely speculative question.

A passing reference is made to Sir Sandford Fleming's humble nature in a letter dated 2 February 1909 from A.O. Wheeler (Topographer) to Mr. Whitcher (Secretary, Geographic Board of Canada): ". . . with regard to Mount Sandford, I may say that I gathered from Sir Sandford Fleming that he preferred not having the "Sir" put in, although personally, I think it would make it more definite in whose honour the peak was named, and would be a greater compliment to Sir Sandford Fleming. It is probably due to his modesty that he wished it left out."[1] It is also possible that Fleming never even thought of naming features or of the implications of such an act, for there is no direct mention of the issue in any of his reports, speeches, or letters.

The only reference applicable to this issue is a short, indirect passage from the diary Fleming kept on his 1872 trek. The statement is actually that of Valad, one of Fleming's principal Métis guides who travelled with the expedition from Lac St. Anne to the ford on the Fraser River. The comment was made as Fleming's party discussed their route up the valley of the "River Myette" [sic]. Fleming noted Valad's words, "He mentioned the old local titles of the mountains on this side, but every passer-by thinks that he has a right to give his own and his friends' names to them over again." Fleming may have noted this conversation in his diary because he was concerned with the naming of geographical features, or out of dismay over the white man's disregard for native heritage.

Fleming's journey preceded the creation of the Geographic Board of Canada (now the Canadian Permanent Committee on Geographical Names), by more than twenty-five years. Not until 1897 was a governing body deemed necessary and were guidelines established to put a halt to what the Geographic Board considered to be abuses. One of these was the replacement of well-established local names, many of which derived from native culture. Interestingly

enough, the concern expressed by Valad over 100 years ago has become one of the prominent guiding forces in the development of the principles and procedures of the Canadian Permanent Committee on Geographical Names.

Sir Sandford Fleming's diary, reports, and maps from his journeys, and those of the C.P.S., provided vital topographical, geological, and geographical information, including distances between features, for generations of explorers and surveyors that followed. Among them were the men of the Dominion Land Survey (D.L.S.) created in 1869. Within a few decades, the future provinces of Manitoba, Saskatchewan, and Alberta were almost completely surveyed into townships and sections. The survey of the boundary between British Columbia and Alberta was begun in 1913 and continued through 1924 with the careful placing of cairns, brass bolts, and officially surveyed monuments marking the division of the territory. People such as J.N. Wallace, the Dominion Land Survey Commissioner, R.W. Cautley, Alberta's Commissioner, and British Columbia's A.O. Wheeler emerged as new actors in Alberta's history.

The eastern boundary of British Columbia was established as early as 1863 by an Act of the British Parliament. The eastern boundary of the British colony, later the Canadian province, from "the Boundary of the United States Northwards," was declared to be the "Rocky Mountains" and "the one hundred and twentieth Meridian of West Longitude." Because the border between Alberta and British Columbia was made up of a natural topographical feature and a straight unalterable line, there was no urgency to survey and mark the boundary on the ground, even after British Columbia joined Canada in 1871. However, in the 1910s, several factors made surveying necessary and advisable.

The discovery of valuable coal deposits required licensing and the collection of royalties, both of which were made difficult by the lack of an officially marked border. The growing value of the forest reserves required strict determination of the jurisdiction of surveyors, timber lessees, fire wardens, and game guardians. A need for knowledge of the passes suitable for wagon roads and pack

---

[1]Geographic Board of Canada Microfiche, Geographical Names Programme, 0133, 3C10.

trails also increased the necessity of a survey. In addition, geographers needed precise border locations to draw accurate maps.

In 1913, commissioners were appointed to represent the two provinces and the Dominion Government during the surveying of the boundary. This survey was to mark the boundary on the ground and also to connect the boundary to the Dominion Lands Survey which had extended from western Ontario to the Rocky Mountain foothills. R.W. Cautley of the Alberta Lands Survey was appointed to represent Alberta and to take charge of the surveying party that would take levels and make a preliminary survey of the boundary in various passes and erect permanent boundary monuments therein. A.O. Wheeler of the British Columbia Land Survey was to represent British Columbia and to take charge of the topographical portion of the survey that concentrated on the delineation of the watershed along the main range from the International Boundary northward to its intersection with the 120th Meridian West. J.N. Wallace of the Dominion Lands Survey represented the Dominion Government until 20 September 1915, when Alberta's representative, Cautley, took over that duty as well. His job was "to visit each pass at such times as may be necessary to enable him to satisfy himself as to the correctness of the work."

To survey the rugged terrain, the surveyor employed the phototopographical method, which maps the natural watershed with greater accuracy than was permitted by any other method. Wheeler was experienced with this, as he had conducted phototopographical surveys of the Kicking Horse Pass-Railway Belt section in the years 1903 to 1906. The Dominion Lands Survey introduced this method in 1885 and developed it over the ensuing years. In 1920, Major E.O. Wheeler, Royal Engineer of the Survey of India, accompanied the expedition in order to study this method. He had been commissioned by his government to study the Canadian applications of this surveying technique in order to apply it within the high mountain areas of his own country.

On 18 June 1913, Dr. Edouard-Gaston Deville, the Surveyor-General of Dominion Lands, gave instructions at the start of the first season of the boundary survey. He listed the survey of the following passes as most urgent: Crowsnest (due to its proximity to mining properties); Vermilion (the site of a motor road); Howse (because of its proximity to timber claims); Kicking Horse, Simpson, and Whiteman (due to their proximity to populated areas.). Because of the short surveying seasons (from mid-June to mid-September), it was not until 1924 that the surveys of the passes and the Peace River area were accomplished. Progress was delayed by late and early snowfalls at the high altitudes, bushfires, difficult terrain, lack of roads and even trails, attacks by bears (fortunately only on equipment and survey monuments), and isolation. (The Commission report states that during winter activities on the Wapiti River, only one snowshoe track was seen, made by a solitary trapper named Osborne.)

The surveyors also named some features in the area of the survey. There were few established names and, in fact, there was some confusion about the names that were being used, not to mention the location of the features themselves. For example, Mount Hooker, originally named by David Douglas in 1827, was recorded as having an altitude of 4877 metres above sea level, despite the fact that Mount Robson, now known to be the highest mountain in the Canadian Rockies, has an altitude of only 4356 metres. In 1893, an expedition measured Mount Hooker at less than 3016 metres. In 1913, members of the Alpine Club (London) gave the name "Scott" to the mountain Douglas named "Hooker." The survey of 1920 under Wheeler was able to map the proper location of this mountain and restore its original name, "Mount Hooker," to official status. Only an extensive survey that accurately defines delineations between mountains can ensure the proper official names for features.

In addition to giving official confirmation to established feature names, the surveyors also put new names to some features that were part of their survey information. The sites of boundary monuments and distinctive features whose identification aided navigation were officialized to ease communication of the survey's information. The surveyors named features according to their physical appearances or after fellow surveyors.

Mount Bridgland (page 31), for example, was named in 1916 after Morrison Parsons Bridgland (1878–1948) of the Dominion Lands Survey. A note in the Report of the Commission states that his maps of the Crownest Forest provided "very adequate assistance" to the Commission. Born in Toronto, he first came west in 1902 as assistant to Wheeler. His impeccable mapping skills and tirelessness made him a renowned topographical surveyor in the Rocky Mountains.

In 1911 Wheeler named another pass, Colonel Pass (page 55), after Colonel Aimé Laussedat (1819–1907), an engineer in the French army who was the originator of the science of phototopography so extensively used for survey purposes in mountainous regions. The "father of photographic surveying," he gave a full exposition of the method in his *Mémorial de l'Officier du Génie*. A pass and creek in the area were also named after him.

In 1924, the survey of the interprovincial boundary was completed, except for about 280 kilometres stretching through the uninhabited country along the northern boundary of the two provinces. The hardy surveyors had triumphed against great hardship and difficulties of all kinds to draw the arbitrary but all-important lines that make order out of nature's flowing patterns.

Surveyors were sometimes also mountain climbers, and some of them organized and supported mountain-climbing associations, which, in turn, contributed considerably to mountain toponymy. The Alpine Club of Canada was founded in 1906. Its founder and first president was Arthur Oliver Wheeler (1860–1945), Dominion Topographic Surveyor and Fellow of the Royal Geographical Society. One of the first Honorary Members of the Club was Edouard-Gaston Deville (1849–1924), the Surveyor-General of Canada. Deville introduced phototopographic surveying into Canada, and Wheeler pioneered its application to the mapping of the Rocky Mountains in 1895.

Phototopographic surveying requires photographing from peak to mountain peak, creating an interrelated series of triangles tied to a previous land survey undertaken by conventional methods (in the Rockies, the base was provided by surveys of the Canadian Pacific Railway). A phototopographic surveyor must of necessity also become a mountaineer, and in the course of his work Wheeler became acquainted with mountaineers such as the Swiss guides who worked for the C.P.R. and others of The Alpine Club (London) and the Appalachian Mountaineering Club. It was from these last that Wheeler conceived the idea of forming the Alpine Club of Canada.

The first stated object of the Club in their 1906 constitution was "promotion of scientific study and exploration of Canadian alpine and glacial regions."[2] The *Canadian Alpine Journal (CAJ)* from volume 1 (1907) to volume 52 (1969) contains special sections of scientific articles, many of them written by Wheeler himself. In 1911, the Alpine Club conducted an expedition to the Mount Robson region of British Columbia and from his own phototopographic survey, Wheeler, the expedition director, produced a "Topographical Map Showing Mount Robson and Mountains of the Continental Divide North of Yellowhead Pass" to accompany the Reports of the Alpine Club of Canada's Expedition 1911. On it appears this note: "The names on this map have not yet been passed by the Geographic Board and are subject to revision." From this it may be seen that the Club, through Wheeler, was aware of the toponymic work of the Geographic Board.

In 1919, the *CAJ* published an article on mountain names written by R. Douglas, M.A., Secretary of the Geographic Board, at the request of the editor of the *CAJ*.[3] The editor appended a note to the effect that the Board welcomed name submissions. It was the editor's intention to make the membership more aware of the work of the

---

[2] *Canadian Alpine Journal*, vol. 1 (1907), p. 39.
[3] *Ibid.*, vol. 10 (1917), pp. 32-7.

Geographic Board. In 1964, again at the request of the editor, the *CAJ* published an article on the principles and procedures of the Canadian Permanent Committee on Geographical Names (C.P.C.G.N.), written by J. Keith Fraser, Secretary of the Committee. Once more the editor urged the membership to submit new names to the C.P.C.G.N.[4] Judging by the number of articles in the *CAJ* directly concerned with names, the membership responded slowly: there were two articles in the 1920s, none in the 1930s or 1940s, three in the 1950s, five in the 1960s, nine in the 1970s, and seven in the 1980s.

In the summer of 1921, Walter Dwight Wilcox, who had been an Honorary Member of the Club since 1909 because of his long acquaintance with the Rockies and because of the books on the mountains he had published between 1896 and 1909, and Alfred L. Castle, a lawyer from Hawaii, made an exploratory trip into the Valley of Hidden Lakes. Wilcox published an article about the trip in the 1923 *CAJ,* in which he stated that he and Castle named the valley and the three lakes in it after Castle's children, Gwendolyn, Donald, and Alfred, and that the names were approved by the Geographic Board.[5] The only reference in Board correspondence is in a letter of 28 November 1922 from Charles D. Walcott, Secretary of the Smithsonian Institution, which includes a photograph of the valley, which he calls Douglas Canyon. Walcott states that Wilcox proposed the name "Valley of the Hidden Lakes," but "as it is a rather long name to be used in a descriptive geological paper, I am using the old name Douglas for it." He also says "I am using the name Gwendolyn for the small glacial lake in the great amphitheatre near the head of Douglas Canyon." (The name has since been changed back to Valley of the Hidden Lakes.) Neither name is mentioned in the Board's reply. All four names in Wilcox's article were approved 31 March 1965 by the C.P.C.G.N. on the recommendation of G.H.L. Dempster, Park Superintendent, Banff National Park.

With the publication of name origins and the development of official naming procedures came the opportunity to accumulate extensive records. The geographical names history of the Alberta foothills and mountains that emerges through those records expresses a collective heritage and an important part of our cultural landscape. Future volumes in this series will treat their respective specific study areas with a similar approach (see map on page xxiv) heralding the uniqueness of each territory. This map provides a province-wide National Topographic System grid for use in orienting the reader to finding a location. The nature of the environment has had a considerable impact on the nature of human activity in the region. On the other hand, individuals and organizations promoting a variety of enterprises have imposed certain clearly defined patterns onto the natural wilderness. In a panorama of stories from Indian occupation, through the fur trade, to the railway, mountain recreation, and resource development, the connection between culture and the environment is revealed in detail by thousands of geographical names.

---

[4] *Ibid.,* vol. 47 (1954), pp. 124-29.
[5] *Ibid.,* vol. 13 (1920), pp. 185 and 187.

EXPLANATORY NOTES

Individual place name entries in this volume may best be understood by considering the following example:

1. **Crowsnest Mountain** (mountain)
2. **82 G/10 - Crowsnest**
3. **2-9-5-W5**
4. **49°42′N 114°35′W**
5. **In the locality of Crowsnest Pass, approximately 52 km west north-west of Pincher Creek.**

6. **This mountain is approximately 2785 m in altitude. There are currently two competing possibilities for the origin of the name . . .**

1. Specialists in geographical nomenclature prefer to think of most place names as being comprised of two parts (both exemplified in the first line of this example): one part is called the "*specific*" (here "Crowsnest"), whereas the other is referred to as the "*generic*" (here "Mountain"). The *generic* identifies the *type* of feature, while the *specific* identifies the name of the feature of that type. Although generics very often form parts of place names (as in "Crowsnest Mountain"), many place names lack them. In this volume, the appropriate generics are always provided in parentheses at the end of the first line. The generics used here are consistent with those found in *Generic Terms in Canada's Geographical Names; Terminology Bulletin 176,* Minister of Supply and Services Canada 1987; Catalogue No. S52-2/176-1987. An asterisk (*) preceding a place name indicates that the name has been rescinded or designates a former locality. A square box (■) at the beginning of an entry indicates that a colour photograph of the feature is to be found at the end of this volume.

2. The National Topographic System Grid Reference is a system that "blocks out" the country using map sheets of increasing scales. The second line of each entry identifies the map sheet corresponding to the described feature — in this case "82 G/10 - Crowsnest." Maps may be obtained from any Maps Alberta outlet in the province.

3. Where available, a legal description (here "2-9-5-W5") is given in the line that follows. It specifies the section, range, township, and meridian.

4. Next is the latitude and longitude location description, e.g. "49°42′N 114°35′W."

5. As a further aid to locating the described feature, the approximate distance to the nearest populated community (as the crow flies) is also provided.

6. The concluding sentence or paragraph of most entries summarizes any available descriptive or historical information concerning the feature or its name. As it happens, the above sample entry shares a *specific* with a number of other entries in the volume: Crowsnest (locality), Crowsnest Creek (creek), Crowsnest Lake (lake), Crowsnest Pass (pass), Crowsnest Ridge (ridge), Crowsnest River (river), and Municipality of Crowsnest Pass (municipality-town). To avoid duplication, the origin information for the common name is usually located under only one of the entries. Since the surrounding features are often named as a result of their proximity to the originally named feature, it seems appropriate to present the origin information under the feature that was named first. Wherever this is the case, the other entries conclude with a cross-reference, e.g. "see Crowsnest Mountain."

PHOTO CREDITS

Feature Name	Number of Negative	Location	Page
Amethyst Lakes and The Ramparts, 1918	V263/NA 71-903	Whyte +	6
Angel Glacier, ca. 1927	V492/PA 283/69	Whyte	6
Athabasca Falls	TH 001	Geog Names	285
Athabasca Glacier, 1914	V263/NA 71-1820	Whyte +	10
Mount Balfour and Waputik Icefield	IPR1-29	Geog Names	285
Barrier Lake	RG 80-7-4	Geog Names	285
Bow Valley and Bow River	B 9730	P.A.A.	29
Brule Lake, 1910	Plate 7 (A 38)	Sp. Coll.	32
Cameron Falls, 1951	PA 3107/1	P.A.A.	39
Castle Mountain and Silver City, 1885	P 6683	P.A.A.	43
The Cave	RG 86-1-22	Geog Names	286
Mount Chephren, n.d.	V263/NA 71-1790	Whyte +	47
Cirrus Mountain	IPR8-14	Geog Names	286
Coral Creek Canyon and Coral Creek	Slide Series #9	Geog Names	286
Corona Ridge	Slide Series #3	Geog Names	287
Crescent Falls	Slide Series #1	Geog Names	287
Crowsnest Mountain	RG 86-1-3	Geog Names	287
"Devil's Head Canyon," n.d.	B 6088	P.A.A.	69
Mount Ernest Ross	Slide Series #5	Geog Names	288

Maps

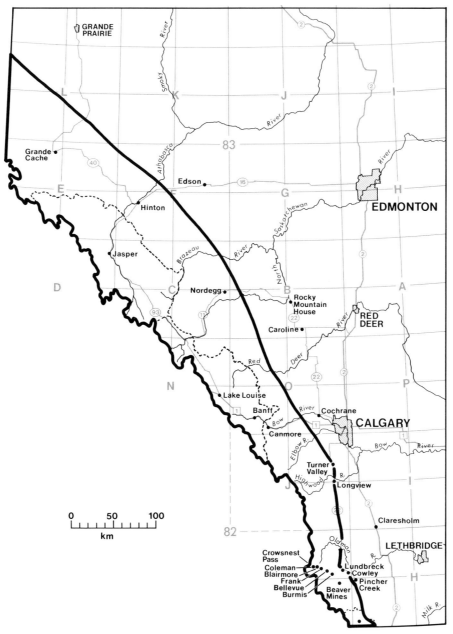

Map showing study areas in the *Place Names of Alberta* series

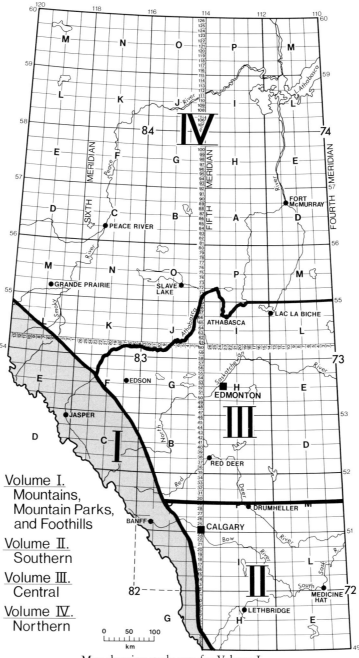

Volume I.
Mountains,
Mountain Parks,
and Foothills

Volume II.
Southern

Volume III.
Central

Volume IV.
Northern

Map showing study area for Volume I

A La Pêche Creek (creek)

83 E/15 - Pierre Greys Lakes
34-55-6-W6
53°48'N 118°48'W
Flows east into Muskeg River, approximately 23 km south-east of Grande Cache.

The name of this creek was officially approved 4 September 1947. A la pêche means "to go fishing" in French.

A La Pêche Lake (lake)

83 E/15 - Pierre Greys Lakes
33-55-6-W6
53°48'N 118°50'W
Approximately 22 km south-east of Grande Cache.

(see A La Pêche Creek)

Abbot Pass (pass)

82 N/8 - Lake Louise
2-28-17-W5
51°22'N 116°17'W
Approximately 55 km west north-west of Banff, on the Alberta-B.C. boundary.

This 2922 m pass was named after Phillip Stanley Abbot, a member of the Appalachian Mountain Club. He fell from the steep mountain slope of Mount Lefroy on 3 August 1896 and died adjacent to the pass.

Abbott Creek (creek)

83 E/15 - Pierre Greys Lakes
8-57-7-W6
53°55'N 119°00'W
Flows north into Peavine Lake, approximately 9 km east north-east of Grande Cache.

This creek was officially named 4 September 1947. It may have been named after Lem Abbott, a resident of the area who was a fine violinist. No other origin information is available.

Aberdeen Glacier (glacier)

82 N/8 - Lake Louise
51°23'12"N 116°14'24"W
Approximately 55 km west north-west of Banff.

The name was officially approved 21 January 1985. (see Mount Aberdeen)

Aberdeen, Mount (mountain)

82 N/8 - Lake Louise
7-28-16-W5
51°23'N 116°14'W
Approximately 55 km west north-west of Banff.

This mountain, which is 3152 m in altitude, was named in 1897 by J.J. McArthur, a D.L.S. member responsible for many of the phototopographical surveys along the railway through the Rocky Mountains. The mountain was named after Lord John Campbell Gordon, the Marquis of Aberdeen, Governor General of Canada from 1893-1898 and a visitor to Lake Louise in 1893.

*Aboukir Island (island)

82 J/11 - Kananaskis Lakes
23-19-9-W5
50°37'N 115°09'W
Approximately 65 km west north-west of Turner Valley.

(see Hogue Island)

Abraham Lake (lake)

83 C/8 - Nordegg
21-38-17-W5
52°17'N 116°22'W
Approximately 24 km south-west of Nordegg.

The name for this lake was chosen from 2000 entries submitted by Alberta students in Grades 1-9 to a contest conducted by the Department of Environmental Affairs in 1972. Proposed by Karen Schauerte (Grade 9-Alder Flats) and Jake Jameson (Grade 4-

Caroline), the name Abraham Lake perpetuates the memories of Silas Abraham and other members of his family. They were Stoney Indians who often hunted and trapped in this area. An Indian burial ground containing graves of Abraham's family was flooded by the man-made lake caused by the Bighorn Dam. The graveyard was moved.

Adair Creek (creek)

82 G/16 - Maycroft
1-12-2-W5
49°58'N 114°09'W
Flows west into Callum Creek, approximately 45 km north-east of Blairmore.

The origin of the name is unknown.

Adam Joachim, Mount (mountain)

83 C/5 - Fortress Lake
52°24'15"N 117°31'35"W
Approximately 65 km south-east of Jasper in Jasper National Park.

The name Mount Adam Joachim was originally proposed in 1968 by James Munroe Thorington (see Thorington Tower) for Adam Joachim (1875-1959). Joachim was born at Berland Lake, Alberta in 1875, a descendant of both the original Iroquois canoemen who came west with the North West Company and Hudson's Bay Company trader Colin Fraser (see Mount Colin). Joachim was educated at Father Lacombe's St. Albert Mission and later studied for three years at a seminary in Montreal. Due to a family crisis, he returned to Alberta and subsequently married a daughter of Ewan Moberly. Joachim was fluent in French, English, Latin and several Indian languages and had little trouble obtaining employment as a guide for trappers, traders and outfitters, when he was not trapping, hunting or fishing himself. Joachim also served as the lead packer for the Alpine Club of Canada when it operated in the Jasper area. As a result, he accompanied numerous prominent mountaineers during the early 1900s as they explored areas previously unseen by white men. (see Joachim Creek)

Adams Creek (creek)

83 E/10 - Adams Lookout
29-54-4-W6
53°41′N 118°33′W
Flows south-east into Berland River, approximately 44 km south-east of Grande Cache.

(see Adams Lookout Mountain)

Adams Lookout Mountain (mountain)

83 E/10 - Adams Lookout
5-55-4-W6
53°43′N 118°34′W
Approximately 40 km east south-east of Grande Cache.

Adam Agnes (born 1900) came to the area in 1911-1912 and settled on what is now locally known as "Agnes Flats." This lookout, 2177 m in altitude, was used by the Forestry Service and was named in 1955.

Adelade Creek (creek)

83 L/3 - Copton Creek
25-59-10-W6
54°08′N 119°21′W
Flows south-east into Copton Creek, approximately 33 km north-west of Grande Cache.

This creek was officially named in 1951 after a trapper named Adelade Joachin who had a trap line nearby.

Adolphus Creek (creek)

83 E/14 - Grande Cache
24-56-8-W6
53°51′N 119°04′W
Flows south-west into Sulphur River, approximately 5 km south-east of Grande Cache.

The origin of the name of this creek is not clear. Some sources claim that it is named after Adolphus Moberly (see Adolphus Lake). Others maintain that it does not have the same origin as the lake. Adolphus Agnes was an early settler of the Grande Cache region. This creek may be named after him. The Agnes family originally came from Lac Ste. Anne about the beginning of the twentieth century.

Adolphus Lake (lake)

83 E/3 - Mount Robson
28-48-8-W6
53°10′N 119°07′W
Approximately 77 km west north-west of Jasper.

Adolphus Moberly was a guide of mixed blood background of Iroquois and white, who guided Coleman and his party to Mount Robson in 1908. A.P. Coleman (see Mount Coleman) named this lake in honour of his guide.

Aeolus, Mount (mountain)

83 E/8 - Rock Lake
34-49-1-W6
53°16′N 118°05′W
Approximately 44 km north of Jasper.

This 2643 m mountain carries the name of the Greek god of the winds. The survey party reached the peak on a windy day.

Afternoon Peak (peak)

83 C/7 - Job Creek
30-38-21-W5
52°18′N 116°59′W
Approximately 100 km south-east of Jasper.

This 3120 m mountain peak was officially named 5 March 1935. Its precise origin is not known, but the name may have been inspired by a good afternoon's climb.

Agnes, Lake (lake)

82 N/8 - Lake Louise
19-28-16-W5
51°25′N 116°15′W
Approximately 55 km north-west of Banff.

This lake was named after Susan Agnes, Baroness Macdonald of Earnscliffe, wife of Prime Minister Sir John A. Macdonald, who visited the lake in 1886.

Aiguille Peak (peak)

82 N/15 - Mistaya Lake
6-33-19-W5
51°48′N 116°40′W
Approximately 105 km north-west of Banff, on the Alberta-B.C. boundary.

The name for this peak, which is 2999 m in altitude, is the French word for needle. It is descriptive of the shape of the peak and was submitted by A.O. Wheeler (see Wheeler Flats) in 1918.

Akamina Lake (lake)

82 G/1 - Sage Creek
11-1-1-W5
49°01′N 114°02′W
Approximately 10 km west south-west of Waterton Park, 3 km north of the Canada-U.S. border.

(see Akamina Pass)

Akamina Pass (pass)

82 G/1 - Sage Creek
1-1-1-W5
49°01′N 114°03′W
Approximately 10 km west south- west of Waterton Park, on the Alberta-B.C. boundary.

Akamina Pass is the most southerly pass through the Rocky Mountains and is 1779 m in altitude. Its name is Indian in origin, meaning "high bench land," referring to the benches near the summit of South Kootenay Pass. The name was applied to a joint astronomical station occupied in 1861 by the British-American Boundary Commission and is referred to in their report as Akamina Camp and Astronomical Station.

Alberta, Mount (mountain)

83 C/6 - Sunwapta
23-38-25-W5
52°18'N 117°28'W
Approximately 77 km south south-east of Jasper.

This mountain, 3619 m in altitude, was named by J.N. Collie (see Collie Creek) in 1899, after Her Royal Highness Princess Louise Caroline Alberta. (see Lake Louise)

Albertine Creek (creek)

83 E/14 - Grande Cache
36-55-8-W6
53°48'N 119°03'W
Flows north-east into Sulphur River, approximately 11 km south south-east of Grande Cache.

The name of this creek was officially adopted 5 March 1953. The origin of the name is unknown.

Alcove Mountain (mountain)

83 D/9 - Amethyst Lakes
16-42-2-W6
52°37'N 118°14'W
Approximately 31 km south south-west of Jasper.

This 2810 m mountain was named by M.P. Bridgland (see Mount Bridgland) in 1916. The mountain is located in a recess behind Ermite Glacier.

Alderson Creek (creek)

82 H/4 - Waterton Lakes
20-1-30-W4
49°03'N 113°58'W

Flows north into Carthew Brook, approximately 50 km south of Pincher Creek.

This creek was named in 1915 by M.P. Bridgland (see Mount Bridgland). It is named after Lieutenant-General E.A.H. Alderson, K.C.B., who commanded the Canadian Expeditionary Force in France from 1915 to 1916.

Alderson Lake (lake)

82 H/4 - Waterton Lakes
1-30-W4
49°02'N 113°59'W
Approximately 50 km south of Pincher Creek, immediately north of Mount Alderson.

(see Alderson Creek)

Alderson, Mount (mountain)

82 H/4 - Waterton Lakes
1-30-W4
49°01'N 113°58'W
Approximately 50 km south of Pincher Creek.

This mountain, which is 2692 m in altitude, was officially named 10 October 1915. (see Alderson Creek)

Alexandra Glaciers (glacier)

82 N/14 - Rostrum Peak
12-25-22,23-W5
51°59'N 117°09'W
Approximately 140 km north-west of Banff.

(see Mount Alexandra)

Alexandra, Mount (mountain)

82 N/14 - Rostrum Peak
3-35-23-W5
51°59'N 117°10'W
Approximately 145 km north-west of Banff, on the Alberta-B.C. boundary.

This mountain, 3388 m in altitude, was named in 1902 by Sir James Outram (see Mount Outram). Queen Alexandra (1844-1925) was consort of King Edward VII. Her father was heir to the Danish throne. Queen Alexandra was an immediate and lasting favourite with the British public.

Alexandra River (river)

83 C/2 - Cline River
3-36-21-W5
52°05'N 116°55'W
Flows east into North Saskatchewan River, approximately 120 km south-east of Jasper, directly below Mount Alexandra.

(see Mount Alexandra)

Alford Creek (creek)

83 B/2 - Caroline
13-36-7-W5
52°05'N 114°53'W
Flows east into Clearwater River, approximately 10 km west of Caroline.

This locally well-established name was officially applied to this feature in 1976. Bert Alford was a trader who often drove his herds of cattle to graze on meadows through which this creek runs. This creek is also locally referred to as Seneca Creek due to the quantity of Seneca roots which are said to grow on the banks and occasionally used in folk medicine.

Alford Lake (lake)

83 B/3 - Tay River
4-36-8-W5
52°04'N 115°05'W
Approximately 24 km south-west of Caroline.

The name for this lake was officially approved in 1976. (see Alford Creek)

Alfred Lake (lake)

82 O/5 - Castle Mountain
8-29-14-W5
51°28'N 115°56'W

Approximately 41 km north-west of Banff.

This lake, which was originally called Lake Alfred, was named by Walter D. Wilcox (see Wilcox Pass) in 1921. One of Wilcox's fellow alpinists, A.L. Castle, had three children, Alfred, Gwendolyn and Donald. It is after Alfred that this feature was named.

Alice, Lake (lake)

82 N/9 - Hector Lake
31-17-W5
51°42′N 116°22′W
Approximately 85 km north-west of Banff.

This lake was named in 1898 by the Reverend Harry P. Nichols (1850-1940), after his wife. Reverend Nichols was rector of Holy Trinity Church in New York from 1899-1902. He was one of the founders of the American Alpine Club and climbed in the area at the turn of the century. Little is known of Alice Nichols.

Allan, Mount (mountain)

82 J/14 - Spray Lakes Reservoir
21-23-9-W5
50°58′00″N 115°12′30″W
Approximately 80 km west of Calgary.

The name for this mountain, which is 2789 m in altitude, was suggested by Mr. Duncan Crockford, in honour of Dr. J.A. Allan (1884-1955), formerly of the Department of Geology at the University of Alberta. Dr. Allan spent much of his time in the Rocky Mountains doing field work, and he also gave a great deal of attention to surveying Alberta's coal resources. The peak was officially named in September 1948.

Allen Creek (creek)

83 E/14 - Grande Cache
1-57-8-W6
53°54′N 119°03′W
Flows south-east into Grande Cache Lake, approximately 5 km east of Grande Cache.

Art Allen was a prominent guide, outfitter and trapper in the Hinton area. He moved to Jasper from Prince George, B.C. in 1942. This creek was named in 1953, possibly after Mr. Allen.

Allen, Mount (mountain)

82 N/8 - Lake Louise
27-16-W5
51°17′N 116°13′W
Approximately 50 km west north-west of Banff, on the Alberta-B.C. boundary.

This mountain, 3310 m in altitude, was named in 1924 after Samuel Evan Stokes Allen (1874-1945), a pioneer climber in the Rockies. Allen came from Philadelphia and visited the Rockies and Selkirks in 1891. He named several features in the Lake Louise area, including Valley of the Ten Peaks.

Allen Bill Pond (pond)

82 J/15 - Bragg Creek
30-22-5-W5
50°54′00″N 114°41′00″W
Approximately 35 km west of Calgary.

This small man-made feature was officially named 7 February 1983, for Allen H. Bill, former editor and outdoor columnist for the *Calgary Herald*. He died 10 September 1970.

Allenby Creek (creek)

82 J/13 - Mount Assiniboine
2-23-12-W5
50°55′N 115°33′W
Flows south-west into Bryant Creek, approximately 100 km west of Calgary.

(see Mount Allenby)

Allenby, Mount (mountain)

82 J/13 - Mount Assiniboine
23-11-W5
50°57′N 115°32′W
Approximately 100 km west of Calgary.

This mountain, which is 2995 m in altitude, was officially named in 1918 after Field Marshal Viscount Edmund Henry Hynman Allenby (1861-1936). He was the Commander of the British Army in Palestine during the Second World War.

Allenby Pass (pass)

82 J/13 - Mount Assiniboine
23-23-12-W5
50°58′N 115°34′W
Approximately 100 km west of Calgary.

(see Mount Allenby)

Allison Creek (creek)

82 G/10 - Crowsnest
2-8-5-W5
49°38′N 114°35′W
Flows south into Crowsnest River, approximately 6 km west south-west of Coleman.

This creek first appears on survey maps of 1914. (see Allison Peak)

Allison Peak (peak)

82 G/10 - Crowsnest
20-9-5-W5
49°45′N 114°39′W
Approximately 16 km north-west of Coleman, on the Alberta-B.C. boundary.

The name Allison Peak was officially approved for this 2643 m peak 22 July 1915. M.P. Bridgland (see Mount Bridgland) submitted the name for the memory of Douglas Allison, a former Royal North West Mounted Policeman who was an early settler in this area. This mountain is also known locally as "The Needle's Eye."

Allsmoke Mountain (mountain)

82 J/10 - Mount Rae
31-20-5-W5
50°45′N 114°41′W

Approximately 30 km west of Turner Valley.

According to some local people, this name originated when a Chinese cook at the Burns Mine near Turner Valley noticed that there was always a haze around the mountain and called it Allsmoke. The mountain is 2109 m in altitude.

Allstones Creek (creek)

83 C/8 - Nordegg
21-38-17-W5
52°17'N 116°22'W
Flows east into Abraham Lake, approximately 28 km south-west of Nordegg.

The well-established name for this creek has been in use locally since at least 1918. The bed of the stream is very rocky, lending appropriateness to the creek's descriptive name.

Allstones Lake (lake)

83 C/8 - Nordegg
14-30-38-17-W5
52°18'05"N 116°25'00"W
Approximately 35 km south-west of Nordegg.

This lake was likely named by local Stoney Indians due to its close proximity to Allstones Creek. The name was officially approved 7 July 1986. (see Allstones Creek)

Alnus Peak (peak)

83 D/8 - Athabasca Pass
36-40-1-W6
52°29'N 118°00'W
Approximately 44 km south of Jasper.

This 2976 m mountain peak was named by A.O. Wheeler (see Wheeler Flats) in 1921 after the alder (Latin, "alnus") tree, likened to the birch tree, which grows on the mountain's sides.

Alpland Creek (creek)

83 C/13 - Medicine Lake
23-45-25-W5
52°54'N 117°31'W
Flows north into the Rocky River, approximately 37 km east of Jasper.

The name for this stream was officially adopted 12 April 1947.

Alpland Creek (creek)

82 N/15 - Mistaya Lake
15-33-21-W5
51°50'N 116°54'W
Flows south into Forbes Creek, approximately 115 km north-west of Banff.

The name for this creek describes the character of a nearby trail and has been in use since at least 1905.

Altrude Creek (creek)

82 O/5 - Castle Mountain
32-26-14-W5
51°16'N 115°56'W
Flows north-east into Bow River, approximately 27 km west north-west of Banff.

The name for this creek was officially approved in 1956. (see Altrude Lakes)

Altrude Lakes (lakes)

82 N/1 - Mount Goodsir
21-26-15-W5
51°14'N 116°03'W
Approximately 35 km west north-west of Banff.

The name for these lakes was approved in 1952. It is said that H.J. Hoffner, in charge of the survey, changed the name of the creek from "Little Vermilion," as instructed. He and his assistant, Robert U. McGuinness, derived the name Altrude from the Latin word, "altus," meaning "high" and the word "pulchritude," meaning "beautiful."

Amber Mountain (mountain)

83 C/13 - Medicine Lake
30-44-27-W5
52°49'N 117°55'W
Approximately 16 km south-east of Jasper.

The descriptive name for this mountain, whose summit is covered with amber-coloured shale, was suggested by M.P. Bridgland (see Mount Bridgland) in 1916.

Ambler Mountain (mountain)

83 E/14 - Grande Cache
15-57-9-W6
53°56'N 119°15'W
Approximately 10 km north-west of Grande Cache.

The origin of the name of this 1905 m mountain is unknown.

Amery, Mount (mountain)

83 C/2 - Cline River
29-35-21-W5
52°02'N 116°58'W
Approximately 120 km south-east of Jasper.

This mountain, 3329 m in altitude, was named after Leopold Charles Maurice Stennett Amery (1873-1955), who climbed the peak in 1929. Amery was an extensive traveller and mountaineer who was also a member of the House of Lords in Britain. He was a persistent advocate of Imperial preference and tariff reform and did much to promote the constitutional advance of colonial territories. He is remembered best for his part in bringing down the Chamberlain Government in 1940.

Amethyst Lakes (lakes)

83 D/9 - Amethyst Lakes
18-43-2-W6
52°42'N 118°16'W
Approximately 22 km south-west of Jasper.

R.W. Cautley (see Mount Cautley) noted in 1921 that the name Amethyst is descriptive of the lakes' beautiful violet-coloured water. The name was made official 8 August 1956.

Amethyst Lakes and The Ramparts, 1918

Ammonite Creek (creek)

83 F/4 - Miette
12-47-27-W5
53°02′N 117°49′W
Flows south-west into Rocky River, approximately 47 km south south-west of Hinton.

The name "Cardioceras Creek" was first suggested by Dr. Mountjoy after the fossil Ammonite Cardioceras which was found in the rocks. Since the Geographic Board considered this name to be difficult to pronounce, the alternate name, Ammonite Creek, was approved instead in 1961. Ammonite is commonly called "snake-stones."

Ancient Wall, The (ridge)

83 E/7 - Blue Creek
51-7-W6
53°27′N 118°53′W
Approximately 80 km north-west of Jasper.

The name for this ridge was officially approved 2 August 1956. The origin of the name is unknown.

Anderson Creek (creek)

83 F/6 - Pedley
15-50-23-W5
53°19′N 117°17′W
Flows east into McLeod River, approximately 64 km south-west of Edson.

This creek was officially named in July 1927 possibly after Harold Anderson, a trapper and homesteader on the McLeod River. He was also involved in census-taking in 1937 and became part of a 1946 survey party which renamed features in the area.

Anderson Peak (peak)

82 G/1 - Sage Creek
2-1-W5
49°08′N 114°04′W
Approximately 15 km north-west of Waterton Park.

This 2652 m peak is named in honour of Lieutenant Samuel Anderson, R.E., the Chief Astronomer of the Second British Boundary Commission of 1872-1876, from Lake of the Woods to the Rockies (see Cameron Lake). Lieutenant Anderson was also the Secretary of the First British Boundary Commission of 1862-1869 surveying from the Strait of Georgia to the Rockies. It was formerly locally called "Millionaires Peak."

Andromeda, Mount (mountain)

83 C/3 - Columbia Icefield
16-37-23-W5
52°10′N 117°13′W
Approximately 95 km south-east of Jasper.

The locally well-established name for this mountain was officially approved in 1959. For years, it has had the name painted on the rock peak finder located in front of the Columbia Icefield Chalet. The name comes from Greek Mythology (wife of Perseus and daughter of Cassiope and King Cepheus) and is the name of a northern constellation. The name was suggested by Major Rex Gibson, former President of the Alpine Club of Canada.

Andy Good Peak (peak)

82 G/10 - Crowsnest
23-7-5-W5
49°34′N 114°35′W
Approximately 10 km south-west of Coleman, on the Alberta-B.C. boundary.

Formerly "Andygood," this peak was officially named Andy Good Peak on 31 December 1962. The 2621 m peak was named for Mr. Andrew Good, Proprietor of the Crows Nest Hotel, which straddled the Alberta-B.C. boundary.

Angel Glacier (glacier)

83 D/9 - Amethyst Lakes
3-43-1-W6
52°41′N 118°04′W
Approximately 23 km south of Jasper.

The name for this glacier describes its fancied resemblance to the most popular conception of the shape of an angel.

Angel Glacier, ca. 1927

Angle Peak (peak)

83 D/9 - Amethyst Lakes
10-42-2-W6
52°36′N 118°12′W
Approximately 32 km south south-west of Jasper.

The name for this 2910 m mountain peak, with its sharp turn in the ridge, was suggested by J. Outram (see Mount Outram) and was officially approved in 1916.

Annette Lake (lake)

82 N/8 - Lake Louise
5-28-16-W5
51°22′N 116°12′W
Approximately 50 km west north-west of Banff.

Named in 1896 by Walter D. Wilcox (see Wilcox Pass), this lake is located in Paradise Valley. Mrs. Annette Astley, a pioneer in her own right, was the wife of the then manager of the Lake Louise Chalet.

Annette Lake (lake)

83 D/16 - Jasper
26-45-1-W6
52°54′N 118°03′W
Approximately 3 km north-east of Jasper.

This attractive lake was named in 1914 by H. Matheson of the Dominion Land Survey after the wife of Colonel S. Maynard Rogers, a Superintendent of Jasper Park.

Antelope Butte (butte)

82 G/16 - Maycroft
28-9-2-W5
49°46′15″N 114°12′25″W
Approximately 35 km north-west of Pincher Creek.

This high, flat-topped hill became known as Antelope Butte at the turn of the century when the Indians drove antelope to the top of the hill to kill them. The name was made official 21 July 1982.

Anthony Creek (creek)

83 E/10 - Adams Lookout
27-53-7-W6
53°36′N 118°57′W
Flows south-west into Sulphur River,

approximately 34 km south south-east of Grande Cache.

The name of this creek is taken from Anthony Hope, an author who wrote *Phroso* and *The Prisoner of Zenda*. Two nearby creeks are named after these two books, i.e., Zenda Creek and Phroso Creek.

Anthozoan Mountain (mountain)

82 N/8 - Lake Louise
10-29-15,16-W5
51°28′N 116°02′W
Approximately 45 km north-west of Banff.

This mountain, which is 2695 m in altitude, was named in 1925 after the type of coral reefs (known as anthozoan) found in the Devonian limestone of the mountain.

Anthracite (locality)

82 O/3 - Canmore
5-26-11-W5
51°12′N 115°30′W
Approximately 6 km east north-east of Banff.

The Canadian Anthracite Company mined anthracite, a nonbituminous variety of coal, in this area from 1886 to 1887, when the sale of coal diminished. Mr. H.W. McNeil, an American entrepreneur, took over the mine in 1891 and kept it operating until the fall of 1904, when it was officially closed. The post office closure followed 7 March 1905, after many former residents of the village of Anthracite moved to Canmore.

Antler Creek (creek)

83 F/3 - Cadomin
28-48-23-W5
53°13′N 117°20′W
Flows north-east into McLeod River, approximately 73 km south-west of Edson.

(see Antler Mountain)

Antler Mountain (mountain)

83 C/13 - Medicine Lake
4-45-25-W5
52°45′N 117°50′W
Approximately 21 km south-east of Jasper.

The origin of the name is unknown.

Apetowun Creek (creek)

83 F/11 - Dalehurst
12-54-23-W5
53°39′N 117°16′W
Flows east into Plante Creek, approximately 55 km west north-west of Edson.

The name for this creek was officially approved in 1946 in preference to "Centre Creek." Apetowun is the Cree Indian word for "it is the middle."

Apex Mountain (mountain)

83 C/4 - Clemenceau Icefield
33-37-27-W5
52°13′N 117°48′W
Approximately 75 km south-east of Jasper, on the Alberta-B.C. boundary.

The name for this mountain describes its appearance.

Apparition Mountain (mountain)

82 O/6 - Lake Minnewanka
2-28-11-W5
51°22′N 115°27′W
Approximately 22 km north north-east of Banff.

The name for this 3002 m mountain was suggested by T.W. Swaddle of Calgary in 1970. Swaddle maintained that the name Apparition (a strange, remarkable, mysterious vision) was concurrent with the "supernatural" idiom which has been long established in the area (e.g., Ghost River, Devil's Head, Phantom Crag, etc.).

Aquila Mountain (mountain)

83 D/9 - Amethyst Lakes
29-43-1-W6
52°44′N 118°07′W
Approximately 16 km south of Jasper.

There was an eagle seen on the peak at the time of naming in 1916. Aquila is the Latin word for "eagle." M.P. Bridgland (see Mount Bridgland) gave this 2825 m mountain its descriptive name.

Arcs, Lac des (lake)

82 O/3 - Canmore
22-24-9-W5
51°03′N 115°11′W
Approximately 12 km east south-east of Canmore.

Arc is the French word for "bow." This lake is an expansion of the Bow River and was named by E. Bourgeau (see Mount Bourgeau) in 1858.

Arctomys Creek (creek)

82 N/15 - Mistaya Lake
8-35-20-W5
51°59′N 116°49′W
Flows east into North Saskatchewan River, approximately 125 km north-west of Banff.

(see Arctomys Peak)

Arctomys Peak (peak)

82 N/15 - Mistaya Lake
23-34-22-W5
51°56′N 116°59′W
Approximately 130 km north-west of Banff.

This 2793 m mountain peak was named by A.O. Wheeler (see Wheeler Flats) in 1919 after the whistling marmots (genus arctomys columbianus) seen in the valley. (see Marmot Mountain)

Arcturus Peak (peak)

83 E/7 - Blue Creek
34-51-5-W6
53°27′N 118°41′W
Approximately 74 km north-west of Jasper.

Arcturus is a character from Greek mythology who was the son of Zeus and Callisto. The name means "Little Bear." Arcturus is one of the five brightest stars in the night sky. This mountain, first named by R.W. Cautley (see Mount Cautley), was officially named 1 May 1934 and is part of the Starlight Range.

Arête, Mount (mountain)

83 C/7 - Job Creek
27-40-21-W5
52°28′N 116°57′W
Approximately 90 km south-east of Jasper.

This mountain, which is 2990 m in altitude, has ragged rock ridges caused by erosion where two neighbouring cirque walls intersect. In physical geography, arête is the term used for the result of such weathering.

Arethusa, Mount (mountain)

82 J/10 - Mount Rae
18-19-7-W5
50°43′N 114°55′W
Approximately 49 km west of Turner Valley.

This mountain, 2912 m in altitude, was named in 1917 after the famous British light cruiser, sunk by a mine on 11 February 1916.

Aries Peak (peak)

82 N/15 - Mistaya Lake
33-32-19-W5
51°47′N 116°38′W
Approximately 100 km north-west of Banff, on the Alberta-B.C. boundary.

This 2996 m mountain peak was named by A.O. Wheeler (see Wheeler Flats) in 1918 after the wild sheep (also known as Aries) seen on the mountain.

Armstrong, Mount (mountain)

82 J/7 - Mount Head
16-6-W5
50°21′N 114°46′W
Approximately 49 km south-west of Turner Valley.

J.D. Armstrong was killed in action 12 April 1917. It is for this member of the Surveyor General's staff that this mountain, 2823 m in altitude, is named.

Arnica Lake (lake)

82 O/4 - Banff
14-26-15-W5
51°13′N 116°00′W
Approximately 31 km west north-west of Banff.

Approved in 1959, the name for this lake is taken from the flowers which border it.

Arris Mountain (mountain)

83 D/16 - Jasper
8-44-2-W6
52°46′N 118°15′W
Approximately 16 km south-west of Jasper.

The name for this 2705 m mountain overlooking the (at that time) unexplored Crescent Valley, is synonymous with arête (see Mount Arête) and is descriptive of the geographical features found there. The name was applied in 1916 by M.P. Bridgland. (see Mount Bridgland)

Arrowhead Lake (lake)

83 D/9 - Amethyst Lakes
21-42-2-W6
52°38′N 118°13′W
Approximately 27 km south south-west of Jasper.

The name for this lake, approved 1 May 1934, is likely descriptive of the shape.

Ashburner Creek (creek)

83 C/5 - Cardinal River
34-43-18-W5
52°45′N 116°32′W
Flows south-east into Brown Creek, approximately 97 km east south-east of Jasper.

The origin for the name of this feature is unknown.

Ashlar Creek (creek)

83 F/4 - Miette
2-49-27-W5
53°12′N 117°50′W
Approximately 31 km south-west of Hinton.

(see Ashlar Ridge)

Ashlar Ridge (ridge)

83 F/4 - Miette
36-48-27-W5
53°11′N 117°48′W
Approximately 32 km south-west of Hinton.

The descriptive name for this ridge, which resembles a smooth wall, was suggested by M.P. Bridgland (see Mount Bridgland) in 1916. Ashlar is a form of masonry.

Assiniboine, Mount (mountain)

82 J/13 - Mount Assiniboine
22-12-W5
50°52′N 115°38′W
Approximately 105 km west of Calgary, on the Alberta-B.C. boundary.

Mount Assiniboine was named in 1885 by Dr. G.M. Dawson (see Tombstone Mountain), Director of the Dominion Geological Survey from 1875-1901. The tribe of Indians, after whom the 3618 m mountain was named, was a branch of the Sioux Nation, which migrated westward to the Great Plains. Their cooking technique of boiling their meat in hide-lined holes in the ground filled with water and hot stones led to their name. The Stoney Indians, or Assiniboines, (the Cree equivalent), hunted in this area.

Assiniboine Pass (pass)

82 J/13 - Mount Assiniboine
23-12-W5
50°56′N 115°36′W
Approximately 100 km west of Calgary, on the Alberta-B.C. boundary.

This pass is 2180 m in altitude. (see Mount Assiniboine)

Association Peak (peak)

82 O/3 - Canmore
36-25-9-W5
51°11′N 115°07′W
Approximately 15 km north-east of Canmore.

The well-established local name for this 2362 m mountain was approved by the Canadian Permanent Committee on Geographical Names 17 March 1967. The origin of the name is unknown.

Aster Creek (creek)

82 J/11 - Kananaskis Lakes
16-19-9-W5
50°36′N 115°11′W
Flows north-east into Foch Creek, approximately 65 km west of Turner Valley.

(see Aster Lake)

Aster Lake (lake)

82 J/11 - Kananaskis Lakes
5-19-9-W5
50°34′N 115°13′W
Approximately 65 km west of Turner Valley.

The name for this lake was proposed by H.G. Bryant (see Bryant Creek) and W.D. Wilcox (see Wilcox Pass). The lake lies above the timberline, and the name is descriptive of the flowers that grow on its banks.

Astoria Pass (pass)

83 D/9 - Amethyst Lakes
13-42-2-W6
52°42′N 118°09′W
Approximately 19 km south of Jasper.

This pass, which was named in 1917, is 2316 m in altitude. It takes its name from the Astoria fur traders who travelled east through the Athabasca Pass in 1814. These traders were named after Fort Astoria, established by the American Fur Co. in 1811 at the mouth of the Columbia River. The fort takes its name from the fur trader and capitalist John Jacob Astor (1763-1848).

Astoria River (river)

83 D/16 - Jasper
13-44-1-W6
52°48′N 118°02′W
Flows east into Athabasca River, approximately 12 km south south-east of Jasper.

(see Astoria Pass)

■ Athabasca Falls (falls)

83 C/12 - Athabasca Falls
31-42-27-W5
52°40′N 117°52′W
Approximately 32 km south south-east of Jasper.

These falls are located on the Athabasca River about 32 km above Jasper. Athabasca is the Cree Indian name meaning "where there are reeds." This refers to the muddy delta of the river where it falls into Lake Athabasca. In 1790, the lake was called "Lake of the Hills" and the river "Great Araubaska." There have been various developments to the name over time which have resulted in the final Athabasca Lake, River, Town, Pass and Falls.

Athabasca Glacier (glacier)

83 C/3 - Columbia Icefield
20-37-23-W5
52°12'N 117°15'W
Approximately 95 km south-east of Jasper.

There have been several other names recorded for the lake and river, including "Lake of the Hills" and "Great Athabasca" (1790), "Lake of the Hills" and "Elk River" (1801), and "Athapascow Lake" and "Athapascow" or "Elk River." (see Athabasca Falls)

Athabasca Glacier, 1914

Athabasca, Mount (mountain)

83 C/3 - Columbia Icefield
15-37-23-W5
52°11'N 117°13'W
Approximately 95 km south-east of Jasper.

This 3491 mountain was named by J.N. Collie (see Collie Creek), who was the first to climb the mountain in 1898. (see Athabasca Falls)

Athabasca Pass (pass)

83 D/8 - Athabasca Pass
26-39-2-W6
52°23'N 118°10'W
Approximately 55 km south of Jasper.

This pass, which is 1737 m in altitude, is on the Alberta-B.C. boundary that makes up part of the Continental Divide. It is the route which the fur traders used to cross the mountains prior to the completion of the railroad. It was discovered in the winter of 1810 by David Thompson (see Thompson Pass). The fur trade route ran from Jasper House (located some 24 km east of the present town of Jasper) to the present site of Jasper, then along the Athabasca River to its junction at Whirlpool River, then up the Whirlpool River to the summit of the Pass and down the Wood River (a tributary to the Columbia River) to Boat Encampment at the big bend of the Columbia River. From there bâteaux brigades took the furs, trade goods and other products to Fort Vancouver at the mouth of the Columbia River. The point at which the trail branched off from the Athabasca River to follow the Whirlpool River and cross the pass is now marked by a monument showing maps and giving a very detailed description of this route. (see Athabasca Falls)

Athabasca River (river)

83 C/3 - Columbia Icefield
37-24-W5
52°13'N 117°14'W
Approximately 75 km south-east of Jasper.

(see Athabasca Falls)

Aura Creek (creek)

82 O/7 - Wildcat Hills
28-27-7-W5
51°20'N 114°56'W
Flows south-west into Waiparous Creek, approximately 36 km north-west of Cochrane.

The origin of the name is unknown.

Aurora Mountain (mountain)

82 J/13 - Mount Assiniboine
21-12-W5
50°50'N 115°33'W
Approximately 100 km west south-west of Calgary, on the Alberta-B.C. boundary.

This 2789 m mountain was named in 1917 after a light cruiser engaged in a battle in the North Sea 24 January 1915.

Avalanche Creek (creek)

83 E/12 - Pauline Creek
21-53-13-W6
53°35'N 119°52'W
Flows south-east into Beaverdam Creek, approximately 61 km south-west of Grande Cache.

(see Avalanche Pass)

Avalanche Pass (pass)

83 E/12 - Pauline Creek
30-53-13-W6
53°37'N 119°55'W
Approximately 60 km west south-west of Grande Cache.

The name of this pass was taken from the Jobe-Phillips map, published in 1915 by the American Geographical Society in its bulletin Vol. XLVII, No. 7. The Jobe-Phillips map is a result of an expedition taken in 1914 by Miss Mary Jobe and her guide, David Phillips. Although the Jobe-Phillips map had some inaccuracies in parts, it was a fair representation of this specific area, and was the only detailed map of its kind published prior to 1922, when the Commission of the Surveyor General was appointed to delimit the boundary between Alberta and British Columbia in this region. The descriptive name, Avalanche Pass, commemorates a snow-slide that occurred close to the pass many years prior to 1918. Some rotting debris of the timber it had swept down at the time served as a reminder of the avalanche.

Avens, Mount (mountain)

82 N/8 - Lake Louise
28-15-W5
51°25′N 116°00′W
Approximately 40 km north-west of Banff.

Named in 1911 by James F. Porter (see Merlin Lake), this 2970 m mountain takes its name from a wildflower of the genus Geum which is found in the vicinity.

Avion Ridge (ridge)

82 G/1 - Sage Creek
2-1-W5
49°10′N 114°07′W
Approximately 20 km north-west of Waterton Park.

"Avion" is the French word applied to any airplane. There is also an Avion in France which is a suburb of Lens. It was taken over by Canadian troops in 1917. This ridge measures 2437 m in altitude and is named because its shape resembles that of an aircraft.

Aye, Mount (mountain)

82 J/13 - Mount Assiniboine
22-12-W5
50°51′N 115°38′W
Approximately 105 km west of Calgary, on the Alberta-B.C. boundary.

This mountain is 3243 m in altitude. The origin of the name is unknown.

Aylmer Canyon (canyon)

82 O/6 - Lake Minnewanka
11-27-11-W5
51°17′N 115°25′W
Approximately 16 km north-east of Banff.

Officially approved 1 March 1904, the name for this canyon was proposed by J. Smith. (see Mount Aylmer)

Aylmer, Mount (mountain)

82 O/6 - Lake Minnewanka
23-27-11-W5
51°19′N 115°27′W
Approximately 19 km north north-east of Banff.

This 3162 m mountain was named in 1890 by J.J. McArthur, D.L.S. after his home town, Aylmer, Quebec.

Aylmer Pass (pass)

82 O/6 - Lake Minnewanka
23-27-11-W5
51°19′N 115°27′W
Approximately 17 km north north-east of Banff.

The Canadian Permanent Committee on Geographical Names approved this name 7 February 1971. (see Mount Aylmer)

Aztec, Mount (mountain)

83 C/6 - Sunwapta
18-40-21-W5
52°27′N 117°01′W
Approximately 88 km south-east of Jasper.

This mountain, north of Brazeau Lake, is approximately 2816 m in altitude. There is a curious zigzag outcrop of red rock near its summit. This motif was a favourite among the Aztecs of Mexico. The name was officially approved 3 July 1946.

Azure Lake (lake)

83 E/6 - Twintree Lake
6-52-7-W6
53°27′N 119°00′W
Approximately 89 km north-west of Jasper.

The name of this lake is likely descriptive of its colour.

Babel Creek (creek)

82 N/8 - Lake Louise
27-27-16-W5
51°20′N 116°10′W
Flows north-west into Moraine Creek, approximately 45 km west north-west of Banff.

(see Mount Babel)

Babel, Mount (mountain)

82 N/8 - Lake Louise
15-27-16-W5
51°18′N 116°10′W
Approximately 45 km west north-west of Banff.

This mountain, which is 3101 m in altitude, is located directly south of the Tower of Babel (mountain). It was named Mount Babel for its fancied resemblance to the feature to its north.

Backus, Mount (mountain)

82 G/8 - Beaver Mines
5-3-W5
49°26′N 114°16′W
Approximately 25 km south-west of Pincher Creek.

This locally well-established name commemorates a homesteader who lived in the area. The mountain is 1815 m in altitude.

Bacon Creek (creek)

83 F/2 - Foothills
19-48-19-W5
53°10′N 116°46′W
Flows north-east into Erith River, approximately 52 km south south-west of Edson.

The name for this creek was suggested by J.B. Beatty, a local resident. The name was officially approved in 1944, but its origin remains unknown.

Badger Pass (pass)

82 O/5 - Castle Mountain
27-28-14-W5
51°25'N 115°53'W
Approximately 34 km north-west of Banff.

Officially approved in 1959, the pass was named for the badger, a nocturnal hibernating animal.

Bailey Creek (creek)

83 C/15 - Cardinal River
21-46-20-W5
52°59'N 116°50'W
Flows south-east into Pembina River, approximately 86 km east north-east of Jasper.

This creek was named after a local rancher and trapper who was shot near the stream. No other information is available.

Baker Creek (creek)

82 N/8 - Lake Louise
32-27-15-W5
51°30'N 116°02'W
Flows south into Bow River, approximately 40 km north-west of Banff.

(see Baker Lake)

Baker Lake (lake)

82 N/8 - Lake Louise
29-15-W5
51°30'N 116°02'W
Approximately 50 km north-west of Banff.

This lake is named after a prospector who worked in the vicinity in 1882 or 1883. No other origin information is available.

Baker, Mount (mountain)

82 N/10 - Blaeberry River
23-31-19-W5
51°40'N 116°35'W
Approximately 90 km north-west of Banff, on the Alberta-B.C. boundary.

This 3172 m mountain was named by Reverend C.L. Noyes and party in 1898. G.P. Baker was a member of The Alpine Club (London) who accompanied members of the Appalachian Mountain Club on an excursion in 1897.

Balcarres, Mount (mountain)

83 C/14 - Mountain Park
24-44-23-W5
52°48'N 117°20'W
Approximately 50 km east south-east of Jasper.

Located opposite Mount Lindsay, this 2897 m mountain was officially named in 1925. (see Mount Lindsay)

Baldy, Mount (mountain)

82 O/3 - Canmore
4-24-8-W5
51°00'40"N 115°02'55"W
Approximately 23 km east south-east of Canmore.

The name for this 2192 m mountain, officially approved 12 September 1984, is descriptive of this feature's flat bare top. (see Barrier Lake)

Balfour Creek (creek)

82 N/9 - Hector Lake
25-30-18-W5
51°35'N 116°24'W
Flows east into Hector Lake, approximately 75 km north-west of Banff.

The Canadian Permanent Committee on Geographical Names approved the name for this creek officially 14 August 1975. (see Mount Balfour)

Balfour Glacier (glacier)

82 N/9 - Hector Lake
30-18-W5
51°34'N 116°26'W
Approximately 75 km north-west of Banff.

(see Mount Balfour)

■ **Balfour, Mount** (mountain)

82 N/9 - Hector Lake
16-30-18-W5
51°34'N 116°28'W
Approximately 75 km north-west of Banff on the Alberta-B.C. boundary.

This mountain, 3272 m in altitude, was named by Sir James Hector (see Mount Hector) in 1859 after Professor John Hitton Balfour, M.D. (1808-1884). He founded the Botanical Society of Edinburgh in 1836 and was a professor of botany at Glasgow University and Dean of the Medical School at the University of Edinburgh.

Balfour Pass (pass)

82 N/9 - Hector Lake
21-30-18-W5
51°36'N 116°28'W
Approximately 80 km north-west of Banff, on the Alberta-B.C. boundary.

(see Mount Balfour)

Balinhard Creek (creek)

83 C/14 - Mountain Park
11-44-24-W5
52°47'N 117°22'W
Flows west into Restless River, approximately 50 km east south-east of Jasper.

(see Mount Balinhard)

Balinhard, Mount (mountain)

83 C/14 - Mountain Park
5-44-23-W5
52°45'N 117°18'W
Approximately 54 km east south-east of Jasper.

This 3130 m mountain was officially named in 1925 after one of the titles held by the Earl of Southesk, Baron Balinhard. Another source (*Alberta History*, Autumn 1988) says that Mount Balinhard was named after one of Southesk's literary works. (see Southesk River)

Ball, Mount (mountain)

82 N/1 - Mount Goodsir
25-15-W5
51°09'N 116°00'W
Approximately 30 km west south-west of Banff.

This 3311 m mountain was named by Sir James Hector (see Mount Hector) in 1858 after John Ball (1818-1889), an Under-Secretary of State for the Colonies who was influential in securing government support for the Palliser Expedition and helped in its organization. Ball was first President of The Alpine Club (London).

Ball Pass (pass)

82 O/4 - Banff
25-15-W5
51°07'N 115°58'W
Approximately 31 km west south-west of Banff.

The name for this pass, which is 2225 m in altitude, was officially approved in 1956. (see Mount Ball)

Ball Range (range)

82 N/1 - Mount Goodsir
25-15-W5
51°10'N 116°01'W
Approximately 35 km west of Banff.

(see Mount Ball)

Banded Peak (peak)

82 J/10 - Mount Rae
28-20-7-W5
50°43'N 114°55'W
Approximately 45 km west of Turner Valley.

The descriptive name for this 2934 m peak has been recorded since at least 1896.

Banff (town)

82 O/4 - Banff
10-35-25-12-W5
51°10'45"N 115°34'15"W
Approximately 92 km west of Calgary.

Lord Strathcona and his colleagues named this town after a town in the county of Banffshire, now Grampian, near his birthplace in Eastern Scotland. Banff is a corruption of "Bunnaimb," meaning "mouth of the river" in the Gaelic language. Apparently, the name was first suggested by Mr. Harry Sandison, former alderman and former chairman of the Parks Board, who was born in the town of Banff in Scotland. The meeting in which the name Banff was decided for this town included personalities such as Dr. James Steward and Mr. H. MacTavish, who were both associated with the Hudson's Bay Company. Other members included George Stephens and William Van Horne of the Canadian Pacific Railway. Also in attendance was Donald Smith, who later became Lord Strathcona, a prominent figure in Canadian history. He was a Canadian High Commissioner in England, a key enterpreneur in the formation of the Canadian Pacific Railway Company, and had also at different times served as a senior Hudson's Bay Company officer and a Member of Parliament. This meeting was called to decide on a name for this town, and it was here it was first called Banff. There are several different Indian names for this townsite. As outlined by Hugh Dempsey, author of *Indian Names for Alberta Communities* Occasional Paper No. 4 (1969), in Blackfoot it is named for the hot springs, for example, "Nato-oh-sis-koom," which means "holy springs." The Stoney Indians call Banff "Minihapa," which means "waterfall place," while the Cree call it "Nipika-Pakitik," both for the falls on Cascade Mountain. In Sarcee, "Tsa-nidza" means "in the mountains." The name Banff was made official 25 December 1905.

Banff National Park (national park)

82 N - Golden
51°30'N 116°15'W

(see Banff)

Bank Creek (creek)

83 L/4 - Kakwa Falls
8-24-58-13-W6
54°02'30"N 119°47'00"W
Flows east into South Kakwa River, approximately 47 km west north-west of Grande Cache.

The name for this creek was officially adopted 4 December 1958. The origin of the name is unknown.

Bankfoot Creek (creek)

82 O/11 - Burnt Timber Creek
14-31-9-W5
51°39'N 115°18'W
Flows south-east into Red Deer River, approximately 55 km north north-east of Banff.

No information regarding the origin of the name is known.

Bankhead (locality)

82 O/4 - Banff
12-26-12-W5
51°12'N 115°32'W
Approximately 3 km north north-east of Banff.

The town of Bankhead grew out of the coal industry. The Canadian Pacific Railway founded the town in 1903. It was named in 1905 by Lord Strathcona (see Banff), after Bankhead, Banffshire, Scotland. This boom town became a ghost town circa 1923.

Barbara Creek (creek)

83 F/12 - Gregg Lake
1-54-26-W5
53°38'N 117°42'W

Flows north-west into Wildhay River, approximately 28 km north north-west of Hinton.

This creek was named 16 July 1922 and was officially confirmed 10 November 1945. Miss Barbara Anne Burrows (1922-) is the daughter of the some time Forest Supervisor at Entrance who was wounded during the Second World War. It is after Miss Burrows, a noted Edmonton musician and critic, that this creek was named.

Barbette Glacier (glacier)

82 N/10 - Blaeberry River
15-32-19-W5
51°44′N 116°36′W
Approximately 95 km north-west of Banff.

(see Barbette Mountain)

Barbette Mountain (mountain)

82 N/10 - Blaeberry River
10-32-19-W5
51°44′N 116°36′W
Approximately 95 km north-west of Banff.

A barbette is a platform in a fort or ship from which guns fire over a parapet without an embrasure. This mountain, 3072 m in altitude has two high platform peaks rising from the mass of the mountain. The descriptive name for the mountain was officially approved in 1918.

Bare Range (range)

82 O/12 - Barrier Mountain
30-13-W5
51°35′N 115°44′W
Approximately 45 km north-west of Banff.

The descriptive name for this range has been in use since at least 1916.

Baril Creek (creek)

82 J/7 - Mount Head
16-5-W5
50°22′N 114°46′W
Flows east into Etherington Creek, approximately 42 km south-west of Turner Valley.

Officially approved 8 December 1943, this creek is named for its proximity to Baril Peak.

Baril Creek (creek)

83 F/7 - Erith
3-51-20-W5
53°22′N 116°52′W
Flows south-east into Lambert Creek, approximately 37 km south-west of Edson.

This creek was named after a timber operator who worked on the stream for several years. The name was officially approved 4 October 1927.

Baril Peak (peak)

82 J/7 - Mount Head
34-15-6-W5
50°18′N 114°45′W
Approximately 52 km south-west of Turner Valley.

M.C.L. Baril, of the Surveyor General's staff, was killed in action 9 November 1915. This 2998 m mountain was named after him.

Barlow, Mount (mountain)

82 N/10 - Blaeberry River
31-20-W5
51°42′N 116°48′W
Approximately 95 km north-west of Banff, on the Alberta-B.C. boundary.

This mountain, which stands 3120 m in altitude, was named by D.B. Dowling (see Dowling Ford) in 1920 after Dr. Alfred Ernest Barlow, D.Sc., F.R.S. (1861-1914), a member of the Geological Survey of Canada

until 1907, when he resigned to become a consulting geologist in Montreal. He and his wife were drowned in the Empress of Ireland disaster in the St. Lawrence River in 1914.

Barnaby Creek (creek)

82 G/8 - Beaver Mines
9-5-3-W5
49°22′N 114°21′W
Approximately 30 km south-west of Pincher Creek.

The name for this creek was taken from Barnaby Ridge. (see Barnaby Ridge)

Barnaby Lake (lake)

82 G/8 - Beaver Mines
31-4-3-W5
49°20′15″N 114°22′50″W
Approximately 34 km south-west of Pincher Creek.

The name was officially applied to this lake 15 February 1978. The lake lies in the watershed of the Castle River at an altitude of approximately 1981 m above sea level. (see Barnaby Ridge)

Barnaby Ridge (ridge)

82 G/8 - Beaver Mines
4-3-W5
49°18′N 114°22′W
Approximately 39 km south-west of Pincher Creek.

This ridge was named officially in 1916, but its precise origin is unknown.

Barnard Dent Group (range)

82 N/10 - Blaeberry River
32-21-W5
51°44′N 116°56′W
Approximately 95 km north-west of Banff, on the Alberta-B.C. boundary.

The origin of the name is unknown.

Barnes Ridge (ridge)

82 O/2 - Jumpingpound Creek
24-5-W5
51°03′N 114°40′W
Approximately 30 km west of Calgary.

This ridge is likely named after Elliott C. Barnes, who came to Alberta in 1905 from North Dakota, U.S.A. He homesteaded on Little Jumpingpound Creek. He died in Calgary in 1938. The name for this ridge was officially approved 12 December 1939.

Barra, Mount (mountain)

83 E/6 - Twintree Lake
2-51-8-W6
53°22′N 119°03′W
Approximately 86 km west north-west of Jasper.

The name for this 2515 m mountain, approved 1 May 1934, may be taken after one of the islands located in the Outer Hebrides, in the Atlantic Ocean west of Scotland. The Hebrides were ceded by Norway to Scotland in the 13th century.

Barricade Mountain (mountain)

83 E/6 - Twintree Lake
2-51-8-W6
53°24′N 119°29′W
Approximately 112 km west north-west of Jasper.

There is a ridge on this 2593 m mountain resembling a barricade. The descriptive name for this mountain was approved in 1924.

■ **Barrier Lake** (lake)

82 O/3 - Canmore
5-24-8-W5
51°01′N 115°04′W
Approximately 22 km east south-east of Canmore.

The lake derives its name from the mountain located at its south end, locally known as "Barrier Mountain" or "The Barrier." This mountain was a barrier to north-south travel along the Kananaskis River before the present road was constructed. The mountain's official name is now Mount Baldy. The name for the lake was approved 1 November 1956.

Barrier Mountain (mountain)

82 O/12 - Barrier Mountain
15-31-12-W5
51°39′N 115°37′W
Approximately 55 km north of Banff.

The name for this 2957 m mountain is descriptive. The mountain forms a ridge which extends north-west as a barrier from the Panther River to the Red Deer River.

Barrier Pond (pond)

82 O/12 - Barrier Mountain
22-31-12-W5
51°40′05″N 115°37′10″W
Approximately 56 km north of Banff.

This lake was officially named 20 October 1983 because of its proximity to Barrier Mountain, which is found immediately to its south.

Barwell, Mount (mountain)

82 J/15 - Bragg Creek
17-21-5-W5
50°47′N 114°39′W
Approximately 40 km south-west of Calgary.

This mountain, which is 1829 m in altitude, was named in 1916 after C.S.W. Barwell of the Dominion Land Survey. He was an assistant to A.O. Wheeler (see Wheeler Flats) on surveys in 1895-1896 and subsequently moved to the Yukon.

Base Line Creek (creek)

83 B/3 - Tay River
29-37-8-W5
52°13′N 115°12′W

Approximately 27 km south-west of Rocky Mountain House.

The name was officially applied to this creek in 1941 and was suggested because the creek has its source on the Tenth Baseline.

Base Line Lake (lake)

83 L/4 - Kakwa Falls
35-60-13-W6
54°14′N 119°51′W
Approximately 62 km north-west of Grande Cache.

The name for this lake was officially adopted 4 December 1958 and is descriptive of the lake's location - it sits on the Sixteenth Baseline. A camera station was located nearby during the Interprovincial Boundary Survey, 1918-24.

Baseline Creek (creek)

83 F/11 - Dalehurst
32-52-23-W5
53°32′N 117°21′W
Flows east into Athabasca River, approximately 61 km west of Edson.

This creek follows along the Fourteenth Baseline. Its descriptive name was made official 27 May 1946.

Baseline Mountain (mountain)

83 B/3 - Tay River
31-36-10-W5
52°08′00″N 115°25′45″W
Approximately 42 km south-west of Rocky Mountain House.

This name, officially adopted 8 November 1978, was given to this mountain because it lies on the Tenth Baseline. The name is locally well-established.

Basilica Mountain (mountain)

83 D/16 - Jasper
22-44-3-W6
52°48′N 118°20′W

Approximately 9 km south south-west of Jasper.

The name for this mountain, 2865 m in altitude, was given by M.P. Bridgland (see Mount Bridgland) in 1916. The mountain has a fancied resemblance to a royal palace law courts or assembly hall, commonly known as a basilica. (see also Curia Mountain)

Bastion Peak (peak)

83 D/9 - Amethyst Lakes
15-43-3-W6
52°43′N 118°20′W
Approximately 26 km south-west of Jasper.

This mountain peak, which is 2970 m in altitude, has a sharp, projecting peak resembling a bastion. The name was suggested by Dr. E. Deville, Director of the Geological Survey of Canada, and was officially approved in 1916.

Bateman Creek (creek)

82 O/2 - Jumpingpound Creek
8-24-5-W5
51°03′N 114°48′W
Flows south-east into Jumpingpound Creek, approximately 40 km west of Calgary.

This creek was named after Tom Bateman, postmaster and pioneer rancher in the area. The name was officially approved 12 December 1939.

Bateman Ridge (ridge)

82 O/2 - Jumpingpound Creek
25-5-W5
51°05′N 114°35′W
Approximately 25 km west of Calgary.

(see Bateman Creek)

Bath Creek (creek)

82 N/8 - Lake Louise
32-28-16-W5
51°27′N 116°12′W

Flows east into Bow River, approximately 55 km north-west of Banff.

The name for this creek originated 20 July 1881 when Major A.B. Rogers of the Canadian Pacific Railway Engineering staff took an involuntary "bath" in the stream when thrown from his horse. The creek had been known formerly as Moore's Creek.

Bath Glacier (glacier)

82 N/9 - Hector Lake
29-17-W5
51°31′N 116°23′W
Approximately 70 km north-west of Banff.

(see Bath Creek)

Bauerman Creek (creek)

82 G/1 - Sage Creek
2-1-W5
49°07′N 114°01′W
Approximately 11 km north-west of Waterton Park.

(see Mount Bauerman)

Bauerman, Mount (mountain)

82 G/1 - Sage Creek
18-2-1-W5
49°08′N 114°08′W
Approximately 17 km north-west of Waterton Park.

Hilary Bauerman (1835-1909) was born in London, England and studied in England and Germany. He travelled widely during his career and was the geologist for the British Boundary Commission of 1858-1862, from the Pacific to the Rockies. It is after this metallurgist, mineralogist and professor of geology that Mount Bauerman, 2377 m in altitude, is named. The name was officially approved 9 January 1917.

Baymar Creek (creek)

82 O/7 - Wildcat Hills
4-27-6-W5
51°16′N 114°48′W
Approximately 25 km west north-west of Cochrane.

The locally well-established name for this creek was officially submitted for approval by D.P. Nichols in 1950 from the name of the oil well near the mouth of the creek.

Bazalgette, Mount (mountain)

83 E/9 - Moberly Creek
5-53-3-W6
53°33′N 118°24′W
Approximately 55 km west north-west of Hinton.

This 2438 m mountain peak was named in honour of Squadron Leader Ian Willoughby Bazalgette, V.C., D.F.C., who was born in Calgary in 1918. He was the only Albertan honoured with the Victoria Cross in the Second World War. In 1949, members of the Geographic Board of Alberta chose to name several unnamed peaks after Alberta's decorated personnel who paid the supreme sacrifice, and Bazalgette was one of these heroes. On 4 August 1944, during a bombing raid on France, Bazalgette and his squadron in their Lancaster bomber suffered heavy anti-aircraft fire. Their mission was a success, but it cost the squadron leader his life. In 1973, Bazalgette's name was entered into the Aviation Hall of Fame.

Beacon Lake (lake)

83 D/9 - Amethyst Lakes
19-41-2-W6
52°32′N 118°16′W
Approximately 39 km south south-west of Jasper.

(see Beacon Peak)

Beacon Peak (peak)
83 D/9 - Amethyst Lakes
20-41-2-W6
52°32'N 118°15'W
Approximately 39 km south south-west of
Jasper.

This 2986 m mountain was named in 1922
by A.O. Wheeler (see Wheeler Flats). The
name is descriptive of the peak's isolated
position.

Beagle Creek (creek)
82 J/1 - Langford Creek
13-2-W5
50°04'N 114°15'W
Flows north into Chaffin Creek,
approximately 45 km west of Claresholm.

The name Beagle Creek was made official 3
December 1941. There is no origin
information available for this locally well-
established name.

Bear Creek (creek)
82 J/8 - Stimson Creek
14-16-4-W5
50°21'N 114°28'W
Flows east into Pekisko Creek,
approximately 38 km south of Turner Valley.

The name for this creek was officially
approved 12 December 1939. The precise
origin of the name is unknown.

Bear Creek (creek)
82 O/11 - Burnt Timber Creek
10-31-19-W5
51°40'N 115°12'W
Flows south into Red Deer River,
approximately 60 km north-east of Banff.

The name for this creek, officially approved
in 1960, is likely descriptive.

Bear Creek (creek)
83 E/13 - Dry Canyon
2-3-58-13-W6
53°59'00"N 119°50'30"W
Flows north into South Kakwa River,
approximately 47 km west north-west of
Grande Cache.

This locally well-established name was
officially approved 4 December 1958. The
name Bear Creek was submitted for approval
by Mr. R. Eben, a local outfitter and guide.

Bear Pond (lake)
82 J/1 - Langford Creek
12-36-14-4-W5
50°14'N 114°26'W
Approximately 90 km south south-west of
Calgary.

The application to have this lake named was
made by the Department of Fish and
Wildlife, after having stocked the lake with
arctic grayling. The need for a name as a
navigational point prompted the proposal.

Bears Hump (cliff)
82 H/4 - Waterton Lakes
26-1-30-W4
49°04'N 113°55'W
Approximately 45 km south of Pincher
Creek.

The Bears Hump is a projecting cliff on the
south-east face of Mount Crandell, about
250 m above the ground. Until recently, the
feature was called "Black Bear Mountain"
and is still identified as such by a number of
local residents. The name was made official 8
June 1971.

Beatty Creek (creek)
83 C/13 - Medicine Lake
14-45-25-W5
52°53'N 117°31'W
Flows north-west into Alpland Creek,
approximately 37 km east of Jasper.

The name for this creek was officially
approved 12 April 1947. Its origin is
uncertain, but it is possibly named after
Vice-Admiral David Beatty, who
commanded ships during the Battle of
Jutland, 1916. (see Jutland Mountain)

Beatty, Mount (mountain)
82 J/11 - Kananaskis Lakes
2-20-10-W5
50°40'N 115°17'W
Approximately 70 km west of Turner Valley.

This mountain, which is 2999 m in altitude,
was named after Admiral Earl Beatty
(1871-1936), who commanded the Grand
Fleet (1916-1919), including the Battle of
Jutland, in the First World War. (see Jutland
Mountain)

Beaudry Lakes (lakes)
82 J/8 - Stimson Creek
34-16-4-W5
50°23'45"N 114°28'00"W
Approximately 30 km south-west of Turner
Valley.

These three small lakes were named after
Oscar Beaudry, who bought the land near
the lakes in 1901. The property was known
as the Eden Valley Ranch and is now part of
the Eden Indian Reserve. The name was
officially approved 5 May 1983.

Beaupre, Mount (mountain)
83 E/2 - Resplendent Creek
11-47-5-W6
53°02'N 118°37'W
Approximately 40 km west north-west of
Jasper.

This mountain, 2778 m in altitude, was named
in 1923 by A.O. Wheeler (see Wheeler Flats).
It was named after a guide of the Canadian
Pacific exploration party of 1872 led by
Sandford Fleming (see Introduction). No
other origin information is available.

Beauty Creek (creek)

83 C/6 - Sunwapta
1-39-24-W5
52°20′N 117°19′W
Flows west into Sunwapta River, over
Stanley Falls, approximately 80 km south
south-east of Jasper.

The name for this creek is descriptive.

Beauvais Lake (lake)

82 G/8 - Beaver Mines
29-5-1-W5
49°25′N 114°06′W
Approximately 15 km south-west of Pincher
Creek.

In 1882, Remi Beauvais came to Pincher
Creek from Union, Oregon by covered
wagon. He brought with him some fine
horses and homesteaded by this lake. It is for
this man that Beauvais Lake, one of the
beauty spots of the foothills, is named.

Beauvais Lake Provincial Park

(provincial park)
82 G/8 - Beaver Mines
5-1-W5
49°25′N 114°06′W
Approximately 16 km west south-west of
Pincher Creek.

(see Beauvais Lake)

Beauvert Lake (lake)

83 D/16 - Jasper
15-45-1-W6
52°53′N 118°03′W
Approximately 1 km east of Jasper.

The name for this lake was suggested by H.
Matheson, D.L.S., in 1914. The lake was
originally called "Horseshoe Lake" because
of its shape, but, to avoid duplication, the
name was changed to describe the "beautiful
green" colouring ranging through every

tone from pale aquamarine to jade and
malachite.

Beaver Bluffs (ridge)

83 E/1 - Snaring
49-1-W6
53°13′N 118°04′W
Approximately 37 km north of Jasper.

This high, steep ridge was named by M.P.
Bridgland (see Mount Bridgland) in 1916
after the numerous beaver found along the
Athabasca River.

Beaver Creek (creek)

82 J/1 - Langford Creek
12-13-4-W5
50°06′N 114°26′W
Flows south-west into Livingstone River,
approximately 60 km west north-west of
Claresholm.

Submitted by B.R. McKay in 1911, Beaver
Creek was officially approved in 1959. The
precise origin of the name is unknown.

Beaver Creek (creek)

82 J/8 - Stimson Creek
11-16-3-W5
50°20′N 114°20′W
Flows south-east into Sheppard Creek,
approximately 38 km south of Turner Valley.

The name for this creek was officially
approved 12 December 1939. No origin
information is available.

Beaver Creek (creek)

83 C/13 - Medicine Lake
52°51′55″N 117°43′15″W
Flows north into Beaver Lake,
approximately 24 km west of Jasper.

This creek is probably named for its
proximity to Beaver Lake. The name is well-
established in local usage. (see Beaver Lake)

Beaver Flat (flat)

83 C/9 - Wapiabi Creek
14-42-17-W5
52°37′N 116°20′W
Approximately 23 km north-west of
Nordegg.

Beaver Flat received its official name in 1940.
The origin of the name is unknown.

Beaver Lake (lake)

83 C/13 - Medicine Lake
8-45-26-W5
52°52′N 117°43′W
Approximately 24 km east of Jasper.

No origin information is known.

Beaver Mines (locality)

82 G/8 - Beaver Mines
10-6-2-W5
49°27′N 114°12′W
Approximately 18 km west south-west of
Pincher Creek.

The Blackfoot name for this locality refers to
the deposit of red ochre found nearby which
the natives used for religious paint. "Estay-
sukta" translates as "where we get paint."
(Dempsey). The post office was opened here
in 1912. The Beaver Coal Mines were
opened in 1905, and it is from these mines
that the town got its name. In 1909 the
Western Coal and Coke Company took an
active interest in the development of the
local coal deposits and sent a crew of twenty-
five men to survey and develop it. The site
grew, as many coal towns did at the time,
but eventually met the fate of dozens like it
after the 1930s. The last mine closed in 1971,
but Beaver Mines was considered a ghost
town long before that.

Beaver Mines Creek (creek)

82 G/8 - Beaver Mines
24-6-2-W5
49°29′N 114°08′W

Approximately 15 km west of Pincher Creek, emptying north-east into Castle River.

(see Beaver Mines)

Beaver Mines Lake (lake)

82 G/8 - Beaver Mines
1-5-3-W5
49°22'N 114°18'W
Approximately 25 km south-west of Pincher Creek.

This lake was officially named in 1958. (see Beaver Mines)

Beaverdam Creek (creek)

82 G/16 - Maycroft
6-11-2-W5
49°53'N 114°15'W
Flows south-west into Bob Creek, approximately 20 km north-east of Blairmore.

This name is likely descriptive. It was officially applied to this feature in 1941. No other origin information is available for this locally well-established name.

Beaverdam Creek (creek)

83 F/3 - Cadomin
11-48-22-W5
53°08'N 117°06'W
Flows north-west into McLeod River, approximately 67 km south-west of Edson.

This creek was named in 1944. The precise origin of the name is unknown.

Beaverdam Creek (creek)

83 E/12 - Pauline Creek
14-53-13-W6
53°34'N 119°48'W
Flows south-east into Pauline Creek, approximately 56 km south-west of Grande Cache.

(see Beaverdam Pass)

Beaverdam Creek (creek)

83 L/3 - Copton Creek
28-59-9-W6
54°08'N 119°18'W
Flows north-west into Copton Creek, approximately 35 km north-west of Grande Cache.

This creek has its source at Beaverdam Pass.

Beaverdam Pass (pass)

83 E/12 - Pauline Creek
17-53-13-W6
53°34'N 119°53'W
Approximately 61 km south-west of Grande Cache, on the Alberta-B.C. boundary.

This pass was named on the Jobe-Phillips map (see Avalanche Pass), as there was evidence of beaver-dams and houses at the time of exploration in 1914. By the time of the Interprovincial Boundary Survey (1918-1924), there was no longer any sign of the beaver, which had either been trapped out or left the area.

Bedson Ridge (ridge)

83 F/5 - Entrance
49-27-W5
53°16'N 117°55'W
Approximately 25 km south-west of Hinton.

This ridge, named in 1916, takes its name from Bedson Railway Station (now Miette), which was likely named for a former Warden of the Stoney Mountain Penitentiary in Manitoba.

Beehive, The (mountain)

82 N/8 - Lake Louise
19-28-16-W5
51°25'N 116°15'W
Approximately 55 km north-west of Banff.

The descriptive name for this 2270 m mountain was applied to this feature by Walter D. Wilcox (see Wilcox Pass) in 1894.

Beehive Creek (creek)

82 J/2 - Fording River
13-5-W5
50°03'N 114°35'W
Flows east into Oldman River, approximately 70 km south-west of Turner Valley.

(see Beehive Mountain)

Beehive Mountain (mountain)

82 J/2 - Fording River
13-5-W5
50°04'N 114°40'W
Approximately 75 km south-west of Turner Valley.

This mountain, which is 2895 m in altitude, was named by Dr. Dawson (see Tombstone Mountain) in 1886. The name for this mountain is descriptive of its appearance.

Beehive Pass (pass)

82 J/2 - Fording River
7-13-5-W5
50°03'50"N 114°40'25"W
Approximately 75 km south-west of Turner Valley.

The pass is adjacent to Beehive Mountain in the High Rock Range. The name was approved by the Historic Sites Board 5 June 1981. (see Beehive Mountain)

Belanger, Mount (mountain)

83 C/12 - Athabasca Falls
11-41-28-W5
52°31'N 117°55'W
Approximately 43 km south of Jasper.

This mountain, 3120 m in altitude, was named in 1921 by D.B. Dowling (see Dowling

Ford). It commemorates André Belanger, who with Lapensée (see Mount Lapensée) accompanied G. Franchère (see Franchère Peak) when he crossed the Athabasca Pass from Astoria in 1814. Belanger and Lapensée drowned in the Athabasca River below Brule Lake 25 May 1814.

Belgium Lake (lake)

82 J/11 - Kananaskis Lakes
24-20-11-W5
50°43′N 115°24′W
Approximately 80 km west of Turner Valley.

The name for this lake was likely chosen for its proximity to Mount Queen Elizabeth and Mount King Albert, which were named in honour of the King and Queen of Belgium. Belgium Lake was officially approved in 1957.

Bell, Mount (mountain)

82 N/8 - Lake Louise
6-27-15-W5
51°17′N 116°06′W
Approximately 40 km west north-west of Banff.

This 2910 m mountain was named during the Alpine Club of Canada Annual Camp held in 1910 on a slope on the east side of the lower of the two Consolation Lakes. Miss Nora Bell, a member of a small climbing party, made the first ascent. Mount Bell replaced the original name "Mount Bellevue."

Bellevue (village)

82 G/9 - Blairmore
20-7-3-W5
49°35′N 114°22′W
Located in the Crowsnest Pass, approximately 5 km south-east of Blairmore.

This village was named in 1905 by Mr. J.J. Fleutôt, who was the originator and

Managing Director of West Canadian Collieries Ltd. His daughter, Elsie, was an early visitor to the Crowsnest Pass. When she saw the view to the west, she exclaimed in French, "Quelle une belle vue," meaning, "what a beautiful view." In English, this became Bellevue, which became the official name for this village, incorporated 1 January 1957. (see Municipality of Crowsnest Pass)

Bellevue Hill (hill)

82 H/4 - Waterton Lakes
2-30-W4
49°07′N 113°55′W
Approximately 40 km south of Pincher Creek.

Bellevue Hill has an altitude of 2112 m. The name was officially applied to this hill 19 May 1943. (see Bellevue)

Benjamin Creek (creek)

82 O/11 - Burnt Timber Creek
4-30-8-W5
51°32′N 115°04′W
Flows east into Fallentimber Creek, approximately 55 km north-west of Cochrane.

Jonas Benjamin was a Stoney Indian chief of the area. The creek was officially named after him in 1917-18.

Bennington Peak (peak)

83 D/9 - Amethyst Lakes
36-42-3-W6
52°39′N 118°18′W
Approximately 29 km south south-west of Jasper.

This 3265 m mountain peak is a peak of Mount Fraser. It was named by A.O. Wheeler (see Wheeler Flats) in 1922 after Bennington, Vermont, where Simon Fraser (see Mount Fraser) was born.

Bergamot Creek (creek)

83 E/14 - Grande Cache
4-58-9-W6
53°59′N 119°17′W
Flows north into Sheep Creek, approximately 14 km north-west of Grande Cache.

This creek was named in 1953 after the wildflower, the Western Wild Bergamot, a native mint of an unusual mauve shade found throughout prairie and thinly wooded areas.

Bergne, Mount (mountain)

82 N/15 - Mistaya Lake
23-32-21-W5
51°46′N 116°52′W
Approximately 110 km north-west of Banff.

Frank Bergne was a member of The Alpine Club (London). He was killed while climbing with A.O. Wheeler (see Wheeler Flats) in 1907 in Switzerland. This 3176 m mountain was named by Wheeler in 1920 after Bergne.

Berland Range (range)

83 E/9 - Moberly Creek
53-4-W6
53°38′N 118°28′W
Approximately 75 km north north-west of Jasper.

This mountain range was officially named 29 July 1955 after the Berland River, which runs through this range. The river was originally "Baptiste River," after Baptiste Berland, an early trader; however, at the time of its naming, there were at least two other rivers of that same name. It therefore became Berland River.

Berry, Mount (mountain)

83 F/3 - Cadomin
7-47-24-W5
53°03′N 117°30′W

Approximately 93 km south-west of Edson.

This mountain, which is 2362 m in altitude was officially named in 1961. (see Berry's Creek)

Berry's Creek (creek)
83 F/3 - Cadomin
33-47-24-W5
53°06′N 117°28′W
Flows north-east into Gregg River, approximately 86 km south-west of Edson.

The name for this creek was suggested by W.H. Miller of the Geological Survey, who performed survey work in the area. It is named after William Oscar Berry (1878-1971), a packer and guide of the area who was associated with H. Mustard and Sons, guides and packers.

Bertha Creek (creek)
82 H/4 - Waterton Lakes
1-30-W4
49°02′N 113°54′W
Approximately 50 km south of Pincher Creek.

This creek was officially named 8 June 1971. (see Bertha Peak)

Bertha Lake (lake)
82 H/4 - Waterton Lakes
9-1-30-W4
49°01′N 113°56′W
Approximately 50 km south of Pincher Creek.

(see Bertha Peak)

Bertha Peak (peak)
82 H/4 - Waterton Lakes
1-30-W4
49°03′N 113°56′W
Approximately 50 km south of Pincher Creek.

The three features, Bertha Creek, Lake and Peak, are said to be named after an early resident of Waterton Lakes National Park. The name was first used on a map of the Crowsnest Forest Reserve by M.P. Bridgland (see Mount Bridgland) in 1914. No details are known about the person after whom these features were named.

Beryl Lake (lake)
83 D/9 - Amethyst Lakes
30-42-1-W6
52°39′N 118°08′W
Approximately 26 km south of Jasper.

This lake is named for the transparent precious stone, beryl, used in this instance as a synonym for beautiful.

Bess, Mount (mountain)
83 E/6 - Twintree Lake
27-50-10-W6
53°21′N 119°23′W
Approximately 103 km west north-west of Jasper.

This mountain, 3216 m in altitude, and the nearby pass were named at the time of an expedition in 1911 by Dr. J. Norman Collie (see Collie Creek) and A.L. Mumm (see Mumm Peak) after Bessie Gunn of Lac Ste. Anne. She became the wife of M.C. McKeen, Member of the Legislative Assembly for Lac Ste. Anne, who held a seat in the United Farmers of Alberta Government during its term (1921-1935).

Bess Pass (pass)
83 E/6 - Twintree Lake
23-50-10-W6
53°20′N 119°21′W
Approximately 102 km west north-west of Jasper.

(see Mount Bess)

Bident Mountain (mountain)
82 N/8 - Lake Louise
27-16-W5
51°17′N 116°08′W
Approximately 45 km west north-west of Banff.

Officially approved 30 May 1904, the name for this feature describes the shape of the mountain, similar to that of a double tooth. The mountain is 3084 m in altitude.

Big Coulee (coulee)
82 G/16 - Maycroft
11-10-2-W5
49°48′N 114°10′W
Approximately 30 km north-east of Blairmore.

Officially approved in 1941, the name for this coulee is probably descriptive. Its precise origin, however, is unknown.

Big Creek (creek)
83 E/16 - Donald Flats
19-56-1-W6
53°51′N 118°07′W
Flows east into Berland Creek, approximately 58 km east of Grande Cache.

The locally well-established name for this creek was officially approved in 1947, but the origin remains unknown.

Big Horn Indian Reserve #144 A (Indian reserve)
83 C/8 - Nordegg
29-39-16-W5
52°23′N 116°15′W
Approximately 11 km south-west of Nordegg.

(see Bighorn Range)

Big Shale Hill (hill)

83 E/12 - Pauline Creek
25-53-12-W6
53°56′N 119°46′W
Approximately 53 km south-west of Grande
Cache.

The name for this hill was officially approved
in 1963 and is likely descriptive of the rock
formation of the hill.

Big Shovel Pass (pass)

83 C/13 - Medicine Lake
52°47′00″N 117°15′00″W
Approximately 18 km south-east of Jasper.

This pass is named to differentiate it from
"Little Shovel Pass." The name is well-
established in local usage and was officially
approved 5 May 1987. (see Shovel Pass)

Bighorn Canyon (canyon)

83 C/8 - Nordegg
25-39-17-W5
52°23′30″N 116°20′59″W
Approximately 21 km west south-west of
Nordegg.

This canyon is nearly 5 km long, and the
Bighorn River plunges into it in two vertical
falls. The name was approved by the
Canadian Permanent Committee on
Geographical Names 19 May 1977. (see
Bighorn Range)

Bighorn Creek (creek)

82 O/12 - Barrier Mountain
17-32-11-W5
51°44′N 115°32′W
Flows south into Red Deer River,
approximately 60 km north of Banff.

(see Bighorn Range)

Bighorn Lake (lake)

82 O/5 - Castle Mountain
29-12-W5
51°27′47″N 115°39′50″W
Approximately 30 km north north-west of
Banff.

This lake was officially named in June 1979
because of the number of Rocky Mountain
Bighorn sheep whose favourite pasturing
ground is in this area.

Bighorn Range (range)

83 C/8 - Nordegg
27-40-17-W5
52°30′N 116°25′W
Approximately 18 km west south-west of
Nordegg.

The name for this range is a translation of an
Indian name. In earlier days, these mountains
were noted hunting grounds full of Rocky
Mountain Sheep or Bighorn Sheep. The name
is first found on Palliser's map of 1865.

Bighorn River (river)

83 C/8 - Nordegg
16-39-16-W5
52°21′N 116°15′W
Flows east into North Saskatchewan River,
approximately 17 km south-west of
Nordegg.

(see Bighorn Range)

Bingley Peak (peak)

83 D/15 - Rainbow
14-45-5-W6
52°52′N 118°37′W
Approximately 36 km west of Jasper.

This mountain peak, which is 2438 m in
altitude, was named in 1863 by Dr. William
Cheadle (1835-1910). The birthplace of this
noted traveller was Bingley, Yorkshire.
Cheadle and Viscount Milton published *The
Northwest Passage by Land* in 1865.

Birdseye Butte (hill)

82 H/4 - Waterton Lakes
1-2-29-W4
49°05′N 113°45′W
Approximately 45 km south south-east of
Pincher Creek.

The name Birdseye Butte was officially
applied to this feature 8 June 1971. No other
origin information for this locally well-
established name is available.

Birdwood Creek (creek)

82 J/14 - Spray Lakes Reservoir
14-21-11-W5
50°47′N 115°26′W
Flows west into Spray River, approximately
90 km west south-west of Calgary.

(see Mount Birdwood)

Birdwood, Mount (mountain)

82 J/14 Spray Lakes Reservoir
18-21-10-W5
50°47′N 115°22′W
Approximately 90 km west south-west of
Calgary.

This 3097 m mountain was named after
Field Marshal Sir William Riddell Birdwood
(1865-1951), who was the first commander of
the Australian and New Zealand Army
Corps (ANZAC). He is especially remembered
for his part in the Gallipoli campaign of the
First World War. He was the author of
Khaki and Gown (1941) and *In My Time*
(1946).

Bishop Creek (creek)

82 J/7 - Mount Head
30-17-6-W5
50°28′N 114°49′W
Flows north-east into Loomis Creek,
approximately 43 km south-west of Turner
Valley.

(see Mount Bishop)

Bishop, Mount (mountain)

82 J/7 - Mount Head
15-17-7-W5
50°26′N 114°52′W
Approximately 48 km south-west of Turner Valley.

This 2850 m mountain was officially named in 1918. Colonel W.A. Bishop, V.C., D.S.O., M.C., a Canadian airman, shot down three enemy planes in one flight in 1917. Colonel W.A. (Billy) Bishop shot down a total of 72 planes during the First World War.

Bison Creek (creek)

82 N/15 - Mistaya Lake
1-33-19-W5
51°53′N 116°41′W
Flows west into Mistaya River, approximately 110 km north-west of Banff.

The origin of the name is unknown.

Bistre, Mount (mountain)

83 E/1 - Snaring
4-49-2-W6
53°12′N 118°13′W
Approximately 38 km north north-west of Jasper.

The descriptive name for this 2346 m mountain was suggested by M.P. Bridgland (see Mount Bridgland) in 1916 from the warm brown colour of the feature.

Black Creek (creek)

82 G/16 - Maycroft
24-11-2-W5
49°56′N 114°08′W
Flows east into Callum Creek, approximately 35 km north-east of Blairmore.

Mr. John Black first came west in 1882 to Helena, Montana and moved two years later to Fort MacLeod, where he took a position with I.G. Baker & Co. He then joined the Hudson's Bay Company for several years until he entered into business for himself. He died in 1899. This creek was named for Black in recognition of his contribution to the early history of southern Alberta.

Black Mountain (mountain)

83 C/8 - Nordegg
20-40-16-W5
52°28′N 116°16′W
Approximately 13 km west of Nordegg.

The name for this mountain, 1930 m in altitude, is descriptive of its appearance.

Black Mountain (mountain)

82 G/16 - Maycroft
2-12-2-W5
49°58′N 114°10′W
Approximately 45 km north-east of Blairmore.

(see Black Creek)

Black Canyon Creek (creek)

83 C/8 - Nordegg
31-39-15-W5
52°24′N 116°09′W
Approximately 10 km south south-west of Nordegg.

The origin of the name is unknown.

Black Cat Mountain (mountain)

83 F/5 - Entrance
20-50-27-W5
53°19′N 117°55′W
Approximately 21 km west south-west of Hinton.

The shape of this mountain at one time resembled a cat with its back arched. After its official naming, a forest fire masked the effect of the mountain's shape, but the name remained.

Black Prince, Mount (mountain)

82 J/11 - Kananaskis Lakes
18-20-9-W5
50°42′N 115°15′W
Approximately 70 km west of Turner Valley.

Officially adopted in 1922, the name for this 2932 m mountain is taken from a British cruiser engaged in the Battle of Jutland, 1916, during the First World War (see Jutland Mountain). The Black Prince was destroyed by the fire of German battleships, and 750 men died.

Black Rock Lake (lake)

82 O/4 - Banff
8-25-14-W5
51°07′N 115°54′W
Approximately 25 km west south-west of Banff.

(see Black Rock Mountain)

Black Rock Mountain (mountain)

82 O/6 - Lake Minnewanka
26-27-9-W5
51°19′N 115°09′W
Approximately 30 km north north-east of Canmore.

This sharp black peak, which rises 2462 m in altitude, is clearly visible from the Cochrane area. The descriptive name is well-established in local usage and was officially approved in 1958.

Blackface Creek (creek)

83 C/14 - Mountain Park
8-45-24-W5
52°52′N 117°27′W
Flows south-west into Rocky River, approximately 43 km east of Jasper.

(see Blackface Mountain)

Blackface Mountain (mountain)

83 C/14 - Mountain Park
28-45-24-W5
52°54′N 117°25′W
Approximately 44 km east of Jasper.

The descriptive name of this mountain, 2867 m in altitude, was officially approved in 1922.

Blackfriars Peak (peak)

83 C/5 - Fortress Lake
52°19′35″N 117°37′00″W
Approximately 60 km south-east of Jasper, in Jasper National Park.

In 1901, Dr. Jean Habel (see Mount Habel) suggested the name Blackfriars for this 3210 m mountain, after the "Schwartze Monche" in the Bernese Oberland; "schwartze" means black and "monche" means friar or monk. In 1921, it was proposed that the name be changed to "Blackmonks" because the name Blackfriars was duplicated nearby in British Columbia. In November of 1921, the name Dais Mountain was approved for the feature. The name Blackfriars was resubmitted by Alfred J. Ostheimer III in honour of Jean Habel's original suggestion. The name Blackfriars Peak for the feature in Alberta is used locally, especially among mountaineers, and appears in the *Climber's Guides*.

Blackhorn Peak (mountain)

83 D/9 - Amethyst Lakes
25-42-2-W6
52°39′N 118°09′W
Approximately 26 km south of Jasper.

The descriptive name for this sharp, black mountain, 3000 m in altitude, was given in 1916 by M.P. Bridgland. (see Mount Bridgland)

Blackrock Mountain (mountain)

83 D/9 - Amethyst Lakes
36-41-3-W6
52°34′N 118°17′W

Approximately 37 km south south-west of Jasper.

The name for this 2910 m mountain is descriptive. A.O. Wheeler (see Wheeler Flats) gave it its name in 1922.

Blackshale Creek (creek)

82 J/11 - Kananaskis Lakes
15-20-9-W5
50°41′35″N 115°10′30″W
Flows south into Smith-Dorrien Creek, approximately 60 km west of Turner Valley.

Officially approved 8 November 1978, Blackshale Creek had previously been a popular local name for the stream due to exposures of large deposits of black shale in the vicinity. The shale dust from these deposits has turned the water in the stream black.

Blackstone Gap (gap)

83 C/10 - George Creek
7-42-18-W5
52°36′N 116°35′W
Approximately 37 km west north-west of Nordegg.

The Blackstone River flows through this gap. The name was officially approved in 1956. (see Blackstone River)

Blackstone River (river)

83 C/16 - Brown Creek
26-44-15-W5
52°50′N 116°07′W
Flows north-east into Brazeau River approximately 39 km north of Nordegg.

The name for this river, which was officially approved in November 1916, is descriptive of an outcrop of dark shale near the head of the stream.

Blairmore (town)

82 G/9 - Blairmore
35-7-4-W5
49°36′N 114°26′W

Approximately 5 km south-east of Coleman.

The McLaren sawmill was built nearby at the turn of the century, and the Blackfoot name, "nitay-stato-ksisokyopi" reflects this, translating as "where they rip logs." (Dempsey). The name Blairmore was derived from the names of two men. Mr. Blair and Mr. More were contractors who laid the Canadian Pacific Railway line through what is now Blairmore. The town was officially given this name 15 November 1898. Other sources contend that the name of the town is after A.G. Blair (1844-1907), the Minister of Railways in the Laurier Government. The town was incorporated 29 September 1911. (see Municipality of Crowsnest Pass)

Blairmore Creek (creek)

82 G/9 - Blairmore
3-8-4-W5
49°36′N 114°28′W
Flows south into Crowsnest River at Blairmore, approximately 6 km south-east of Coleman.

(see Blairmore)

Blairmore Range (range)

82 G/9 - Blairmore
7-4-W5
49°35′N 114°25′W
Approximately 1 km east of Blairmore.

(see Blairmore)

Blakiston Creek (creek)

82 H/4 - Waterton Lakes
31-1-29-W4
49°05′N 113°51′W
Flows east into the Dardanelles, approximately 45 km south of Pincher Creek.

Blakiston Creek was once known as both "Blakiston Brook" and "Pass Creek." (see Mount Blakiston)

Blakiston Falls (falls)

82 G/1 - Sage Creek
15-2-1-W5
49°07′N 114°03′W
Approximately 13 km north-west of
Waterton Park.

The name Blakiston Falls was officially
applied to this feature by the Canadian
Permanent Committee on Geographical
Names 27 April 1972. (see Mount Blakiston)

Blakiston, Mount (mountain)

82 G/1 - Sage Creek
2-1-W5
49°05′N 114°02′W
Approximately 10 km north-west of
Waterton Park.

Lieutenant Thomas Blakiston of the Royal
Artillery was attached to the Palliser
Expedition (see Palliser Range) as a magnetic
observer. He named this 2910 m mountain
after himself in 1858 during his exploration
of the North and South Kootenay passes. In
1861, he organized an expedition up the
Yangtze River in China and went 1448 km
further than any Westerner had before him.
He always had an interest in nature, which
led him to compile a catalogue on birds
while he was in Japan on another of his
many expeditions.

Blanchard Creek (creek)

83 C/15 - Cardinal River
1-45-19-W5
52°51′N 116°37′W
Flows north-west into Brazeau River,
approximately 94 km east of Jasper.

The origin of the name is unknown.

Blane, Mount (mountain)

82 J/11 - Kananaskis Lakes
29-20-8-W5
50°43′N 115°04′W

Approximately 55 km west of Turner Valley.

Sir C.R. Blane was the commander of the
battleship Queen Mary, destroyed by
German gunfire at the Battle of Jutland (see
Jutland Mountain) in May 1916. The
mountain, 2993 m in altitude, was officially
named Mount Blane in 1922 in honour of
this commander.

Blind Canyon (canyon)

82 H/4 - Waterton Lakes
22-3-30-W4
49°13′N 113°59′W
Approximately 20 km north of Waterton
Park.

The name Blind Canyon was officially
approved for this feature in 1943. The name
is a locally well-established one, but its
precise origin is unknown.

Block Lakes (lakes)

82 O/5 - Castle Mountain
12-28-14-W5
51°22′N 115°50′W
Approximately 30 km north north-west of
Banff.

(see Block Mountain)

Block Mountain (mountain)

82 O/5 - Castle Mountain
11-28-14-W5
51°22′N 115°51′W
Approximately 35 km north north-west of
Banff.

The name for this squarish mountain, 2935
m in altitude, was officially approved in
1958. It is descriptive of the mountain's
shape.

Bloodroot Creek (creek)

83 E/14 - Grande Cache
11-57-10-W6
53°54′N 119°22′W

Flows north into Neighbor Creek,
approximately 16 km west of Grande Cache.

No information regarding the origin is
known.

Blue Creek (creek)

83 E/7 - Blue Creek
22-50-5-W6
53°20′N 118°38′W
Flows south-east into Snake Indian River,
approximately 63 km north-west of Jasper.

The name for this creek is likely descriptive
of its colour and was officially approved 2
August 1956.

Blue Lake (lake)

83 F/5 - Entrance
8-52-26-W5
53°29′N 117°48′W
Approximately 16 km north-west of Hinton.

The name of this lake is likely descriptive of
its appearance.

Blue Range (range)

82 J/13 - Mount Assiniboine
21-12-W5
50°48′N 115°33′W
Approximately 100 km west south-west of
Calgary, on the Alberta-B.C. boundary.

The descriptive name for this range is due to
its blue appearance when seen from a
distance.

Blue Ridge (ridge)

82 J/10 - Mount Rae
18-4-W5
50°32′N 114°30′W
Approximately 20 km south-west of Turner
Valley.

Officially approved in 1943, the name for
this ridge comes from the fact that the ridge
is covered with blue spruce which from a

distance and certain vantage point appears to be a bluish colour.

Blue Grouse Basin (basin)

82 G/1 - Sage Creek
13-2-1-W5
49°07′N 114°09′W
Approximately 45 km south of Pincher Creek.

The Blue Grouse is a large, slate-coloured bird found only in the foothills and mountains of Alberta. The intriguing annual movement of the Blue Grouse includes wintering at high altitudes and migrating to the lower slopes for the breeding season. The name was officially approved by the Canadian Permanent Committee on Geographical Names 27 April 1972.

Blue Grouse Pass (pass)

83 E/10 - Adams Lookout
35-52-7-W6
53°32′30″N 118°55′00″W
Approximately 38 km south south-east of Grande Cache.

Emil Moberly was a native trapper and hunter and when interviewed in 1978 said that this pass had been known by this name for at least 50 years. It was named because of the large number of Blue Grouse which inhabit the area. The name was officially approved 15 January 1979. (see Blue Grouse Basin)

Bluerock Creek (creek)

82 J/10 - Mount Rae
14-19-6-W5
50°36′N 114°44′W
Flows south-east into Sheep River, approximately 33 km west south-west of Turner Valley.

(see Bluerock Mountain)

Bluerock Mountain (mountain)

82 J/10 - Mount Rae
6-20-6-W5
50°40′N 114°48′W
Approximately 37 km west of Turner Valley.

The name of this mountain, 2789 m in altitude, is descriptive of its appearance.

Bluff Mountain (mountain)

82 G/9 - Blairmore
12-8-4-W5
49°38′N 114°25′W
Approximately 3 km north-east of Blairmore, in the Blairmore Range.

This 2145 m mountain was named in 1902. Its name is descriptive of a high steep bank or cliff. It was officially named by M.P. Bridgland (see Mount Bridgland) in 1915.

Bob Creek (creek)

82 G/16 - Maycroft
30-10-2-W5
49°52′N 114°15′W
Flows south into the Oldman River, approximately 30 km north-east of Blairmore.

This stream was named after a miner known as "Old Bob," who opened up a coal mine in the area in the early 1880s. The name was officially approved 29 July 1943.

Bogart, Mount (mountain)

82 J/14 - Spray Lakes Reservoir
31-22-9-W5
50°55′N 115°14′W
Approximately 75 km west of Calgary.

Dr. Donald Bogart Dowling's (see Dowling Ford) work in coal, petroleum and natural gas development warranted the naming of this 3144 m mountain after him.

Bolton Creek (creek)

83 L/2 - Bolton Creek
3-61-4-W6
54°15′N 118°31′W
Approximately 57 km north-east of Grande Cache.

M.E. Nidd, a field topographer, worked in this area and obtained the local names for features in the vicinity in 1947. Bolton Creek was one of these names, but its precise origin is unknown.

Bolton, Mount (mountain)

82 J/7 - Mount Head
16-6-W5
50°20′N 114°48′W
Approximately 52 km south-west of Turner Valley.

L.E.S. Bolton, D.L.S., was a member of the Surveyor General's staff at Ottawa. He was killed in action in the First World War, in June 1916. This mountain, 2706 m in altitude, was named in his honour.

Bonhomme, Roche (mountain)

83 C/13 - Medicine Lake
10-46-28-W5
52°57′N 117°57′W
Approximately 11 km north-east of Jasper.

This mountain, which stands 2495 m in altitude, has a curious shape: it resembles a man's face. The French word for this mountain literally translates to the English equivalent, "old chap" or "good fellow." It is mentioned in Grant's book *Ocean to Ocean* (1878). (see Grant Pass)

Bonnet Glacier (glacier)

82 O/5 - Castle Mountain
28-14-W5
51°27′N 115°55′W
Approximately 36 km north-west of Banff.

(see Bonnet Peak)

Bonnet Peak (peak)

82 O/5 - Castle Mountain
34-28-14-W5
51°25′N 115°52′W
Approximately 36 km north-west of Banff.

The name for this 3136 m mountain was applied in 1890 because of the summit's striking resemblance to a bonnet.

Boom Creek (creek)

82 N/1 - Mount Goodsir
27-26-15-W5
51°15′N 116°02′W
Flows east into Altrude Creek, approximately 35 km west north-west of Banff.

(see Boom Lake)

Boom Lake (lake)

82 N/8 - Lake Louise
31-26-15-W5
51°16′N 116°05′W
Approximately 48 km west north-west of Banff.

This lake was named in 1908. The name is considered descriptive of the driftwood dammed against a shoal resembling a lumberman's boom. The other features in the area take their names from this lake.

Boom Mountain (mountain)

82 N/8 - Lake Louise
30-26-15-W5
51°15′N 116°04′W
Approximately 45 km west north-west of Banff.

This mountain stands 2760 m in altitude. (see Boom Lake)

Border Ranges (range)

82 G/8 - Beaver Mines
4-4-W5
49°20′N 114°30′W

Approximately 30 km south-west of Pincher Creek.

The name for this range was officially approved 17 March 1967. Its name is descriptive of its location along the boundary between Alberta and British Columbia.

Bosche Range (range)

83 E/8 - Rock Lake
49-1-W6
53°15′N 118°05′W
Approximately 42 km north of Jasper.

See Roche à Bosche, which terminates the range, and gives it its name.

Bosche, Roche à (mountain)

83 E/1 - Snaring
11-49-1-W6
53°12′N 118°02′W
Approximately 40 km north of Jasper.

The meaning of "bosche" is not known. It may be a corruption of the French word for "a hump," which is "bosse." The name for this 2123 m mountain has been in use since 1873, and is mentioned in Grant's (see Grant Pass) *Ocean to Ocean* (1878).

Bosporus (narrows)

82 H/4 - Waterton Lakes
1-30-W4
49°03′N 113°54′W
Approximately 50 km south of Pincher Creek.

This feature joining the Upper and Middle Waterton Lakes was named after the Bosporus, the strait joining the Sea of Marmara and the Black Sea. The name was officially approved in 1943.

Boswell, Mount (mountain)

82 H/4 - Waterton Lakes
1-29-W4
49°01′N 113°52′W

Approximately 55 km south of Pincher Creek.

Mount Boswell, 2400 m in altitude, was named in 1917 after an obscure veterinary surgeon associated with the British Boundary Commission. The mountain was previously known as "Street Mountain," named after Jack Street, who lost his life in an avalanche.

Bosworth, Mount (mountain)

82 N/8 - Lake Louise
29-17-W5
51°28′N 116°20′W
Approximately 65 km north-west of Banff.

This 2771 m mountain was named in 1904 after G.M. Bosworth, fourth Vice President of the Canadian Pacific Railway.

Botten (railway point)

83 L/2 - Bolton Creek
14-60-5-W6
54°11′N 118°37′W
Approximately 48 km north-east of Grande Cache.

This station on the Alberta Resources Railway was officially named 8 May 1969, after Private Cyril Botten, who was killed in action December 1943.

Boulder Pass (pass)

82 N/8 - Lake Louise
17-29-15-W5
51°29′N 116°05′W
Approximately 50 km north-west of Banff.

The name for this pass is likely descriptive of its appearance.

Boule Range (range)

83 E/8 - Rock Lake
50-1-W6
53°20′N 118°02′W

Approximately 54 km north of Jasper.

(see Roche Boule)

Boule, Roche (mountain)

83 F/5 - Entrance
24-50-28-W5
53°20′N 117°57′W
Approximately 24 km west south-west of
Hinton.

This 2204 m mountain has a descriptive
name, since it is shaped like a bowl or a ball
at its summit. "Roche Boule" is the French
name meaning "ball rock." Prior to 1916,
the mountain was erroneously named
"Bullrush Mountain."

Boulton Creek (creek)

82 J/11 - Kananaskis Lakes
25-19-9-W5
50°38′N 115°08′W
Flows west into Lower Kananaskis Lake,
approximately 60 km west of Turner Valley.

This creek may be named after Tom
Boulton, an old-time rancher who lived at
Blacktail P.O. on Willow Creek.

Boundary Lake (lake)

83 L/12 - Nose Creek
14-65-14-W6
54°37′N 120°00′W
Approximately 97 km south-west of Grande
Prairie, on the Alberta-B.C. boundary.

The name is descriptive of the lake's location
on the boundary.

Bourgeau Lake (lake)

82 O/4 - Banff
17-25-13-W5
51°08′N 115°47′W
Approximately 17 west south-west of Banff.

(see Mount Bourgeau)

Bourgeau, Mount (mountain)

82 O/4 - Banff
17-25-13-W5
51°08′N 115°48′W
Approximately 16 km west south-west of
Banff.

This 2930 m mountain was named by Sir
James Hector (see Mount Hector) of the
Palliser Expedition (see Palliser Range), after
Eugène Bourgeau (1813-1877), the
expedition's botanist. Bourgeau was a
charming Frenchman, known for his sober
and economical habits. His frank good
nature gained him friends wherever he
travelled. The nearby lake was also named
for him and was officially approved 16
January 1912. (see Massive Range)

Bovin Lake (lake)

82 G/1 - Sage Creek
19-3-1-W5
49°13′N 114°08′W
Approximately 25 km north-west of
Waterton Park.

The meadows surrounding this small lake
were the pasture ground for large herds of
horses in the days when the natives roamed
the area. The grounds are now a rich big
game region.

Bow Falls (falls)

82 O/4 - Banff
25-12-W5
51°10′N 115°35′W
Approximately 1 km south-east of Banff.

(see Bow River)

Bow Glacier (glacier)

82 N/10 - Blaeberry River
31-18-W5
51°39′N 116°31′W
Approximately 85 km north-west of Banff.

(see Bow River)

Bow Lake (lake)

82 N/9 - Hector Lake
23-31-18-W5
51°40′N 116°27′W
Approximately 85 km north-west of Banff.

(see Bow River)

Bow Pass (pass)

82 N/9 - Hector Lake
32-18-W5
51°43′N 116°30′W
Approximately 90 km north-west of Banff.

This pass is approximately 2042 m in
altitude. (see Bow River)

Bow Peak (peak)

82 N/9 - Hector Lake
32-31-17-W5
51°38′N 116°23′W
Approximately 75 km north-west of Banff.

(see Bow River)

Bow Range (range)

82 N/8 - Lake Louise
28-17-W5
51°21′N 116°16′W
Approximately 55 km west north-west of
Banff, along the Alberta-B.C. boundary.

(see Bow River)

Bow River (river)

82 N/10 - Blaeberry River
31-18-W5
51°39′N 116°30′W
At source.

The nearby glacier, lake, pass and peak take
their names from this river which rises at
the Bow Glacier and eventually becomes
part of the South Saskatchewan River system,
some 507 kilometres in length. Its name is a

translation of the Cree "manachaban sipi" meaning "the place from which bows are taken," according to J.B. Tyrrell. This is a reference to the availability along its course of wood suitable for making bows, a fact of which the Indians took full advantage. Bow Lake itself, however, was known as "Minismeimme" in Stoney and "Askowioosipi" in Cree, both of which translate to "cold water lake." These names, while recorded, were superseded by the name Bow Lake.

Bow Valley (provincial park)

82 O/3 - Canmore
29-24-8-W5
51°05'N 115°05'W
Approximately 9 km east of Canmore.

(see Bow Pass)

Bow Valley and Bow River

Bowlen, Mount (mountain)

82 N/8 - Lake Louise
16-27-16-W5
51°18'N 116°11'W
Approximately 50 km west north-west of Banff, on the Alberta-B.C. boundary.

Mount Bowlen, which stands 3072 m in altitude, may be clearly seen from Moraine Lake. The mountain was named in 1958 after Dr. J.J. Bowlen (1876-1959). Dr. Bowlen was honoured as a result of a lifetime of

service, both on a provincial and national level. Pioneer farmer, rancher and sheepman, he served in the Alberta Legislature as Liberal Member for the City of Calgary from 1931-1944 and was Leader of the Opposition for two years. As the first westerner to be appointed to an executive position with the organization, he was named a member of the Board of Governors of the Canadian Broadcasting Corporation in 1947. He served as Lieutenant-Governor of Alberta from 1950-1959.

Brachiopod Lake (lake)

82 N/8 - Lake Louise
10-29-15-W5
51°29'N 116°02'W
Approximately 45 km north-west of Banff.

(see Brachiopod Mountain)

Brachiopod Mountain (mountain)

82 N/8 - Lake Louise
10-29-15-W5
51°29'N 116°03'W
Approximately 45 km north-west of Banff.

Named by J.F. Porter, an alpinist from Chicago, this mountain has west slopes which are literally covered with fine samples of brachiopods and fossil corals of the Devonian Period. The name was officially approved 6 February 1912.

Bragg Creek (creek)

82 J/15 - Bragg Creek
12-23-5-W5
50°47'N 114°34'W
Flows east into Elbow River, approximately 30 km west of Calgary.

The name for this creek was officially approved 6 October 1949. (see Bragg Creek)

Bragg Creek (hamlet)

82 J/15 - Bragg Creek
12-23-5-W5
50°57'N 114°35'W

Approximately 30 km west of Calgary.

Albert W. Bragg, the earliest settler in the area, homesteaded here in 1894 and the hamlet was ostensibly named after him. The post office opened here in 1911. The name for this hamlet was officially approved 6 October 1949.

Bragg Creek Park (provincial park)

82 J/15 - Bragg Creek
12-23-5-W5
50°56'N 114°59'W
Approximately 30 km west of Calgary.

(see Bragg Creek)

Braithwaite, Mount (mountain)

83 E/14 - Grande Cache
23-55-9-W6
53°46'N 119°14'W
Approximately 15 km south south-west of Grande Cache.

This mountain, which is 2134 m in altitude, was officially named in 1953 after Edward Ainslie Braithwaite, M.D. (1862-1949). He was born in England and came to Canada in 1884 when he joined the North West Mounted Police to perform medical duties in the force. This pioneer doctor was appointed Coroner and Medical Health Officer under the Northwest Territories Government in 1896.

Brandy Creek (creek)

83 E/10 - Adams Lookout
29-54-7-W6
53°42'N 119°00'W
Flows west into Sulphur River, approximately 23 km south south-east of Grande Cache.

The origin of the name is unknown.

Brazeau (railway point)

83 C/8 - Nordegg
27-40-15-W5
52°28'N 116°05'W

Located within the town of Nordegg.

(see Brazeau Range)

Brazeau Lake (lake)

83 C/6 - Sunwapta
2-40-22-W5
52°25′N 117°05′W
Approximately 84 km south-east of Jasper.

(see Brazeau Range)

Brazeau Mine Pit (lake)

83 C/8 - Nordegg
32-40-15-W5
51°29′N 116°07′W
Approximately 4 km west north-west of
Nordegg.

This small lake is a former mine pit that has
filled with water and is being stocked with
fish (see Brazeau Range). The name was
made official 5 May 1987.

Brazeau, Mount (mountain)

83 C/11 - Southesk Lake
23-41-24-W5
52°33′N 117°21′W
Approximately 87 km west of Nordegg.

This mountain stands 3470 m in altitude.
(see Brazeau Range)

Brazeau Range (range)

83 B/5 - Saunders
2-40-14-W5
52°24′N 115°55′W
Approximately 5 km east of Nordegg.

This range and the other surrounding
features are named after Joseph E. Brazeau, a
native of Missouri who entered the fur trade
in 1830. He was an employee of the
Hudson's Bay Company serving as clerk and
postmaster at Edmonton, Jasper and Rocky
Mountain House from 1852-1864. This
linguist was of great help to the Palliser

Expedition and was spoken of with highest
praise.

Bread Creek (creek)

82 O/14 - Limestone Mountain
32-33-8-W5
51°52′N 115°07′W
Flows east into James River, approximately
85 km north-east of Banff.

The name for this creek was officially
approved 14 August 1941. During a forest
fire in 1919, bread for all the camps was
baked at a camp located near this stream.

Breaker Mountain (mountain)

82 N/15 - Mistaya Lake
32-19-W5
51°46′N 116°39′W
Approximately 100 km north-west of Banff,
on the Alberta-B.C. boundary.

This 3058 m mountain was named by A.O.
Wheeler (see Wheeler Flats) in 1918. The
name describes the snow formations on the
mountain.

Breccia Creek (creek)

83 F/4 - Miette
28-46-26-W5
53°00′N 117°44′W
Flows south into Rocky River, approximately
47 km south south-west of Hinton.

This creek takes its name from the amount
of rock breccia found near its mouth.
Breccia is a rock of angular stone cemented
by lime. The name was approved 10 August
1960.

Breeding Valley (valley)

82 J/1 - Langford Creek
12-2-W5
50°02′N 114°12′W
Approximately 45 km west of Claresholm.

Officially approved 3 December 1941, this
feature was named after William Patterson
Breeding. He was a farmer in the area
around 1910. Eventually Mr. Breeding sold
his section to the Johnson brothers and
moved north. He died suddenly of lead
poisoning after his move north.

Brett, Mount (mountain)

82 O/4 - Banff
25-25-14-W5
51°10′N 115°49′W
Approximately 18 km west south-west of
Banff.

This mountain, 2984 m in altitude, was
named after the Honourable Robert George
Brett, M.D. (1851-1929). This pioneer
resident of Banff was born in Ontario. He
was the surgeon assigned to the C.P.R.
construction project in 1886. Dr. Brett was
the Lieutenant Governor of Alberta,
1915-1925. (see Massive Range)

Brewster Creek (creek)

82 O/4 - Banff
13-25-12-W5
51°09′N 115°39′W
Flows north-west into Healy Creek,
approximately 10 km south-west of Banff.

This creek was named for James (Jim) Irvine
Brewster (1882-1947), a well-known camp
outfitter from Banff. He arrived in the area
in 1888, before Banff was declared a townsite
and previous to the time when the railway
grade was complete. He accepted a contract
to provide railway ties and set up a tie-camp
and logging operation with John Gerome
Healy (see Healy Pass Lakes) along the
stream. Jim Brewster and his brother Bill,
with the assistance of their father John,
began guiding horseback trail trips in 1892.

Brewster Glacier (glacier)

82 J/13 - Mount Assiniboine
23-11-W5
50°58′N 115°32′W

Approximately 100 km west of Calgary, the source of Brewster Creek.

Mr. Jim Brewster was the first recorded visitor to the glacier. (see Brewster Creek)

Brewster, Mount (mountain)

82 O/4 - Banff
30-26-12-W5
51°14'N 115°39'W
Approximately 9 km north-west of Banff.

This 2859 m mountain was officially named 1 October 1929. The name was suggested by Mr. Tom Wilson (see Mount Wilson) after Mr. John Brewster, a pioneer dairyman in the area, and father of the original Brewster family of Banff.

Brewster Rock (mountain)

82 O/4 - Banff
24-13-W5
51°05'N 115°45'W
Approximately 17 km south-west of Banff.

The name for this 2697 m mountain formerly referred to a large boulder located on the shoulder of Lookout Mountain. Local usage referred the name to the whole feature and it was officially named 3 July 1968. Apparently, Mr. Jim Brewster (see Brewster Creek) would accompany a group of people who had plans to ski, hike and climb in the "Sunshine Valley" area. He would go only as far as this rock and then stop, choosing to wait for the return of the others.

Brewster's Wall (vertical valley headwall)

83 E/7 - Blue Creek
52-7-W6
53°28'10"N 118°53'20"W
Approximately 85 km north-west of Jasper.

The locally well-established name for this feature was officially approved 20 October 1983. It is named for its proximity to the old Brewster camp used by Fred and Jack

Brewster of Jasper in their outfitting businesses. (see Brewster Creek)

Bridal Veil Falls (falls)

83 C/3 - Columbia Icefield
37-22-W5
52°10'55"N 117°03'05"W
Approximately 100 km south-east of Jasper.

The descriptive name for these falls was officially approved 28 February 1980.

Bridgland Creek (creek)

82 O/14 - Limestone Mountain
16-32-10-W5
51°45'N 115°21'W
Flows south-east into James River, approximately 65 km north north-east of Banff.

The name for this creek was officially approved 17 December 1941. (see Mount Bridgland)

Bridgland, Mount (mountain)

83 D/15 - Rainbow
16-46-4-W6
52°57'N 118°32'W
Approximately 32 km west north-west of Jasper.

This 2930 m mountain was named in 1918 after Morrison Parsons Bridgland (1878-1948). He was born and educated in Toronto and had a background in mathematics and physics. He first came west in 1902 as an assistant to A.O. Wheeler (see Wheeler Flats). Bridgland was a fine mountaineer and was a founding member of the Alpine Club of Canada. His mapping skills were impeccable, and his maps were used as guides for alpinists who climbed after him. He was a surveyor for the Dominion Land Survey and performed extensive topographical surveys in the Rocky Mountains.

Broad Creek (creek)

83 E/9 - Moberly Creek
36-53-3-W6
53°37'N 118°18'W
Flows north into Little Berland River, approximately 85 km north north-west of Jasper.

This creek was named 15 April 1946 by M.E. Nidd, a field topographer in the area. The origin of the name is unknown.

Brock, Mount (mountain)

82 J/11 - Kananaskis Lakes
32-20-8-W5
50°44'N 115°05'W
Approximately 55 km west north-west of Turner Valley.

This 2902 m mountain was officially named in 1922 after Rear Admiral Osmond de Beauvoir Brock, who sailed on the Princess Royal in the First World War Battle of Jutland. (see Jutland Mountain)

Brokenleg Lake (lake)

82 O/3 - Canmore
4-26-8-W5
51°11'N 115°04'W
Approximately 23 km north-east of Canmore.

The name for this feature is apparently the translation of a native name, the story of which is unknown.

Brown Creek (creek)

83 C/9 - Wapiabi Creek
21-43-16-W5
52°43'N 116°16'W
Flows south-east into Blackstone River, approximately 31 km north north-west of Nordegg.

The origin of the name is unknown.

Brown Creek (creek)

83 F/5 - Entrance
3-50-27-W5
53°17'N 117°52'W
Flows east into Brule Lake, approximately 20 km south-west of Hinton.

This creek may be named after Sandy Brown, a former forest ranger in the area. No other information is available.

Brown, Mount (mountain)

83 D/8 - Athabasca Pass
27-39-2-W6
52°22'N 118°14'W
Approximately 57 km south south-west of Jasper.

This mountain, which is 2799 m in altitude, was named by David Douglas (see Mount Douglas) in 1827 in honour of Robert Brown (1773-1858). Brown was a Scottish botanist who contributed substantially to the knowledge of plant morphology, embryology and geography by his work done on the flora of Australia. Brown published a few pamphlets on his findings in various areas, but little else is known about him.

Brule (hamlet)

83 F/5 - Entrance
10-50-27-W5
53°18'N 117°52'W
Approximately 19 km south south-west of Hinton.

Brule was founded as a mining town in 1912 by the Mackenzie and Mann interests of the Canadian Northern Railway. The mine was closed in 1928. (see Brule Lake)

Brule Hill (hill)

83 F/5 - Entrance
50-27-W5
53°18'N 117°53'W
Approximately 21 km west south-west of Hinton.

The name of this hill was officially approved in 1956. (see Brule-hamlet)

Brule Lake (lake)

83 F/5 - Entrance
2-50-27-W5
53°17'N 117°51'W
Approximately 17 km west south-west of Hinton.

"Brûlé" is French for "burnt." The name likely refers to the burnt timber found along the shores of the lake by the early settlers.

Brule Lake, 1910

Brussels Peak (peak)

83 C/12 - Athabasca Falls
10-41-27-W5
52°31'N 117°48'W
Approximately 44 km south south-east of Jasper.

This mountain peak, which is 3161 m in altitude, was named in 1922 after Captain Fryatt's ship (see Fryatt Creek). No other information is available.

Bryan Creek (creek)

83 F/3 - Cadomin
16-49-21-W5
53°14'N 117°00'W

Flows east into Embarras River, approximately 53 km south-west of Edson.

This creek flows through what was the property of the Bryan Coal Company Ltd. The Bryan mine was open until 1923. It was an underground project, spearheaded by James Bryan, a well-known northern fur trader.

Bryant Creek (creek)

82 J/14 - Spray Lakes Reservoir
10-22-11-W5
50°52'N 115°27'W
Flows south-east into the Spray River, approximately 95 km west of Calgary.

This creek was named after Mr. H.G. Bryant of Philadelphia, who, along with W.D. Wilcox (see Wilcox Pass) and L.J. Steele, attempted to scale some of the mountains in this area around 1899.

Bryant Creek (creek)

82 O/2 - Jumpingpound Creek
9-24-6-W5
51°02'N 114°47'W
Flows south into Jumpingpound Creek, approximately 40 km west of Calgary.

(see Mount Bryant)

Bryant, Mount (mountain)

82 J/15 - Bragg Creek
25-22-8-W5
50°54'N 114°59'W
Approximately 60 km west of Calgary.

Alfred Harold Bryant (1882-1934) came from England and after spending two years in Africa, came to Canada in 1905 and homesteaded in the Jumpingpound area for several years. In 1916 he was offered the position of forest ranger, which he held until his death. The name for this mountain, 2629 m in altitude, was officially approved 6 October 1949.

Bryant, Mount (mountain)

83 F/4 - Miette
3-47-26-W5
53°01′N 117°43′W
Approximately 45 km south of Hinton.

This 2621 m mountain was officially named in 1960 after Frank Bryant, former Warden of Jasper Park, Chief Warden at Waterton Lakes Park and Superintendent of Kootenay National Park.

Bryce, Mount (mountain)

83 C/3 - Columbia Icefield
52°03′N 117°20′W
Approximately 92 km south-west of Jasper.

This 3507 m mountain was named in 1898 after late Viscount James Bryce, then president of the Alpine Club (London) and British Ambassador at Washington 1907 to 1912.

Buchanan Peak (peak)

82 H/4 - Waterton Lakes
19-1-30-W4
49°03′N 113°59′W
Approximately 50 km south of Pincher Creek.

Senator William A. Buchanan had a very direct interest in the development of Waterton Lakes National Park and had a summer residence in the park area for several years. It was for his service and public effort for the park that this mountain peak was named after him. Its official approval date was 8 June 1971. (see Buchanan Ridge)

Buchanan Ridge (ridge)

82 H/4 - Waterton Lakes
1-30-W4
49°02′N 114°00′W
Approximately 50 km south of Pincher Creek.

This ridge was named after the Honourable William Asbury Buchanan, (1876-1954). He was educated at Trenton, Warworth and Brighton, all in Ontario. At the age of 16, he became a "printer's devil" with the *Peterborough Examiner*. He later joined the staff of the *Peterborough Review* as a reporter and then worked for the *Toronto Evening Telegram*. In 1905 he arrived in Lethbridge and by 1907, he founded the daily paper, *The Lethbridge Herald*. In 1909 he was elected Member of the Legislative Assembly for Lethbridge City. He served in the House of Commons from 1911 to 1921 and in 1925 was made a senator. Throughout his political career he was an active newspaperman. On 17 May 1949, he was made an Honorary Doctor of Laws of the University of Alberta in recognition of his distinguished service.

Buck Lake (lake)

83 C/12 - Athabasca Falls
22-41-26-W5
52°33′N 117°39′W
Approximately 46 km south south-east of Jasper.

The origin of this lake's name is unknown.

Bull Creek (creek)

82 J/9 - Turner Valley
16-19-18-2-W5
50°32′25″N 114°15′19″W
Flows north-east into Highwood River, approximately 2 km north of Longview.

The name was annotated properly on the 1903 Land Survey map, but prior to 29 July 1986, the name Bull Creek was attached to the wrong feature. The origin for the now current Bull Creek is not known precisely. It may be named after Charlie Bull, a rancher who lived around Priddis.

Bull Creek (creek)

82 O/11 - Burnt Timber Creek
8-31-8-W5
51°38′N 115°06′W

Flows south-east into Red Deer River, approximately 65 km north-west of Cochrane.

Officially approved in 1956, this name may be taken from the bull elk found along the creek.

Bull Creek Hills (hill)

82 J/7 - Mount Head
24-17-5-W5
50°25′30″N 114°34′20″W
Approximately 75 km south-west of Calgary.

Bull Creek Hills is a group of high foothills, with a maximum elevation of 2150 m above sea level. The locally well-established name was officially approved 28 February 1980. The hills were so named because they are at the head of Bull Creek.

Buller Creek (creek)

82 J/14 - Spray Lakes Reservoir
20-22-10-W5
50°53′N 115°21′W
Flows north-west into Spray Lakes Reservoir, approximately 85 km west of Calgary.

The name Buller Creek was rescinded 3 October 1975 because of flooding due to the building of the Three Sisters Dam. The entire feature was not inundated, however, and the name was reinstated 22 June 1976. (see Mount Buller)

Buller, Mount (mountain)

82 J/14 - Spray Lakes Reservoir
27-22-10-W5
50°54′N 115°19′W
Approximately 80 km west of Calgary.

This 2805 m mountain was named in 1922 in honour of Lieutenant Colonel H.C. Buller of the Princess Patricia's Canadian Light Infantry. He was killed in the First World War.

Bulyea, Mount (mountain)

82 N/10 - Blaeberry River
4-32-21-W5
51°43'N 116°55'W
Approximately 115 km north-west of Banff,
on the Alberta-B.C. boundary.

This 3304 m mountain was named in 1920
after the Honourable George Hedley Vickers
Bulyea (1859-1926), the first Lieutenant
Governor of Alberta (1905-1915).

Burke, Mount (mountain)

82 J/7 - Mount Head
15-4-W5
50°17'N 114°31'W
Approximately 45 km south-west of Turner
Valley.

Mount Burke measures 2542 m in altitude
and its name was officially approved in 1919.
It is named after Denis Charles Burke, a
rancher and forest ranger from the area. He
was a member of the North West Mounted
Police from 1896 to 1901.

Burleigh Creek (creek)

83 E/15 - Pierre Greys Lakes
8-57-4-W6
53°54'N 118°33'W
Flows north into Lone Teepee Creek,
approximately 37 km east of Grande Cache.

This creek was officially named in 1947 after
J.H. Burleigh, a former forest ranger who
almost lost his pack outfit in muskeg on the
bank of this creek.

Burmis (locality)

82 G/9 - Blairmore
13-7-3-W5
49°33'N 114°17'W
Located on Highway 3, approximately 5 km
south-east of Bellevue.

The name Burmis is a combination of the
parts of the first two surnames on a petition
sent to the federal government to change the
name of the "flag-stop" from Livingstone in
1906. Robert H. Burns was a rancher located
13 km east of Livingstone station, and Jack
Kemmis was a rancher located north-west of
Cowley. It is from the names of these two
ranchers that the name Burmis was born.

Burney, Mount (mountain)

82 J/11 - Kananaskis Lakes
20-20-8-W5
50°43'N 115°03'W
Approximately 55 km west of Turner Valley.

Vice-Admiral Sir Cecil Burney was the
commander of the ship Marlborough in the
1916 Battle of Jutland (see Jutland
Mountain). This 2934 m mountain was
officially named in 1922 in his honour.

Burns Creek (creek)

82 J/10 - Mount Rae
22-19-7-W5
50°37'N 114°53'W
Flows north-east into Sheep River,
approximately 43 km west of Turner Valley.

(see Mount Burns)

Burns Lake (lake)

83 J/10 - Mount Rae
4-17-19-7-W5
50°36'N 114°57'W
Approximately 47 km west of Turner Valley.

The name of this lake was officially
approved 5 May 1987 because it is located on
the upper reaches of Burns Creek. The name
Burns Lake also enjoys established local
usage. (see Mount Burns)

Burns, Mount (mountain)

82 J/10 - Mount Rae
26-19-7-W5
50°38'N 114°52'W
Approximately 41 km west of Turner Valley.

Once described by Alberta's Premier
Brownlee as "a grand pioneer, a true
Canadian gentleman, a philanthropist and a
prince of good fellows," Patrick Burns
(1856-1937) contributed greatly to the life of
Calgary and of Canada. He was born in
Oshawa, Ontario, and eventually made his
way to Calgary where he quickly became
one of the best known ranching men of the
West. Burns built up a large meat packing
business. He was called to the Senate in 1931
and remained a senator until he resigned in
1936. This 2936 m mountain was officially
named after him in 1922.

Burnt Timber Creek (creek)

82 O/11 - Burnt Timber Creek
4-31-8-W5
51°38'N 115°06'W
Flows north-east into Red Deer River,
approximately 60 km north-east of Banff.

Officially approved in 1956, the name for
this creek is likely descriptive of the
appearance of the area through which it
flows.

Burstall Creek (creek)

82 J/14 Spray Lakes Reservoir
15-21-10-W5
50°47'N 115°18'W
Flows north-east into Mud Lake,
approximately 85 km west south-west of
Calgary.

(see Mount Burstall)

Burstall Lake (lake)

82 J/14 - Spray Lakes Reservoir
16-21-10-W5
50°47'N 115°20'W
Approximately 85 km west south-west of
Calgary.

(see Mount Burstall)

Burstall, Mount (mountain)

82 J/14 - Spray Lakes Reservoir
9-21-10-W5
50°46′N 115°19′W
Approximately 85 km west south-west of Calgary.

This 2760 m mountain was officially named in 1918, after Lieutenant-General Sir E.H. Burstall, who commanded Canadian troops in the First World War.

Burstall Pass (pass)

82 J/14 - Spray Lakes Reservoir
50°45′40″N 115°22′35″W
Approximately 90 km west south-west of Calgary.

The name for this feature was officially approved 21 January 1985. (see Mount Burstall)

Burton Creek (creek)

82 G/16 - Maycroft
1-12-2-W5
49°58′N 114°09′W
Flows west into Callum Creek, approximately 45 km north-east of Blairmore.

This creek is named after Frederick Alfred Burton, a prominent rancher in the early 1900s.

Bury Ridge (ridge)

83 E/9 - Moberly Creek
53-4-W6
53°33′N 118°28′W
Approximately 84 km north north-west of Jasper.

Named in 1948 after Major W.B. Bury, D.S.O., this ridge was formerly known as "Minny Ridge." Major Bury, of Edmonton, was killed in action in the Second World War.

Busby Lake (lake)

83 E/8 - Rock Lake
6-51-1-W6
53°22′10″N 118°08′45″W
Approximately 55 km north of Jasper.

This lake was officially named 20 October 1983 after George Busby, a trapper who was active in the area circa 1900. The name was locally well-established prior to its gaining official status.

Bush Pass (pass)

82 N/15 - Mistaya Lake
6-33-21-W5
51°48′N 116°58′W
Approximately 120 km north-west of Banff, on the Alberta-B.C. boundary.

This pass, 2396 m in altitude, is located at the head of Bush River, British Columbia, which is so called because its banks are covered with impenetrable undergrowth.

Butterwort Creek (creek)

83 F/4 - Miette
26-47-28-W5
53°06′N 117°59′W
Flows north into Talbot Lake, approximately 47 km south-west of Hinton.

The name for this creek was officially applied in 1960 and is taken from the wildflower Pinguicula Vulgaris or Common Butterwort, found in nearby wooded areas and foothills.

Buttress Lake (lake)

83 D/9 - Amethyst Lakes
32-42-1-W6
52°39′N 118°07′W
Approximately 24 km south of Jasper.

The descriptive name for this lake was applied by M.P. Bridgland (see Mount Bridgland) in 1916 because of the numerous cliffs around it.

Buttress Mountain (mountain)

83 D/16 - Jasper
28-46-2-W6
52°59′N 118°13′W
Approximately 10 km north north-west of Jasper.

This 2685 m mountain was named by M.P. Bridgland (see Mount Bridgland) in 1916. (see Buttress Lake)

Byng, Mount (mountain)

82 J/13 - Mount Assiniboine
21-11-W5
50°50′N 115°31′W
Approximately 100 km west south-west of Calgary.

This mountain, which is 2940 m in altitude, was officially named in 1918. (see Byng Pass)

Byng Pass (pass)

83 E/7 - Blue Creek
15-50-7-W6
53°19′N 118°57′W
Approximately 76 km north-west of Jasper.

This pass was named in 1922 in honour of Viscount Julian Hedworth George Byng (1862-1935), who was a frequent visitor to Jasper Park. Lord Byng of Vimy served in the Boer War from 1899 to 1902, and the First World War in 1914 to 1915. He was appointed to the command of the Canadian Corps in 1916 and was promoted to the rank of General in 1917 (the year Canadian troops took Vimy Ridge) until the end of the war. He became Governor General of Canada in 1921.

Byron Creek (creek)

82 G/9 - Blairmore
9-7-3-W5
49°33′N 114°21′W
Flows north into Crowsnest River, approximately 9 km south-east of Blairmore.

The name Byron Creek was applied to this
feature in 1905. It commemorates a lone
prospector, Frank Byron, who came to the
Crowsnest Pass area from the United States
prior to the building of the railroad looking
for gold and any other fortune he could
find.

Byron Hill (hill)

82 G/9 - Blairmore
34-6-3-W5
49°31′N 114°20′W
Approximately 13 km south south-east of
Blairmore.

(see Byron Creek)

Byron Creek Falls (falls)

82 G/9 - Blairmore
11-9-7-3-W5
49°32′50″N 114°21′00″W
Approximately 9 km south-east of
Blairmore.

The name for these falls was officially
approved 21 July 1982. (see Byron Creek)

Cabin Creek (creek)

82 G/9 - Blairmore
33-8-1-W5
49°42′N 114°05′W
Flows west into Oldman River, approximately 13 km north-east of Lundbreck.

This locally well-established name was approved in 1942. The origin of the name is unknown.

Cabin Creek (creek)

83 D/16 - Jasper
8-45-1-W6
52°52′N 118°06′W
Flows south into Miette River, approximately 1 km south of Jasper.

(see Cabin Lake)

Cabin Creek (creek)

83 E/16 - Donald Flats
22-55-3-W6
53°46′N 118°21′W
Flows east into Berland River, approximately 54 km east south-east of Grande Cache.

The name of this creek has been in local usage since at least 1935. The origin of the name is unknown.

Cabin Lake (lake)

83 D/16 - Jasper
18-45-1-W6
52°53′N 118°07′W
Approximately 3 km west of Jasper.

This lake and the nearby creek were named in 1914 by H. Matheson of the Dominion Land Survey. The water supply for Jasper is obtained from this source. No other information about the origin of this name is available.

Cabin Ridge (ridge)

82 J/1 - Langford Creek
28-12-4-W5
50°02′N 114°29′W

Approximately 65 km west of Claresholm.

The highest part of this ridge is 2181 m in altitude. The origin of the name is unknown.

Cache Creek (creek)

82 J/2 - Fording River
2-13-5-W5
50°03′N 114°35′W
Flows north-east into Oldman River, approximately 75 km south-west of Turner Valley.

This name was adopted at the Geographic Board of Canada meeting 1 May 1934. It is a translation of the Cree Indian word "astachi-kuwin," meaning storehouse.

Cache Lake (lake)

83 F/5 - Entrance
17-52-26-W5
53°29′N 117°52′W
Approximately 17 km north-west of Hinton.

Cache Lake gets its name because A.H. Hawkins of the Dominion Land Survey, who ran the Fifteenth and Sixteenth Baselines, had a supply of iron pins and food packed into a cache here.

Cache Percotte Creek (creek)

83 F/5 - Entrance
30-51-24-W5
53°26′N 117°32′W
Flows north into Athabasca River, approximately 2 km north-east of Hinton.

The name for this creek was made official in 1944. Its precise origin is uncertain. One story, however, relates that in 1870 Indians in the Jasper area, stricken with smallpox ("percotte" in French) camped near this creek while waiting for one of them to go to the mission at Lac Ste. Anne for food and medicine. By the time the medicine arrived, all or most of the group had died. About half of the Indian population of Alberta died of the smallpox epidemic during that period.

Cadomin (hamlet)

83 F/3 - Cadomin
5-47-23-W5
53°02′N 117°20′W
Approximately 85 km south-west of Edson.

The name Cadomin is a contraction of Canadian Dominion Mine, the coal developing company. The name was coined by F.L. Hammond, the first president of the Cadomin Coal Company. The village was established in 1913.

Cadomin Creek (creek)

83 C/14 - Mountain Park
30-46-23-W5
52°59′N 117°20′W
Flows west into McLeod River, approximately 51 km east north-east of Jasper.

(see Cadomin)

Cadomin Mountain (mountain)

83 C/14 - Mountain Park
28-46-23-W5
52°59′N 117°17′W
Approximately 55 km east north-east of Jasper.

(see Cadomin)

Cairn Lakes (lakes)

83 C/11 - Southesk Lake
43-22-W5
52°42′25″N 117°08′57″W
A group of lakes, located approximately 68 km east south-east of Jasper.

(see Cairn River)

Cairn River (river)

83 C/11 - Southesk Lake
31-42-21-W5
52°38′N 116°58′W
Flows south-east into Southesk River,
approximately 64 km west north-west of
Nordegg.

The precise origin of the name is unknown.
The term "cairn" refers to a pyramid of
rough stones, used as a marker or memorial.

Cairngorm (mountain)

83 D/16 - Jasper
2-46-2-W6
52°56′N 118°12′W
Approximately 10 km north-west of Jasper.

Cairngorm is Gaelic for "blue-green
mountain." There is a mountain range in
Scotland called Cairngorm Mountains. The
name for this feature was applied in 1916 by
M.P. Bridgland, D.L.S. (see Mount Bridgland).
The mountain measures 2610 m in altitude.

Caldron Creek (creek)

82 N/10 - Blaeberry River
51°41′50″N 116°32′10″W
Flows east into Peyto Creek, approximately
90 km north-west of Banff.

The name for this creek was officially approved
21 January 1985. (see Caldron Lake)

Caldron Lake (lake)

82 N/10 - Blaeberry River
36-31-19-W5
51°42′N 116°34′W
Approximately 90 km north-west of Banff.

This lake was named by A.O. Wheeler (see
Wheeler Flats) in 1918. The kettle-shaped
rock depression is reminiscent of a caldron,
leading to its descriptive name.

Caldron Peak (peak)

82 N/10 - Blaeberry River
51°42′05″N 116°34′40″W

Approximately 90 km north-west of Banff.

(see Caldron Lake)

Caledonia Lake (lake)

83 D/16 - Jasper
12-45-2-W6
52°52′N 118°09′W
Approximately 6 km west of Jasper.

(see Caledonia Mountain)

Caledonia Mountain (mountain)

83 D/15 - Rainbow
10-46-5-W6
52°57′N 118°40′W
Approximately 39 km west north-west of
Jasper.

The name "New Caledonia" was applied to
the area between the Rocky Mountains and
the Coast Range, and between 51°30′ and
57°00′, by Simon Fraser (see Mount Fraser)
in 1806. Now part of British Columbia, the
area was traversed by fur traders and
members of the Hudson's Bay Company.
The mountain likely takes its name from
Caledonian Valley, where the Miette and
Fraser rivers flow, the only remnant of the
"province" of New Caledonia.

Calumet Creek (creek)

83 E/3 - Mount Robson
8-49-8-W6
53°13′N 119°08′W
Flows west into Smoky River, approximately
79 km north-west of Jasper.

(see Calumet Peak)

Calumet Peak (peak)

83 E/6 - Twintree Lake
30-49-7-W6
53°16′N 119°01′W
Approximately 86 km west north-west of
Jasper.

The name Calumet River originates from
the name of the old fort called Pierre au
Calumet, which is shown on Franklin's map
of 1823. This mountain, which is 2977 m in
altitude, was named in 1923. The term
Calumet is the translation from the Cree,
"ospwâgan," for "peace pipe."

Calumet Ridge (ridge)

83 E/6 - Twintree Lake
49-8-W6
53°15′N 119°07′W
Approximately 83 km west north-west of
Jasper.

The name for this feature was officially
approved in 1956. (see Calumet Peak)

Calypso Creek (creek)

83 E/14 - Grande Cache
25-55-10-W6
53°47′N 119°20′W
Flows north-west into Smoky River,
approximately 18 km south-west of Grande
Cache.

The name for this creek, which was
officially approved in 1953, is taken from the
small orchid which grows in moist to dry
coniferous woods, commonly under pine
trees, known as the Calypso or Venus'
Slipper.

Cambrai, Mount (mountain)

82 N/15 - Mistaya Lake
18-33-22-W5
51°50′N 116°58′W
Approximately 120 km north-west of Banff,
on the Alberta-B.C. boundary.

This mountain, which stands 3134 m in
altitude, was named in 1920 after Cambrai, a
town in Flanders in France. It is located
approximately 60 km south south-east of
Lille. During the First World War, in
November 1917, Cambrai was the site of the
first tank attack of the war. Allied leaders

wanted a success on the Western Front to raise morale and tie German troops down in France. The battle plan was based on the use of tanks in a surprise onslaught, instead of the usual intense artillery bombardment. Begun on 20 November as a raid, the attack by the Third British Army, which included Canadian formations, was initially successful and the bells of London were pealed in celebration of the victory. The generals changed it to a full-scale offensive but did not allow enough troops to maintain the success. On 30 November, the German troops counterattacked, themselves launching an attack without advance artillery bombardment and pushed the front line almost back to the positions held by the opposing armies prior to 20 November. Canadian troops entered the town of Cambrai 9 October 1918.

Cameron Creek (creek)

82 H/4 - Waterton Lakes
22-1-30-W4
49°03′N 113°54′W
Flows south-east into Upper Waterton Lake, approximately 50 km south of Pincher Creek.

This creek was formerly known as Cameron Brook. (see Cameron Lake)

Cameron Falls (falls)

82 H/4 - Waterton Lakes
1-30-W4
49°03′N 113°55′W
Approximately 50 km south of Pincher Creek.

Within a short walk to the west of the hamlet of Waterton Park is Cameron Falls, one of the most interesting cascades in the Rockies. The waterfall takes its name from Cameron Lake, named after General D.R. Cameron. (see Cameron Lake)

Cameron Falls, 1951

Cameron Lake (lake)

82 G/1 - Sage Creek
3-1-1-W5
49°01′N 114°04′W
Approximately 10 km west south-west of Waterton Park.

This lake was named in 1916 after Captain Donald Roderick Cameron (1834-1921), a Scottish-born officer nominated by Prime Minister Sir John A. Macdonald as the British Commissioner on the International Boundary Commission of 1872-1876, Lake of the Woods to the Rockies. The creek, falls and mountain take their names from this lake.

Cameron Mountain (mountain)

82 G/1 - Sage Creek
24-1-1-W5
49°02′N 114°00′W
Approximately 7 km south-west of Waterton Park.

This 2565 m mountain, formerly also known as Cameronian Mountain, is a part of Buchanan Ridge. (see Buchanan Ridge and Cameron Lake)

Camp Creek (creek)

83 B/5 - Saunders
16-40-11-W5
52°27′N 115°31′W
Flows south into North Saskatchewan River, approximately 38 km east of Nordegg.

The locally well-established name for this creek was officially approved in 1941. The origin of the name is unknown.

Camp Creek (creek)

82 G/16 - Maycroft
35-10-3-W5
49°52′N 114°18′W
Flows south into Oldman River, approximately 30 km north north-east of Blairmore.

Officially named in 1943, this creek is probably called Camp Creek after the cow-camp of the Kew Ranch which was located on this creek.

Campion, Mount (mountain)

83 E/9 - Moberly Creek
19-53-3-W6
53°36′N 118°26′W
Approximately 83 km north north-west of Jasper.

This mountain, which is 2484 m in altitude, was named in 1949 in honour of Corporal George Campion, M.M. of Edmonton, who was killed in action in the Second World War.

Campus Creek (creek)

83 D/9 - Amethyst Lakes
35-42-2-W6
52°40′N 118°11′W
Flows west into Astoria River, approximately 24 km south south-west of Jasper.

"Campus" is the Latin word meaning "field." When seen from above, Campus Creek and Pass look like wide open meadows.

Campus Pass (pass)

83 D/9 - Amethyst Lakes
19-42-1-W6
52°38'N 118°08'W
Approximately 27 km south of Jasper.

(see Campus Creek)

Canary Creek (creek)

83 B/4 - Cripple Creek
7-36-14-W5
52°04'N 115°59'W
Flows north-east into Hummingbird Creek, approximately 45 km south-east of Nordegg.

The locally well-established name for this creek was made official 29 September 1976. Its precise origin is unknown, but local forest rangers suggest it may be after the wild canaries seen along its course.

Caniche Peak (peak)

83 D/16 - Jasper
33-43-3-W6
52°45'N 118°22'W
Approximately 24 km south-west of Jasper.

Caniche is the French word for "poodle." This mountain peak, 2552 m in altitude, resembles a poodle's head. The descriptive name for this peak was officially approved in 1922.

Canmore (town)

82 O/3 - Canmore
32-24-10-W5
51°05'N 115°21'W
Approximately 17 km south-east of Banff.

The first divisional point for the Canadian Pacific Railway west of Calgary was completed in 1884 and named in honour of

King Malcolm III, Canmore, Scotland, who defeated Macbeth and reigned as King of Scots from 1057-1093. The name Canmore was likely chosen by Donald A. Smith, one of the Canadian Pacific Railway pioneers. "Ceann mor" is Gaelic for "Big Head." The Stoney Indian name for Canmore, "too-wup-chinchin-koodibee," (Dempsey) translates as "shooting at a young spruce tree." Young Indian boys practised by shooting at a young tree while camping on the Canmore flats.

Canmore Creek (creek)

82 O/3 - Canmore
29-24-10-W5
51°05'N 115°22'W
Flows east into Bow River, approximately 1 km south-west of Canmore.

(see Canmore)

Canoe Pass (pass)

83 D/8 - Athabasca Pass
9-40-2-W6
52°25'N 118°14'W
Approximately 51 km south south-west of Jasper.

This pass is named after the Canoe River, located in British Columbia. David Thompson (see Thompson Pass) wintered at the mouth of Canoe River from January to April 1811 and built a canoe in which he descended the Columbia River.

Canyon Creek (creek)

83 C/15 - Cardinal River
36-44-19-W5
52°50'N 116°37'W
Flows north-east into Blanchard Creek, approximately 94 km east of Jasper.

The locally well-established name for this creek was officially approved in 1940. The creek flows in a long, deep and narrow canyon and there is a high falls in the creek.

Canyon Creek (creek)

82 J/15 - Bragg Creek
15-22-6-W5
50°52'N 114°45'W
Flows south into Elbow River, approximately 40 km west of Calgary.

The name for this creek is locally well-established and was made official in 1939. The name is descriptive of the area in which the creek flows.

Canyon Creek (creek)

83 F/6 - Pedley
23-52-24-W5
53°30'N 117°26'W
Flows south-east into Athabasca River, approximately 67 km west of Edson.

The descriptive name for this creek is locally well-established and was made official in 1939.

Cap Ridge (ridge)

83 F/5 - Entrance
49-26-W5
53°15'N 117°41'W
Approximately 16 km south south-west of Hinton.

The name for this ridge was suggested by the Geological Survey in 1945 and was officially approved 21 April of the same year. John Capotinski (alias Cap) operated a sawmill nearby, and this mountain ridge is named after him.

Capitol Creek (creek)

83 F/4 - Miette
27-47-27-W5
53°06'N 117°52'W
Flows south-west into Makwa Creek, approximately 41 km south south-west of Hinton.

(see Capitol Mountain)

Capitol Mountain (mountain)

83 F/4 - Miette
10-48-27-W5
53°08′N 117°50′W
Approximately 37 km south south-west of Hinton.

This 2438 m mountain was given its name in 1916 by M.P. Bridgland (see Mount Bridgland). It has an imposing position south-east of Roche Miette.

Capricorn Glacier (glacier)

82 N/15 - Mistaya Lake
21-34-29-W5
51°46′N 116°38′W
Approximately 100 km north-west of Banff.

The area surrounding the glacier is frequented by mountain goats. Capricorn is Latin for goat and, thus, the name for this feature is taken from the name of this animal.

Capricorn Lake (lake)

82 N/15 - Mistaya Lake
28-32-19-W5
51°46′N 116°37′W
Approximately 100 km north-west of Banff.

(see Capricorn Glacier)

Carbondale Hill (hill)

82 G/8 - Beaver Mines
33-5-3-W5
49°26′N 114°21′W
Approximately 12 km west south-west of Beaver Mines.

This hill stands at 1805 m in altitude. (see Carbondale River)

Carbondale River (river)

82 G/8 - Beaver Mines
14-6-3-W5
49°28′N 114°19′W
Approximately 9 km west north-west of Beaver Mines.

This river, the west branch of Castle River, was officially named Carbondale River in 1918 after the coal mine on the river. Captain Palliser (see Introduction) called it Railway River due to the striking advantage its "levels" offered for the entry of a railway into the mountains.

Carcajou Creek (creek)

83 E/6 - Twintree Lake
33-50-9-W6
53°17′N 119°14′W
Flows north-east into Smoky River, approximately 90 km west north-west of Jasper.

(see Carcajou Pass)

Carcajou Pass (pass)

83 E/3 - Mount Robson
17-49-9-W6
53°14′N 119°16′W
Approximately 90 km north-west of Jasper.

To avoid duplication of already established names, the native word for "wolverine" was chosen for this pass. The name Carcajou Pass was officially approved in 1925.

Carconte Creek (creek)

83 E/14 - Grande Cache
7-57-7-W6
53°54′N 119°01′W
Flows south-east into Susa Creek, approximately 6 km east north-east of Grande Cache.

One of the original Iroquois to arrive in Jasper was named Dominick Karayinter. He arrived in 1814, sent from Montreal by the North West Company. He was later known as Carconte. It may be after him that this creek was named.

Cardinal Hills (hills)

83 C/15 - Cardinal River
27-45-19-W5
52°50′N 116°40′W
Approximately 97 km east of Jasper.

The name for this hill was officially approved in 1939. (see Cardinal River)

Cardinal, Mount (mountain)

83 C/14 - Mountain Park
6-45-23-W5
52°51′N 117°18′W
Approximately 52 km east of Jasper.

This mountain, which is 2515 m in altitude, was officially named in 1922. (see Cardinal River)

Cardinal Pass (pass)

83 C/14 - Mountain Park
4-44-23-W5
52°51′N 117°17′W
Approximately 53 km east of Jasper.

(see Cardinal River)

Cardinal River (river)

83 C/15 - Cardinal River
8-45-18-W5
52°52′N 116°35′W
Flows east into Brazeau River, approximately 100 km east of Jasper.

This river is named after Jacques Cardinal, a local fur trader, whose grave is located on the bank of the stream. The name was made official 6 March 1918. The other features in the area take their names from this river.

Carnarvon Creek (creek)

82 J/7 - Mount Head
10-17-6-W5
50°24′N 114°45′W
Flows north-east into McPhail Creek, approximately 44 km south-west of Turner Valley.

Herbert Henry Howard Melyneux, 4th Earl of Carnarvon, (1831-1890) was educated at Eton and Oxford, graduating with first-class honours. He entered the House of Lords in 1854 and was Under-Secretary for the Colonies in 1858-1859. The mountain in B.C. was named after him by A.M. Burgess, once Deputy Minister of the Interior for B.C. The creek and lake take their names from this mountain.

Carnarvon Lake (lake)

82 J/7 - Mount Head
30-16-6-W5
50°22′N 114°49′W
Approximately 46 km south-west of Turner Valley.

(see Carnarvon Creek)

Carrot Creek (creek)

82 O/3 - Canmore
23-25-11-W5
51°08′N 115°27′W
Flows south-west into Bow River, approximately 10 km south-east of Banff.

The name for this creek was officially approved in 1956, but its origin is unknown.

Carson Creek (creek)

83 E/9 - Moberly Creek
24-52-4-W6
53°30′N 118°27′W
Flows south into Wildhay River, approximately 74 km north north-west of Jasper.

Mr. Carson was a prominent Irishman of the area. This creek was named on the Forestry Branch of Alberta map of 1920. This name was officially approved for this stream in 1946.

Carthew Creek (creek)

82 H/4 - Waterton Lakes
1-30-W4
49°03′N 113°56′W

Flows north-east into Cameron Creek, approximately 45 km south of Pincher Creek.

Formerly "Carthew Brook," this creek takes its name from the mountain of the same name. The name Carthew Creek was officially approved 8 June 1971. (see Mount Carthew)

Carthew Lakes (lakes)

82 H/4 - Waterton Lakes
17-1-30-W4
49°02′N 113°59′W
Approximately 50 km south of Pincher Creek.

(see Mount Carthew)

Carthew, Mount (mountain)

82 G/1 - Sage Creek
13-1-1-W5
49°02′N 114°00′W
Approximately 8 km west of Waterton Park.

William Morden Carthew was R.W. Cautley's assistant in the Dominion Land Survey and climbed this mountain to set a signal for his colleague. Carthew was killed in Ypres 1 June 1916.

Cascade Mountain (mountain)

82 O/5 - Castle Mountain
26-12-W5
51°16′N 115°35′W
Approximately 5 km north of Banff.

The name for this mountain, which is 2998 m in altitude, was taken from the translation of a native name meaning "mountain where the water falls." Sir James Hector (see Mount Hector) abbreviated the term to Cascade Mountain in 1858.

Cascade River (river)

82 O/3 - Canmore
28-25-11-W5
51°10′N 115°29′W

Flows south into Bow River, approximately 7 km east of Banff.

The name was approved in 1911. (see Cascade Mountain)

Cascade Rock (peak)

82 J/13 - Mount Assiniboine
22-12-W5
50°55′N 115°35′W
Approximately 100 km west of Calgary.

(see Cascade Mountain)

Casket Creek (creek)

83 E/13 - Dry Canyon
16-56-13-W6
53°51′N 119°52′W
Flows north-east into Sheep Creek, approximately 50 km west of Grande Cache.

(see Casket Mountain)

Casket Lake (lake)

83 E/13 - Dry Canyon
30-55-13-W6
53°47′N 119°56′W
Approximately 54 km west south-west of Grande Cache.

(see Casket Mountain)

Casket Mountain (mountain)

83 E/13 - Dry Canyon
1-56-14-W6
53°48′N 119°57′W
Approximately 55 km west south-west of Grande Cache.

The rock formation at the summit of this 2231 m mountain resembles a casket. Casket was the name preferred to "Coffin" in 1925, the year this mountain and the surrounding features were officially named.

Casket Pass (pass)

83 E/13 - Dry Canyon
30-55-13-W6
53°47′N 119°56′W
Approximately 54 km west south-west of
Grande Cache.

(see Casket Mountain)

Castelets, The (mountain)

83 C/3 - Columbia Icefield
35-22-W5
52°04′N 117°08′W
Approximately 110 km south-east of Jasper.

This mountain has three pinnacles, each
resembling small castles. The descriptive
name was officially approved in 1920.

Castilleja Lake (lake)

82 N/9 - Hector Lake
25-29-15-W5
51°31′N 116°06′W
Approximately 55 km north-west of Banff.

The name for this small lake near the pass at
the head of the Spray River reflects the
profuse growth of the wildflower known as
Indian Paintbrush (castilleja). Two American
alpinists, W.D. Wilcox (see Wilcox Pass) and
H.G. Bryant suggested the name Castilleja
Lake, after the flower.

Castle Mountain (locality)

82 O/5 - Castle Mountain
32-26-14-W5
51°16′N 115°55′W
Approximately 25 km west north-west of
Banff.

This locality was named in 1883 after the
nearby mountain. (see Castle Mountain)

Castle Mountain (mountain)

82 O/5 - Castle Mountain
27-14-W5
51°19′00″N 115°57′00″W

Approximately 30 km west north-west of
Banff.

This 2862 m mountain was named by Sir
James Hector (see Mount Hector) in 1857
for its fortress-like appearance. In 1946 the
name was changed to Mount Eisenhower in
honour of General Dwight D. Eisenhower
(1890-1969), Supreme Commander of Allied
Forces in Europe during the Second World
War. Bowing to public pressure, the name
was officially changed back to Castle
Mountain 15 October 1979.

Castle Mountain and Silver City, 1885

Castle Peak (peak)

82 G/8 - Beaver Mines
8-4-2-W5
49°17′N 114°14′W
Approximately 30 km south-west of Pincher
Creek.

The origin of this feature's name remains
unknown.

Castleguard Meadows (meadows)

83 C/3 - Columbia Icefield
36-23-W5
52°07′00″N 117°12′00″W
Approximately 105 km south-east of Jasper.

The name Castleguard Meadows was
approved 21 January 1985. (see Castleguard
Mountain)

Castleguard Mountain (mountain)

83 C/3 - Columbia Icefield
35-23-W5
52°07′N 117°15′W
Approximately 100 km south-east of Jasper.

This 3090 m mountain was named by A.O.
Wheeler in 1919, owing to its castellated
appearance and the fact that it rises as a
guardian over the southern portion of the
Columbia Icefield.

Castleguard River (river)

83 C/3 - Columbia Icefield
35-22-W5
52°02′N 117°07′W
Flows south-east into Alexandra River,
approximately 115 km south-east of Jasper.

(see Castleguard Mountain)

Castor Creek (creek)

83 E/5 - Holmes River
25-51-12-W6
53°26′N 119°39′W
Flows north into Jackpine River,
approximately 122 km north-west of Jasper.

The name was approved in 1925. Castor is
the Latin word for "beaver." The creek flows
from the base of Mount Holmes, which was
formerly known as "Mount Beaver."

Cat Creek (creek)

82 J/7 - Mount Head
2-17-6-W5
50°24′N 114°43′W
Flows south into Highwood River,
approximately 42 km south-west of Turner
Valley.

This locally well-established name was officially applied to this feature 8 December 1940. The origin of the name is unknown.

Catacombs Creek (creek)

83 C/5 - Fortress Lake
34-40-27-W5
52°30′N 117°47′W
Flows north into Lick Creek, approximately 45 km south-east of Jasper.

(see Catacombs Mountain)

Catacombs Mountain (mountain)

83 C/5 - Fortress Lake
12-40-27-W5
52°26′N 117°45′W
Approximately 55 km south-east of Jasper.

This mountain, which is 3330 m in altitude, was named in 1921 by A.O. Wheeler. The feature has an alcove formation, likened to an underground burial tomb containing such recesses.

Cataract Creek (creek)

82 J/2 - Fording River
35-16-5-W5
50°24′N 114°35′W
Flows north into Highwood River, approximately 36 km south-west of Turner Valley.

The origin of the name is unknown.

Cataract Creek (creek)

83 C/2 - Cline River
6-37-20-W5
52°09′N 116°50′W
Flows south-east into Cline River, approximately 115 km south-east of Jasper.

(see Cataract Peak)

Cataract Pass (pass)

83 C/3 - Columbia Icefield
35-37-22-W5
52°13′N 117°03′W
Approximately 105 km south-east of Jasper.

The name was officially approved 5 March 1935. (see Cataract Creek)

Cataract Peak (peak)

82 N/9 - Hector Lake
17-31-15-W5
51°39′N 116°05′W
Approximately 65 km north-west of Banff.

This mountain peak is 3333 m in altitude and takes its name from the locally named falls in the vicinity, which have a descriptive name.

Caudron Creek (creek)

82 G/9 - Blairmore
30-8-3-W5
49°41′N 114°24′W
Flows west into Gold Creek, approximately 9 km north north-east of Blairmore.

(see Caudron Peak)

Caudron Peak (peak)

82 G/9 - Blairmore
4-8,9-3-W5
49°42′N 114°21′W
Approximately 13 km north of Bellevue.

Caudron Peak stands at 2547 m in altitude. The origin of the name is unknown.

Cautley, Mount (mountain)

82 J/13 - Mount Assiniboine
22-12-W5
50°54′N 115°34′W
Approximately 100 km west of Calgary.

This 2880 m mountain was officially named in 1917 after Richard William Cautley (1873-1953), a member of the Dominion Land Survey and the Commissioner representing Alberta on the Inter-Provincial Boundary Commission.

▪ Cave, The (cave)

82 G/10 - Crowsnest
8-8-5-W5
49°38′05″N 114°38′25″W
Approximately 10 km west of Coleman.

The spring which enters the north side of Crowsnest Lake is one of the main sources of the Crowsnest River and issues from a cave-like gash in the cliff. The name ''The Cave'' is therefore descriptive of this crevasse-like opening in the cliff.

Cave Mountain (mountain)

82 J/13 - Mount Assiniboine
23-12-W5
50°56′N 115°35 W Approximately 100 km west of Calgary.

This 2651 m mountain was officially named in 1916. The mountain received its name likely due to the presence of a cave in it.

Cavell Lake (lake)

83 D/9 - Amethyst Lakes
15-43-1-W6
52°42′N 118°04′W
Approximately 20 km south of Jasper.

(see Mount Edith Cavell)

Caw Creek (creek)

83 L/4 - Kakwa Falls
13-59-11-W6
54°06′N 119°31′W
Flows west into South Kakwa River, approximately 35 km north-west of Grande Cache.

This creek was named for Captain Bruce Edward Ashton Caw, M.C. and Bar, of Vegreville. He was killed in the Second World War.

Caywood Creek (creek)

83 F/5 - Entrance
23-50-27-W5
53°20′N 117°31′W
Flows east into Brule Lake, approximately 17 km west south-west of Hinton.

The origin of the name is unknown.

Celestine Lake (lake)

83 E/1 - Snaring
35-48-1-W6
53°11′N 118°03′W
Approximately 35 km north of Jasper.

The name for this lake was chosen in 1917 to reflect the feature's beauty. The word celestine is derived from the French, "céleste," meaning "heavenly," and from the Latin, "caelestis" (caelum heaven). The term was used at the time of the naming as a synonym for "beautiful."

Centre Creek (creek)

83 C/15 - Cardinal River
22-46-19-W5
52°59′N 116°41′W
Flows south-east into Pembina River approximately 95 km east north-east of Jasper.

(see Centre Peak)

Centre Creek (creek)

83 F/5 - Entrance
31-51-24-W5
53°27′N 117°31′W
Flows south-east into Athabasca River, approximately 2 km north-east of Hinton.

The name for this creek was officially approved in 1944. The origin for the locally well-established name of this creek is unknown.

Centre Mountain (mountain)

83 C/13 - Medicine Lake
31-44-27-W5
52°50′N 117°55′W
Approximately 14 km east south-east of Jasper.

This mountain is located between the heads of two valleys halfway between Excelsior and Amber mountains. The name was applied to the feature in 1916 by M.P. Bridgland. (see Mount Bridgland)

Centre Pass (pass)

83 E/2 - Resplendent Creek
34-46-5-W6
53°01′N 118°40′W
Approximately 41 km west north-west of Jasper.

This pass is the central passage of Miette Pass. The Miette Pass is a gap of about 5 km between high mountains of the main divide, in which there are three distinct passages separated by two ridges. It was deemed appropriate to name them separately. (see Roche Miette)

Centre Peak (peak)

82 G/9 - Blairmore
9-9-3-W5
49°43′N 114°21′W
Part of the Livingstone Range, approximately 15 km north of Bellevue.

This 2547 m peak is centred between the heads of two valleys.

Chaba Glacier (glacier)

83 C/4 - Clemenceau Icefield
23-37-26-W5
52°14′N 117°41′W
Approximately 75 km south-east of Jasper.

A.P. Coleman was a geologist born in eastern Canada in 1852. He was elected a Fellow of the Royal Society of Canada and

was associated with many universities across the country. He was the author of two books: *The Canadian Rockies* (1912), and *Ice Ages, Recent and Modern* (1926). Of the Chaba River and Glacier, Coleman wrote, "there were endless beaver dams and trees cut by the beaver along its course, we named it Chaba River, from the Stoney word for beaver."

Chaba Icefield (icefield)

83 C/4 - Clemenceau Icefield
38-27-W5
52°13′N 117°45′W
Approximately 75 km south south-east of Jasper.

(see Chaba Glacier)

Chaba Peak (peak)

83 C/4 - Clemenceau Icefield
29-37-26-W5
52°12′N 117°40′W
Approximately 80 km south-east of Jasper on the Alberta-B.C. boundary.

This 3212 m mountain was named by A.P. Coleman in memory of a Stoney Indian called Job Beaver who hunted in this area. The name was officially approved 1 March 1921. This feature was formerly called "Pimple Mountain," because of the physical shape of the feature but the name was changed because of the peak's proximity to the glacier of the same name. (see Chaba Glacier)

Chaba River (river)

83 C/5 - Fortress Lake
3-40-26-W5
52°25′N 117°39′W
Flows north into Athabasca River, approximately 60 km south-east of Jasper.

(see Chaba Glacier)

Chaffen Creek (creek)

82 J/1 - Langford Creek
23-13-2-W5
50°04′N 114°15′W
Flows east into South Willow Creek,
approximately 45 km west north-west of
Claresholm.

Chaffen Creek and Ridge are named for the
Chaffin family who came from Oregon to
homestead the area near Chimney Rock around
1905. The name Chaffen Creek was officially
approved 28 October 1959. The spelling of
the official name has been corrupted.

Chaffen Ridge (ridge)

82 J/1 - Langford Creek
28-12-3-W5
50°02′N 114°21′W
Approximately 55 km west of Claresholm.

This ridge was located behind the Chaffin
family's home. (see Chaffen Creek)

Chain Lakes (lake)

82 J/1 - Langford Creek
14,15-2-W5
50°14′N 114°13′W
Approximately 50 km north-west of
Claresholm.

The name was officially approved 3
December 1941. (see Chain Lakes Reservoir)

Chain Lakes Provincial Park (park)

82 J/1 - Langford Creek
34-14-2-W5
50°12′N 114°11′W
Approximately 50 km north-west of
Claresholm.

The name was officially approved 22 April
1970. (see Chain Lakes Reservoir)

Chain Lakes Reservoir (reservoir)

82 J/1 - Langford Creek
3-15-2-W5
50°14′N 114°12′W

Approximately 50 km north-west of
Claresholm.

The name is descriptive in that there were
originally four sloughs in a narrow valley
joined by marshy ground with only one foot
difference in elevation along the four or five
miles of length from north to south. The
natives called the body of water "Long
Lake." After the dams were built at the
north and south ends, the "chain of lakes"
became one body of water.

Chak Peak (mountain)

83 D/9 - Amethyst Lakes
19-43-1-W6
52°43′N 118°08′W
Approximately 19 km south south-west of
Jasper.

This mountain, which is 2774 m in altitude,
bears the native name for "eagle." It was
given the name by M.P. Bridgland (see
Mount Bridgland), and it was officially
approved in 1916.

Chalk Lake (lake)

83 D/9 - Amethyst Lakes
8-41-2-W6
52°31′N 118°13′W
Approximately 40 km south south-west of
Jasper.

This lake is fed by a glacial torrent carrying a
whitish boulder clay silt which at times
turns the lake a milky-white colour. The
descriptive name was applied to the lake in 1921.

Champion Lakes (lakes (2))

82 J/15 - Cluny
12-26-21-5-W5
50°49′00″N 114°36′37″W
Approximately 45 km south-west of Calgary.

A name was requested by the Department of
Fish and Wildlife for these two small lakes
for stocking identification purposes. After
much research, a suitable name was found,

taken after a prominent family of the area
west of Priddis. At the turn of the century,
John Norman Champion came to Canada
from England as a young man to homestead.
Mr. Champion was the first lookout man
with the Alberta Forest Service on Moose
Mountain. He was a forest warden for many
years and his son, Dexter, who now resides
in Hinton, was one of the first wardens in
the Kananaskis Valley.

Chance Creek (creek)

83 F/3 - Cadomin
33-48-21-W5
53°11′N 117°01′N
Flows south-east into Embarras River,
approximately 58 km south-west of Edson.

The origin of the name is unknown.

Channel Ridge (ridge)

82 J/9 - Turner Valley
3-19-4-W5
50°34′N 114°28′W
Approximately 20 km south of Calgary.

This ridge was named after Billy Channell,
son of the Lord Chief Justice of England. He
lived near the mouth of Highwood River in
the 1880s. The spelling of his name has since
been altered.

Chapel Rock (butte)

82 G/9 - Blairmore
7-9-2-W5
49°43′15″N 114°15′15″W
Approximately 16 km north of Lundbreck.

This 1600 m rounded butte was formerly
called "Chapel Butte." The name Chapel
Rock was officially approved 5 May 1983 to
conform with established local usage. The
origin of the name is unknown.

Chapman Creek (creek)

83 E/15 - Pierre Greys Lakes
21-56-5-W6
53°51′N 118°41′W

Flows east into Muskeg River, approximately 29 km east south-east of Grande Cache.

The name of this creek was made official in 1947. It may be named after Charlie Chapman, a former forest ranger in the area.

Charles Stewart, Mount (mountain)

82 O/3 - Canmore
28-25-10-W5
51°09′N 115°20′W
Approximately 8 km north north-east of Canmore.

The name Mount Charles Stewart was officially adopted 7 February 1928 for the highest peak of the Fairholme Range, 2809 m in altitude. The mountain was named in honour of the late Honourable Charles Stewart (1868-1946), federal Minister of the Interior. He was also federal Minister of Mines and Superintendent General of Indian Affairs from 1921-1930. In 1936, he was appointed Chairman of the International Joint Commission.

Charlton, Mount (mountain)

83 C/12 - Athabasca Falls
15-42-25-W5
52°36′N 117°30′W
Approximately 44 km south-east of Jasper.

The name for this 3217 m mountain was officially approved 23 November 1911. It was suggested by Mrs. Mary T.S. Schäffer in memory of H.R. Charlton, then General Advertising Agent for the Grand Trunk Pacific Railway.

Chephren Lake (lake)

82 N/15 - Mistaya Lake
17-33-19-W5
51°50′N 116°39′W
Approximately 105 km north-west of Banff.

(see Mount Chephren)

Chephren, Mount (mountain)

82 N/15 - Mistaya Lake
24-33-20-W5
51°51′N 116°41′W
Approximately 105 km north-west of Banff.

This 3266 m mountain was was originally known as "Pyramid Mountain," but was changed in order to avoid confusion with the Pyramid Mountain, which is located near Jasper (Mount Chephren is part of the Waputik Range). J.M. Thorington, an American alpinist and author from Philadelphia, noted that the western summit of this mountain was called White Pyramid for its shape. Chephren, or Khafre, was the fourth pharoah of the Fourth Dynasty of Egypt. He built the second of the Three Great Pyramids of Egypt. His father Khufu, or Cheops, built the Great Pyramid and reigned between 2540-2514 B.C. Thorington liked the association of Pyramid Mountain with the Great Pyramid and came up with the name.

Mount Chephren, n.d.

Cherry Hill (hill)

82 G/8 - Beaver Mines
6-4-W5
49°27′N 114°27′W
Approximately 17 km south of Blairmore.

The origin of the name is unknown.

Chester Creek (creek)

82 J/14 - Spray Lakes Reservoir
15-21-10-W5
50°49′N 115°16′W
Approximately 80 km west south-west of Calgary.

(see Mount Chester)

Chester Lake (lake)

82 J/14 - Spray Lakes Reservoir
26-21-10-W5
50°49′N 115°16′W
Approximately 80 km west south-west of Calgary.

(see Mount Chester)

Chester, Mount (mountain)

82 J/14 - Spray Lakes Reservoir
25-21-10-W5
50°48′N 115°16′W
Approximately 80 km west south-west of Calgary.

This mountain, 3054 m in altitude, was officially named in 1922 after the cruiser Chester, a ship engaged in the Battle of Jutland.

Chetamon Lake (lake)

83 E/1 - Snaring
28-47-2-W6
53°03′N 118°11′W
Approximately 21 km north north-west of Jasper.

(see Chetamon Mountain)

Chetamon Mountain (mountain)

83 E/1 - Snaring
11-47-2-W6
53°03′N 118°13′W
Approximately 23 km north north-west of Jasper.

Chetamon is the Stoney Indian word for "squirrel." Two rocks on the mountain resemble the small animal. M.P. Bridgland (see Mount Bridgland) suggested the descriptive name for this 2504 m mountain in 1916.

Chetang Ridge (ridge)

83 E/3 - Mount Robson
48-8-W6
53°11′N 119°04′W
Approximately 74 km west north-west of Jasper.

This ridge was named in 1912 by C.D. Walcott. Charles D. Walcott Jr., the son of Charles D. Walcott, Secretary of the Smithsonian Institute, was a member of a joint A.C.C.-Smithsonian expedition to Mt. Robson in 1912. "Chetang" is the Stoney Indian word for "hawk." The name was officially approved 3 December 1912.

Cheviot Creek (creek)

83 C/14 - Mountain Park
4-46-23-W5
52°56′N 117°17′W
Flows north into McLeod River, approximately 54 km east of Jasper.

(see Cheviot Mountain)

Cheviot Mountain (mountain)

83 C/14 - Mountain Park
25-45-24-W5
52°54′N 117°21′W
Approximately 50 km east of Jasper.

H.M. Thornton, the General Manager of the Mountain Park Coal Company Limited, named this 2720 m mountain. It reminded him of the Cheviot Hills located along the border of England and Scotland, his former home. This mountain is many times larger than the feature for which it was named, but it resembles its shape.

Chevron Mountain (mountain)

83 D/9 - Amethyst Lakes
30-42-1-W6
52°38′N 118°07′W
Approximately 26 km south of Jasper.

A chevron may be defined as a bent bar with an inverted "V" shape. This mountain, which stands 2835 m in altitude, is double pointed. It was given this descriptive name in 1916 by M.P. Bridgland. (see Mount Bridgland)

Chicken Creek (creek)

83 L/6 - Chicken Creek
6-61-10-W6
54°16′N 119°28′W
Flows south-west into Kakwa River, approximately 48 km north-west of Grande Cache.

The origin of this creek's name is unknown.

Chief Creek (creek)

83 F/3 - Cadomin
29-47-21-W5
53°05′N 117°02′W
Flows south into Beaverdam Creek, approximately 67 km south-west of Edson.

The name for this creek was made official in 1944 and was suggested by J.R. Matthews, a local resident. The origin of the name is unknown.

Childear Mountain (mountain)

83 E/11 - Hardscrabble Creek
17-54-9-W6
53°40′15″N 119°18′00″W
Approximately 30 km south-west of Grande Cache.

The name for this 2485 m mountain was the winning entry in the Geographical Naming Contest for elementary school children to commemorate the International Year of the Child in 1979. The name is a contraction of the words Child and year. It was officially approved 14 April 1980.

Chilver Lake (lake)

82 O/3 - Canmore
13-24-8-W5
51°04′N 115°04′W
Approximately 21 km east of Canmore.

The origin of the name is unknown.

Chimney Creek (creek)

83 C/15 - Cardinal River
16-44-20-W5
52°47′N 116°51′W
Flows north-east into Thistle Creek, approximately 83 km east south-east of Jasper.

The origin of the name is unknown.

Chimney Ridge (ridge)

82 J/1 - Langford Creek
12-2-W5
50°01′N 114°14′W
Approximately 35 km west of Claresholm.

The descriptive name for this ridge refers to a notable chimney (i.e., a narrow cleft by which a cliff can be climbed) on one of its faces.

Chinamans Peak (peak)

82 O/3 - Canmore
24-24-11-W5
51°03′50″N 115°24′00″W
Approximately 3 km south-west of Canmore.

The well-established local name for this 2680 m mountain peak commemorates an incident in 1886 in which a Chinese mine employee took a bet of fifty dollars that he could climb the mountain within six hours. He successfully did so, and placed a flag on the top. The name was officially recognized July 1980 on the basis of strong local usage and its historical significance.

*Chiniki (former locality)

82 O/2 - Jumpingpound Creek
34-25-6-W5
51°10′N 114°46′W

---

*denotes rescinded name or former locality.

Approximately 40 km north-west of Calgary.

(see Chiniki Creek)

Chiniki Creek (creek)

82 O/2 - Jumpingpound Creek
33-25-6-W5
51°11'N 114°48'W
Flows north-east into Bow River, approximately 40 km north-west of Calgary.

Officially approved 12 December 1939, this locally well-established name is taken from the Stoney Indian Chief who signed Treaty Number 7 of September 1877 at Blackfoot Crossing as "Cheneka" or "John."

Chiniki Lake (lake)

82 O/2 - Jumpingpound Creek
4-25-7-W5
51°06'N 114°36'W
Approximately 50 km west north-west of Calgary.

(see Chiniki Creek)

Chinook Creek (creek)

83 L/12 - Nose Creek
19-66-13-W6
54°44'N 119°57'W
Flows north into Wapiti River, approximately 90 km south-west of Grande Prairie.

Chinook Ridge crosses the Alberta-B.C. boundary, and forms the watershed between Mistanusk and Chinook creeks. The creek is named from the ridge, and both have been in use since the Alberta-B.C. Boundary Survey, 1918-1924.

Chinook Lake (lake)

83 L/12 - Nose Creek
17-65-13-W6
54°38'N 119°56'W

Approximately 92 km south-west of Grande Prairie.

(see Chinook Creek)

Chinook Peak (peak)

82 G/10 - Crowsnest
7-5-W4
49°38'05"N 114°38'25"W
Approximately 10 km west of Coleman.

Chinook Peak, which is 2591 m in altitude, was named officially in 1962. Dr. R.A. Price, in charge of a party of the Geological Survey of Canada, was mapping this area in 1960 and found it necessary to name some of the peaks of the Flathead Range. Residents of the area watch this peak during cold winters allegedly for indications of a coming chinook.

Chipman Creek (creek)

82 G/8 - Beaver Mines
35-5-1-W5
49°26'N 114°02'W
Approximately 9 km south-west of Pincher Creek.

The origin of the name is unknown.

Chocolate Mountain (mountain)

83 C/6 - Sunwapta
28-40-22-W5
52°28'N 117°05'W
Approximately 80 km south-east of Jasper.

The origin of the name of this mountain, which is 3049 m in altitude, is unknown.

Chown Creek (creek)

83 E/6 - Twintree Lake
21-50-9-W6
53°20'N 119°15'W
Flows south-east into Smoky River, approximately 94 km west north-west of Jasper.

(see Mount Chown)

Chown Glacier (glacier)

83 E/6 - Twintree Lake
51-10-W6
53°23'N 119°23'W
Approximately 102 km west north-west of Jasper.

(see Mount Chown)

Chown, Mount (mountain)

83 E/6 - Twintree Lake
16-51-10-W6
53°24'N 119°23'W
Approximately 107 km west north-west of Jasper.

This 3381 m mountain is the central and highest of the Resthaven Group. It was named in 1912 by H.A. Stevens after Reverend Samuel Dwight Chown, D.D., LL.D. (1853-1933). Chown was an advocate of church union and was General Superintendent of the Methodist Church from 1914-1925. He was instrumental in forming what is now the United Church of Canada.

Christian Peak (peak)

82 N/14 - Rostrum Peak
34-22-W5
51°56'36"N 117°05'53"W
Approximately 140 km north-west of Banff, on the Alberta-B.C. boundary.

The name for this 3390 m peak on Mount Lyell is named after Christian Kaufmann. The peaks on Mount Lyell were named for five prominent mountaineering guides originally brought out by the C.P.R. from Switzerland, and who later became residents of Canada. The five were Edward, Ernest and Walter Feuz, Rudolph Aemmer and Christian Kaufmann. They are part of what was called the "Swiss Guide Group."

Christie, Mount (mountain)

83 C/12 - Athabasca Falls
15-41-27-W5
52°32'N 117°48'W

Approximately 43 km south south-east of Jasper.

This mountain, which is 3103 m in altitude, was named by Sir James Hector (see Mount Hector) in 1859 after William Joseph Christie. After serving the Hudson's Bay Company as a clerk, Christie was appointed chief trader in charge of the Swan River District in 1854. He became a chief factor in 1858, and he retired in 1870, settling in Brockville, Ontario. He played a part as a commissioner in the negotiation of Indian Treaties in 1874-1876.

Christie Mine Ridge (ridge)

82 G/8 - Beaver Mines
10-5-1-W5
49°22′N 114°04′W
Approximately 15 km south-west of Pincher Creek.

The Christie Coal Mine, established by Andrew Christie, was the first mine to be opened along Pincher Creek. From 1911-1927, coal was extracted on the north face of this ridge from a 16 hectare coal base of the Christie Mine. The name was officially approved 15 February 1978.

Christine Lake (lake)

83 D/16 - Jasper
17-45-2-W6
52°53′N 118°14′W
Approximately 11 km west of Jasper.

This lake was named in 1917 by H. Matheson of the Dominion Land Survey, but its origin is unknown.

Chrome Lake (lake)

83 D/9 - Amethyst Lake
33-42-2-W6
52°40′N 118°14′W
Approximately 25 km south-west of Jasper.

The name for this lake describes the curious yellow brilliancy of the water.

Chungo Creek (creek)

83 C/9 - Wapiabi Creek
19-43-16-W5
52°43′N 116°19′W
Flows east into Blackstone River, approximately 32 km north north-west of Nordegg.

This creek was named in 1910 by George Mallock to replace the name "Trail Creek." Nothing is known of Mr. Mallock. Chungo is the Stoney Indian word signifying "trail."

Chungo Gap (gap)

83 C/10 - George Creek
33-42-19-W5
52°40′N 116°41′W
Approximately 46 km west north-west of Nordegg.

(see Chungo Creek)

Chushina Ridge (ridge)

83 E/3 - Mount Robson
48-8-W6
53°07′N 119°04′W
Approximately 71 km west north-west of Jasper.

Chushina is the Stoney Indian word signifying "small" and was thought to be descriptive when in 1912 W.D. Walcott Jr., a member of a joint A.C.C.-Smithsonian expedition to Mount Robson, named this feature.

Cinema Lake (lake)

83 C/3 - Columbia Icefield
32-35-23-W5
52°03′N 117°15′W
Approximately 110 km south-east of Jasper.

This lake received its descriptive name because it particularly shows the reflective properties of water. The name was officially approved in 1961.

Cinquefoil Creek (creek)

83 F/4 - Miette
26-47-28-W5
53°05′N 118°00′W
Flows north-west into Jacques Creek approximately 47 km south-west of Hinton.

(see Cinquefoil Mountain)

Cinquefoil Mountain (mountain)

83 F/4 - Miette
14-47-28-W5
53°03′N 118°00′W
Approximately 50 km south south-west of Hinton.

The cinquefoil, a five petalled, bright gold shrub, is a type of rose which grows in the valley below this 2259 m mountain. The small yellow wildflower grows on exposed slopes in high altitudes of the Rocky Mountains. The creek is named after the mountain. M.P. Bridgland (see Mount Bridgland) gave this mountain its name in 1916.

Circus Creek (creek)

83 D/16 - Jasper
52°46′01″N 118°07′20″W
Approximately 12 km south-west of Jasper.

This creek was named for its proximity to Circus Valley. (see Circus Valley)

Circus Valley (valley)

83 D/16 - Jasper
1-44-2-W6
52°46′N 118°09′W
Approximately 14 km south south-west of Jasper.

A circus is a ring or roundabout, and since this oblong valley is enclosed on three sides, it resembles a circus tent. The name is thus likely descriptive of the feature's appearance.

Cirque Lake (lake)

82 N/15 - Mistaya Lake
4-33-19-W5
51°48′N 116°38′W
Approximately 100 km north-west of Banff.

(see Cirque Peak)

Cirque Peak (peak)

82 N/9 - Hector Lake
36-31-18-W5
51°42′N 116°25′W
Approximately 85 km north-west of Banff.

The descriptive name for this mountain peak, 2993 m in altitude, was approved in 1909. The semi-circular formation of limestone on the peak gives the impression of a huge amphitheatre.

■ **Cirrus Mountain** (mountain)

83 C/2 - Cline River
17-37-21-W5
52°10′N 116°59′W
Approximately 105 km south-east of Jasper.

The name "Mount Huntington" was originally submitted for this 3270 m mountain, but Cirrus Mountain was recommended by the National Parks Branch and was eventually approved in November 1935. Cirrus clouds are feathery wisps of ice crystals that form at high altitudes, and many may be seen around this peak.

Citadel Lake (lake)

82 O/4 - Banff
51°01′10″N 115°43′20″W
Approximately 20 km south-west of Banff.

The name for this lake was officially approved 21 January 1985. (see Citadel Peak)

Citadel Pass (pass)

82 O/4 - Banff
24-13-W5
51°01′N 115°43′W
Approximately 21 km south-west of Banff.

(see Citadel Peak)

Citadel Peak (peak)

82 O/4 - Banff
24-13-W5
51°01′N 115°44′W
Approximately 21 km south-west of Banff.

This mountain peak, which is 2610 m in altitude, was named by A.O. Wheeler after its fortress-like shape.

Clairvaux Creek (creek)

83 D/16 - Jasper
14-45-3-W6
52°53′N 118°18′W
Flows north into Miette River, approximately 15 km west of Jasper.

(see Mount Clairvaux)

Clairvaux Glacier (glacier)

83 D/16 - Jasper
7-44-3-W6
52°46′N 118°24′W
Approximately 25 km south-west of Jasper.

(see Mount Clairvaux)

Clairvaux, Mount (mountain)

83 D/16 - Jasper
20-44-3-W6
52°48′N 118°25′W
Approximately 24 km west south-west of Jasper.

Clairvaux is the French word for "clear valleys," which describes the situation of this 2690 m mountain. The other features take their name from this mountain, located at the head of a clear valley.

Clark Creek (creek)

83 C/10 - George Creek
35-43-18-W5
52°36′N 116°31′W
Flows north into Brown Creek, approximately 44 km north-west of Nordegg.

The origin of the name is unknown.

Clark Range (range)

82 G/1 - Sage Creek
35-2-2-W5
49°09′N 114°10′W
Approximately 15 km west of Waterton Park.

This range, which is located south of North Kootenay Pass, was named after Captain William Clark (1770-1835), a soldier and explorer born in Caroline County, Virginia. The now well-known Lewis and Clark Expedition began at St. Louis, Missouri 14 May 1804 and ended in the same city 23 September 1806. Clark's function during the journey was drawing maps and illustrating the wildlife of the region.

Clarks Crossing (river crossing)

83 E/14 - Grande Cache
24-55-10-W6
53°46′N 119°21′W
Approximately 20 km south-west of Grande Cache.

Stan Clark was born in Ontario and was Superintendent of the Rocky Mountain Forest Reserve in 1912. Clark became a forestry supervisor, game commissioner and rancher who was instrumental in establishing the Athabasca Forest Reserve. This river crossing was named in 1953, probably after Mr. Clark.

Clearwater Lake (lake)

82 N/16 - Siffleur River
25-32-16-W5
51°46′N 116°09′W

Approximately 75 km north-west of Banff.

(see Clearwater River)

Clearwater Mountain (mountain)

82 N/9 - Hector Lake
32-16-W5
51°45′N 116°17′W
Approximately 85 km north-west of Banff.

This 3275 m mountain takes its name from the nearby stream. (see Clearwater River)

Clearwater Pass (pass)

82 N/9 - Hector Lake
32-16-W5
51°43′N 116°17′W
Approximately 80 km north-west of Banff.

(see Clearwater River)

Clearwater River (river)

83 B/7 - Rocky Mountain House
52°22′N 114°57′W
Flows north-west into North Saskatchewan River, approximately 2 km south-west of Rocky Mountain House.

The descriptive name for this river has been in local use since at least 1814. It was officially approved 5 July 1951. The other features in the area take their name from this river.

Cliff Creek (creek)

82 J/10 - Mount Rae
8-19-6-W5
50°36′N 114°47′W
Flows north-east into Sheep River, approximately 37 km west of Turner Valley.

The name for this creek is likely descriptive of its location near a cliff.

Cliff Mountain (mountain)

83 E/1 - Snaring
28-47-2-W6
53°05′N 118°13′W

Approximately 25 km north north-west of Jasper.

The steep cliffs along each face of this 2743 m mountain suggested its descriptive name to M.P. Bridgland (see Mount Bridgland) in 1916.

Climax Creek (creek)

83 C/13 - Medicine Lake
24-45-25-W5
52°53′N 117°30′W
Flows south into Rocky River, approximately 43 km east of Jasper.

(see Climax Mountain)

Climax Mountain (mountain)

83 C/14 - Mountain Park
29-45-24-W5
52°54′N 117°25′W
Approximately 44 km east of Jasper.

The summit of this mountain protrudes above a long ridge. The descriptive name for the mountain, which stands 2823 m in altitude, was officially approved 5 March 1935.

Cline, Mount (mountain)

83 C/2 - Cline River
5-36-19-W5
52°05′N 116°41′W
Approximately 130 km south-east of Jasper.

This mountain, 3361 m in altitude, was named in 1902, after Michael Cline (Klyne, Clyne, Klein, Kline), who was born in 1781, and was an employee of the North West Company and Hudson's Bay Company. This fur trader and voyageur was also a postmaster at Jasper in 1834. The peak was named by J.N. Collie. (see Collie Creek)

Cline Pass (pass)

83 C/2 - Cline River
6-38-21-W5
52°14′25″N 117°00′05″W

Approximately 100 km south-east of Jasper.

This pass was originally named in 1935, but the feature which was described by the former location is impassable. The present location was officially approved 20 October 1983. (see Mount Cline)

Cline River (river)

83 C/1 - Whiterabbit Creek
13-36-18-W5
52°08′N 116°29′W
Flows north-west into Abraham Lake, approximately 107 km west south-west of Rocky Mountain House.

(see Mount Cline)

Cline River Canyon (canyon)

83 C/2 - Cline River
4-37-18-W5
52°09′N 116°30′W
Approximately 110 km west south-west of Rocky Mountain House.

(see Mount Cline)

Cline River Settlement (locality)

83 C/1 - Whiterabbit Creek
22-37-18-W5
52°12′N 116°28′W
Approximately 105 km west south-west of Rocky Mountain House.

The name for this settlement was approved by the Canadian Permanent Committee on Geographical Names 21 March 1975. (see Mount Cline)

Clitheroe, Mount (mountain)

83 D/9 - Amethyst Lakes
16-43-2-W6
52°43′N 118°14′W
Approximately 23 km south south-west of Jasper.

This mountain, which stands 2747 m in altitude, was named by M.P. Bridgland (see

Mount Bridgland) in 1916. It takes its name from Clitheroe, a municipal borough of Lancashire, England, home of the Clitheroe Royal Grammar School. The name is derived from an Old English (450-1150 A.D. - a combination of partly Celtic and Germanic influences) term, "clyder," and an Old Norse (a Germanic language of medieval Scandinavia) term, "haugr." Combined, the name means "hill of loose stones," and is descriptive of the large deposits of limestone in the area.

Cloister Mountains (mountains)

83 C/2 - Cline River
20-37-20-W5
52°13′N 116°49′W
Approximately 115 km south-east of Jasper.

This group resembles "four fine cathedrals with splendid walls and buttresses on the south-west," according to A.P. Coleman (see Mount Coleman), the alpinist who named it. Another mountain, called Minster Mountain, lies across the main fork of Cataract River, and it was named as a match to this one.

Cloudy Ridge (ridge)

82 H/4 - Waterton Lakes
3-30-W4
49°11′N 113°59′W
Approximately 35 km south of Pincher Creek.

This ridge, which is 2587 m in altitude, was named for its distinctive shape.

Clyde Creek (creek)

83 F/13 - Hightower Creek
21-55-25-W5
53°46′N 117°40′W
Flows north into Wildhay River, approximately 40 km north of Hinton.

This creek was named in 1945, when a Clydesdale horse allegedly fell into the stream.

Coal Coulee (coulee)

82 J/9 - Turner Valley
9-18-2-W5
50°30′15″N 114°13′40″W
Runs north-east into Highwood River, approximately 60 km south of Calgary.

During the late 1800s local ranchers were able to obtain coal from the open seams exposed along the coulee. A number of mines were located in the coulee later on. The "Coal Trail" ran from this coulee, through Longview, to High River.

Coal Creek (creek)

82 J/10 - Mount Rae
30-19-4-W5
50°38′114°32′W
Flows north-east into Sheep River, approximately 18 km west of Turner Valley.

In 1916, Dave Blacklock opened a mine at the mouth of this creek, which has numerous outcroppings of coal along its banks. The name was officially approved 6 September 1951.

Coal Ridge (ridge)

83 L/4 - Kakwa Falls
11-60-14-W6
54°10′N 119°59′W
Approximately 63 km north-west of Grande Cache.

There are many deposits and seams of coal found in the area. This ridge has been known as Coal Ridge since the Alberta-B.C. Boundary Survey, 1918-1924, but the name did not become official until 1959.

Coal Valley (locality)

83 F/2 - Foothills
26-47-20-W5
53°05′N 116°48′W
Approximately 60 km south south-west of Edson.

Coal Valley was a typical coal branch town when the coal industry was alive. The Coal Valley Mining Company installed a Cable-Way Excavator in 1930. There was no underground mining at this site, which was established in 1923. The town made its final closing in 1955. The final scars of strip mining are all that remain of this town.

Coalspur (locality)

83 F/3 - Cadomin
33-48-21-W5
53°11′N 117°01′W
Approximately 58 km south-west of Edson.

The site of this present ghost town was important for its coal mines as well as for its position as a railway centre and headquarters for the Brazeau Forest Reserve. Located on the coal spur at Yellowhead, the name for the hamlet eventually was run together as one word, yielding Coalspur.

Coast Creek (creek)

83 C/15 - Cardinal River
28-44-19-W5
52°50′N 116°42′W
Flows north-west into Brazeau River, approximately 91 km east of Jasper.

The origin of the name is unknown.

Cobblestone Creek (creek)

83 E/1 - Snaring
8-47-1-W6
53°02′N 118°07′W
Flows east into Athabasca River, approximately 17 km north of Jasper.

The descriptive name for this creek is locally well-established.

Cockscomb Mountain (mountain)

82 O/4 - Banff
21-26-13-W5
51°14′N 115°44′W
Approximately 13 km north-west of Banff.

The descriptive name for this 2776 m mountain was officially approved 4 May 1921. The outline of the summit resembles the comb of a cock.

Cold Creek (creek)

83 F/5 - Entrance
25-50-26-W5
53°20′N 117°40′W
Flows north-west into Maskuta Creek, approximately 6 km south south-west of Hinton.

No information regarding the origin of the name is known.

Cold Sulphur Spring (spring)

83 E/1 - Snaring
9-47-1-W6
53°03′N 118°05′W
Approximately 18 km north of Jasper.

Small quantities of sodium chloride, calcium bicarbonate, magnesium sulphide and hydrogen sulphate give the water of this spring its sour smell and the spring itself its descriptive name.

Coleman (town)

82 G/9 - Blairmore
8-8-4-W5
49°38′N 114°30′W
Located in the Crowsnest Pass, approximately 5 km north-west of Blairmore.

The town was named in 1904 by Mr. A.C. Flumerfelt, the President of the International Coal and Coke Company after his youngest daughter, Florence Coleman Flumerfelt. Coleman was incorporated in September of 1910.

Coleman Glacier (glacier)

83 E/3 - Mount Robson
48-8-W6
53°10′N 119°03′W

Approximately 70 km west north-west of Jasper.

(see Mount Coleman)

Coleman Lake (lake)

83 C/2 - Cline River
52°07′55″N 116°56′55″W
Approximately 115 km south-east of Jasper.

The name of this lake was officially approved 21 January 1985. (see Mount Coleman)

Coleman, Mount (mountain)

83 C/2 - Cline River
27-36-21-W5
52°07′N 116°55′W
Approximately 115 km south-east of Jasper.

This 3135 m mountain was named in 1902 after Arthur Philemon Coleman (1852-1939). He was a professor, originally from Toronto. Coleman was one of the first alpinists to begin climbing and exploring the Rockies in 1889. This charter member of the Alpine Club of Canada also attempted to conquer Mount Robson in 1908.

Colin Creek (creek)

83 F/4 - Miette
1-47-28-W5
53°01′N 117°58′W
Flows north into Jacques Creek, approximately 53 km south south-west of Hinton.

(see Mount Colin)

Colin, Mount (mountain)

83 F/4 - Miette
34-46-28-W5
53°00′N 117°59′W
Approximately 55 km south south-west of Hinton.

This mountain, which is 2687 m in altitude, was named in 1859 by Sir James Hector of

the Palliser Expedition (see Mount Hector) after Colin Fraser (1805-1867). Fraser was attached to the Hudson's Bay Company and was in charge of Jasper House from 1835-1849. The other features in the area take their name from this mountain.

Colin Range (range)

83 C/13 - Medicine Lake
46-28-W5
52°58′N 117°57′W
Approximately 14 km north-east of Jasper.

Colin Range stands 2687 m in altitude. (see Mount Colin)

Colin Ridge (ridge)

83 C/13 - Medicine Lake
46-27-W5
52°58′N 117°53′W
Approximately 17 km north-east of Jasper.

This ridge has been measured at 2687 m in altitude. (see Mount Colin)

Coliseum Mountain (mountain)

83 C/9 - Wapiabi Creek
41-15-W5
52°31′N 116°05′W
Approximately 5 km north of Nordegg.

Martin Nordegg (see Nordegg) mentions this 1981 m mountain in his memoirs (pp. 154 & 186). The mountain resembles a coliseum, a rounded arena commonly built in Ancient Rome. The name was officially proposed in 1941.

Collie Creek (creek)

83 E/9 - Moberly Creek
20-52-1-W6
53°30′N 118°07′W
Flows south-east into Wildhay River, approximately 70 km north of Jasper.

In 1910 and 1911, Professor John Norman Collie and A.L. Mumm (see Mumm Peak) were accompanied by two other members of The Alpine Club (London) for the purpose of mountaineering and adventure into the area and along the Continental Divide. Collie's other alpine endeavours took him to the Alps and the Himalayas. He and Hermann Woolley discovered and named the Columbia Icefield in 1898. This creek was named after J.N. Collie.

Collier Peak (peak)

82 N/8 - Lake Louise
51°23′25″N 116°18′00″W
Approximately 60 km west north-west of Banff, on the Alberta-B.C. boundary.

Recommended by the British Columbia Representative for Geographical Names, D.F. Pearson, 6 June 1984, the name for this mountain peak was officially approved 21 January 1985. The origin of the name is unknown.

Colonel Pass (pass)

83 E/2 - Resplendent Creek
23-47-6-W6
53°03′N 118°46′W
Approximately 51 km west north-west of Jasper.

Colonel Peak, Pass and Creek were so named by A.O. Wheeler in 1911 after Colonel Aimé Laussedat (1819-1907), an engineer in the French Army who was the originator of the science of phototopography so extensively used for survey purposes in the mountainous regions of the country. The "father of photography surveying," he gave a full exposition of the method in his *Mémorial de l'Officier du Génie* (1898?).

Colt Creek (creek)

83 C/9 - Wapiabi Creek
35-42-15-W5
52°40′N 116°04′W
Flows north into Nordegg River, approximately 22 km north of Nordegg.

This creek was officially named in 1941. The origin of the name is unknown.

Colt Creek (creek)

83 E/16 - Donald Flats
34-57-1-W6
53°58′N 118°21′W
Flows east into Horse Creek, approximately 70 km east north-east of Grande Cache.

This tributary to Horse Creek was officially named in 1947. The precise origin of the name is unknown.

Columbia Glacier (glacier)

83 C/3 - Columbia Icefield
8-37-24-W5
52°10′N 117°25′W
Approximately 90 km south-east of Jasper.

(see Mount Columbia)

Columbia Icefield (icefield)

83 C/3 - Columbia Icefield
37-24-W5
52°10′N 117°20′W
Approximately 95 km south-east of Jasper.

The Columbia Icefield is the centre of the greatest accumulation of ice in the Rocky Mountains. With its outlet glaciers the Columbia Icefield covers an area of nearly 337 km² of which 130 km² are more than 2591 m above sea level in the area of accumulation, usually called the "névé." The Columbia Icefield is the source of three great rivers — the Athabasca (1231 m), the Saskatchewan (1939 m), and the Columbia (1947 m), flowing respectively to the Arctic Ocean, Hudson Bay and the Pacific Ocean. (see Mount Columbia)

Columbia, Mount (mountain)

83 C/3 - Columbia Icefield
1-37-25-W5
52°09′N 117°27′W

Approximately 90 km south-east of Jasper, on the Alberta-B.C. boundary.

The summit of this 3747 m mountain is the highest point in Alberta. It takes its name from the Columbia River that flows between Oregon and Washington. In 1972, Captain Robert Tray (1755-1806) an American sea captain, noticing the movement of the water, decided to explore, and, at high tide, with all sails set, successfully crossed the dangerous bar located at the river's mouth. He named the river after his ship, "Columbia."

Committee Punch Bowl (lake)

83 D/8 - Athabasca Pass
26-39-2-W6
52°22′N 118°11′W
Approximately 55 km south of Jasper, on the Alberta-B.C. boundary.

Alexander Ross, a fur trader and author, crossed the Athabasca Pass with Sir George Simpson, future Governor-in-Chief of Rupert's Land, in 1825. A mere four years earlier, the North West Company and the Hudson's Bay Company had amalgamated under the Hudson's Bay title. This circular basin, which measures about 18 m in diameter was dubbed the Committee Punch Bowl, referring to the Governing Committee of the Hudson's Bay Company. A bottle of wine was shared among Simpson and his companions. This was to be a tribute always paid to this place when a nabob of the fur trade passed by. This was the common meeting-ground between Hudson's Bay Company factors of the interior and those of the coast.

Commonwealth Creek (creek)

82 J/14 - Spray Lakes Reservoir
13-33-21-10-W5
50°49′55″N 115°20′20″W
Flows north into Smuts Creek, approximately 85 km west south-west of Calgary.

The creek is named Commonwealth Creek because of its proximity to Commonwealth Peak. The name was officially approved 12 September 1984. (see Commonwealth Peak)

Commonwealth Lake (lake)

82 J/14 - Spray Lakes Reservoir
1-29-21-10-W5
50°49′25″N 115°20′40″W
Approximately 85 km west south-west of Calgary.

This lake is located on the north slope of Commonwealth Peak and was officially named 12 September 1984. (see Commonwealth Peak)

Commonwealth Peak (peak)

82 J/14 - Spray Lakes Reservoir
20-21-10-W5
50°47′25″N 115°20′55″W
Approximately 85 km west south-west of Calgary.

Officially named 12 March 1979, this 2774 m peak commemorates the XI Commonwealth Games held in Edmonton from August 3-12, 1978. The name of this peak is particularly significant because the peak is located in the British Military Group of twenty-seven mountains named for Allied Generals and British Battleships of the First World War.

Comoy Lake (lake)

83 E/2 - Resplendent Lake
53°10′35″N 118°40′00″W
Approximately 50 km west north-west of Jasper.

The precise origin of this locally well-established name is unknown.

Compass Creek (creek)

83 L/12 - Nose Creek
24-64-14-W6
54°33′N 119°59′W
Flows north-east into Mistanusk Creek,

approximately 98 km south-west of Grande Prairie.

There is a Compass Hill in British Columbia, the name of which describes a well-defined spur of the main ridge, and a noticeable landmark. The name was suggested by the Boundary Commission. This creek likely takes its name from the hill, which is 1613 m in altitude.

Compression Ridge (ridge)

82 J/15 - Bragg Creek
22-7-W5
50°53′N 114°58′W
Approximately 55 km west of Calgary.

The name for this feature was officially approved 6 October 1949 and is likely descriptive of the tectonic movements which shaped the ridge.

Condor Peak (peak)

82 O/13 - Scalp Creek
33-32-14-W5
51°48′N 115°55′W
Approximately 75 km north-west of Banff.

The name for this peak is suggestive of mountain eagles and was proposed by R.W. Cautley of the Dominion Land Survey.

Cone Mountain (mountain)

82 J/14 - Spray Lakes Reservoir
22-11-W5
50°53′N 115°28′W
Approximately 90 km west of Calgary.

This mountain, 2896 m in altitude, is shaped like a cone. Its descriptive name was made official in 1915.

Confederation, Mount (mountain)

83 C/5 - Fortress Lake
31-39,40-25-W5
52°24′N 117°36′W

Approximately 60 km south-east of Jasper.

The name for this 2969 m mountain was suggested by American climber A.J. Ostheimer III. The name commemorates the Fathers of Confederation and was made official in 1961.

Conical Peak (peak)

82 N/16 - Siffleur River
2-32-18-W5
51°48′N 116°27′W
Approximately 95 km north-west of Banff.

The name for this 2840 m mountain peak is descriptive of its appearance.

Conifer Creek (creek)

83 D/16 - Jasper
10-45-2-W6
52°52′N 118°12′W
Flows north-east into Miette River, approximately 8 km west of Jasper.

This creek runs through an area which is thickly forested with evergreens. Near the bottom of the creek there is a small waterfall. Its descriptive name was applied in 1916 by M.P. Bridgland. (see Mount Bridgland)

Connelly Creek (creek)

82 G/9 - Blairmore
36-7-2-W5
49°36′N 114°08′W
Flows east into Crowsnest River, approximately 2 km north-east of Lundbreck.

Robert Connelly and five of his brothers came to the Pincher Creek area in the 1880s. Originally from Ireland, these early home-steaders were an important factor in the development of Bellevue and its surrounding area. Connelly Creek is named after the Connelly brothers and was officially approved 7 February 1905.

Connop Creek (creek)

82 J/15 - Bragg Creek
33-22-5-W5
50°55'N 114°38'W
Flows north into Elbow River,
approximately 35 km west of Calgary.

The Connop family arrived in Bragg Creek
in 1919, built their home, and started
carving a ranch out of the wilderness.
Sydney Mosley Connop (1885-1964) came
from England and earned a fine reputation
as a social and political force. This creek was
officially named after him 6 October 1949,
but the name was well-established in local
usage many years prior to its official naming.

Consolation Lakes (lake)

82 N/8 - Lake Louise
14-27-16-W5
51°19'N 116°09'W
Approximately 45 km west north-west of
Banff.

(see Consolation Valley)

Consolation Pass (pass)

82 N/8 - Lake Louise
27-16-W5
51°17'N 116°07'W
Approximately 45 km west north-west of
Banff.

(see Consolation Valley)

Consolation Valley (valley)

82 N/8 - Lake Louise
27-16-W5
51°19'N 116°09'W
Approximately 45 km west north-west of
Banff.

This valley was named by Walter D. Wilcox
(see Wilcox Pass) and Ross Peacock in 1899,
as they were "very much pleased with the
place." This valley is very much in contrast
to the neighbouring valley which is locally

called "Desolation Valley," named by S.E.S.
Allen, and now known as the Valley of the
Ten Peaks.

Consort Mountain (mountain)

83 E/1 - Snaring
35-46-3-W6
53°01'N 118°21'W
Approximately 23 km north-west of Jasper.

This mountain is 2883 m in altitude. The
origin of the name is unknown.

Conway Creek (creek)

82 N/15 - Mistaya Lake
8-33-20-W5
51°49'N 116°47'W
Flows north-west into Howse River,
approximately 110 km north-west of Banff.

(see Mount Conway)

Conway Glacier (glacier)

82 N/15 - Mistaya Lake
32-20,21-W5
51°46'N 116°50'W
Approximately 110 km north-west of Banff.

(see Mount Conway)

Conway Group (range)

82 N/15 - Mistaya Lake
32-20-W5
51°45'N 116°49'W
Approximately 105 km north-west of Banff.

(see Mount Conway)

Conway, Mount (mountain)

82 N/15 - Mistaya Lake
20-32-20-W5
51°46'N 116°47'W
Approximately 105 km north-west of Banff,
on the Alberta-B.C. boundary.

William Martin Conway, 1st Baron of
Allington, Kent (1856-1937) was an English

art historian and mountaineer. This
mountain was named after him in 1902.
Conway climbed in the Himalayas, the
Andes and the Alps and was President of
The Alpine Club (London) from 1902-1904.
He wrote several books, including *The First
Crossing of Spitsbergen* (1897), and what he
considered his best book, *The Crowd in
Peace and War* (1915).

Copper Lake (lake)

82 O/5 - Castle Mountain
32-26-14-W5
51°16'N 115°55'W
Approximately 26 km west north-west of
Banff.

The name for this lake was officially
approved by the Canadian Permanent
Committee on Geographical Names 16
September 1965. (see Copper Mountain)

Copper Mountain (mountain)

82 O/4 - Banff
16-16-14-W5
51°13'N 115°54'W
Approximately 23 km west north-west of
Banff.

This 2795 m mountain was named by Dr.
G.M. Dawson of the Geological Survey of
Canada after the copper found near its
summit by prospectors Joe Healy and J.S.
Dennis.

Coppermine Creek (creek)

82 H/4 - Waterton Lakes
4-2-30-W4
49°06'N 113°57'W
Flows south into Blakiston Creek,
approximately 45 km south of Pincher
Creek.

Outcroppings of copper mineral may be
seen along this stream's banks above the
present highway. An old cabin and
abandoned mine indicate an attempt was

made to mine the copper many years ago. The name Coppermine Creek was officially applied to this feature 8 June 1971.

Copton Creek (creek)

83 L/6 - Chicken Creek
3,10-61-9-W6
54°16′N 119°15′W
Flows north into Kakwa River, approximately 43 km north-west of Grande Cache.

The origin of the name is unknown.

Copton Pass (pass)

83 E/13 - Dry Canyon
57-11-W6
53°57′N 119°35′W
Approximately 32 km west north-west of Grande Cache.

The name for this pass was approved in 1958, but the origin of the name remains unknown.

Copton Ridge (ridge)

83 L/3 - Copton Creek
60-10-W6
54°10′N 119°22′W
Approximately 37 km north-west of Grande Cache.

No information regarding the origin of the name is known.

Coral Creek (creek)

83 C/2 - Cline River
9-37-19-W5
52°10′N 116°30′W
Flows south-east into Cline River, approximately 110 km west south-west of Rocky Mountain House.

This creek was named by A.P. Coleman (see Mount Coleman) "after the many fossil corals among its gravels."

Coral Lake (lake)

83 C/7 - Job Creek
9-21-38-W5
52°17′00″N 116°47′00″W
Approximately 110 km south-east of Jasper.

This locally well-established name was made official 29 July 1986. (see Coral Creek)

■ Coral Creek Canyon (canyon)

83 C/2 - Cline River
9-37-18-W5
52°10′N 116°31′W
Approximately 110 km west south-west of Rocky Mountain House.

(see Coral Creek)

Cordonnier, Mount (mountain)

82 J/11 - Kananaskis Lakes
30-18-9-W5
50°33′N 115°14′W
Approximately 70 km west south-west of Turner Valley, on the Alberta-B.C. boundary.

This 3021 m mountain was officially named in 1918 after a French general. Victor Louis Emilien Cordonnier distinguished himself at Mangiennes 10 August 1914 and in the heavy fighting of the IV Army in the Ardennes during the Second World War.

Corkscrew Mountain (mountain)

83 B/3 - Tay River
23-35-10-W5
52°01′N 115°19′W
Approximately 41 km south-west of Caroline.

This locally well-established name was applied to this feature in 1941. The name is probably descriptive of the shape of the mountain.

Cornwall, Mount (mountain)

82 J/10 - Mount Rae
32-20-7-W5
50°44′N 114°57′W

Approximately 46 km west of Turner Valley.

This mountain, which is 2970 m in altitude, was officially named in 1922 after the cruiser Cornwall, engaged in the Battle of Jutland.

Cornwell, Mount (mountain)

82 J/7 - Mount Head
15-6-W5
50°18′N 114°47′W
Approximately 53 km south-west of Turner Valley.

This 2972 m mountain was officially named in 1918 in honour of John Travers Cornwell, V.C., a boy hero on the crew of the H.M.S. Chester during the battle of Jutland. He died from wounds he incurred during action.

Corona Creek (creek)

82 N/15 - Mistaya Lake
18-35-19-W5
52°00′N 116°33′W
Flows north into North Saskatchewan River, approximately 115 km north-west of Banff.

This creek rises from a mountain whose peak resembles the shape of a crown. Corona is the Latin word for crown.

Corona Creek (creek)

83 C/2 - Cline River
18-35-18-W5
52°00′N 116°33′W
Flows north into North Saskatchewan River, approximately 120 km west south-west of Rocky Mountain House.

The origin of the name is unknown.

■ Corona Ridge (ridge)

82 N/15 - Mistaya Lake
34-19-W5
51°56′N 116°34′W
Approximately 110 km north-west of Banff.

The Canadian Permanent Committee on Geographical Names approved the name for this ridge in June 1971. (see Corona Creek)

Coronach Creek (creek)

83 F/4 - Miette
2-49-28-W5
53°12'N 118°00'W
Flows west into Athabasca River, approximately 37 km south-west of Hinton.

(see Coronach Mountain)

Coronach Mountain (mountain)

83 E/1 - Snaring
22-49-1-W6
53°14'N 118°03'W
Approximately 40 km north of Jasper.

This mountain, which is 2462 m in altitude, was originally named because of the howling of coyotes, the sound of which resembled a funeral cry. Perhaps there was a spelling error in the original form of the name for this mountain, because "coranach" is the Gaelic word for a funeral dirge, while "coronach" means a crown, rich in chaplets.

Coronation Mountain (mountain)

82 N/15 - Mistaya Lake
3-33-21-W5
51°48'N 116°54'W
Approximately 115 km north-west of Banff.

This 3170 m mountain was named 9 August 1901 by J.N. Collie (see Collie Creek), as it was the coronation day of King Edward and Queen Alexandra.

Coronet Creek (creek)

83 C/11 - Southesk Lake
6-42-24-W5
52°36'N 117°26'W
Flows north into Maligne Lake, approximately 93 km west north-west of Nordegg.

(see Coronet Mountain)

Coronet Glacier (glacier)

83 C/11 - Southesk Lake
11-41-24-W5
52°31'N 117°21'W
Approximately 88 km west of Nordegg.

(see Coronet Mountain)

Coronet Mountain (mountain)

83 C/11 - Southesk Lake
10-41-24-W5
52°31'N 117°23'W
Approximately 89 km west of Nordegg.

Mountaineer Howard Palmer gave this mountain its name in 1923. The mountain, 3152 m in altitude, resembles the shape of a crown.

Corral Creek (creek)

82 N/8 - Lake Louise
14-28-16-W5
51°24'N 116°08'W
Flows south into Bow River, approximately 50 km north-west of Banff.

During the construction of the Canadian Pacific Railway, Major Hurd, a member of the construction staff, arranged with Dave McDougall, of Morley, to deliver four head of beef steers to the crew. This move was to mollify workers because there were signs of a strike in the air. A large corral was built at the head of the creek to hold the steers. The stream became locally known from then on as Corral Creek.

Corral Creek (creek)

82 J/8 - Stimson Creek
18-15-3-W5
50°16'N 114°24'W
Flows north-east into Willow Creek, approximately 46 km south of Turner Valley.

The name was officially approved 28 October 1959. The origin of the name is unknown.

Corral Creek (creek)

83 E/1 - Snaring
21-47-1-W6
53°04'N 118°05'W
Flows east into Vine Creek, approximately 22 km north of Jasper.

Named since 1915, this creek likely had a horse corral on its banks at one time.

Corral Creek (creek)

83 E/14 - Grande Cache
29-55-9-W6
53°47'N 119°18'W
Flows south-east into Smoky River, approximately 16 km south-west of Grande Cache.

The name for this creek was officially approved in 1953. The origin of the name is unknown.

Corral Creek (creek)

83 F/6 - Pedley
14-52-22-W5
53°29'N 117°07'W
Flows east into McLeod River, approximately 47 km west south-west of Edson.

The name for this creek was suggested by Harold Anderson (see Anderson Creek) because of a large stock corral erected on the flats at the mouth of the creek. The official approval date for the name Corral Creek was 21 December 1944.

Cory, Mount (mountain)

82 O/4 - Banff
12-26-13-W5
51°12'N 115°42'W
Approximately 9 km west north-west of Banff.

This mountain, which is 2802 m in altitude, was named in 1923 after William Wallace Cory (1865-1943), who was the Deputy

Minister of the Interior from 1905-1930. The feature has a "hole-in-the-wall" in it, which is easily seen from the highway.

Cory Pass (pass)

82 O/4 - Banff
51°12'15"N 115°40'30"W
Approximately 8 km west north-west of Banff.

The name of Corry Pass was officially approved 21 January 1985. (see Mount Cory)

Corydalis Creek (creek)

83 E/14 - Grande Cache
30-57-9-W6
53°58'N 119°19'W
Flows north into Sheep Creek, approximately 15 km north-west of Grande Cache.

This creek was officially named in 1953 after the wildflower of the same name.

Costigan, Mount (mountain)

82 O/6 - Lake Minnewanka
12-27-10-W5
51°17'N 115°18'W
Approximately 22 km north north-east of Canmore.

This mountain, 2980 m in altitude, was officially named 1 March 1904, after the Honourable John Costigan (1835-1916), a frequent visitor to the area. He held many government positions including federal Minister of Inland Revenue (1882-1892), Secretary of State (1892-1894) and Minister of Marine and Fisheries (1894-1896). He was called to the Senate in 1907.

Côté Creek (creek)

83 E/13 - Dry Canyon
18-56-12-W6
53°50'N 119°56'W
Flows south-east into Sheep Creek, approximately 43 km west of Grande Cache.

This creek was officially named 3 October 1957. (see Mount Côté)

Côté, Mount (mountain)

83 E/13 - Dry Canyon
56-14-W6
53°53'N 120°00'W
Approximately 48 km west of Grande Cache, on the Alberta-B.C. boundary.

The name of this mountain, which is 2391 m in altitude, was officially approved 3 November 1925. The Honourable Jean Leon Côté, C.E., M.E.D.L.S. (1867-1924), of Côté and Pearson Land Surveyors and Engineers, was born in Quebec and made his way west to Edmonton in 1904. He was elected to the Alberta Legislature for Athabasca in 1909 and was appointed Provincial Secretary 5 September 1918.

Cottonwood Creek (creek)

83 D/16 - Jasper
22-45-1-W6
52°54'N 118°04'W
Flows east into Athabasca River, approximately 2 km north north-east of Jasper.

Cottonwood Creek was named in 1914 by H. Matheson, D.L.S. The origin of the name is unknown.

Cottonwood Slough (intermittant lake)

83 D/16 - Jasper
52°53'45"N 118°05'30"W
Approximately 2 km north of Jasper.

This slough was named for its proximity to Cottonwood Creek. (see Cottonwood Creek)

Cougar Creek (creek)

82 J/10 - Mount Rae
23-20-7-W5
50°43'N 114°52'W

Flows north-west into Elbow River, approximately 41 km west of Turner Valley.

(see Cougar Mountain)

Cougar Mountain (mountain)

82 J/10 - Mount Rae
3-20-7-W5
50°41'N 114°54'W
Approximately 43 km west of Turner Valley.

This 2863 m mountain was officially named prior to 1928 after the cougar, or mountain lion, frequently found on its slopes.

Coulthard, Mount (mountain)

82 G/10 - Crowsnest
13-7-5-W5
49°33'N 114°34'W
Approximately 10 km south south-west of Coleman.

This 2642 m peak was named after Mr. Coulthard, a prominent mining engineer who was General Manager of the West Canadian Coal Company in 1910.

Cow Creek (creek)

82 G/9 - Blairmore
6-8-1-W5
49°37'N 114°08'W
Approximately 4 km north-east of Lundbreck.

No information regarding the origin of the name is known.

Cow Lake (lake)

83 B/3 - Tay River
12-38-8-W5
52°15'N 115°01'W
Approximately 16 km south-west of Rocky Mountain House.

This name was officially approved in 1951. The origin of the name is unknown.

Cow Juicer Creek (creek)

82 G/16 - Maycroft
19-11-3-W5
49°55′15″N 114°23′30″W
Flows south-west into Oldman River, approximately 35 km north-east of Blairmore.

It is said that during the drought of 1919, Vic Robinson moved his cows up to the Gap and milked them there. The name for this stream was officially approved 21 July 1982.

Cow Juicer Pass (pass)

82 G/16 - Maycroft
28-11-3-W5
49°56′35″N 114°20′50″W
Approximately 38 km north-east of Blairmore.

(see Cow Juicer Creek)

Cowley (village)

82 G/9 - Blairmore
21-7-1-W5
49°34′N 114°05′W
On Highway 3, approximately 6 km east south-east of Lundbreck.

There are several accounts of how Cowley acquired its name, but in all of them the credit is given to Mr. F. Godsal, a prominent land owner in the district. One account states that Mr. Godsal had as a visitor Lord Cowley from England and that the village was named after him in honour of his stay. Another, and the most commonly believed, is that one evening Mr. Godsal, a poetic soul, observing his cattle stringing out to water at the big spring was reminded of the line from Gray's "Elegy," "The lowing herd winds slowly o'er the lea." Lea or ley, was the Anglo-Saxon word for pasture, and so when asked to give the new town a name, he named it Cowley. The third version is that the new village was named after Mr. Godsal's home in England, which was Cowley, Oxford.

Cowlick Creek (creek)

83 E/14 - Grande Cache
13-56-8-W6
53°50′N 119°03′W
Flows north-west into Sulphur River, approximately 8 km south-east of Grande Cache.

Salt, sulphur and minerals may be found near this creek to which cows would come to lick. The name Cowlick Creek was officially approved for this feature in 1947.

Cox Hill (hill)

82 J/15 - Bragg Creek
27-23-7-W5
50°59′N 114°55′W
Approximately 50 km west of Calgary.

This foothill was named in 1896 by A.O. Wheeler after an assistant of his named Cox. The name was officially approved 6 October 1949.

Coxhill Creek (creek)

82 O/2 - Jumpingpound Creek
8-24-6-W5
51°02′N 114°48′W
Flows north into Jumpingpound Creek, approximately 40 km west of Calgary.

Officially named 12 December 1939, this creek may take its name from nearby Cox Hill. (see Cox Hill)

Coyote Creek (creek)

82 G/16 - Maycroft
29-10-2-W5
49°51′N 114°14′W
Flows south into Oldman River, approximately 30 km north-east of Blairmore.

The origin of the name is unknown.

Crandell Lake (lake)

82 H/4 - Waterton Lakes
32-1-30-W4
49°05′N 113°58′W

Approximately 45 km south of Pincher Creek.

Edward Henry Crandell (1859-1944) was a pioneer Calgary businessman who worked the oil wells near this lake and the adjacent mountain. It is this Mr. Crandell after whom this lake is named. (see Mount Crandell)

Crandell, Mount (mountain)

82 H/4 - Waterton Lakes
1-30-W4
49°05′N 113°56′W
Approximately 45 km south of Pincher Creek.

Until a few years ago the mountain which is now known as Mount Crandell was called "Black Bear Mountain" and is still called that today by many people. The 2381 m mountain lies between Blakiston and Cameron Creeks, east of the oil wells worked by Edward Henry Crandell of Calgary.

Crescent Creek (creek)

83 B/4 - Cripple Creek
13-37-14-W5
52°10′N 115°35′W
Flows north-west into Fall Creek, approximately 46 km south-east of Nordegg.

The name for this rocky, crescent-shaped stream was officially approved in July 1941.

Crescent Creek (creek)

83 D/16 - Jasper
19-44-2-W6
52°49′N 118°17′W
Flows north-west into Meadow Creek, approximately 15 km south-west of Jasper.

The name for this creek describes its semi-circular course.

Crescent Creek (creek)

83 E/9 - Moberly Creek
26-53-4-W6
53°36′N 118°28′W
Flows north-east into Moon Creek,
approximately 86 km north north-west of
Jasper.

This creek was named in 1946 by M.E.
Nidd, a field topographer in the area. The
name is likely descriptive of the creek's flow
pattern.

■ **Crescent Falls** (falls)

83 C/8 - Nordegg
29-39-17-W5
52°23′N 116°21′W
Approximately 21 km west south-west of
Nordegg.

The name for these falls is descriptive of the
shape of the crest.

Cripple Creek (creek)

83 B/5 - Saunders
11-38-14-W5
52°16′N 115°52′W
Flows north-east into North Ram River,
approximately 28 km south-east of Nordegg.

The origin for the locally well-established
name of this creek is unknown.

Cromwell, Mount (mountain)

83 C/6 - Sunwapta
52°16′10″N 117°23′30″W
Approximately 80 km south-east of Jasper,
in Jasper National Park.

The name Mount Cromwell was first
suggested in 1967 by Dr. James Munroe
Thorington for Oliver Eaton Cromwell
(died 1987; *CAJ* 1987 p. 60), a member of the
French, British and American Alpine Clubs.
Cromwell was a wealthy American who
began climbing in the Canadian Rockies in
1928. Cromwell and Thorington were

climbing partners and were members of the
team that first climbed this 3330 m
mountain in 1936. The name Mount
Cromwell was officially approved 7
November 1984 and is well-known among
climbing fraternities and is used extensively
by Park Wardens and visitors.

Crooked Creek (creek)

83 C/8 - Nordegg
6-39-16-W5
52°19′N 116°17′W
Flows north into North Saskatchewan River,
approximately 22 km south-west of
Nordegg.

The name for this creek is descriptive of its
path of flow.

Crooked Creek (creek)

83 C/15 - Cardinal River
23-46-19-W5
52°59′N 116°39′W
Flows north into Pembina River,
approximately 96 km east north-east of
Jasper.

The locally well-established name for this
creek was in use prior to 1927 and is likely
descriptive of the stream's flow.

Crowfoot Glacier (glacier)

82 N/9 - Hector Lake
11-31-18-W5
51°38′N 116°25′W
Approximately 80 km north-west of Banff.

(see Crowfoot Mountain)

Crowfoot Mountain (mountain)

82 N/9 - Hector Lake
11-31-18-W5
51°38′N 116°27′W
Approximately 80 km north-west of Banff.

The glacier resembles a crow's foot along the
side of this 3050 m mountain and it is from

this that the feature derived its descriptive
name. The name Crowfoot Mountain was
made official in 1959.

Crowsnest (locality)

82 G/10 - Crowsnest
2-9-5-W5
49°38′N 114°41′W
Approximately 52 km west north-west of
Pincher Creek.

(see Crowsnest Mountain)

Crowsnest Creek (creek)

82 G/10 - Crowsnest
7-8-5-W5
49°37′N 114°39′W
Flows north into Crowsnest River,
approximately 54 km west north-west of
Pincher Creek.

(see Crowsnest Mountain)

Crowsnest Lake (lake)

82 G/10 - Crowsnest
8-8-5-W5
49°38′N 114°38′W
Approximately 51 km west north-west of
Pincher Creek.

(see Crowsnest Mountain)

■ **Crowsnest Mountain** (mountain)

82 G/10 - Crowsnest
2-9-5-W5
49°42′N 114°35′W
In the locality of Crowsnest Pass,
approximately 52 km west north-west of
Pincher Creek.

This mountain is approximately 2785 m in
altitude. There are currently two competing
possibilities for the origin of the name
Crowsnest. The first possibility, which
appeared in *Maclean's Magazine* in 1928, is
that around 1852 or 1854, a party of
Blackfoot Indians managed to cut off a
group of raiding Crow Indians who had

tried to escape westward through the pass. They were allegedly cornered and slaughtered at the base of a mountain which was then called Crow's Nest in memory of the occasion. Since First Nations groups rarely left written records, this version of the origin of the name is difficult to substantiate in documents. The other possibility, and the one for which documentary evidence can be more readily supplied, pertains to the nesting of the crows or ravens near the base of the peak. The name on the Palliser Expedition preliminary report refers to it as "lodge des Corbeaux." Translated literally, it means "the nest of the crow or raven." The first mention of the name in historic documents was made by Captain Blakiston of the Palliser Expedition, who noted under the date 15 December 1858, "I have not mentioned the existence of two other passes across this portion of the mountains called the Crow-Nest and Flathead Passes . . . the Crow-Nest Pass of which I have marked the general direction on the plan follows up the Crow-Nest River . . . by report of the natives it is a very bad road and seldom used." Blakiston did not explain why the Indians considered this to be such a poor road.

Crowsnest Pass (pass)

82 G/10 - Crowsnest
12-8-6-W5
52°38′N 114°40′W
Approximately 55 km north-west of Pincher Creek, on the Alberta-British Columbia boundary.

The pass is approximately 1246 m in altitude. (see Crowsnest Mountain)

Crowsnest Ridge (ridge)

82 G/10 - Crowsnest
8-8-5-W5
52°38′N 114°40′W
Approximately 50 km north-west of Pincher Creek.

(see Crowsnest Mountain)

Crowsnest River (river)

82 G/9 - Blairmore
34-7-1-W5
49°36′N 114°03′W
Flows east into Oldman River, approximately 15 km north north-west of Pincher Creek.

(see Crowsnest Mountain)

Crypt Falls (falls)

82 H/4 - Waterton Lakes
4-1-29-W4
49°01′N 113°50′W
Approximately 55 km south south-east of Pincher Creek.

(see Crypt Lake)

Crypt Lake (lake)

82 H/4 - Waterton Lakes
4-1-29-W4
49°00′N 113°50′W
Approximately 55 km south south-east of Pincher Creek.

A crypt is an undercroft or basement of a church; the word is derived from the Greek, *krupto*, meaning, "to hide." The lake is hidden high in a mountain of Waterton Lakes National Park. Both the lake and the falls have, thus, an appropriately descriptive name which was officially approved 8 June 1971.

Cummings Creek (creek)

82 J/7 - Mount Head
17-15-5-W5
50°16′N 114° 39′W
Flows north into Cataract Creek, approximately 52 km south-west of Turner Valley.

Officially approved 8 December 1943, this creek carries the name of George Cummings, the Manager of the Two Dot Ranch from 1943-1959. It may be been

named after Mr. Cummings himself or for one of his forefathers.

Cumnock, Mount (mountain)

83 E/1 - Snaring
26-48-2-W6
53°10′N 118°11′W
Approximately 34 km north north-west of Jasper.

This 2460 m mountain was named in 1916 by M.P. Bridgland (see Mount Bridgland) after Cumnock, Ayrshire (now Strathclyde), Scotland. It means "meeting of waters."

Curator Lake (lake)

83 C/13 - Medicine Lake
44-27-W5
52°47′35″N 117°51′50″W
Approximately 18 km south-east of Jasper.

This alpine lake was officially named 5 May 1987 for its proximity to Curator Mountain. (see Curator Mountain)

Curator Mountain (mountain)

83 C/13 - Medicine Lake
9-44-27-W5
52°47′N 117°50′W
Approximately 19 km south-east of Jasper.

This 2622 m mountain was named in 1916 by M.P. Bridgland (see Mount Bridgland) for its position as "custodian" of Shovel Pass.

Curia Mountain (mountain)

83 D/16 - Jasper
15-44-3-W6
52°48′N 118°20′W
Approximately 19 km south south-west of Jasper.

Named in 1916 by M.P. Bridgland (see Mount Bridgland), this mountain resembles the shape of a Roman senate house, or curia. The Roman tradition states that Romulus divided the people into three political "tribes," which each had families of "curiae." One of the principal buildings of Ancient Rome was the Curia, and along with the Basilica and Rostrum, the religious, political and judicial lives of the people were housed. The three features, Basilica Mountain, Curia Mountain (2835 m) and Rostrum Hill, surround the valley known as "The Forum."

Currie Creek (creek)

82 J/14 - Spray Lakes Reservoir
33-21-11-W5
50°50′N 115°27′W
Flows north-east into Spray Lakes Reservoir, approximately 90 km west of Calgary.

(see Mount Currie)

Currie, Mount (mountain)

82 J/14 - Spray Lakes Reservoir
29-21-11-W5
50°48′N 115°29′W
Approximately 95 km west south-west of Calgary.

This 2810 m mountain was named in 1918 after General Sir Arthur William Currie, K.C.B., K.C.M.G. (1875-1933). This soldier and educator was appointed to the command of the Canadian Army in June of 1917. From 1920 until his death, he was principal and vice-chancellor of McGill University.

Cuthead Creek (creek)

82 O/5 - Castle Mountain
18-28-12-W5
51°24′N 115°41′W
Flows south-east into Cascade River, approximately 26 km north north-west of Banff.

The name for this creek is a translation of a Stoney Indian name. The legend behind the name refers to a Cree Indian who had eloped with a Stoney woman. He was overtaken and beheaded near this stream.

Cuthead Lake (lake)

82 O/5 - Castle Mountain
3-29-13-W5
51°27′N 115°45′W
Approximately 33 km north north-west of Banff.

(see Cuthead Creek)

Cutoff Creek (creek)

83 C/9 - Wapiabi Creek
15-42-17-W5
52°37′N 116°02′W
Flows south into Blackstone River, approximately 25 km north of Nordegg.

The origin of the name is unknown.

Cutoff Creek (creek)

83 B/4 - Cripple Creek
16-35-11-W5
52°37′N 116°02′W
Approximately 52 km south-west of Caroline.

The origin of this locally well-established name is unknown.

Cutpick Hill (hill)

83 L/3 - Copton Creek
5-59-8-W6
54°04′N 119°10′W
Approximately 23 km north north-east of Grande Cache.

The origin for the name of this hill, which is 1819 m in altitude, is unknown. The name has been in use since at least 1946.

Cutt Lake (lake)

83 D/16 - Jasper
52°54′05″N 118°19′45″W
Approximately 13 km west north-west of Jasper.

Staff of a local fish hatchery and Norman Woody, a Jasper Park Warden, stocked this

small lake with cutthroat trout in 1962 and named the lake after the fish. The locally well-established name was officially approved 5 May 1987.

Cutthroat Creek (creek)

82 J/8 - Stimson Creek
33-17-4-W5
50°28′40″N 114°25′05″W
Flows north-east into Flat Creek, approximately 75 km south south-west of Calgary.

The name Cutthroat was proposed by the Alberta Fisheries biologist for Kananaskis Country, to reflect the significant cutthroat trout population in the stream. During Fisheries inventory work it was discovered that this creek is one of the more important streams in the Highwood River watershed for cutthroat trout. The name was officially approved 5 December 1984.

Cyclamen Ridge (ridge)

82 J/2 - Thunder Lake
13-5-W5
50°04′N 114°33′W
Approximately 55 km south south-west of Turner Valley.

The name Cyclamen Ridge was proposed by M.P. Bridgland (see Mount Bridgland) 30 June 1915. This ridge is approximately 2256 m in altitude and is named for the wildflower of the same name.

Cyclone Mountain (mountain)

82 N/9 - Hector Lake
17-30-15-W5
51°34′N 116°04′W
Approximately 55 km north-west of Banff.

The name for this 3050 m mountain is descriptive of the storm that was raging on its peak at the time of its naming.

Daigle Lake (lake)

82 G/8 - Beaver Mines
14,23-5-1-W5
49°23'N 114°02'W
Approximately 12 km south-west of Pincher Creek.

Edmund Daigle and his wife Hermaine homesteaded in this area, across from the lake named in their honour. Edmund raised good Hereford cattle and was also a councillor for the Municipal District of Castle River until it was absorbed into the larger unit. Daigle Lake is the second largest lake in this area, after Beauvais Lake.

Dais Mountain (mountain)

83 C/5 - Fortress Lake
22-38-26-W5
52°17'N 117°37'W
Approximately 75 km south-east of Jasper.

This mountain, which is 3300 m in altitude, dominates Chaba Valley. It acts as a raised table or platform, making its name appropriately descriptive.

Daisy Creek (creek)

82 G/16 - Maycroft
24-10-4-W5
49°51'N 114°25'W
Flows north into Racehorse Creek, approximately 25 km north of Blairmore.

The precise origin of the name is unknown.

Dalhousie, Mount (mountain)

83 C/10 - George Creek
15-42-21-W5
52°36'N 116°56'W
Approximately 61 km west north-west of Nordegg.

This mountain, which is 2438 m in altitude, was named in 1859 after the 11th Earl of Dalhousie (1801-1874). He was the Secretary for War during the Crimean War and the

nephew of the 9th Earl, who was Governor General of Canada. The name was proposed by James Carnegie, 9th Earl of Southesk. (see Southesk River)

Daly, Mount (mountain)

82 N/9 - Hector Lake
25-29-18-W5
51°31'N 116°24'W
Approximately 70 km north-west of Banff, on the Alberta-B.C. boundary.

This 3152 m mountain was named in 1898 by Professor Charles E. Fay (see Mount Fay) after Chief Justice Charles P. Daly, President of the American Geographical Society, 1864-1899.

Damon Creek (creek)

82 G/16 - Maycroft
12-11-2-W5
49°54'N 114°09'W
Flows west into Callum Creek, approximately 35 km north-east of Blairmore.

The origin of this locally well-established name is unknown.

Daniel Creek (creek)

83 L/6 - Chicken Creek
5-61-9-W6
54°15'N 119°18'W
Flows north into Kakwa River, approximately 42 km north north-west of Grande Cache.

This creek was officially named in 1951 after a trapper named Daniel Joachin who had a trap line nearby.

Daniels Flats (flats)

83 L/3 - Copton Creek
24-58-8-W6
54°02'N 119°03'W
Approximately 15 km north north-east of Grande Cache.

D~E

These flats were named 4 May 1950, possibly after Daniel Joachin. (see Daniel Creek)

Dardanelles (channel)

82 H/4 - Waterton Lakes
1-29-W4
49°04'N 113°51'W
Approximately 5 km north-west of Waterton Park.

There is a narrows leading from the Mediterranean Sea into the Black Sea in Asia. This passage and the passage of water between Middle and Lower Waterton Lakes which may be found in Waterton Lakes National Park, are both called the Dardanelles. The name was officially adopted by the Geographic Board of Canada on 4 May 1943.

Darnell Lake (lake)

82 O/2 - Jumpingpound Creek
25-24-6-W5
51°04'N 114°43'W
Approximately 35 km west of Calgary.

(see Darnell Ridge)

Darnell Ridge (ridge)

82 O/2 - Jumpingpound Creek
25-6-W5
51°05'N 114°41'W
Approximately 35 km west of Calgary.

The name "Darnell" was suggested by D.A. Nichols, a member of the Geological Survey of Canada. Mr. Darnell was a local rancher in the area. The name was officially approved 12 December 1939.

Darrah, Mount (mountain)

82 G/7 - Upper Flathead
15-6-5-W5
49°28′N 114°35′W
Approximately 20 km south-west of
Coleman, on the Alberta-B.C. boundary.

This 2755 m mountain was once known as
Gable Mountain. It was renamed in 1917.
Captain Darrah was the astronomer attached
to the British Boundary Commission of
1858-1862, from the Pacific to the Rockies.

Davey Creek (creek)

83 E/14 - Grande Cache
2-56-9-W6
53°49′N 119°13′W
Flows south-east into Smoky River,
approximately 10 km south-west of Grande
Cache.

This creek may be named after Bert Davey, a
bookkeeper in the area. No other
information is available about the origin of
the name.

David Creek (creek)

82 N/15 - Mistaya Lake
20-33-20-W5
51°51′N 116°48′W
Flows south into Howse River,
approximately 110 km north-west of Banff.

(see Mount David)

David Creek (creek)

83 E/16 - Donald Flats
34-57-3-W6
53°58′N 118°21′W
Flows east into Little Smoky River,
approximately 53 km east north-east of
Grande Cache.

This creek was officially named in 1947
allegedly after a local trapper. No other
information is available.

David Lake (lake)

82 N/15 - Mistaya Lake
19-33-20-W5
51°51′N 116°50′W
Approximately 115 km north-west of Banff.

(see Mount David)

David, Mount (mountain)

82 N/15 - Mistaya Lake
18-33-20-W5
51°50′N 116°49′W
Approximately 115 km north-west of Banff.

This 2780 m mountain, the nearby creek
and lake were named in 1920 after David
Thompson (1770-1857), who travelled
through adjacent Howse Pass in 1806-1807.
This explorer and geographer was
apprenticed to the Hudson's Bay Company
in 1784. In 1797, he transferred his services
to the North West Company, in which he
later became a partner. In 1812, he compiled
a map of western Canada. From 1816-1826
he was employed in surveying the border
line between Canada and the United States.

Daybreak Peak (peak)

83 E/8 - Rock Lake
32-51-3-W6
53°26′N 118°24′W
Approximately 67 km north north-west of
Jasper.

The origin of the name is unknown.

De Smet Range (range)

83 E/1 - Snaring
48-2-W6
53°10′N 118°13′W
Approximately 40 km north north-west of
Jasper.

(see Roche de Smet)

De Smet, Roche (mountain)

83 E/1 - Snaring
17-48-1-W6
53°08′N 118°05′W
Approximately 29 km north of Jasper.

This 2539 m mountain was named with
great ceremony of muskets and loud hurrahs
by the Iroquois Indians working in the fur
trade in the Jasper area in 1846. It was
named in honour of Belgian missionary
Pierre-Jean de Smet (1801-1873). He laboured
among the Indians in the north-western
States and western Canada, crossing the
Rockies by way of Cross River and
Whiteman Pass; wintered at Edmonton in
1845. He then recrossed the mountains via
the Athabasca Pass in 1846.

De Veber, Mount (mountain)

83 E/12 - Pauline Creek
26-54-12-W6
53°41′N 119°38′W
Approximately 40 km south-west of Grande
Cache.

This 2573 m mountain was named after the
late Honourable Leverett George de Veber,
M.D. (1849-1925), a physician and surgeon
who served the Royal North West Mounted
Police for some years. He was elected the
first president of the Southern Alberta
Medical Association in 1909. He was part of
Premier Rutherford's cabinet, having been
elected as the M.L.A. for Lethbridge in 1905,
and the following year was summoned to the
Senate of Canada, where he served until his
death.

De Wind, Mount (mountain)

83 E/9 - Moberly Creek
13-53-4-W6
53°35′N 118°27′W
Approximately 81 km north north-west of
Jasper.

This 2438 m mountain was named in 1948
after Second Lieutenant Edmund De Wind,

V.C. (1883-1918). De Wind was a native of Ireland who came to Canada in 1910. On 21 March 1918 he was awarded the Victoria Cross for his conspicuous bravery and self-sacrifice in the First World War.

Dead Horse Meadows (meadows)

83 L/4 - Kakwa Falls
29-59-13-W6
54°08′N 119°55′W
Approximately 63 km west north-west of Grande Cache.

The name for these meadows appears on a field engineer's Report dated 21 September 1954. The precise origin is not clear: either one or a number of horses were left unattended at this place, and eventually died.

Dead Man Flat (flat)

82 O/3 - Canmore
13-24-10-W5
51°02′N 115°17′W
Approximately 8 km south-east of Canmore.

The name was officially approved 7 August 1958. (see Dead Man's Flats)

Dead Man's Flats (locality)

82 O/3 - Canmore
13-24-10-W5
51°02′25″N 115°15′50″W
Approximately 7 km south-east of Canmore.

Two origins for this name follow: 1. Two brothers, John and François Marrett, operated a dairy on the flat. In May 1904, François borrowed a double-edged axe from a neighbour and killed his brother with it. He was taken into custody and found guilty but insane and was committed to the Ponoka Institute where he remained for the rest of his life. 2. Around the turn of the century, two or three Indians were trapping beaver in the area. They spied the Park Warden approaching them and knew there was no time to escape. They quickly smeared themselves with beaver blood and played

dead until the warden ran for help. When the warden left, the Indians got up and took their beaver pelts home.

Dead Mans Pass (pass)

82 G/10 - Crowsnest
6-9-5-W5
49°42′N 114°40′W
Approximately 14 km north-west of Coleman.

According to legend, a group of Indians were chased by forty members of the United States cavalry. The Indians knew this area very well and enticed the Americans into the pass, where some of the Indians stayed off the trail and hid behind the trees to wait for the unsuspecting cavalrymen. Once the Americans realized they were trapped, they turned and fled, only to be faced by a blockade of huge timbers. The Indians slaughtered all members of the cavalry, and since that time the pass has been called "Dead Mans Pass."

Dean Pass (pass)

83 C/11 - Southesk Lake
52°39′30″N 117°02′10″W
Approximately 73 km east south-east of Jasper.

This pass was named after Norman Woody's son, Dean. The name is well-established in local usage.

Death Valley Creek (creek)

82 J/10 - Mount Rae
20-20-4-W5
50°43′N 114°32′W
Flows north into Ware Creek, approximately 17 km west of Turner Valley.

Officially approved 6 September 1951, this creek was so named because many years ago someone brought in a herd of horses from the prairies. The horses ate themselves to death on the lush grass in the valley. The creek is also known locally as "Sinnott

Creek" after Henry Thomas Sinnott, who homesteaded near Kew in 1901.

Deception Creek (creek)

83 C/14 - Mountain Park
36-44-24-W5
52°50′N 117°22′W
Flows south-east into Medicine Tent River, approximately 45 km east south-east of Jasper.

The origin of the name is unknown.

Deception Mountain (mountain)

83 C/14 - Mountain Park
10-45-24-W5
52°52′N 117°24′W
Approximately 45 km east of Jasper.

Deception Mountain stands 2819 m in altitude. The origin of the name is unknown.

Deception Pass (pass)

82 N/8 - Lake Louise
29-15-W5
51°30′N 116°04′W
Approximately 50 km north-west of Banff.

No information regarding the origin of the name is known.

Decoigne (locality)

83 D/16 - Jasper
16-45-3-W6
52°53′N 118°22′W
Approximately 20 km west of Jasper.

This locality was named in 1926 after Francois Décoigne, a yellow haired trapper who was in charge of Jasper's House on Brule Lake. (see Roche Tête)

Deep Creek (creek)

83 B/5 - Cripple Creek
5-40-13-W5
52°25′N 115°50′W

Flows south into North Saskatchewan River, approximately 17 km south-east of Nordegg.

The locally well-established name for this creek was made official in 1941. It is likely descriptive of the creek's depth.

Deep Creek (creek)

82 J/8 - Stimson Creek
10-17-4-W5
50°25′N 114°28′W
Flows south-east into Highwood River, approximately 30 km south-west of Turner Valley.

The name Deep Creek was officially applied to this feature 12 December 1939 and is likely descriptive.

Deer Creek (creek)

83 E/7 - Blue Creek
3-51-4-W6
53°22′N 118°30′W
Flows south-east into Snake Indian River, approximately 59 km north north-west of Jasper.

Hunters in search of big game frequently used this creek which leads into rich hunting territory. It is likely named because of the deer in the area.

Deer Creek (creek)

83 E/10 - Adams Lookout
20-52-5-W6
53°31′N 118°43′W
Flows east into Rock Creek, approximately 50 km south-east of Grande Cache.

The origin of the name is unknown.

Deer Creek Flats (flat)

82 O/11 - Burnt Timber Creek
18-31-8-W5
51°39′N 115°07′W
Approximately 60 km north-east of Banff.

No information regarding the origin of the name is known.

Deer Lake (lake)

82 H/4 - Waterton Lakes
4-1-28-W4
49°00′N 113°43′W
Approximately 55 km south south-east of Pincher Creek.

The origin of the name is unknown.

Deerlick Creek (creek)

83 F/3 - Cadomin
23-48-23-W5
53°09′N 117°15′W
Flows north into McLeod River, approximately 72 km south-west of Edson.

The origin of the name is unknown.

Delorme Creek (creek)

83 E/14 - Grande Cache
2-56-9-W6
53°49′N 119°13′W
Flows north-west into Smoky River, approximately 11 km south south-west of Grande Cache.

(see Delorme Pass)

Delorme Pass (pass)

83 E/11 - Hardscrabble Creek
1-55-9-W6
53°43′00″N 119°12′45″W
Approximately 18 km south south-west of Grande Cache.

Louie Delorme, a native old-timer, said that this pass at the head of Delorme Creek was named for his family and had been known by that name for many years. The name was officially approved 1 January 1979.

Delphinium Creek (creek)

83 F/4 - Miette
6-47-26-W5
53°01′N 117°47′W

Flows south-west into Rocky River, approximately 48 km south south-west of Hinton.

The name for this creek was changed in 1961 to avoid duplication from "Larkspur Creek" to its present name, Delphinium Creek. Wild delphiniums grow profusely near the stream's headwaters.

Delta Creek (creek)

82 N/10 - Blaeberry River
18-32-18-W5
51°45′N 116°32′W
Flows north-east into Mistaya River, approximately 95 km north-west of Banff.

(see Delta Glacier)

Delta Glacier (glacier)

82 N/10 - Blaeberry River
11-32-19-W5
51°44′N 116°35′W
Approximately 95 km north-west of Banff.

A.O. Wheeler (see Wheeler Flats) named this glacier, the source of the Delta Creek, in 1918 after the wide delta found at the lower end of the creek.

Deltaform Mountain (mountain)

82 N/8 - Lake Louise
27-16-W5
51°18′N 116°14′W
Approximately 50 km west north-west of Banff, on the Alberta-B.C. boundary.

This 3424 m mountain was named by Walter D. Wilcox (see Wilcox Pass) in 1897 because of the similarity of its form to the Greek letter "Δ" — delta.

Dent, Mount (mountain)

82 N/15 - Mistaya Lake
32-22-W5
51°45′N 116°58′W

Approximately 115 km north-west of Banff, on the Alberta-B.C. boundary.

Clinton Thomas Dent, M.A., M.C., F.R.C.S., after whom this 3267 m mountain was named in 1899, was President of The Alpine Club (London), 1886-1889. He was first and foremost a surgeon, but he had a great interest in mountaineering.

Deome Creek (creek)

83 L/2 - Bolton Creek
15-58-4-W6
54°00′N 118°32′W
Flows north into Simonette River, approximately 40 km east north-east of Grande Cache.

This creek was officially named in 1947 after a trapper named Deome Findley who had a trap line nearby.

Derr Creek (creek)

83 D/16 - Jasper
16-45-3-W6
52°53′N 118°23′W
Flows south into Miette River, approximately 20 km west of Jasper.

This creek was named in 1898 by J. McEvoy of the Geological Survey of Canada after S. Derr, a packer and guide from Edmonton.

Desolation Creek (creek)

83 E/11 - Hardscrabble Creek
8-53-9-W6
53°33′N 119°17′W
Flows north-east into Smoky River, approximately 37 km south south-west of Grande Cache.

(see Valley of the Ten Peaks, formerly, Desolation Valley)

Devil's Bite (pass - gap)

82 J/8 - Stimson Creek
1-16-4-W5
50°18′50″N 114°25′30″W

Approximately 35 km south of Turner Valley.

The locally well-established name for this feature is descriptive — it looks like a "bite" has been taken out of the hills. The name was officially approved 5 May 1983.

Devils Gap (gap)

82 O/6 - Lake Minnewanka
27-9-W5
51°16′N 115°12′W
Approximately 25 km north north-east of Canmore.

The Devils Gap was the route by which Sir George Simpson (see Simpson Pass) entered the mountains in 1858 during his expedition around the world.

Devils Head (mountain)

82 O/6 - Lake Minnewanka
31-27-9-W5
51°21′N 115°16′W
Approximately 27 km north-east of Banff.

The name for this 2997 m mountain is a translation of the Cree Indian name, "we-ti-kwos-ti-kwan;" in Stoney, "si-ham-pa." It was described by Sir George Simpson (see Simpson Pass) as having a "rude resemblance to an upturned face." The name "Devils Head" was officially approved 3 March 1902.

"Devil's Head Canyon," n.d.

Devils Thumb (mountain)

82 N/8 - Lake Louise
24-28-17-W5
51°25′N 116°15′W
Approximately 55 km north-west of Banff.

All information about this 2459 m mountain maintains that it is descriptive. This explanation is supported by S.E.S. Allen (see Mount Allen), who spent the years 1891-1894 in the vicinity of Lake Louise. He writes in his published articles, *Explorations Among the Watershed Rockies of Canada* (1894?) about Devils Thumb: "In a short visit to the beautiful Lake Louise, south of Laggan, I was also such impressed by the majestic peak reflecting in its clear surface, whose avalanches were continually heard and seen as they plunged with a roar of thunder down the mountain cliffs from overhanging ice walls. With the ascent of a pinnacle which I named the Devil's Thumb . . . rising from Lake Agnes above Lake Louise and well seen from the chalet, I ended my climbing for the summer." This incident occurred in 1891. The name was officially approved 25 October 1910. W.D. Wilcox (see Wilcox Pass) describes Devils Thumb as a rough crag on an abrupt precipice.

Devon Lakes (lake)

82 N/9 - Hector Lake
32-16-W5
51°43′N 116°14′W
Approximately 80 km north-west of Banff.

(see Devon Mountain)

Devon Mountain (mountain)

82 N/9 - Hector Lake
6-31-16-W5
51°43′N 116°15′W
Approximately 75 km north-west of Banff.

There is a well-developed Devonian geological formation here. The descriptive name for this 3004 m mountain was officially adopted in 1923.

Devona (locality)

83 F/4 - Miette
15-48-22-W5
53°09′N 118°00′W
Approximately 30 km north of Jasper.

The Canadian National Railway opened a station here in 1915. The locality is likely named after the Devonian geological formations nearby.

Devona Flats (flats)

83 E/1 - Snaring
48-1-W6
53°09′N 118°00′W
Approximately 30 km north of Jasper.

(see Devona)

Diadem Creek (creek)

83 C/6 - Sunwapta
14-39-24-W5
52°21′N 117°21′W
Flows north-east into Sunwapta River, approximately 76 km south south-east of Jasper.

The Canadian Permanent Committee on Geographical Names approved this name 18 October 1976. (see Diadem Peak)

Diadem Peak (peak)

83 C/6 - Sunwapta
31-38-24-W5
52°19′N 117°25′W
Approximately 77 km south south-east of Jasper.

Diadem is a poetical word for "crown." This mountain peak, which measures 3371 m in altitude, is crowned by snow measuring about 30 m in depth. J.N. Collie (see Collie Creek) named the peak in 1898 after climbing it.

Dickson Coulee (coulee)

82 J/8 - Stimson Creek
16-4-W5
50°22′N 114°27′W
Flows north into Pekisko Creek, approximately 35 km south-west of Turner Valley.

Dickson Coulee is named after Jonas Dickson, the first settler to the area. The name was officially approved 12 December 1939.

Dinosaur Creek (creek)

83 L/5 - Two Lakes
18-62-13-W6
54°22′N 119°56′W
Flows north into Narraway River, approximately 62 km north-west of Grande Cache.

(see Dinosaur Ridge)

Dinosaur Ridge (ridge)

83 L/5 - Two Lakes
35-61-14-W6
54°19′N 119°59′W
Approximately 77 km west north-west of Grande Cache.

The name for this ridge, which is 1691 m in altitude, was proposed by R.W. Cautley (a Commissioner on the Provincial Boundary Survey, 1918-1924) on 18 October 1922, for the skyline resemblance of the ridge to the shape of a dinosaur. The creek takes its name from the ridge.

Dip Slope Mountain (mountain)

82 N/9 - Hector Lake
36-31-16-W5
51°43′N 116°08′W
Approximately 75 km north-west of Banff.

The name for this 3125 m mountain was officially approved by the Canadian Permanent Committee on Geographical

Names 24 August 1971. The absolute conformity of the south-west slope to the dip of the strata led Mr. W.L. Putnam to suggest the descriptive name.

Dirtywater Lake (lake)

82 J/8 - Stimson Creek
17-15-2-W5
50°15′N 114°14′W
Approximately 45 km south of Turner Valley.

This little lake is muddy around the edges and is never clear. This appropriately descriptive name was officially approved 12 December 1939.

Disaster Point (point)

83 F/4 - Miette
25-48-28-W5
53°11′N 117°58′W
Approximately 38 km south-west of Hinton.

Dr. E. Deville, once Surveyor General, named this point because Sir Sandford Fleming's brandy flask was broken on a rock here. C.P.R. surveyor Fleming (see Introduction) was exploring a possible route for the Canadian Pacific Railway in 1872 when this "disaster" occurred. Another version for the origin of this name suggests that several horses lost their footing on the smooth sloping surfaces of the rocks, and met their death from the fall.

Divergence Creek (creek)

83 D/9 - Amethyst Lakes
34-41-1-W6
52°35′N 118°03′W
Flows north-west into Whirlpool River, approximately 35 km south of Jasper.

(see Divergence Peak)

Divergence Peak (peak)

83 C/5 - Fortress Lake
6-41-28-W5
52°30′N 118°00′W

Approximately 40 km south-east of Jasper.

Named by A.O. Wheeler (see Wheeler Flats) in 1921, this 2827 m mountain peak lies at an angle on the Alberta-B.C. boundary line.

Divide Creek (creek)

82 O/12 - Barrier Mountain
19-31-13-W5
51°40′N 115°51′W
Flows south-east into Red Deer River, approximately 60 km north-west of Banff.

The origin of the name is unknown.

Divide Pass (pass)

82 O/13 - Scalp Creek
51°45′10″N 115°55′40″W
Approximately 70 km north-west of Banff.

The name for this pass was officially approved 20 January 1985. (see Divide Creek)

Division Mountain (mountain)

82 N/14 - Rostrum Peak
2-34-22-W5
51°53′30″N 117°01′00″W
Approximately 130 km north-west of Banff, on the Alberta-B.C. boundary.

The descriptive name for this 3030 m mountain was approved in 1920.

Dizzy Creek (creek)

83 B/5 - Saunders
5-40-13-W5
52°25′N 115°50′W
Flows north into North Saskatchewan River, approximately 18 km south-east of Nordegg.

This creek was officially named in 1941 for its proximity to a nearby hill locally known as "Dizzy Hill."

Doctor Creek (creek)

83 E/9 - Moberly Creek
27-52-1-W6
53°31′N 118°03′W
Flows south-east into Wildhay River, approximately 71 km north of Jasper.

The origin of the name for Doctor Creek is unknown.

Dog Creek (creek)

83 C/8 - Nordegg
40-16-W5
52°27′N 116°11′W
Flows south-east into Shunda Creek, approximately 25 km west of Nordegg.

No information regarding the origin of the name is known.

Dogrib Creek (creek)

82 O/11 - Burnt Timber Creek
34-30-11-W5
51°37′N 115°28′W
Flows south into Panther River, approximately 50 km north of Banff.

The origin of the name is unknown.

Dolly Lakes (lakes)

83 E/7 - Blue Creek
53°17′40″N 118°38′00″W
Approximately 55 km north-west of Jasper.

The name of these two lakes is probably derived from the species of fish called Dolly Varden, a char native to north-west America and north-east Asia. The locally well-established name has been in use since the early 1900s. It was officially approved 5 May 1987.

Dolomite Creek (creek)

82 N/9 - Hector Lake
12-33-18-W5
51°49′N 116°25′W

Flows north into Siffleur River, approximately 90 km north-west of Banff.

(see Dolomite Peak)

Dolomite Pass (pass)

82 N/9 - Hector Lake
31-31-17-W5
51°42′N 116°23′W
Approximately 80 km north-west of Banff.

(see Dolomite Peak)

Dolomite Peak (peak)

82 N/9 - Hector Lake
20-31-17-W5
51°40′N 116°22′W
Approximately 80 km north-west of Banff.

This 2782 m mountain peak was named in 1897 by Messrs. Fay (see Mount Fay), J.N. Collie (see Collie Creek), Dixon and Thompson (see Thompson Pass) due to its resemblance to the Dolomites, a range of the Alps, located in northeastern Italy, between the Valleys of Adige and Piave.

Dome Glacier (glacier)

83 C/3 - Columbia Icefield
24-37-24-W5
52°12′N 117°17′W
Approximately 90 km south-east of Jasper.

The local name for the mountain on which this glacier sits is descriptively called "The Dome." The glacier takes its name from that feature.

Dominion Prairie (prairie)

83 D/16 - Jasper
17-45-3-W6
52°53′N 118°23′W
Approximately 22 km west of Jasper.

The origin of the name is unknown.

Donald Flats (flats)

83 E/16 - Donald Flats
31-55-1-W6
53°47′N 118°07′W
Approximately 68 km east south-east of
Grande Cache.

The well-established local name for these
flats was officially approved in 1947. Its
earliest record dates back to 1920.

Donald Lake (lake)

82 O/5 - Castle Mountain
17-29-14-W5
51°29′N 115°56′W
Approximately 42 km north north-west of
Banff.

This lake, which was originally called Lake
Donald, was named by Walter D. Wilcox
(see Wilcox Pass) in 1921. One of Wilcox's
companions, A.L. Castle, had three children:
Gwendolyn, Alfred and Donald Castle. This
feature is named after Donald. (see
Gwendolyn Lake and Alfred Lake)

Donald Lake (lake)

83 E/16 - Donald Flats
4-58-1-W6
53°59′N 118°05′W
Approximately 71 km east north-east of
Grande Cache.

This lake was formerly known as "Baptiste
Lake;" however, Donald Lake was the name
substituted for it because it was better
known as such. The name was officially
approved in 1947.

Door Jamb Mountain (mountain)

82 O/3 - Canmore
35-24-9-W5
51°05′N 115°08′W
Approximately 15 km east of Canmore.

The name for this mountain, which is 1996
m in altitude, was approved by the Canadian

Permanent Committee on Geographical
Names 4 August 1965. It is descriptive in
that the mountain stands at one side of the
entrance to the mountains, and the sideposts
of a door are called "jambs."

Dormer Lake (lake)

82 O/11 - Burnt Timber Creek
4-28-30-11-W5
51°35′27″N 115°29′40″W
Approximately 53 km north of Banff.

The name for this lake was officially
approved 29 July 1986. (see Dormer River)

Dormer Mountain (mountain)

82 O/12 - Barrier Mountain
36-30-12-W5
51°37′N 115°34′W
Approximately 50 km north of Banff.

This mountain is 2766 m in altitude. (see
Dormer River)

Dormer Pass (pass)

82 O/5 - Castle Mountain
51°26′35″N 115°34′10″W
Approximately 40 km north of Banff.

The locally well-established name for this
pass is taken from the feature's proximity to
Dormer Mountain and River. (see Dormer
River)

Dormer River (river)

82 O/12 - Barrier Mountain
33-30-11-W5
51°37′N 115°30′W
Flows north-east into Panther River,
approximately 50 km north of Banff.

The name for this river and the nearby
mountain is taken from the fact that the
ridges above the valley terminate in right
angles like dormer windows.

Dorothy Creek (creek)

83 C/16 - Brown Creek
36-43-18-W5
52°45′N 116°29′W
Flows north-east into Brown Creek,
approximately 43 km north-west of
Nordegg.

The origin of the name is unknown.

Dorothy Lake (lake)

83 D/16 - Jasper
16-45-2-W6
52°53′N 118°14′W
Approximately 10 km west of Jasper.

This lake was named in 1914 by H.
Matheson, D.L.S., but no other information
about the origin of the name is known.

Douai Mountain (mountain)

82 N/14 - Rostrum Peak
3-35-23-W5
51°59′N 117°11′W
Approximately 145 km north-west of Banff
on the Alberta-B.C. boundary.

This 3120 m mountain was named in 1920
after the celebrated fortified town in northeastern France in commemoration of its
occupation on 18 October 1918 by the
Canadians and other Allied troops during
the First World War. Douai, the site of a
castle, belonged to the courts of Flanders in
the Middle Ages. There are still many buildings
in the area which date back to the 16th-18th
centuries, although they are now damaged.

Douglas Creek (creek)

82 O/12 - Barrier Mountain
5-30-14-W5
51°32′N 115°56′W
Flows north-west into Red Deer River,
approximately 55 km north-west of Banff.

(see Mount Douglas)

Douglas Creek (creek)

83 E/10 - Adams Lookout
4-55-6-W6
53°43'N 118°50'W
Flows north into Muskeg River,
approximately 25 km south-east of Grande
Cache.

(see Mount Douglas)

Douglas Lake (lake)

82 O/12 - Barrier Mountain
5-30-14-W5
51°32'N 115°56'W
Approximately 50 km north-west of Banff.

(see Mount Douglas)

Douglas, Mount (mountain)

82 O/12 - Barrier Mountain
31-29-14-W5
51°32'N 115°58'W
Approximately 50 km north-west of Banff.

This impressive mountain, which is 3235 m
in altitude, was named in 1884 by Dr. G.M.
Dawson (see Tombstone Mountain), after
David Douglas (1798-1834), a Scottish
botanist and naturalist who crossed the
Rockies in 1827. His journal, entitled
*Journal During Travels in North America,
1823-27,* was published in 1914. (see Mount
St. Bride)

Dowling Ford (ford)

83 C/10 - George Creek
21-42-20-W5
52°38'N 116°49'W
Approximately 53 km west north-west of
Nordegg.

Donaldson Bogart Dowling (1858-1925), an
engineer in the Geological Survey
Department of Canada, was in the area
surveying coal fields of the Rocky
Mountains in 1906-1907. It is after Mr.
Dowling that the Canadian Permanent

Committee on Geographical Names
approved the name for this feature 22 June
1976. (see Lusk Creek)

Draco Peak (peak)

83 E/5 - Holmes River
14-52-12-W6
53°29'N 119°40'W
Approximately 125 km north-west of Jasper.

This 2587 m mountain was named "Dragon
Mountain" in 1917 after its distinctive shape.
But it was duplicating another nearby
feature's name, so "Dragon" was changed to
Draco, the Latin word for "dragon."

Dragon Creek (creek)

83 C/5 - Fortress Lake
27-40-26-W5
52°29'N 117°40'W
Flows east into Athabasca River,
approximately 50 km south-east of Jasper.

(see Dragon Peak)

Dragon Lake (lake)

83 C/5 - Fortress Lake
52°28'15"N 117°39'45"W
Approximately 50 km south-east of Jasper,
in Jasper National Park.

Dragon Lake is named for its proximity to
Dragon Peak and Dragon Creek, although it
is not part of the Dragon Creek drainage
system. The name Dragon Lake has been
used locally since the 1940s by the Jasper
Park Warden Service.

Dragon Peak (peak)

83 C/5 - Fortress Lake
17-40-26-W5
52°27'N 117°43'W
Approximately 55 km south-east of Jasper.

The descriptive name for this mountain
peak, 2940 m in altitude, was applied in 1921
by A.O. Wheeler (see Wheeler Flats). At the

time of naming the rock shape near the
summit was said to resemble a dragon.

Drawbridge Peak (peak)

83 D/9 - Amethyst Lakes
14-43-3-W6
52°42'N 118°19'W
Approximately 27 km south-west of Jasper.

The name for this mountain, which is 2720
m in altitude, was adopted 1 May 1943. It is
probably associated in meaning to the range
to which it is connected: "The Ramparts,"
and the surrounding peaks - Dungeon Peak,
Bastion Peak, The Turret, etc.

Drinnan Creek (creek)

83 F/4 - Miette
31-48-24-W5
53°11'N 117°30'W
Flows north-east into Gregg River,
approximately 27 km south of Hinton.

This creek, which was named in 1925, takes its
name from R.G. Drinnan who had a coal
prospect camp on the stream at one time.
Drinnan was Manager of the Cadomin Mine
and later became the Director of the Luscar
Collieries.

Drinnan, Mount (mountain)

83 F/4 - Miette
6-48-25-W5
53°07'N 117°39'W
Approximately 24 km south south-west of
Hinton.

This mountain is 2242 m in altitude. (see
Drinnan Creek)

Dromore Creek (creek)

83 F/4 - Miette
6-47-26-W5
53°02'N 117°48'W
Flows north into Rocky River, approximately
47 km south south-west of Hinton.

(see Mount Dromore)

Dromore, Mount (mountain)

83 C/13 - Medicine Lake
6-46-27-W5
52°57′N 117°53′W
Approximately 16 km north-east of Jasper.

This mountain was named by M.P. Bridgland (see Mount Bridgland) in 1916 after the town of Dromore, County Down, Ireland. The word "dromore" is Gaelic for "great ridge."

Drum Creek (creek)

82 G/9 - Blairmore
16-7-3-W5
49°34′N 114°21′W
Flows south-east into Crowsnest River, approximately 8 km south-east of Blairmore.

The origin of the name is unknown.

Drummond Creek (creek)

83 C/14 - Mountain Park
23-46-24-W5
52°59′N 117°22′W
Flows north-east into Whitehorse Creek, approximately 49 km east south-east of Jasper.

(see Mount Drummond)

Drummond Glacier (glacier)

82 N/9 - Hector Lake
28-30-15-W5
51°36′N 116°03′W
Approximately 60 km north-west of Banff.

The name Drummond Glacier was officially approved 25 February 1965. (see Mount Drummond)

Drummond Lake (lake)

82 O/12 - Barrier Mountain
19-30-14-W5
51°35′N 115°59′W

Approximately 55 km north-west of Banff.

(see Mount Drummond)

Drummond, Mount (mountain)

82 N/9 - Hector Lake
26-30-15-W5
51°36′N 116°01′W
Approximately 60 km north-west of Banff.

This 3148 m mountain was named in 1884 by Dr. G.M. Dawson (see Tombstone Mountain) after Thomas Drummond, the Assistant Naturalist to Franklin's Second Arctic Expedition, 1825-1827.

Dry Canyon (canyon)

83 E/13 - Dry Canyon
16-56-11-W6
53°50′N 119°35′W
Approximately 30 km west south-west of Grande Cache.

The Canadian Permanent Committee on Geographical Names approved the name for this canyon in January 1963. It is likely descriptive of the arid nature of the area.

Dry Creek (creek)

82 J/1 - Langford Creek
22-14-4-W5
50°11′N 114°29′W
Flows east into Livingstone River, approximately 95 km south south-west of Calgary.

The origin of the name is unknown.

Dry Creek (creek)

83 B/3 - Tay River
9-38-10-W5
52°15′N 115°21′W
Flows north into Lick Creek, approximately 33 km south-west of Rocky Mountain House.

This locally well-established name was officially applied to this feature in 1941. The

name may be descriptive but its precise origin is unknown.

Drystone Creek (creek)

83 F/4 - Miette
22-49-27-W5
53°14′N 117°51′W
Flows north-west into Brule Lake, approximately 27 km south-west of Hinton.

The origin for the locally well-established name of this creek is unknown. The name has been in use since 1904.

Drywood Mountain (mountain)

82 G/8 - Beaver Mines
34-3-1-W5
49°16′N 114°03′W
Approximately 25 km south-west of Pincher Creek.

(see Drywood Creek)

Dummy Creek (creek)

83 F/3 - Cadomin
28-48-21-W5
53°10′N 117°00′W
Flows west into Embarras River, approximately 59 km south-west of Edson.

The origin of the name is unknown.

Dungarvan Creek (creek)

82 H/4 - Waterton Lakes
24-3-29-W4
49°14′N 113°46′W
Flows east into Waterton River, approximately 40 km south south-east of Pincher Creek.

The name was officially approved 19 May 1946. (see Mount Dungarvan)

Dungarvan, Mount (mountain)

82 H/4 - Waterton Lakes
2-30-W4
49°09′N 113°58′W

Approximately 35 km south of Pincher Creek.

"Dungarvan" is the Irish description of a rough and broken mountain. It is also a seaport and harbour in Ireland in the county of Waterford. It comes from "dungarbhain," meaning "Garvan's fortress." The name Mount Dungarvan for this 2575 m mountain was officially approved 19 May 1943.

Dungeon Peak (peak)

83 D/9 - Amethyst Lakes
12-43-3-W6
52°41'N 118°17'W
Approximately 26 km south south-west of Jasper.

The surveyors who christened The Ramparts thought of it as a castellated range and bestowed upon the peaks names relating to medieval times - The Turret, Bastion, Redoubt, Dungeon, etc. This peak measures 3130 m in altitude.

Dunn Lake (lake)

83 F/12 - Gregg Lake
13-7-54-25-W5
53°39'20"N 117°41'05"W
Approximately 30 km north of Hinton.

The name "Scout Lake" was adopted 9 May 1985. At one time there was a scout camp on the shores of this lake but the name fell out of usage. Due to various representations by the public who did not support the name, a proposal to change the name was submitted. Local fisherman, Art Dunn, was an enthusiastic angler and outdoors-man who was alleged to have known every good fishing hole in the area. This lake was his favourite. Art Dunn died in 1983. The name was officially approved for this feature 5 May 1987.

Dutch Creek (creek)

82 G/16 - Maycroft
8-11-3-W5
49°54'N 114°23'W

Flows east into Oldman River, approximately 30 km north of Blairmore.

The name for this creek is taken after an old prospector known as "Dutch" Bernhardt. The name was officially approved in 1941. No other information is available about this man.

Dutch Creek (creek)

83 B/5 - Saunders
34-39-14-W5
52°24'N 115°56'W
Flows south-east into North Saskatchewan River, approximately 13 km south-east of Nordegg.

The local history of the name of this creek states that a Dutch prospector was reputed to have found some very valuable minerals in this stream. He was murdered by his partner and from that time, the creek has been called Dutch Creek.

Dyson Creek (creek)

82 J/10 - Mount Rae
34-19-5-W5
50°39'N 114°37'W
Flows north-east into Sheep River, approximately 24 km west of Turner Valley.

The name for this creek was officially approved 6 September 1951. (see Mount Dyson)

Dyson, Mount (mountain)

82 J/10 - Mount Rae
7-19-4-W5
50°36'N 114°33'W
Approximately 22 km west of Turner Valley.

This mountain, which is 1768 m in altitude, was named after a local rancher. The name was officially approved 6 September 1951 after several years of local usage.

Eagle Coulee (coulee)

82 G/16 - Maycroft
21-10-2-W5
49°51'N 114°13'W
Approximately 30 km north-east of Blairmore.

The origin for this locally well-established name is unknown.

Eagle Creek (creek)

82 O/12 - Barrier Mountain
4-32-11-W5
51°42'N 115°38'W
Flows south into Red Deer River, approximately 63 km north of Banff.

The origin for the locally well-established name of this creek is not known. It was submitted by Mel Kraft of the Alberta Department of Fish and Wildlife, who wished to establish a management program for the sport fishing facility. The name was officially approved 5 May 1987.

Eagle Lake (lake)

82 O/11 - Burnt Timber Creek
11-32-11-W5
51°44'N 115°28'W
Approximately 60 km north of Banff.

The origin of the name is unknown.

Eagle Mountain (mountain)

82 O/4 - Banff
3-25-13-W5
51°06'N 115°45'W
Approximately 15 km south-west of Banff.

Eagle Mountain is 2820 m in altitude. The name for this feature was officially approved in 1958, but its origin is unknown.

Eagles Nest Creek (creek)

83 E/10 - Adams Lookout
22-52-4-W6
53°31'N 118°32'W

Flows north-east into Wildhay River, approximately 58 km south-east of Grande Cache.

(see Eagles Nest Pass)

Eagles Nest Pass (pass)

83 E/7 - Blue Creek
7-52-4-W6
53°28′30″N 118°34′00″W
Approximately 75 km north north-west of Jasper.

C.P.R. surveyor Walter Moberly named this pass after following the eagles through the mountains to the opening. The name became increasingly well-established locally and was officially approved 15 January 1979. The elevation of the narrow mountain pass measures 2103 m in altitude.

East Glacier (glacier)

82 N/15 - Mistaya Lake
28-33-21-W5
51°52′N 116°55′W
Approximately 120 km north-west of Banff.

The name for this glacier is likely descriptive of its location, but the reference point is not known.

East Crowsnest Creek (creek)

82 G/10 - Crowsnest
36-7-6-W5
49°36′N 114°42′W
Flows north into Crowsnest Creek, approximately 14 km west south-west of Coleman.

This creek was named for its proximity as a tributary to Crowsnest Creek. (see Crowsnest Mountain)

East Elk Pass (pass)

82 J/11 - Kananaskis Lakes
3-19-8-W5
50°34′N 115°03′W

---

*denotes rescinded name or former locality.

Approximately 55 km west south-west of Turner Valley, on the Alberta-B.C. boundary.

The Canadian Permanent Committee on Geographical Names approved this name 13 September 1966. (see Elk Range)

*East Kootenay (former locality)

82 G/10 - Crowsnest
9-8-5-W5
49°38′10″N 114°36′55″W
Approximately 120 km west of Lethbridge.

No information regarding the origin of the name is known.

East Lyell Glacier (glacier)

82 N/14 - Rostrum Peak
27-34-22-W5
51°57′N 117°03′W
Approximately 135 km north-west of Banff.

(see Mount Lyell)

East Rice Glacier (glacier)

83 C/3 - Columbia Icefield
52°00′02″N 117°12′04″W
Approximately 110 km south-east of Jasper, immediately south-east of Mount Spring-Rice.

This glacier was named for its proximity to Mount Spring-Rice. (see Mount Spring-Rice)

Easy Creek (creek)

83 B/5 - Saunders
29-38-13-W5
52°18′N 115°49′W
Flows east into Nice Creek, approximately 27 km south-east of Nordegg.

The name for this stream describes the travelling conditions along the creek.

Eaton Creek (creek)

83 E/14 - Grande Cache
12-56-9-W6
53°49′N 119°12′W
Flows south-east into Smoky River, approximately 9 km south-west of Grande Cache.

There is some evidence to suggest that this creek may have been named after a member of the Eaton family, who arrived at Hinton in 1926 to homestead.

Ebon Peak (peak)

82 N/15 - Mistaya Lake
29-32-19-W5
51°47′N 116°38′W
Approximately 100 km north-west of Banff, on the Alberta-B.C. boundary.

The name for this peak, which is 2910 m in altitude, was applied by A.O. Wheeler (see Wheeler Flats) in 1918. An ebony-coloured line may be distinctly seen amid a line of snow-covered peaks.

Eccles Pond (pond)

83 F/11 - Dalehurst
8-53-22-W5
53°33′N 117°13′W
Approximately 50 km west of Edson.

This pond was originally constructed as part of a sawmill operation for the purpose of washing logs. A man named Eccles operated the sawmill. The name was officially approved 12 March 1976.

Echo Creek (creek)

82 O/4 - Banff
25-12-W5
51°11′N 115°25′W
Flows north-east into Bow River, approximately 1 km west of Banff.

The name for this creek was officially approved in 1958, but its origin remains unknown.

Eden Peak (peak)

83 C/4 - Clemenceau Icefield
52°11'30"N 117°44'00"W
Approximately 80 km south south-east of Jasper, on the Alberta-B.C. boundary.

This 3180 m mountain was named by Jean Habel (see Mount Habel) in 1901, and the name was officially approved 5 May 1987. The origin of the name is unknown.

Edith Lake (lake)

83 D/16 - Jasper
26-45-1-W6
52°54'N 118°02'W
Approximately 4 km north north-east of Jasper.

H.A. McColl was a General Superintendent of the Grand Trunk Pacific Railway who was in charge of construction at the time of this lake's naming in 1914. H. Matheson, D.L.S. gave this lake its name after Edith McColl, the wife of H.A. McColl.

Edith, Mount (mountain)

82 O/4 - Banff
7-26-12-W5
51°12'N 115°40'W
Approximately 7 km west north-west of Banff.

This 2554 m mountain was named by J. Smith and Mr. L. Stewart while on a trip over the pass in 1887-88, after Mrs. J.F. Orde (née Edith Cox) from Ottawa, who visited Banff with Lady MacDonald in 1886.

Edith Cavell, Mount (mountain)

83 D/9 - Amethyst Lakes
2-43-1-W6
52°40'N 118°03'W
Approximately 24 km south of Jasper.

This mountain, which is 3363 m in altitude, was officially named Mount Edith Cavell in March, 1916. Edith Louise Cavell (1865-1915) was a Norfolk heroine whom some dub a martyr. She saw herself as "a nurse who tried to do her duty." She returned to Brussels from Swardeston (her hometown) in 1907 and by 1914 was put in charge of a pioneer training school engaged in helping soldiers trapped behind enemy lines to rejoin their armies. This was regarded as treason by the German authorities, and Edith Cavell was sentenced to death by a firing squad. Her defence was that as a nurse, her duty was to save lives. This is what she was doing by concealing and smuggling away hunted men. She was buried in a spot called "Life's Green," just outside of Norwich Cathedral.

Edna Lake (lake)

83 E/1 - Snaring
23-47-1-W6
53°04'N 118°02'W
Approximately 22 km north of Jasper.

The origin of the name is unknown.

Edward Peak (peak)

82 N/14 - Rostrum Peak
34-22-W5
51°57'35"N 117°05'40"W
Approximately 135 km north-west of Banff.

(see Christian Peak)

Edworthy Falls (water falls)

82 J/10 - Mount Rae
2-20-8-W5
50°39'50"N 114°59'50"W
Located on the Elbow River, approximately 45 km west south-west of Turner Valley.

These two small waterfalls on the Upper Elbow River are named after George Edworthy, (1899-1978) who discovered them in 1929 while on a pack trip into the Kananaskis region. George Edworthy at that time referred to the falls as "Desolation Falls." The names "Desolation Falls" and "Desolation Flats" did not, however, become established in local usage. The name Edworthy Falls was made official 29 July 1986.

Egypt Lake (lake)

82 O/4 - Banff
4-25-14-W5
51°05'N 115°54'W
Approximately 25 km south-west of Banff.

The origin of the name is unknown.

Eiffel Lake (lake)

82 N/8 - Lake Louise
19-27-16-W5
51°20'N 116°14'W
Approximately 50 km west north-west of Banff.

(see Eiffel Peak)

Eiffel Peak (peak)

82 N/8 - Lake Louise
30-27-16-W5
51°20'N 116°14'W
Approximately 50 km west north-west of Banff.

The descriptive name for this mountain, 3084 m, was officially applied to this feature in 1908, from a map of A.O. Wheeler's (see Wheeler Flats). There is a huge tower rising for about 305 m at the summit of the mountain, suggestive of the Eiffel Tower in Paris, France.

*Eisenhower, Mount (mountain)

82 O/5 - Castle Mountain
9-27-14-W5
51°18'00"N 115°54'00"W
Approximately 28 km north-west of Banff.

(see Castle Mountain)

---

*denotes rescinded name or former locality.

Eisenhower Peak (peak)
82 O/5 - Castle Mountain
27-14-W5
51°18′00″N 115°55′20″W
Approximately 28 km north-west of Banff.

This peak at the east end of Castle Mountain was named in honour of General Dwight D. Eisenhower, Commander of the Allied Forces in Europe. The name was adopted 15 October 1979, the same date as the name Mount Eisenhower was officially changed back to Castle Mountain. (see Castle Mountain)

Elbow Falls (falls)
82 J/15 - Bragg Creek
8-16-22-6-W5
50°52′10″N 114°46′00″W
Approximately 50 km west of Calgary.

The name for these falls was officially approved 20 October 1983. (see Little Elbow River)

Elbow Lake (lake)
82 J/11 - Kananaskis Lakes
26-19-8-W5
50°38′20″N 115°25′00″W
Approximately 50 km west of Turner Valley.

This lake forms the headwaters of the Elbow River. The name Elbow Lake has been in local usage for at least 60 years and was made official 8 November 1978.

Elbow Pass (pass)
82 J/11 - Kananaskis Lakes
26-19-8-W5
50°38′00″N 115°40′00″W
Approximately 50 km west of Turner Valley.

The name was officially approved 8 November 1978. (see Elbow Lake)

Elbow River (river)
82 J/10 - Mount Rae
1-21-7-W5
50°45′00″N 114°50′20″W
Approximately 62 km south-west of Calgary.

This river was named after the elbow-like curve it goes through approximately 8 km south of Calgary.

Eldon (locality)
82 N/8 - Lake Louise
28-27-15-W5
51°20′N 116°03′W
Approximately 35 km west north-west of Banff.

The Canadian Pacific Railway station was named Eldon in 1883. The name may be taken after John Scott, 3rd Earl of Eldon. There is an English village in Derbyshire called Eldon Hill as well as an Eildon Hall located in Banffshire, Scotland. The precise origin for the name of this locality is not known.

Elephas Mountain (mountain)
83 D/9 - Amethyst Lakes
2-42-3-W6
52°36′N 118°20′W
Approximately 36 km south-west of Jasper.

Named by A.O. Wheeler (see Wheeler Flats) in 1922, this mountain is 2940 m in altitude. Its fancied resemblance in shape to an elephant's head inspired the name elephas, the Latin word for elephant.

Elk Creek (creek)
83 B/4 - Cripple Creek
33-35-12-W5
52°03′N 115°39′W
Flows south into Clearwater River, approximately 56 km south-east of Nordegg.

The origin for this locally well-established name is unknown.

Elk Creek (creek)
83 C/15 - Cardinal River
5-46-18-W5
52°56′N 116°35′W
Flows east into Pembina River, approximately 101 km east of Jasper.

(see Elk Range)

Elk Lake (lake)
82 O/5 - Castle Mountain
8-27-12-W5
51°17′N 115°39′W
Approximately 14 km north north-west of Banff.

The name for this lake was officially approved in 1959 and is likely taken from the local elk or wapiti.

Elk Pass (pass)
82 J/11 - Kananaskis Lakes
4-19-8-W5
50°35′N 115°04′W
Approximately 55 km west south-west of Turner Valley.

(see Elk Range)

Elk Range (range)
82 J/10 - Mount Rae
18-7-W5
50°32′N 114°59′W
Approximately 52 km west of Turner Valley.

The mountains are at the head of Elk River, in B.C., which is so called from the number of elk found near it.

Elkwoods Fen (fen)
82 J/11 - Kananaskis Lakes
1-20-9-W5
50°39′30″N 115°07′00″W
Approximately 60 km west of Turner Valley.

The name Elkwoods Fen was selected and officially approved 18 September 1978 because of its proximity to the Elkwoods Campground. The area is locally called "Muskeg Creek," but since this feature is not true muskeg (i.e., no sphagnum moss or black spruce) nor a true creek (i.e., no flow of water), the name "fen" was selected.

Elliot Peak (peak)

83 C/1 - Whiterabbit Creek
26-36-18-W5
52°08′N 116°29′W
Approximately 125 km north-west of Banff.

Elliot Barnes, the son of a local rancher from North Dakota, climbed this 2872 m mountain in 1906 when he was eight years old. The mountain was officially named Elliot Peak in 1907, replacing "Sentinel Peak," given by A.P. Coleman (see Coleman Peak), which was a duplication of a name given to another peak.

Elpoca Creek (creek)

82 J/11 - Kananaskis Lakes
6-20-8-W5
50°39′55″N 115°05′20″W
Flows from the west slope of Elpoca Mountain, west into Pocaterra Creek, approximately 60 km west of Turner Valley.

The name was officially approved 18 September 1978. (see Elpoca Mountain)

Elpoca Falls (falls)

82 J/11 - Kananaskis Lakes
5-20-8-W5
50°39′50″N 115°04′30″W
Located on Elpoca Creek, approximately 55 km west of Turner Valley.

(see Elpoca Mountain)

Elpoca Mountain (mountain)

82 J/11 - Kananaskis Lakes
3-20-8-W5
50°40′N 115°01′W
Approximately 55 km west of Turner Valley.

The name "Elpoca" is derived from the names Elbow River and Pocaterra Creek (see Pocaterra Creek). This mountain, 3029 m in altitude, is at the head of these two streams, and its name was officially approved in 1920.

Elusive Pass (pass)

83 C/12 - Athabasca Falls
52°31′05″N 117°29′05″W
Approximately 55 km south-east of Jasper.

The pass is deceiving in that it appears to be closer than it actually is.

Elysium Mountain (mountain)

83 D/16 - Jasper
10-46-3-W6
52°57′N 118°22′W
Approximately 20 km west north-west of Jasper.

In Greek mythology, elysium is the supposed state or abode of the blessed after death. This mountain, which is 2446 m in altitude, overlooks fine meadows, the sight of which alludes to "Elysian fields." M.P. Bridgland, D.L.S. (see Mount Bridgland) gave this mountain its allegorically descriptive name in 1916.

Elysium Pass (pass)

83 D/16 - Jasper
11-46-3-W6
52°57′N 118°20′W
Approximately 19 km west north-west of Jasper.

(see Elysium Mountain)

Embarras (locality)

83 F/7 - Erith
8-50-20-W5
53°18′N 116°53′W
Approximately 44 km south-west of Edson.

(see Embarras River)

Embarras River (river)

83 F/7 - Erith
5-52-18-W5
53°27′N 116°37′W
Flows north-east into McLeod River, approximately 19 km south-west of Edson.

The locally well-established name for this river was approved in 1927 and is probably associated with the French word for an obstruction, because the river is obstructed by driftwood.

Emerald Bay (bay)

82 H/4 - Waterton Lakes
23-1-30-W4
49°03′N 113°54′W
Approximately 1 km west of Waterton Park.

Emerald Bay lies on the north-west shore of Upper Waterton Lake. Its name is descriptive of the gem-like colour of the water in the bay. The name was officially applied 18 January 1974.

Emerald Lake (lake)

82 G/10 - Crowsnest
8-8-5-W5
49°37′20″N 114°38′30″W
Approximately 9 km west of Coleman.

The name Emerald Lake was approved for this feature in 1977. Old-timers in the area have said that the name is descriptive of the lake's colour and has been used for over seventy years. At one time a man named Steward built a cabin beside the lake, and his efforts to have the lake renamed "Steward Lake" were rigorously opposed by local residents.

Emerson Creek (creek)

82 J/8 - Stimson Creek
3-17-3-W5
50°24'N 114°20'W
Flows east into Pekisko Creek, approximately 30 km south of Turner Valley.

This creek was officially named 12 December 1939 after George Emerson (1841-1920). Mr. Emerson was born and raised in Danville, Quebec. He headed west to Iowa, where he homesteaded until 1868, when rumours of gold drew him north. He worked for several years with the Hudson's Bay Company and then worked with Tom Lynch as a rancher until he sold his cattle and brand in 1914. He was one of the most highly esteemed men in the area and was an original member of the Stock Association.

Emerson Lakes (lakes)

83 F/11 - Dalehurst
7-55-21-W5
53°43'30"N 117°07'00"W
Approximately 47 km west north-west of Edson.

The name was first suggested for official approval by R.J. Paterson, Director of Surveys, 25 November 1968. This group of lakes was originally called "Seven Lakes" by the natives, but the name Emerson had since then become locally well-established. The name was officially approved 21 March 1978, but its origin is unknown.

Emigrants Mountain (mountain)

83 D/16 - Jasper
2-46-3-W6
52°57'N 118°19'W
Approximately 17 km west north-west of Jasper.

This mountain, which is 2553 m in altitude, was named by the Geographic Board of Canada in 1916 in commemoration of the over 30,000 miners who travelled north from the California goldfields to the Cariboo Mountains. The mountain range is located in British Columbia, between the Fraser River to the north, and the North Thompson River to the south. This range was home to the Cariboo Gold Rush (1860-1863), British Columbia's most famous gold rush. A group of about 150 settlers, known as the Overlanders of 1862, travelled from Ontario to the interior of B.C. using Red River carts and packhorses. With the aid of Indian guides, they passed over the Rockies, losing only 6 of the members on the rivers. Most of the group joined the many other "Cariboo Gold Seekers," inspired by rumours of large fortunes to be made in a few months, travelling great distances to find wealth. Few found wealth in the form of gold, but many settled in the interior of B.C.

Emir Creek (creek)

83 F/4 - Miette
35-47-28-W5
53°06'N 117°59'W
Flows north-west into Talbot Lake, approximately 46 km south-west of Hinton.

(see Emir Mountain)

Emir Mountain (mountain)

83 F/4 - Miette
6-47-27-W5
53°01'N 117°55'W
Approximately 51 km south-west of Hinton.

An emir is a Saracen or Arab prince or governor. This mountain was named in 1916 by M.P. Bridgland (see Mount Bridgland) because of its prominence.

End Mountain (mountain)

82 O/3 - Canmore
1-26-9-W5
51°12'N 115°08'W
Approximately 20 km north-east of Canmore.

The name for this mountain, which is 2420 m in altitude, is descriptive: it is at the "end" of the range. The name was officially approved 16 January 1912. J.J. McArthur, D.L.S., who surveyed along the railway belt in 1885, may have named this feature, as he named many features on the Alberta-B.C. boundary.

Endless Chain Ridge (ridge)

83 C/6 - Sunwapta
40-25-W5
52°30'N 117°27'W
Approximately 50 km south-east of Jasper.

This low rocky ridge stretches for miles in an unbroken line along the Athabasca River. The descriptive name for the ridge was applied by Mrs. Mary T.S. Schäffer in 1907.

Engadine, Mount (mountain)

82 J/14 - Spray Lakes Reservoir
15-22-10-W5
50°52'N 115°18'W
Approximately 85 km west of Calgary

This 2970 m mountain was officially named in 1922 after a cruiser engaged in the Battle of Jutland. (see Jutland Mountain)

Engelhard, Mount (mountain)

83 C/6 - Sunwapta
52°16'30"N 117°24'20"W
Approximately 80 km south-east of Jasper, in Jasper National Park.

Officially approved 7 November 1984, the 3270 m Mount Engelhard was named for Georgia Engelhard (died October 1986), wife of Eaton Cromwell (see Mount Cromwell).

Although an American, she was a member of the Alpine Club of Canada and one of the most prominent climbers in Canada during the 1930s. She was author of the children's book, *Peterli and the Mountain*.

Entrance (locality)

83 F/5 - Entrance
1-51-26-W5
53°22′N 117°41′W
Approximately 6 km west south-west of Hinton.

The name for this former hamlet was chosen because of its location: in 1915 it was the first Canadian National Railway station east of the entrance to Jasper National Park.

Entry Creek (creek)

83 C/2 - Cline River
1-37-19-W5
52°09′N 116°35′W
Flows north into Cline River, approximately 115 km west south-west of Rocky Mountain House.

This creek is the first stage of the route into the mountains. The name Entry is, thus, descriptive of its location.

Eon Mountain (mountain)

82 J/13 - Mount Assiniboine
22-12-W5
50°50′N 115°38′W
Approximately 105 km west south-west of Calgary, on the Alberta-B.C. boundary.

It is said that the name given to this 3310 m mountain in 1917 is probably descriptive of the ages that elapsed during the slow elevation of the mountain.

Epaulette Creek (creek)

82 N/15 - Mistaya Lake
6-34-19-W5
51°53′N 116°41′W
Flows north-east into Mistaya River, approximately 110 km north-west of Banff.

The Canadian Permanent Committee on Geographical Names approved this name in September 1971. (see Epaulette Mountain)

Epaulette Lake (lake)

82 N/15 - Mistaya Lake
36-33-20-W5
51°52′N 116°42′W
Source of Epaulette Creek, approximately 110 km north-west of Banff.

The Canadian Permanent Committee on Geographical Names approved this name in September 1971. (see Epaulette Mountain)

Epaulette Mountain (mountain)

82 N/15 - Mistaya Lake
34-33-20-W5
51°52′N 116°46′W
Approximately 110 km north-west of Banff.

Formerly known as "Pyramid Mountain," this 3095 m mountain has a fancied resemblance in shape to a shoulder ornament on a uniform. It was officially named in 1961.

Erasmus, Mount (mountain)

82 N/15 - Mistaya Lake
33-34-21-W5
51°58′N 116°55′W
Approximately 125 km north-west of Banff.

This 3265 m mountain was named after Peter Erasmus (1833-1931), a noted guide and trader. He was Dr. James Hector's (see Mount Hector) special assistant from 1858-1859, and the last survivor of the Palliser Expedition. He later became an official with the Department of Indian Affairs on the Blackfoot Reserve.

Erebus Mountain (mountain)

83 D/9 - Amethyst Lakes
19-42-2-W6
52°38′N 118°17′W
Approximately 30 km south-west of Jasper.

This 3119 m mountain was named in 1916 by M.P. Bridgland (see Mount Bridgland). There is a dark rock precipice on the mountain which faces north-east: the Greek word for "darkness" is "erebus"; hence, the name for this feature is descriptive of the north-east face of the mountain.

Eremite Creek (creek)

83 D/9 - Amethyst Lakes
33-42-2-W6
52°40′N 118°14′W
Flows north into Chrome Lake, approximately 26 km south-west of Jasper.

(see Eremite Mountain)

Eremite Glacier (glacier)

83 D/9 - Amethyst Lakes
42-2-W6
52°38′N 118°14′W
Approximately 29 km south-west of Jasper.

(see Eremite Mountain)

Eremite Mountain (mountain)

83 D/9 - Amethyst Lakes
20-42-2-W6
52°38′N 118°15′W
Approximately 30 km south-west of Jasper.

The descriptive name for this 2910 m mountain was applied in 1916 by M.P. Bridgland (see Mount Bridgland). The remote position of this mountain reminds one of an eremite, a hermit. The other features take their names from this mountain.

Eric Lake (lake)

83 E/10 - Adams Lookout
22-54-7-W6
53°38′N 118°54′W
Approximately 30 km south south-east of Grande Cache.

No information regarding the origin of the name is known.

Erickson Creek (creek)

83 F/3 - Cadomin
21-47-21-W5
53°04′N 117°01′W
Flows south-west into Beaverdam Creek, approximately 68 km south-west of Edson.

The local name for this creek was officially approved in 1944 and was suggested by a local resident, J.R. Matthews. The origin of the name is unknown.

Erith (locality)

83 F/7 - Erith
14-51-19-W5
53°25′N 116°40′W
Approximately 26 km south-west of Edson.

(see Erith River)

Erith River (river)

83 F/2 - Foothills
28-49-18-W5
53°15′N 116°35′W
Flows north into Embarras River, approximately 26 km south-west of Edson.

There was a railway station named Erith on the Canadian National Railway line near this river. The next station along the line was named Weald. Both Weald and Erith are names of places in Kent, England.

Ermatinger, Mount (mountain)

83 D/8 - Athabasca Pass
9-40-1-W6
52°25′N 118°02′W
Approximately 51 km south of Jasper.

This 3060 m mountain was named by A.O. Wheeler (see Wheeler Flats) in 1921 after Edward Ermatinger (1797-1876). Ermatinger served with the Hudson's Bay Company from 1818-1828, after taking his education in England. He moved to St. Thomas, Ontario, where he became merchant, banker and postmaster after his retirement from the Hudson's Bay Company. This mountain was originally known as "Horn Mountain."

Ernest Peak (peak)

82 N/14 - Rostrum Peak
34-22-W5
51°57′26″N 117°06′12″W
Approximately 135 km north-west of Banff.

This peak is 3511 m in altitude. (see Christian Peak)

■ **Ernest Ross, Mount** (mountain)

83 C/1 - Whiterabbit Creek
36-18-W5
52°06′10″N 116°27′10″W
Approximately 125 km north-west of Banff.

Although this 2454 m mountain was named in 1969, official approval was not given until 20 June 1980. The Rocky Mountain House Chamber of Commerce requested that a mountain along the David Thompson Highway be named after Mr. Ernest Ross, an early pioneer who contributed greatly to the completion of the highway. Ross was born in Ontario in 1883 and came to Alberta in 1903. He began trail-blazing when he made his first trip west to Saunders in 1928. The Rocky Mountain House Chamber of Commerce awarded Ernest Ross the honour of "Citizen of the Year" in 1963, the year he died.

Ernst Creek (creek)

82 G/16 - Maycroft
26-10-3-W5
49°51′N 114°18′W
Flows north into the Oldman River, approximately 30 km north-east of Blairmore.

This creek is named after Ernest Ernst, a prospector in the area who located coal below The Gap.

Errington Hill (hill)

83 F/5 - Entrance
9-50-27-W5
53°18′N 117°53′W
Approximately 21 km south-west of Hinton.

This hill may possibly be named after Joe Errington, a well-known pioneer of the area. He was the controller of the Blue Diamond Coal Company in 1912. No other information is available.

Erris, Mount (mountain)

82 G/15 - Tornado Mountain
7-11-5-W5
49°54′N 114°41′W
Approximately 30 km north-west of Coleman, on the Alberta-B.C. boundary.

Officially named by M.P. Bridgland (see Mount Bridgland) in 1915, this 2820 m mountain is named after Erris, a prominent headland on the west coast of Ireland in County Mayo.

Error Lake (lake)

83 C/15 - Cardinal River
26-45-18-W5
52°54′N 116°31′W
Approximately 105 km east of Jasper.

The origin of the name is unknown.

Escarpment River (river)

82 N/16 - Siffleur River
21-34-17-W5
51°56′N 116°21′W
Flows north-west into Siffleur River, approximately 100 km north-west of Banff.

The name is likely descriptive of the river's steep banks.

Esplanade Mountain (mountain)

83 E/1 - Snaring
25-47-2-W6
53°05′N 118°10′W

Approximately 23 km north north-west of Jasper.

The descriptive name for this flat-topped ridge was applied to the mountain in 1916 by M.P. Bridgland. (see Mount Bridgland)

Estella, Mount (mountain)

83 D/16 - Jasper
3-44-2-W6
52°46′N 118°12′W
Approximately 15 km south-west of Jasper.

The name for this 3069 m mountain was intended to be, but is not, the Spanish word for "rock." It was officially named Mount Estella, nonetheless, in 1916.

Etherington Creek (creek)

82 J/7 - Mount Head
29-16-5-W5
50°23′N 114°39′W
Flows north into Highwood River, approximately 41 km south-west of Turner Valley.

The name for this creek was officially approved 8 December 1943. (see Mount Etherington)

Etherington, Mount (mountain)

82 J/7 - Mount Head
15-6-W5
50°16′N 114°45′W
Approximately 55 km south-west of Turner Valley.

This 2877 m mountain was officially named in 1918 in honour of Colonel Frederick Etherington, C.M.G., who served during the First World War with the Canadian Army Medical Corps.

Eunice Creek (creek)

83 F/3 - Cadomin
23-48-23-W5
53°09′N 117°14′W

Flows north-west into McLeod River, approximately 72 km south-west of Edson.

This creek was named by a member of the Dominion Land Survey's party in 1926, and the name became incorporated on Survey maps. The origin of the name is unknown.

Evan's Creek (creek)

83 E/9 - Moberly Creek
31-53-2-W6
53°38′N 118°16′W
Flows north-west into Little Berland River, approximately 85 km north of Jasper.

Records indicate that a nearby trail of the same name was named after Mr. Evan Moberly in 1946 at the same time as this creek's naming. There is no information available about an Evan Moberly, but there was, however, a Ewan Moberly (1860-1918/ flu epidemic) who was quite well-known in the area. Ewan and his brother John were the offspring of Henry John Moberly and Suzanne Cardinal (see Moberly Flats). Ewan married Madeleine Finlay; they had several children, including Fresine Moberly, who became the first wife of Adam Joachim (see Joachim Creek). John married Marie Joachim.

Evan-Thomas Creek (creek)

82 J/14 - Spray Lakes Reservoir
26-22-9-W5
50°54′N 115°09′W
Flows north into Kananaskis River, approximately 70 km west of Calgary.

(see Mount Evan-Thomas)

Evan-Thomas, Mount (mountain)

82 J/14 - Spray Lakes Reservoir
7-21-8-W5
50°46′N 115°04′W
Approximately 70 km west south-west of Calgary.

This 3097 m mountain was named in 1922 after Rear Admiral H. Evan-Thomas

(1862-1928), who fought in the Battle of Jutland. (see Jutland Mountain)

Evans, Mount (mountain)

83 D/8 - Athabasca Pass
18-40-1-W6
52°26′N 118°07′W
Approximately 49 km south of Jasper.

The name for this 3210 m mountain was suggested by G.E. Howard in 1914, after Captain E.R.G. Evans, R.N., second in command of the British Antarctic Expedition, 1910-1912 and commander of it after the death of Captain Scott (see Mount Scott) in 1912.

Evelyn Creek (creek)

83 C/13 - Medicine Lake
52°46′10″N 117°41′00″W
Flows north-east into Maligne River, approximately 30 km east south-east of Jasper.

This creek was so named because it flows north from the locally named "Evelyn Pass" and Evelyn Lakes, both named after Evelyn, Duchess of Devonshire, who visited Jasper in 1920. The name was officially approved 5 May 1987.

Evelyn Lakes (lakes)

83 C/12 - Athabasca Falls
52°39′50″N 117°44′20″W
Approximately 34 km south-east of Jasper.

(see Evelyn Creek)

Excelsior Creek (creek)

83 C/13 - Medicine Lake
15-45-27-W5
52°53′N 117°50′W
Flows north into Maligne River, approximately 13 km east of Jasper.

(see Excelsior Mountain)

Excelsior Mountain (mountain)

83 C/13 - Medicine Lake
5-45-27-W5
52°51'N 117°53'W
Approximately 13 km east south-east of
Jasper.

This 2744 m mountain was named in 1916
by W.P. Hinton (see Hinton). The name is
descriptive of the feature's height.

Exshaw (hamlet)

82 O/3 - Canmore
22-24-9-W5
51°03'N 115°09'W
Approximately 15 km east south-east of
Canmore.

This community was named by Sir Sandford
Fleming, one-time C.P.R. surveyor (see
Introduction) and director of the cement
company, Western Canada Cement and Coal
Company. It is named after his son-in-law,
Lord Exshaw, who helped form the
company.

Exshaw Creek (creek)

82 O/3 - Canmore
23-24-9-W5
51°03'N 115°10'W
Flows south-east into Bow River,
approximately 15 km east of Canmore.

(see Exshaw)

Exshaw Mountain (mountain)

82 O/3 - Canmore
26-24-9-W5
51°04'N 115°09'W
Approximately 16 km south-east of
Canmore.

The name of this mountain, which is 1783
m in altitude, was officially approved by the
Canadian Permanent Committee on
Geographical Names 19 July 1965. (see
Exshaw)

Fairfax Lake (lake)

83 C/15 - Cardinal River
17-46-18-W5
52°58′N 116°34′W
Approximately 101 km east north-east of
Jasper.

This lake was originally named "Lohoar
Lake" after Jimmy Lohoar, who was
drowned in 1916 while attempting to save
the lives of two boys. He was in a boat with
them when it capsized. He told them to stay
with the boat while he attempted to swim to
shore, but he drowned on the way. The boys
were rescued. The name for the lake was
changed, but the origin of the name Fairfax
is unknown.

Fairholme Range (range)

82 O/3 - Canmore
28-25-10-W5
51°10′N 115°20′W
Approximately 5 km north-east of Canmore.

This range was named in 1859 by Captain
Palliser (see Palliser Range), after his sister,
Grace, who married William Fairholme, of
Greenknowe, Berwickshire, Scotland, 15
June 1853. (see Orient Point)

Fairview Mountain (mountain)

82 N/8 - Lake Louise
17-28-16-W5
51°24′N 116°13′W
Approximately 55 km north-west of Banff.

Walter D. Wilcox (see Wilcox Pass) named
Fairview Mountain in 1894, to describe the
magnificent view from its 2744 m summit.

Fall Creek (creek)

82 O/3 - Canmore
15-24-10-W5
51°02′30″N 115°18′15″W
Flows east into Stewart Creek,
approximately 6 km south-east of Canmore.

The origin of the name is unknown.

Fall Creek (creek)

83 B/5 - Saunders
18-38-11-W5
52°16′N 115°32′W
Flows north into Ram River, approximately
45 km south-west of Rocky Mountain House.

The name is well-known locally because of
the waterfalls found along the creek.

Falls Creek (creek)

83 L/4 - Kakwa Falls
NW 15-2-59-12-W6
From 54°01′30″N 119°38′30″W
to 54°04′45″N 119°41′00″W
Flows north-west into South Kakwa River,
approximately 42 km west north-west of
Grande Cache.

This creek is likely named for its proximity
to Kakwa Falls. (see Kakwa Falls)

Falls, Lake of (lake)

83 C/2 - Cline River
10-36-19-W5
52°05′N 116°37′W
Approximately 120 km west south-west of
Rocky Mountain House.

The name for this lake is appropriately
descriptive.

Famm Creek (creek)

83 E/13 - Dry Canyon
56-11-W6
53°51′N 119°35′W
Flows south-east into Sheep Creek,
approximately 30 km west of Grande Cache.

The name was officially approved 3 December
1958. The origin of the name is unknown.

Farbus Mountain (mountain)

82 N/14 - Rostrum Peak
36-34-23-W5
51°58′N 117°08′W

Approximately 140 km north-west of Banff,
on the Alberta-B.C. boundary.

This mountain, which is 3150 m in altitude,
was named in 1920. The name commemorates
the village of Farbus, located on the eastern
slope of Vimy Ridge in France, where Canadian
troops fought during the First World War.

Farquhar, Mount (mountain)

82 J/2 - Fording River
14-6-W5
50°12′N 114°44′W
Approximately 60 km south-west of Turner
Valley.

This 2905 m mountain was officially named
in 1917 after Lieutenant-Colonel F.D.
Farquhar, the Commander of the Princess
Patricia's Canadian Light Infantry. He was
killed during the First World War.

Fatigue Creek (creek)

82 O/4 - Banff
31-24-12-W5
51°05′N 115°39′W
Flows north into Brewster Creek,
approximately 11 km south-west of Banff.

(see Fatigue Mountain)

Fatigue Lakes (lakes)

82 O/4 - Banff
51°02′15″N 115°41′40″W
Approximately 17 km south south-west of
Banff.

Fatigue Lakes were so named because of
their proximity to Fatigue Mountain (see
Fatigue Mountain). These lakes are not part
of the Fatigue Creek drainage system.

Fatigue Mountain (mountain)

82 O/4 - Banff
12-24-13-W5
51°02′N 115°42′W
Approximately 18 km south south-west of
Banff.

This mountain, which is 2959 m in altitude,
was named in 1888. The name refers to the
physical experience of the explorer W.S.
Drewry, D.L.S. (see Tallon Peak) when
climbing it.

Fatigue Pass (pass)

82 O/4 - Banff
24-13-W5
51°01′N 115°41′W
Approximately 20 km south south-west of
Banff.

(see Fatigue Mountain)

Faulk Creek (creek)

83 E/14 - Grande Cache
33-55-10-W6
53°47′N 119°25′W
Flows south-east into Muddywater River,
approximately 22 km west south-west of
Grande Cache.

The origin of the name is unknown.

Fault Creek (creek)

83 E/9 - Moberly Creek
24-52-4-W6
53°30′N 118°26′W
Flows south into Wildhay River,
approximately 75 km north north-west of
Jasper.

The origin for the locally well-established
name of this creek, which was approved in
1946, is unknown.

Fay Glacier (glacier)

82 N/8 - Lake Louise
51°18′05″N 116°10′00″W

Approximately 45 km west north-west of
Banff.

The name was officially approved 21 January
1985. (see Mount Fay)

Fay, Mount (mountain)

82 N/8 - Lake Louise
10-27-16-W5
51°18′N 116°09′W
Approximately 45 km west north-west of
Banff, on the Alberta-B.C. boundary.

This 3235 m mountain was officially named
1 March 1904 after Professor Charles E. Fay
(1846-1930). Professor Fay was founder and
first president of both the Appalachian
Mountain Club (1878, 1881, 1893, 1905) and
the American Alpine Club (1902-1908,
1917-1919). An enthusiastic mountaineer, he
first came to the Canadian Rockies in 1890.

Felix Creek (creek)

83 F/11 - Dalehurst
4-16-55-23-W5
53°44′45″N 117°21′45″W
Flows north into Oldman Creek,
approximately 61 km west north-west of
Edson.

This creek was officially named in 1946 after
Felix Paul Plante (1893-?), a homesteader and
trapper of the area. He acted as a guide at
Jasper for fifteen years with Fred Brewster.
By 1927, Felix opened his own business as a
guide and outfitter and operated for another
twenty-five years. Felix married Caroline
Moberly (?-1979) circa 1922. (see Joachim
Creek)

Felton Creek (creek)

83 F/6 - Pedley
2-50-23-W5
53°17′N 117°17′W
Flows north-west into McLeod River,
approximately 64 km south-west of Edson.

The origin of the locally well-established
name of this creek, which was officially
approved 21 December 1944, is unknown.

Festubert Mountain (mountain)

82 G/1 - Sage Creek
31-1-1-W5
49°05′N 114°07′W
Approximately 18 km west north-west of
Waterton Park.

The 2522 m mountain is named after the
village east of Bassée, France, where
Canadian troops fought in 1915.

Fetherstonhaugh Creek (creek)

83 E/13 - Dry Canyon
34-55-12-W6
53°48′N 119°41′W
Flows north-east into Muddywater River,
approximately 38 km west south-west of
Grande Cache.

(see Mount Fetherstonhaugh)

Fetherstonhaugh, Mount (mountain)

83 E/12 - Pauline Creek
18-55-13-W6
53°44′N 119°54′W
Approximately 50 km west south-west of
Grande Cache.

This 2088 m mountain was named in 1925
after W.S. Fetherstonhaugh, who was the
Division Engineer for the Canadian
National Railway out of Calgary. He
worked in this area in 1905 and 1906.

Fetherstonhaugh Pass (pass)

83 E/12 - Pauline Creek
4-55-13-W6
53°43′N 119°51′W
Approximately 52 km west south-west of
Grande Cache.

(see Mount Fetherstonhaugh)

Fickle Lake (lake)

83 F/7 - Erith
31-51-19-W5
53°27′N 116°46′W
Approximately 28 km south-west of Edson.

This lake was named after the original white settler who lived on the lake.

Fiddle Pass (pass)

83 F/4 - Miette
6-47-24-W5
53°02′N 117°30′W
Approximately 44 km south of Hinton.

The name for this pass was officially approved in 1960. (see Fiddle Range)

Fiddle Peak (peak)

83 F/4 - Miette
33-48-26-W5
53°11′N 117°45′W
Approximately 30 km south south-west of Hinton.

The name was officially approved in 1960. (see Fiddle Range)

Fiddle Range (range)

83 F/4 - Miette
49-26-W5
53°12′N 117°45′W
Approximately 29 km south south-west of Hinton.

The two origins of the name of this range and the river are as follows: the first suggests that when the wind strikes the mountains in this range from a certain direction, it sounds like a #4 fiddle string. Another version describes the shape of the river as resembling a violin. The name dates back to 1846, when Father de Smet (see Roche de Smet) refers to the river as "Violin." In 1865 Palliser (see Palliser Range) called the range Fiddle Range. The name was officially approved in March 1917.

Fiddle River (river)

83 F/4 - Miette
15-49-27-W5
53°13′N 117°51′W
Flows north into Brule Lake, approximately 29 km south-west of Hinton.

(see Fiddle Range)

Fifi, Mount (mountain)

82 O/4 - Banff
26-13-W5
51°13′N 115°41′W
Approximately 10 km north-west of Banff.

The origin of the name is unknown for this 2621 m mountain.

Findley Creek (creek)

83 E/15 - Pierre Greys Lakes
14-57-6-W6
53°56′N 118°46′W
Flows south-east into Muskeg River, approximately 24 km east of Grande Cache.

This creek was named in 1947 after Deome Findley, who had a trap line on the stream.

Fir Creek (creek)

82 J/7 - Mount Head
16-5-W5
50°24′N 114°34′W
Approximately 34 km south-west of Turner Valley.

This locally well-established name was officially approved 8 December 1943. It is likely descriptive of the area through which the creek flows.

First Range (range)

83 C/8 - Nordegg
22-38-18-W5
52°18′N 116°28′W
Approximately 32 km south-west of Nordegg.

This is the first mountain range encountered when travelling west from Nordegg. Its name is therefore appropriately descriptive.

Fish Butte (butte)

82 J/15 - Bragg Creek
5-23-4-W5
50°55′N 114°32′W
Approximately 25 km west of Calgary.

The name was officially approved 6 October 1949. The origin of the name is unknown.

Fish Creek (creek)

83 F/5 - Entrance
25-51-25-W5
53°26′N 117°32′W
Flows south-east into Athabasca River, approximately 1 km north of Hinton.

The locally well-established name of this creek was approved in 1944, but its origin remains unknown.

Fish Lake (lake)

82 G/8 - Beaver Mines
26-4-1-W5
49°20′N 114°00′W
Approximately 17 km south-west of Pincher Creek.

This lake has been known as Fish Lake since 29 August, 1884, when it abounded in trout weighing from 8 to 10 pounds.

Fish Lakes (lakes)

82 N/9 - Hector Lake
2-31-16-W5
51°39′N 116°11′W
Approximately 70 km north-west of Banff.

No information regarding the origin of the name is known.

Fisher Creek (creek)

82 J/16 - Priddis
5-21-3-W5
50°45′N 114°23′W
Flows east into Threepoint Creek, approximately 29 km south-west of Calgary.

This creek was named after Joseph Fisher, a rancher who came to the Millarville area around 1883. He was one of the first settlers and did considerable work in irrigation in the vicinity.

Fisher Peak (peak)

82 J/14 - Spray Lakes Reservoir
27-21-8-W5
50°49′N 115°02′W
Approximately 65 km west south-west of Calgary.

This mountain is 3053 m in altitude. (see Fisher Range)

Fisher Range (range)

82 J/14 - Spray Lakes Reservoir
22-8-W5
50°53′N 115°02′W
Approximately 65 km west south-west of Calgary.

This range was named by Captain Palliser (see Palliser Range), probably after the family name of someone who accompanied him on a hunting excursion near New Orleans in 1847.

Fitzsimmons Creek (creek)

82 J/7 - Mount Head
31-16-5-W5
50°23′N 114°41′W
Flows north-east into Strawberry Creek, approximately 44 km south-west of Turner Valley.

The name for this creek was officially approved 8 December 1943. Charlie Fitzsimmons was the camp boss for Lineham's Lumber Company (see Lineham Creek and Mount Lineham) and it is likely after him that the creek was named. He came to Alberta from Manitoba in 1905-1906, already an expert logger. This creek eventually runs into the Highwood River.

Flapjack Creek (creek)

83 C/15 - Cardinal River
35-44-21-W5
52°50′N 116°57′W
Flows north-east into Ruby Creek, approximately 77 km east of Jasper.

The origin of the name is unknown.

Flapjack Lake (lake)

83 C/15 - Cardinal River
27-44-21-W5
52°49′N 116°57′W
Approximately 75 km east of Jasper.

No origin information is available.

Flare Creek (creek)

83 B/3 - Tay River
8-37-9-W5
52°10′10″N 115°14′55″W
Approximately 30 km south-west of Rocky Mountain House.

The name of this creek, which was officially adopted in 1983, is descriptive of the creek's proximity to the flare stack of the Ram River Gas Plant.

Flat Creek (creek)

82 J/7 - Mount Head
36-17-4-W5
50°28′45″N 114°25′05″W
Flows east into Highwood River, approximately 23 km south-west of Turner Valley.

This creek was erroneously named "Trap Creek" 12 December 1939. The name "Trap Creek" has been shown on maps in the area since at least 1914. The residents of the Highwood River area enthusiastically maintained that the local name was Flat Creek, and therefore the name Flat Creek was officially reinstated for this feature 9 July 1980. The topography through which this creek flows is relatively flat, especially

near the mouth, and the name is likely descriptive of this.

Flat Pond (pond)

82 J/14 - Spray Lakes Reservoir
36-22-9-W5
50°54′N 115°07′W
Approximately 70 km west of Calgary.

The origin of this well-established local name is not known, but it may be descriptive of the area surrounding the pond. The name was officially approved in October 1977.

Flat Ridge (ridge)

83 C/6 - Sunwapta
16-39-22-W5
52°21′N 117°07′W
Approximately 87 km south-east of Jasper.

The name for this ridge was officially approved 5 March 1935. It is likely descriptive.

Flat Top Creek (creek)

83 L/5 - Two Lakes
22-62-11-W6
54°22′N 119°34′W
Flows north-west into Nose Creek, approximately 60 km north north-west of Grande Cache.

This creek has its source in a flat-topped mountain. The descriptive name was proposed by A.M. Floyd, a field topographer, circa 1946.

■ Flathead Range (range)

82 G/10 - Crowsnest
34-6-5-W5
49°31′N 114°33′W
Approximately 10 km south-west of Coleman.

A portion of this range makes up the Alberta-B.C. boundary. The name Flathead

Range, approved 3 October 1904, is taken from the Flathead Indians of Western Montana.

Flints Park (valley)

82 O/5 - Castle Mountain
22-28-13-W5
51°25′N 115°44′W
Approximately 29 km north north-west of Banff.

The origin of the name is unknown.

Flints Peak (peak)

82 O/5 - Castle Mountain
28-13-W5
51°27′N 115°45′W
Approximately 32 km north north-west of Banff.

No information regarding the origin of the name of this 2950 m mountain peak is known.

Flood Creek (creek)

83 E/14 - Grande Cache
7-58-7-W6
54°00′N 119°02′W
Flows north-east into Muskeg River, approximately 14 km north north-east of Grande Cache.

A possible cloudburst in the surrounding mountains may have caused a sudden flooding of this stream at the time of its naming.

Flood Mountain (mountain)

83 E/14 - Grande Cache
31-57-7-W6
53°58′N 119°02′W
Approximately 11 km north north-east of Grande Cache.

This mountain stands 1814 m in altitude. (see Flood Creek)

Flume Creek (creek)

83 L/5 - Two Lakes
18-62-13-W6
54°22′N 119°57′W
Flows south-east into Narraway River, approximately 63 km north-west of Grande Cache.

The descriptive name for this creek was proposed 18 October 1922 by R.W. Cautley (a Commissioner of the Provincial Boundary Survey, 1918-1924), due to the steep gradient and outbanks on both sides of the stream. A flume is defined as an "artificial channel conveying water etc. for industrial use; ravine with stream" (*Concise Oxford Dictionary* 1982). The name was suggested as an alternative to "canyon" or "cutbank."

Fly Creek (creek)

82 G/16 - Maycroft
32-10-4-W5
49°52′N 114°23′W
Approximately 30 km north of Blairmore.

This creek is named for its proximity to Fly Hill from whose eastern watershed it flows. (see Fly Hill)

Fly Hill (hill)

82 G/16 - Maycroft
2-11-4-W5
49°52′N 114°27′W
Approximately 30 km north of Blairmore.

Fly Hill is a well-established local name for this feature, which measures 2000 m in altitude. The origin of the name is unknown.

Foch Creek (creek)

82 J/11 - Kananaskis Lakes
21-19-9-W5
50°36′N 115°11′W
Flows north into Hidden Lake, approximately 65 km west south-west of Turner Valley.

(see Mount Foch)

Foch, Mount (mountain)

82 J/11 - Kananaskis Lakes
2-19-9-W5
50°34′N 115°09′W
Approximately 65 km west south-west of Turner Valley.

This 3180 m mountain was officially named in 1918. It commemorates Marshal Ferdinand Foch (1851-1929), Generalissimo of the Allied Forces 1918 and hero of the Battle of Marne. Marshal Foch received many honours for his role in the First World War.

Folding Mountain (mountain)

83 F/4 - Miette
11-49-26-W5
53°12′N 117°42′W
Approximately 25 km south south-west of Hinton.

The descriptive name for this 2844 m mountain was applied by James McEvoy in 1898. It was officially approved 3 October 1911 and describes the "folding" due to severe compressions of the rock strata composing the mountain.

Folding Mountain Creek (creek)

83 F/4 - Miette
24-48-26-W5
53°09′N 117°40′W
Flows south into Drinnan Creek, approximately 31 km south of Hinton.

Officially approved in 1960, the name for this creek is taken from the nearby mountain. (see Folding Mountain)

Font Creek (creek)

82 G/1 - Sage Creek
15-3-2-W5
49°13′N 114°12′W
Approximately 25 km north-west of Waterton Park.

(see Font Mountain)

Font Mountain (mountain)

82 G/1 - Sage Creek
4-3-2-W5
49°10′N 114°13′W
Approximately 25 km north-west of Waterton Park, on the Alberta-B.C. boundary.

A font is defined, among other things, as "a basin holding water for baptism, a basin for holy water." There is a font-like formation immediately west of this 2353 m feature, and it may be for this likeness that the mountain is named.

Foothills (locality)

83 F/2 - Foothills
24-47-20-W5
53°04′N 116°47′W
Approximately 62 km south south-west of Edson.

Foothills was a coal branch town that opened in 1913. Prior to this, it was known as "Mudge." It stayed alive longer than most of the coal branch towns like it, not closing until 1958. Its name is descriptive in that it is located in the foothills of the mountains.

Forbes Creek (creek)

82 N/15 - Mistaya Lake
8-33-20-W5
51°49′N 116°50′W
Flows west into Howse River, approximately 110 km north-west of Banff.

(see Mount Forbes)

Forbes, Mount (mountain)

82 N/15 - Mistaya Lake
29-33-21-W5
51°52′N 116°55′W
Approximately 120 km north-west of Banff.

This 3612 m mountain was named by Sir James Hector (see Mount Hector) after Professor Edward Forbes (1815-1854), a British naturalist who made an important

analysis of the distribution of plant and animal life in the British Isles. The name was officially approved 6 February 1912.

Forbidden Creek (creek)

82 O/13 - Scalp Creek
1-34-13-W5
51°33′N 115°43′W
Flows north into Clearwater River, approximately 80 km north north-west of Banff.

The origin of the name is unknown.

Forbidden Lake (lake)

82 O/13 - Scalp Creek
5-33-13-W5
52°48′N 115°48′W
Approximately 80 km north north-west of Banff.

This lake is located at the headwaters of Forbidden Creek. Its name was officially approved 5 May 1987.

Ford Creek (creek)

82 J/15 - Bragg Creek
24-21-7-W5
50°48′N 114°51′W
Flows south-east into Elbow River, approximately 50 km west south-west of Calgary.

Officially approved 6 October 1949, the name for this creek may be taken from Henry Ford (1866-1933), an early rancher of the Priddis district, originally from Portland, Oregon.

Fording River Pass (pass)

82 J/7 - Mount Head
16-6-W5
50°19′N 114°47′W
Approximately 52 km south-west of Turner Valley.

This pass is a narrow crossing of the continental watershed at an altitude of 2263 m. It was originally named Fording River Pass by George Dawson (see Tombstone

Mountain) of the Geological Survey of Canada in 1884 because he and his party forded it several times and also because it lies at the head of the eastern source of the Fording River. The original name was officially restored 15 January 1979, replacing the name "Fording Pass."

Forget, Mount (mountain)

83 E/12 - Pauline Creek
17-54-12-W6
53°40′N 119°44′W
Approximately 47 km south-west of Grande Cache.

This 2121 m mountain was named in 1925 after the Honourable Amedée Emmanuel Forget (1847-1923), who came to Banff from Quebec in 1910. He was Assistant Commissioner for Indian Affairs for Manitoba and the North West Territories in 1888. He became Lieutenant Governor of the North West Territories in 1898-1905. Forget was summoned to the Canadian Senate 2 May 1911.

Forgetmenot Mountain (mountain)

82 J/15 - Bragg Creek
4-21-6-W6
50°46′N 114°48′W
Approximately 50 km south-west of Calgary.

This 2332 m mountain rises south-east of the Powder Face Ridge in the vicinity of Kananaskis Country. There is an abundance of Alpine Forget-me-not flowers found in the vicinity of the peak. This showy perennial blossom is bright blue with a yellow centre and measures about 3/16-inch across. It grows in dense clusters and is easily recognizable, as it resembles very closely the garden variety of Forget-me-not. The name was officially approved 6 October 1949.

Forgetmenot Pass (pass)

83 E/13 - Dry Canyon
16-55-13-W6
53°45′N 119°53′W

Approximately 52 km west south-west of Grande Cache.

(see Forgetmenot Mountain)

Forgetmenot Pond (pond)

82 J/15 - Bragg Creek
24-21-7-W5
50°48′10″N 114°50′30″W
Approximately 50 km west south-west of Calgary.

This very small man-made body of water was named for its proximity to Forgetmenot Ridge. The name was officially approved 12 September 1984.

Forgetmenot Ridge (ridge)

82 J/15 - Bragg Creek
21-6-W5
50°47′N 114°48′W
Approximately 50 km south-west of Calgary.

(see Forgetmenot Mountain)

Fortalice Mountain (mountain)

83 D/16 - Jasper
15-44-2-W6
52°47′N 118°13′W
Approximately 13 km south-west of Jasper.

This outlying peak, which is 2835 m in altitude, has a descriptive name, since it resembles a small fort. It was named by M.P. Bridgland (see Mount Bridgland) in 1916.

Fortress, The (mountain)

82 J/14 - Spray Lakes Reservoir
31-21-9-W5
50°50′N 115°14′W
Approximately 83 km west south-west of Calgary.

This 3000 m mountain was once known as "The Tower," but because there is a Tower Mountain at 50°53′N 115°36′W whose name dates back to at least 1885, the name that

was officially applied to this feature was "The Fortress," to avoid duplication. It is an appropriately descriptive name.

Fortress Lake (lake)

82 J/14 - Spray Lakes Reservoir
2-6-22-9-W5
50°51′N 115°15′W
Approximately 83 km west south-west of Calgary.

The name for this lake was officially approved 5 May 1987. (see The Fortress)

Fortress Mountain (mountain)

83 C/5 - Fortress Lake
29-39-26-W5
52°24′N 117°43′W
Approximately 60 km south-east of Jasper, on the Alberta-B.C. boundary.

This 3020 m mountain has a fancied resemblance to a fortress. A.P. Coleman (see Mount Coleman) gave the feature its descriptive name in 1892. The nearby pass takes its name from the mountain.

Fortress Pass (pass)

83 C/5 - Fortress Lake
16-39-26-W5
52°22′N 117°42′W
Approximately 65 km south-east of Jasper, on the Alberta-B.C. boundary.

(see Fortress Mountain)

Fortune, Mount (mountain)

82 J/14 - Spray Lakes Reservoir
26-22-11-W5
50°53′N 115°25′W
Approximately 80 km west south-west of Calgary.

Officially named in 1922, this mountain, which is 2819 m in altitude, commemorates a destroyer which was sunk at the Battle of Jutland. (see Jutland Mountain)

Forty-Five Mile Creek (creek)

83 B/3 - Tay River
16-37-10-W5
52°11′00″N 115°22′25″W
Approximately 38 km south-west of Rocky Mountain House.

This creek, officially named 19 February 1979 at the request of several residents of the Rocky Mountain House area, is descriptive. The creek is the 45 mile point on the Dominion Telegraph Lines pack trail from Shunda Ranger Station to Swan Lake, which was in use in the 1930s.

Forty Mile Creek (creek)

82 O/4 - Banff
35-25-12-W5
51°11′N 115°35′W
Flows south-east into Echo Creek, approximately 2 km north north-west of Banff.

Officially approved in 1958, the precise origin for the name of this creek is unknown.

Forty Mile Summit (pass)

82 O/5 - Castle Mountain
51°20′25″N 116°44′05″W
Approximately 25 km north of Banff, at the head of Forty Mile Creek.

This pass was named for its location at the head of Forty Mile Creek. (see Forty Mile Creek)

Forty-One Mile Creek (creek)

83 E/9 - Moberly Creek
19-52-3-W6
53°30′N 118°25′W
Flows south into Wildhay River, approximately 73 km north north-west of Jasper.

Officially approved in 1946, the name is likely descriptive of the feature's length.

Forum, The (valley)

83 D/16 - Jasper
23-44-3-W6
52°48′N 118°19′W
Approximately 18 km west south-west of
Jasper.

The Forum of Ancient Rome was the centre
of the city. This valley is surrounded by
Basilica Mountain, Curia Mountain and
Rostrum Hill, all of which carry the names
of the principal buildings of the Roman
Empire's social system. (see Curia Mountain)

Forum Peak (peak)

82 G/1 - Sage Creek
4-1-1-W5
49°00′N 114°04′W
Approximately 13 km south-west of Waterton
Park, on the Alberta-B.C. boundary.

The name for this 2415 m peak is taken
from a small lake in B.C. of the same name.
The mountain is hollowed out in a series of
very striking circular amphitheatres, whose
curving walls embrace a number of lakes of
various sizes. The Boundary Commission
bestowed the name Forum Lake on one of
these lakes located just west of the watershed.

Fossil Falls (falls)

82 J/11 - Kananaskis Lakes
9-19-9-W5
50°35′N 115°12′W
Approximately 65 km west south-west of
Turner Valley.

The narrow valley in which the creek and
these falls may be found is rich in fossils.
The descriptive name was suggested prior to
1905 by H.B. Bryant and Walter D. Wilcox.
(see Wilcox Pass)

Fossil Mountain (mountain)

82 N/9 - Hector Lake
21-29-15-W5
51°30′N 116°03′W

Approximately 50 km north-west of Banff.

Numerous fossils may be found in the limestone
on the slopes of this 2946 m mountain. Its
name was officially approved in 1908.

Four Point Creek (creek)

83 C/6 - Sunwapta
22-38-22-W5
52°17′N 117°04′W
Flows south-east into Brazeau Creek,
approximately 90 km south south-east of
Jasper.

This creek was officially named by the
Canadian Permanent Committee on
Geographical Names 18 October 1976. The
name Four Point is appropriate since the
junction of four trails near its confluence
with the Brazeau River has historically been
known as such.

Fox Creek (creek)

82 J/11 - Kananaskis Lakes
18-19-8-W5
50°37′N 115°06′W
Flows north into Boulton Creek,
approximately 60 km west south-west of
Turner Valley.

(see Mount Fox)

Fox Creek (creek)

83 E/9 - Moberly Creek
4-55-2-W6
53°43′N 118°14′W
Flows north-east into Little Berland River,
approximately 94 km north of Jasper.

The locally well-established name for this
creek was made official in 1946. The precise
origin of the name is unknown.

Fox, Mount (mountain)

82 J/11 - Kananaskis Lakes
36-18-9-W5
50°34′N 115°07′W

Approximately 60 km west south-west
of Turner Valley, on the Alberta-B.C.
boundary.

This 2973 m mountain was named by
Captain Palliser (see Palliser Range) in 1859.
It is probably named after Lieutenant-
General C.R. Fox, who was on the Council
of the Royal Geographic Society in 1900.

Franchère Peak (peak)

83 D/9 - Amethyst Lakes
20-43-1-W6
52°43′N 118°06′W
Approximately 18 km south of Jasper.

This 2812 m mountain was named in 1917
after Gabriel Franchère, the author of
*Relation d'un voyage à la Coté du Nord-Ouest
de l'Amérique Septentrionale* (1820). This was
the first published description of a journey
through the Rockies by way of the
Athabasca River. (see Mount Lapensée and
Mount Belanger)

Francis Peak (peak)

83 L/4 - Kakwa Falls
58-13-W6
54°03′20″N 119°56′10″W
Approximately 63 km west north-west of
Grande Cache.

This 2406 m mountain peak was named 6
February 1926 after Private Francis Loren
May. Private May was killed in action at the
Third Battle of Ypres, 3 June 1916. Previous
to the War, Private May was employed as a
member of a Topographical Survey Party,
under the direction of a Mr. George
McMillan, D.L.S. The mountain on which
the two peaks, George and Francis are
situated, commemorates the two brothers,
both killed in the First World War. The
names were suggested by their uncle, V.
Ernest May, Chief Map Draughtsman,
Topographical Survey, 20 May 1925.

Francis Peak Creek (creek)

83 L/4 - Kakwa Falls
12-59-13-W6
54°05'N 119°49'W
Fows north-east into Kakwa River,
approximately 42 km north-west of Grande
Cache.

The name was approved in 1959. (see Francis
Peak)

Francis Peak Creek Falls (falls)

83 L/4 - Kakwa Falls
2-9-59-13-W6
54°04'40"N 119°52'55"W
Approximately 141 km south south-west of
Grande Prairie.

Francis Peak Creek Falls is a small waterfall,
or chute located on Francis Peak Creek. (see
Francis Peak)

■ **Frank** (village)

82 G/9 - Blairmore
31-7-3-W5
49°36'N 114°24'W
Located on Highway 3, approximately 2 km
south-east of Blairmore.

The village was incorporated 3 September 1901.
Mr. S.W. Gebo came to Alberta in 1900
looking for coal property and found some
choice land around this site. In 1901, he sent
for a Mr. A.L. Frank in Butte, Montana, and
together they formed the Canadian American
Company, which started immediately to
develop the coal deposits. The grand opening
of the town occurred 10 September 1901. By
1903, the town of Frank was well built up. A
tremendous rock slide crashed down from
Turtle Mountain 29 April 1903. This slide
destroyed part of the town and took seventy
lives. (see the Municipality of Crowsnest Pass)

Frank Lake (lake)

82 G/9 - Blairmore
30-7-3-W5
49°35'20"N 114°23'50"W

Approximately 3 km south of Frank.

This lake was formed by the Frank Slide (the
Crowsnest River was diverted). Local legend
has it that the block of stone from Turtle
Mountain came down intact and gouged out
the lake bed on impact. The name was
officially applied 21 July 1982.

Frank Slide (rock slide)

82 G/9 - Blairmore
31-7-3-W5
49°35'00"N 114°24'00"W
Immediately east of the village of Frank.

Approximately 90 million tons of limestone
rock fell from Turtle Mountain in about 100
seconds 29 April 1903. Frank Slide is one of
the largest rock slides that have occurred in
the past few thousand years. (see Frank; also
see photograph of Turtle Mountain)

Franks Creek (creek)

83 E/14 - Grande Cache
30-55-7-W6
53°46'N 119°02'W
Flows north-east into Sulphur River,
approximately 14 km south south-east of
Grande Cache.

The name was officially approved 5 March
1953. The origin of the name is unknown.

Fraser Glacier (glacier)

83 D/9 - Amethyst Lakes
30-42-2-W6
52°39'N 118°17'W
Approximately 30 km south-west of Jasper.

(see Mount Fraser)

Fraser, Mount (mountain)

83 D/9 - Amethyst Lakes
36-42-3-W6
52°39'N 118°18'W
Approximately 29 km south-west of Jasper.

Simon Fraser (1776-1862) was a fur trader
and explorer who discovered the Fraser
River in British Columbia. Fraser was
apprenticed as a clerk with the North West
Company in 1792 and was made a partner in
1801. The river which is named after him
was mistakenly thought by him to be the
Columbia River. He retired from his charge
of the Athabasca Department in 1818. This
3269 m mountain is comprised of
Bennington, McDonell and Simon Peaks.
The name was officially approved in 1917.

Fred Creek (creek)

83 E/9 - Moberly Creek
26-53-1-W6
53°36'N 118°03'W
Flows north-east into Teitge Creek,
approximately 82 km north of Jasper.

This creek was named in 1946 after a forest
ranger of the district. No other information
is available.

Frederick Lake (lake)

82 O/2 - Jumpingpound Creek
35-24-6-W5
51°05'N 114°44'W
Approximately 35 km west north-west of
Calgary.

This lake was officially named 12 December
1939 after a local land owner.

French Creek (creek)

82 J/14 - Spray Lakes Reservoir
14-21-10-W5
50°47'N 115°18'W
Flows east into Smith-Dorrien Creek,
approximately 85 km west south-west of
Calgary.

(see Mount French)

French, Mount (mountain)

82 J/11 - Kananaskis Lakes
27-20-10-W5
50°44'N 115°18'W

Approximately 70 km west north-west of Turner Valley.

John Denton Pinkstone French (1852-1925), who during the first 16 months of the First World War served as field marshall in command of the British Forces on the Western Front, was named first Earl of Ypres in 1922. This 3234 m mountain was named after J.D.P. French by M.P. Bridgland (see Mount Bridgland) in 1915. The name was made official in 1918.

Freshfield Creek (creek)

82 N/15 - Mistaya Lake
8-33-20-W5
51°48'N 116°49'W
Flows west into Howse River, approximately 110 km north-west of Banff.

(see Mount Freshfield)

Freshfield Glacier (glacier)

82 N/15 - Mistaya Lake
33-21-W5
51°47'N 116°53'W
Approximately 115 km north-west of Banff.

(see Mount Freshfield)

Freshfield Icefield (icefield)

82 N/15 - Mistaya Lake
32-21-W5
51°45'N 116°54'W
Approximately 110 km north-west of Banff, on the Alberta-B.C. boundary.

(see Mount Freshfield)

Freshfield, Mount (mountain)

82 N/10 - Blaeberry River
17-32-21-W5
51°44'N 116°57'W
Approximately 115 km north-west of Banff, on the Alberta-B.C. boundary.

This 3336 m mountain was named in 1897

by Stutfield and Collie of the Alpine Club after Douglas William Freshfield (1845-1934). This noted British climber was a member of the Alpine Club as well as Secretary and President of the Royal Geographical Society. He was chairman of the Society of authors from 1908 to 1909. He advocated the recognition of geography as a separate discipline in schools. The name for the mountain was officially approved 3 January 1911.

Fresnoy Mountain (mountain)

82 N/14 - Rostrum Peak
35-23-W5
52°00'N 117°13'W
Approximately 120 km south-east of Jasper, on the Alberta-B.C. boundary.

This 3240 m mountain was named in 1920 after Fresnoy, in the department of Aisne in France. Canadian troops captured Fresnoy 13 April 1917, and it was in commemoration of this First World War event that this name was chosen for the feature.

Frog Rocks (rock)

83 C/12 - Athabasca Falls
52°42'N 117°36'W
Approximately 35 km south-east of Jasper.

The precise origin of the name is unknown.

Fryatt Creek (creek)

83 C/12 - Athabasca Falls
34-41-27-W5
52°35'N 117°49'W
Flows south into Athabasca River, approximately 37 km south south-east of Jasper.

This creek and the nearby mountain were named in 1921 after Captain Charles Algernon Fryatt (1872-1916), a merchant seaman and war hero who met his death during the First World War, when he was shot by the Germans.

Fryatt Lake (lake)

83 C/12 - Athabasca Falls
52°31'20"N 117°51'50"W
Approximately 40 km south south-east of Jasper.

This lake was named for its proximity to Mount Fryatt and Fryatt Creek. (see Mount Fryatt).

Fryatt, Mount (mountain)

83 C/12 - Athabasca Falls
23-41-28-W5
52°33'N 117°54'W
Approximately 38 km south south-east of Jasper.

Mount Fryatt is 3361 m in altitude. (see Fryatt Creek)

Fullerton, Mount (mountain)

82 J/15 - Bragg Creek
31-21-7-W5
50°49'N 114°58'W
Approximately 60 km west of Calgary.

This mountain, which is 2728 m in altitude, was officially named 6 October 1949 after Ernest Redpath "Jake" Fullerton (1882-1975). He lived near Calgary in his youth and later helped on his father's homestead. He earned his nickname from a struggle with one of his father's bulls named Jake. A boxer and a blacksmith, "Jake" was a rancher in his own right and set up his own ranch in the Bragg Creek area in 1914.

Gable Mountain (mountain)

82 O/12 - Barrier Mountain
18-31-12-W5
51°39'N 115°42'W
Approximately 55 km north north-west of Banff.

The name of this 2928 m mountain was suggested by J. Outram (see Mount Outram) in 1916. Officially approved in 1919, the name describes this long ridge with gradual slopes resembling the gable of a house.

Galatea Creek (creek)
82 J/14 - Spray Lakes Reservoir
15-22-9-W5
50°52′N 115°10′W
Flows east into Kananaskis River, approximately 75 km west of Calgary.

(see Mount Galatea)

Galatea Lakes (lakes (2))
82 J/14 - Spray Lakes Reservoir
13-22-10-W5
50°52′00″N 115°16′25″W
Approximately 80 km west of Calgary.

The name for these lakes was officially approved 7 February 1983. (see Mount Galatea)

Galatea, Mount (mountain)
82 J/14 - Spray Lakes Reservoir
2-22-10-W5
50°51′N 115°16′W
Approximately 80 km west of Calgary.

The British cruiser H.M.S. Galatea was the first ship to open fire in the First World War Battle of Jutland (see Jutland Mountain). This 3185 m mountain was named in 1922 after this flagship of the 1st Light Cruiser Squadron.

Galwey Brook (brook)
82 H/4 - Waterton Lakes
21-2-29-W4
49°08′N 113° 51′W
Approximately 12 km north-east of Waterton Park.

The name was officially approved for this brook 8 July 1954. (see Mount Galwey)

Galwey, Mount (mountain)
82 H/4 - Waterton Lakes
2-30-W4
49°07′N 113°57′W
Approximately 40 km south of Pincher Creek.

This 2377 m mountain was named after Lieutenant Galwey, R.E., who was the Assistant Astronomer for the British Boundary Commission in its expedition from Lake of the Woods to the Rockies, 1872-1876 (see Cameron Lake). The name was officially approved 19 May 1943.

Gap, The (pass)
82 G/16 - Maycroft
10-3-W5
49°52′N 114°22′W
Approximately 30 km north of Blairmore.

It is said that a Dutchman named Cootz was probably the first white man to have gone into The Gap. The Oldman River forces its way through the Livingstone Range here and there is a conspicuous gap in the mountains at this point.

Gap Creek (creek)
83 B/5 - Saunders
35-39-14-W5
52°24′N 115°54′W
Flows north-west into North Saskatchewan River, approximately 14 km south-east of Nordegg.

The name for this creek was suggested due to its proximity to "The Gap" on the North Saskatchewan River. It was officially approved in 1941.

Gap Lake (lake)
82 O/3 - Canmore
20-24-9-W5
51°03′N 115°14′W

Approximately 10 km east south-east of Canmore.

(see The Gap)

Gap Lake (lake)
83 B/5 - Saunders
5-39-13-W5
52°20′N 115°49′W
Approximately 22 km south-east of Nordegg.

(see Gap Creek)

Gap Mountain (mountain)
82 J/11 - Kananaskis Lakes
33-19-8-W5
50°39′20″N 115°03′10″W
Approximately 55 km west of Turner Valley.

The mountain, which stands 2675 m in altitude, is situated just north of the break between the Opal and Elk Ranges, locally called "The Gap." The descriptive name Gap Mountain was officially applied 8 November 1978.

Gardiner Creek (creek)
82 G/8 - Beaver Mines
35-5-4-W5
49°26′N 114°27′W
Flows north into Carbondale River, approximately 18 km west south-west of Beaver Mines.

The origin of this locally well-established name is unknown.

Gargoyle Mountain (mountain)
83 E/1 - Snaring
36-47-2-W6
53°06′N 118°10′W
Approximately 26 km north north-west of Jasper.

A "gargoyle" is described as "a grotesque spout, usually in the form of a human or animal mouth, head or body, projecting from a gutter" (*Concise Oxford Dictionary* 1982). A stream flows from the base of this mountain, 2693 m in altitude, as from a gargoyle. The descriptive name for this feature was given by M.P. Bridgland (see Mount Bridgland) in 1916.

Gargoyle Valley (valley)

82 O/4 - Banff
51°12′30″N 115°40′30″W
Approximately 8 km west north-west of Banff.

The name for this valley was officially approved 21 January 1985. (see Gargoyle Mountain)

Garonne Creek (creek)

83 D/16 - Jasper
23-46-1-W6
52°59′N 118°03′W
Flows west into Athabasca River, approximately 12 km north of Jasper.

The term Garonne is Gaelic for "the rough stream." This creek runs through a rough canyon with high rocky walls.

Garth, Mount (mountain)

82 N/15 - Mistaya Lake
3-33-21-W5
51°48′N 116°55′W
Approximately 115 km north-west of Banff.

This mountain, which is 3030 m in altitude, was named in 1920 after John McDonald of Garth (circa 1774-1860). Born in Scotland, John McDonald came to Canada in the service of the North West Company in 1791. He served in western Canada almost continuously until his posting to Fort

Astoria, in present day Oregon, U.S.A., in 1813. He retired in 1815.

Gass, Mount (mountain)

82 J/2 - Fording River
13-6-W5
50°08′N 114°44′W
Approximately 70 km south-west of Turner Valley.

L.H. Gass of the Dominion Land Survey was the Assistant to J.A. Calder on surveys in the Railway Belt (1913-1915). He was killed in action in the First World War in 1917, and this 2865 m mountain was named for him.

Gaunce, Mount (mountain)

83 E/9 - Moberly Creek
13-53-3-W6
53°35′N 118°18′W
Approximately 79 km north north-west of Jasper.

Squadron Leader Lionel M. Gaunce, D.F.C. (1915-1941) was born at Lethbridge, Alberta, and served as Corporal in the Militia at Edmonton from 1933-1935. He went through Civil Flying School, Austy, Coventry, England from 6 January 1936 to 9 March 1936. He earned the Distinguished Flying Cross in 1940 and was killed 19 November 1941. This mountain was officially named in 1949 in memory of this flight commander.

Gauthier Creek (creek)

83 E/16 - Donald Flats
1-58-3-W6
53°59′N 118°19′W
Flows south-east into Little Smoky River, approximately 55 km east north-east of Grande Cache.

This creek was officially named in 1947, likely after Julian Gauthier, a local trapper.

Gec, Mount (mountain)

83 C/6 - Sunwapta
7-39-24-W5
52°20′N 117°27′W
Approximately 74 km south south-east of Jasper.

The name "Gec" was derived from the initials of the first three mountaineers to climb this mountain in July, 1948. To come up with the name, George Harr (G), Ellen Wilts (E) and Chuck Wilts (C) combined the first letter of their Christian names. During the approval process in May, 1961, the spelling was accidentally altered and the name approved was Mount Gee. This name, however, suggested something more spectacular than this peak, and the original name, Mount Gec, was approved 7 November 1984 for this 3130 m mountain.

*Geikie (former locality)

83 D/16 - Jasper
34-45-2-W6
52°52′N 118°16′W
Approximately 12 km west of Jasper.

This locality took its name from Mount Geikie, located in British Columbia. Sir Archibald Geikie (1835-1924) was an eminent Scottish geologist who was the Director of the Geological Survey of Scotland in 1867. Along with several other positions, Geikie was the Director of the Museum of Practical Geology in Great Britain from 1882 to 1901.

Gendarme Mountain (mountain)

83 E/3 - Mount Robson
11-49-9-W6
53°13′N 119°12′W
Approximately 84 km west north-west of Jasper.

Named in 1911, this 2922 m mountain is imagined to stand on guard. Its name, "gendarme," is the French word for a police constable.

---

*denotes rescinded name or former locality.

Gentian Creek (creek)

83 E/14 - Grande Cache
32-55-9-W6
53°47′N 119°17′W
Flows north into Smoky River, approximately 15 km south-west of Grande Cache.

This creek was named in 1953 after the wildflower, the Northern Gentian, distinguished by its bottle or vase-shaped corollas. It is widely distributed and found abundantly in moist and grassy places.

George Creek (creek)

83 C/10 - George Creek
1-42-19-W5
52°36′N 116°36′W
Flows north into Blackstone River, approximately 38 km west north-west of Nordegg.

The name for this creek was proposed by H.L. Seymour of the Dominion Land Survey in 1907 and was officially approved 4 November 1910. It commemorates George Bernstein (1857-1924), a German book and newspaper publisher. He was a co-founder and officer of the German Voluntary Automobile Corps and was President of the German Development Company on Blackstone River.

George Peak (peak)

83 L/4 - Kakwa Falls
58-13-W6
54°02′N 119°54′W
Approximately 62 km west north-west of Grande Cache.

This peak on Mount May is 2412 m in altitude, and was named after Lieutenant George Geoffrey May. (see Francis Peak)

Geraldine Lakes (lake)

83 C/12 - Athabasca Falls
34-41-28-W5
52°34′N 117°57′W

Approximately 35 km south of Jasper.

The origin of the name is unknown.

Ghost Lakes (lakes)

82 O/6 - Lake Minnewanka
32-26-9-W5
51°16′N 115°13′W
Approximately 21 km north north-east of Canmore.

These lakes are the flooded area formed by the Ghost Dam. (see Ghost River)

Ghost River (river)

82 O/2 - Jumpingpound Creek
13-26-6-W5
51°13′N 114°43′W
Flows south into Bow River, approximately 50 km north-west of Calgary.

This river was named prior to the mid-19th century by the Indians due to the area's connection with death and ghosts. A battle took place near the joining of the Bow and Ghost rivers in which the Cree slew many of the (unnamed) enemy. The bodies were buried in the wood at the top of Deadman Hill (see Deadman Hill). The river is named "Deadman" on Palliser's 1860s maps due to its proximity to this hill. According to Erasmus, (see Mount Erasmus) after the battle, a ghost was seen going up and down this river picking up the skulls of the dead who had been killed by the Cree, thus giving the river its present name.

Giant Steps (cascades)

82 N/8 - Lake Louise
27-16-W5
51°21′N 116°14′W
Approximately 55 km west north-west of Banff.

The name for this feature, which has a high step-like formation with a series of waterfalls, is descriptive.

Gibbon Pass (pass)

82 O/4 - Banff
6-26-14-W5
51°11′N 115°58′W
Approximately 28 km west of Banff.

This 2256 m pass was officially named in 1958 after John Murray Gibbon LL.D. (1875-1952), a journalist and author who was brought to Canada by the Canadian Pacific Railway in 1913. The name Gibbon Pass has been in local usage since at least 1929 by Tom Wilson and the Trail Riders of the Canadian Rockies.

Gibraltar Mountain (mountain)

82 J/10 - Mount Rae
12-19-7-W5
50°35′N 114°50′W
Approximately 40 km west of Turner Valley.

This 2665 m mountain was named from the fancied resemblance to the famous Rock of Gibraltar. It is locally known as "Sheer Cliff."

Gibralter Rock (peak)

82 J/13 - Mount Assiniboine
22-12-W5
50°55′N 115°35′W
Approximately 100 km west of Calgary.

This 2870 m peak was named Gibralter Rock from the fancied resemblance to the famous Rock of Gibraltar.

Gilgit Mountain (mountain)

82 N/10 - Blaeberry River
12-32-21-W5
51°43′N 116°51′W
Approximately 110 km north-west of Banff, on the Alberta-B.C. boundary.

Approved in 1961, the name for this 3090 m feature was taken from a mountain in the district of Kashmir, in north-west India.

Ginger Hill (hill)

82 G/8 - Beaver Mines
22-6-3-W5
49°29′N 114°20′W
Approximately 10 km west north-west of
Beaver Mines.

The name was officially approved in 1974.
The origin of the name is unknown.

Girouard, Mount (mountain)

82 O/3 - Canmore
24-26-11-W5
51°13′N 115°25′W
Approximately 13 km north-east of Banff.

Named in 1904 and officially approved 16
January 1912, this 2995 m mountain
honours Colonel Sir Eduard Percy Girouard,
K.C.M.G., D.S.O. (1867-1932), who served as
a railway builder in north and south Africa
during the expansion of the British Empire
there in the 1890s and 1900s. He later
became a colonial administrator. He was the
Director General of Munitions Supply in
1915.

Girvan Creek (creek)

83 E/8 - Rock Lake
36-49-1-W6
53°15′N 118°01′W
Flows south-west into Moosehorn Creek,
approximately 44 km north of Jasper.

The name for this short branch of
Moosehorn Creek is the Gaelic word for
"the short stream." The creek may also be
named after Peter Girvan, a local packer and
outfitter.

Glacier Creek (creek)

83 E/13 - Dry Canyon
58-13-W6
53°59′N 119°52′W
Flows north-east into Goat Creek,
approximately 49 km west north-west of
Grande Cache.

Donald Phillips (see Mount Phillips) gave
this creek its descriptive name in the winter
of 1911-1912. The stream originates in a
nearby glacier.

Glacier Lake (lake)

82 N/15 - Mistaya Lake
14-34-21-W5
51°55′N 116°51′W
Approximately 120 km north-west of Banff.

This lake was named by Sir James Hector
(see Mount Hector). The lake is fed by a
glacial stream, and its name is appropriately
descriptive.

Glacier Pass (pass)

83 E/7 - Blue Creek
23-51-6-W6
53°25′55″N 118°45′25″W
Approximately 85 km north-west of Jasper.

The precise origin for the name of this pass
is not known. It is, however, likely
descriptive of the area in which the pass is
found.

Glacier Peak (peak)

82 N/8 - Lake Louise
51°21′05″N 116°17′05″W
Approximately 55 km west north-west of
Banff.

This peak measures 3283 m in altitude. The
name was officially approved 21 January 1985.
The precise origin of the name is unknown.

Glacier River (river)

82 N/15 - Mistaya Lake
15-34-20-W5
51°55′N 116°53′W
Flows east into Glacier Lake, approximately
120 km north-west of Banff.

This river drains many glaciers. (see Glacier
Lake)

Gladstone Creek (creek)

82 G/8 - Beaver Mines
25-5-2-W5
49°24′N 114°09′W
Approximately 17 km south-west of Pincher
Creek.

William Shanks Gladstone was born in
Montreal 29 December 1832. In 1848, he
signed up with the Hudson's Bay Company
as a carpenter and apprentice boat builder.
He was nicknamed "Old Glad" during his
association with the Hudson's Bay
Company and was known as such until his
death in 1911. He was one of the builders of
Fort Whoop-up. In a sawpit near the creek
named in honour of him, "Old Glad" rip-
sawed lumber and made the windows and
doors for Fort Macleod. The name
Gladstone Creek was made official in 1916.

Gladstone, Mount (mountain)

82 G/8 - Beaver Mines
29-4-2-W5
49°19′N 114°13′W
Approximately 28 km south-west of Pincher
Creek.

This 2370 m mountain is located north of
Turret Mountain. (see Gladstone Creek)

Glasgow Creek (creek)

82 J/15 - Bragg Creek
12-21-7-W5
50°46′N 114°51′W
Flows north-east into Elbow River,
approximately 55 km west south-west of
Calgary.

(see Mount Glasgow)

Glasgow, Mount (mountain)

82 J/15 - Bragg Creek
5-21-7-W5
50°45′N 114°56′W
Approximately 60 km west south-west of
Calgary.

This mountain, which is 2935 m in altitude, was named in 1922 after a battle cruiser engaged in the Battle of Jutland. (see Jutland Mountain)

Glendowan, Mount (mountain)

82 G/1 - Sage Creek
34-2-1-W5
49°11′N 114°03′W
Approximately 17 km north-west of Waterton Park.

This mountain was named by M.P. Bridgland (see Mount Bridgland) in 1915. Mount Glendowan rises 2673 m above sea level and is the namesake of the range in Ireland in the county of Donegal of the same name. Glendowan means "deep glen."

Gloomy Creek (creek)

83 B/4 - Cripple Creek
12-37-12-W5
52°10′N 115°35′W
Flows north into Fall Creek, approximately 51 km south-west of Rocky Mountain House.

The name for this creek was approved in July 1941 and is descriptive in that it flows through a dark, narrow valley with steeply sloping hills on each side.

Gloria Lake (lake)

82 J/13 - Mount Assiniboine
16-22-12-W5
50°51′N 115°36′W
Approximately 100 km west of Calgary.

This lake was named by A.O. Wheeler (see Wheeler Flats) for its magnificent colour in 1917.

Gloria, Mount (mountain)

82 J/13 - Mount Assiniboine
22-12-W5
50°51′N 115°36′W

Approximately 100 km west of Calgary.

Mount Gloria stands 2908 m in altitude. The precise origin of the name is unknown. (see Gloria Lake)

Goat Cliffs (cliff)

83 E/14 - Grande Cache
2-58-8-W6
53°59′N 119°06′W
Approximately 10 km north of Grande Cache.

The Rocky Mountain Goat is found among these cliffs. The name is likely descriptive.

Goat Creek (creek)

82 G/7 - Upper Flathead
29-6-4-W5
49°30′N 114°27′W
Approximately 15 km south of Coleman.

The origin of the name of this creek is unknown, but may come from the fact that Rocky Mountain Goats inhabit this area.

Goat Creek (creek)

82 O/3 - Canmore
5-25-11-W5
51°06′N 115°30′W
Flows north-west into Spray River, approximately 10 km west of Canmore.

(see Goat Range)

Goat Creek (creek)

83 L/5 - Two Lakes
21-62-13-W6
54°23′N 119°53′W
Flows south-east into Narraway River, approximately 75 km north-west of Grande Cache.

This creek was named officially in 1947 due to the large population of mountain goats which used to inhabit the area.

Goat Lake (lake)

82 G/1 - Sage Creek
32-2-1-W5
49°10′N 114°05′W
Approximately 18 km north-west of Waterton Park.

The name of this lake may have arisen from the abundance of mountain goats in the western mountains. There is a range of mountains nearby which also may have given the lake its present name.

Goat Lake (lake)

82 O/5 - Castle Mountain
2-29-14-W5
51°27′N 115°51′W
Approximately 36 km north-west of Banff.

(see Goat Range)

Goat Pond (reservoir)

82 O/3 - Canmore
1-24-11-W5
51°01′00″N 115°24′00″W
Approximately 9 km south south-west of Canmore.

The name for this reservoir comes from Goat Creek, which is one of its sources. It was officially approved 7 February 1983.

Goat Range (range)

82 O/3 - Canmore
24-11-W5
51°00′N 115°25′W
Approximately 9 km south-west of Canmore.

This mountain range, which is 2822 m in altitude, was named by Captain Palliser (see Palliser Range) and is the translation of the Indian word, "wap-u-tik."

Goat's Eye (cave)

82 O/4 - Banff
25-13-W5
51°07′N 115°44′W
Approximately 14 km south-west of Banff.

The descriptive name for this cave was officially approved in 1958.

Gold Creek (creek)

82 G/9 - Blairmore
30-7-3-W5
49°36′N 114°24′W
Flows south into Crowsnest River, approximately 1 km south of Frank.

The British Columbia Gold Fields, Ltd. had visions of finding gold along this creek. The representatives for this company were J.J. Fleutot and C. Remy, who prospected for gold in this creek.

Golden Lake (lake)

83 D/16 - Jasper
52°53′55″N 118°19′40″W
Approximately 16 km west of Jasper.

Fish Hatchery staff and Norman Woody, a local warden, stocked High Serrin Golden Trout eggs in this lake in 1962. The stocking, however, was unsuccessful. The lake is named for these trout. The name was officially approved 5 May 1987.

Golden Eagle Peak (peak)

82 N/15 - Mistaya Lake
21-33-21-W5
51°51′N 116°56′W
Approximately 120 km north-west of Banff.

This mountain peak, 3048 m in altitude, was named in 1919 by A.O. Wheeler (see Wheeler Flats). There were, at the time of naming, a number of Golden Eagles seen in the vicinity of the peak.

Goldenrod Creek (creek)

83 E/14 - Grande Cache
33-55-9-W6
53°47′N 119°16′W
Flows north into Smoky River, approximately 14 km south-west of Grande Cache.

This creek was named in 1953 after the wildflower, the Goldenrod. There are ten species of this flower found in Alberta. The Tall Smooth Goldenrod is commonly found in woodlands and river flats in August and September.

Goldeye Lake (lake)

83 C/8 - Nordegg
14-40-16-W5
52°27′N 116°11′W
Approximately 8 km west south-west of Nordegg.

This lake was officially named 17 December 1963. The Goldeye is a freshwater fish known in Alberta and is a popular sport and commercial fish in some areas. It has a bluish dorsal and is silver or white laterally. Its iris is yellowish or goldish, and its fins are spotless. This lake supports an abundance of Goldeye.

Gong Glacier (glacier)

83 C/6 - Sunwapta
52°21′57″N 117°28′50″W
Approximately 75 km south south-east of Jasper.

Officially approved 3 August 1987. (see Gong Lake)

Gong Lake (lake)

83 C/5 - Fortress Lake
30-39-25-W5
52°23′N 117°35′W
Approximately 65 km south-east of Jasper.

This lake has a peculiar shade of brownish-green and is held in by the sides of a hanging valley resembling a gong.

Gong Peak (peak)

83 C/6 - Sunwapta
39-25-W5
52°22′N 117°27′W
Approximately 72 km south south-east of Jasper.

Named for its proximity to Gong Lake (see Gong Lake). The peak is 3120 m in altitude.

Gonika Creek (creek)

83 C/8 - Nordegg
11-40-16-W5
52°24′N 116°12′W
Flows south-east into Haven Creek, approximately 10 km south-west of Nordegg.

The name for this creek is the Stoney Indian word for "snow." It was made official in 1932.

Gorge Creek (creek)

82 J/10 - Mount Rae
29-19-5-W5
50°39′N 114°39′W
Flows south-east into Sheep River, approximately 26 km west of Turner Valley.

Officially approved 6 September 1951, the name "Gorge Creek" is descriptive of the land through which the creek flows.

Gorge Creek (creek)

83 F/11 - Dalehurst
25-53-23-W5
53°37′N 117°16′W
Flows north-east into Athabasca River, approximately 55 km west of Edson.

This creek flows through a gorge to reach the Athabasca River. Its descriptive name was officially approved in 1946.

Gorman, Mount (mountain)

83 L/4 - Kakwa Falls
60-14-W6
54°11′N 120°00′W
Approximately 65 km north-west of Grande Cache, on the Alberta-B.C. boundary.

This mountain, which is 2340 m in altitude, was named in 1925 after A.O. Gorman, of the Dominion Land Survey. Gorman was employed on both subdivision resurveys and base line surveys. During the Mines and Technical Surveys of the Alberta-B.C. Boundary Surveys, he was Assistant Surveyor General, and also was Chief of the Legal Surveys Division in the early 1900s.

Gould Dome (mountain)

82 G/15 - Tornado Mountain
21-11-5-W5
49°55′N 114°38′W
Approximately 35 km north-west of Coleman.

This 2894 m rounded mountain was named by Captain Thomas Blakiston (see Mount Blakiston) in 1858. He named it after John Gould (1804-1881), a distinguished British naturalist.

Grande Mountain (mountain)

83 E/14 - Grande Cache
15-57-8-W6
53°56′N 119°07′W
Approximately 5 km north of Grande Cache.

This mountain is 1987 m in altitude. (see Grande Cache)

Grande Cache (town)

83 E/14 - Grande Cache
3-57-8-W6
53°53′N 119°08′W
Approximately 115 km north-west of Hinton.

The word "cache" often refers to a place where supplies have been deposited on a raised platform out of reach of wild animals. The French meaning, however, is "a hiding place." Early fur traders often used this storage technique to protect stored goods from potential marauders. Ignace Giasson was a fur trader, formerly of the North West Company, who later became involved with the Hudson's Bay Company and was stationed in the Peace River and Athabasca districts. His travels up the Smoky River on at least one occasion led him into British Columbia. On his return, he was loaded down with more furs than he could carry, and stored them in a large (grande) cache. Grande Cache was incorporated as a *new town* in September 1966. A new town is the term recently used for a settlement that develops suddenly as a result of one resource or economic activity. Grande Cache grew because of its coal industry. The post office opened in August, 1969. As various other activities such as trading and industry opened up, the status of Grande Cache changed; it was officially recognized as the *town* of Grande Cache on 1 September 1983. A town is a more permanent establishment than a settlement.

Grande Cache Ford (river crossing)

83 E/14 - Grande Cache
5-57-8-W6
53°54′N 119°09′W
Approximately 5 km south-west of Grande Cache.

(see Grande Cache)

Grande Cache Lake (lake)

83 E/14 - Grande Cache
1-57-8-W6
53°54′N 119°03′W
Approximately 5 km west of Grande Cache.

(see Grande Cache)

Grande Cache Valley (valley)

83 E/14 - Grande Cache
57-7-W6
53°55′N 119°02′W
Approximately 6 km east north-east of Grande Cache.

(see Grande Cache)

Grant Pass (pass)

83 E/2 - Resplendent Creek
11-47-6-W6
53°03′N 118°45′W
Approximately 47 km west north-west of Jasper.

Named in 1924, this pass takes its name from George Munro Grant (1853-1902), a Presbyterian Minister from Halifax who served as secretary on an expedition with Sandford Fleming (see Introduction) in 1872. The account of this exploration which sought a possible route for a transcontinental railway is recorded in his book *Ocean to Ocean* (1873).

Grass Pass (pass)

82 J/7 - Mount Head
11-17-5-W5
50°24′50″N 114°25′15″W
Approximately 35 km south-west of Turner Valley.

This is a well-known descriptive name for the low pass at the head of Wileman Creek. The name was officially approved 28 February 1980. It has an elevation of approximately 1850 m.

Grassi Lakes (lakes)

82 O/3 - Canmore
24-24-11-W5
51°04′N 115°24′W
Approximately 85 km west of Calgary.

According to records, the name Grassi Lakes, after Lawrence Grassi, well-known mountaineer and guide, was submitted to the Canadian Board on Geographical Names for approval in a letter dated 9 March 1927 from Mr. E. Mallabone, President, and Mr. W.M. Ramsay, of the Canmore Advisory Council. The Board officially adopted the name 3 May 1927.

Grassy Mountain (mountain)

82 G/9 - Blairmore
1-9-4-W5
49°42′N 114°25′W
Approximately 12 km north of Blairmore.

M.P. Bridgland (see Mount Bridgland) suggested the descriptive name of Grassy Mountain for this feature, which is 2065 m in altitude. The name was officially approved in July 1915, but the mountain has since been exploited for coal, and its name is no longer as appropriate as it once was.

Grassy Ridge (hill)

82 O/5 - Castle Mountain
51°24′00″N 115°39′00″W
Approximately 30 km north-west of Banff.

Grassy Ridge is a descriptive name, used locally and by the Banff Park Warden Service.

Grassy Ridge (ridge)

82 G/15 - Tornado Mountain
11-5-W5
49°55′N 114°33′W
Approximately 30 km north of Coleman.

Grassy Ridge was the name officially applied to this ridge in 1916. The name is descriptive of the ridge's appearance. The highest point measures 1979 m.

Grassy Ridge (ridge)

83 E/1 - Snaring
48-2-W6
53°07′N 118°10′W
Approximately 26 km north north-west of Jasper.

This descriptive name was officially applied to this ridge in 1916 by M.P. Bridgland (see Mount Bridgland). The ridge has a grassy shoulder on it.

Grave Brook (brook)

82 G/16 - Maycroft
16-10-2-W5
49°49′N 114°13′W
Flows east into Tetley Creek, approximately 30 km north-east of Blairmore.

The origin of the name is unknown.

Grave Creek (creek)

83 C/15 - Cardinal River
18-45-20-W5
52°53′N 116°53′W
Flows south-east into Cardinal River, approximately 77 km east of Jasper.

This creek was named in 1916 by C.M. Edwards, who claimed there was evidence of two Stoney Indian graves at the edge of the small prairie at the mouth of the stream.

Gravenstafel Ridge (ridge)

82 G/8 - Beaver Mines
4-4-W5
49°19′N 114°26′W
Approximately 32 km south of Blairmore, on the Alberta-B.C. boundary.

The name was officially applied to this ridge 6 November 1917 when the Canadian Corps, fighting in Belgium in the First World War, were succeeding, though with great casualties, in capturing Passchendaele and nearby Gravenstafel Ridge from German soldiers. The next day, after a loss of 15,000 troops, the Canadians captured the ridge and its 5 square kilometres of mud.

Graveyard Flats (flat)

83 C/2 - Cline River
52°04′30″N 116°55′10″W

Approximately 120 km south-west of Jasper.

The name for this flat was officially approved 21 January 1985. Mary T.S. Schäffer mentioned that the name suggested itself from the quantities of bear, sheep, goat and buffalo bones which were strewed over the ground.

Graveyard Flats, 1924

Graveyard Lake (lake)

83 F/5 - Entrance
17-52-26-W5
53°30′N 117°48′W
Approximately 18 km north-west of Hinton.

An old Indian burial ground lies in the timber on the east side of this creek.

Grayling Creek (creek)

83 L/12 - Nose Creek
30-65-11-W6
54°39′N 119°38′W
Flows north-west into Nose Creek, approximately 78 km south-west of Grande Prairie.

The name for this creek has been in local use since at least 1941. J.V. Butterworth, a field topographer, suggested the name Grayling Creek be officially adopted. A grayling is a silver-grey freshwater fish with a long high dorsal fin. Perhaps these fish swim in abundance in this stream.

Great West Ridge (ridge)

82 O/11 - Burnt Timber Creek
32-9-W5
51°45'N 115°10'W
Approximately 65 km north-east of Banff.

The name for this ridge was officially approved 25 November 1941. The Great West Lumber Company was in operation from 1907 until 1915, when a flood washed away most of the company's logs, which were stacked on the banks of the James River. Some floated as far as eight hundred kilometres away. Remains of two of the original lumbering camps were found at the base of the ridge, and it was named to commemorate the company's existence.

Green Creek (creek)

82 G/9 - Blairmore
8-8-3-W5
49°38'N 114°23'W
Flows west into Gold Creek, approximately 4 km north-east of Frank.

This name has been in local usage since at least 1908. Its precise origin is unknown, but its name is likely descriptive.

Green Mountain (mountain)

82 J/10 - Mount Rae
22-19-5-W5
50°37'N 114°37'W
Approximately 25 km west of Turner Valley.

Green Mountain is 1844 m in altitude. The name was officially approved 6 September 1951. The origin of the name is unknown.

Greenfeed Creek (creek)

82 J/8 - Stimson Creek
11-16-4-W5
50°20'N 114°27'W
Flows east into Pekisko Creek, approximately 40 km south-west of Turner Valley.

Paddy Green was a concert pianist whose love for music inspired him to order a piano to be transported to his sod-roofed cabin. The piano slipped off the load en route and went crashing over a bank along the south fork of this stream, and there it remained for many impromptu concerts performed by Paddy Green. This creek is named for the eccentric and quiet man, who came to the area as mysteriously as he left it. The name was officially approved 12 December 1939, but should possibly be "Greenford Creek."

Greenock, Mount (mountain)

83 E/1 - Snaring
4-48-1-W6
53°06'N 118°05'W
Approximately 25 km north of Jasper.

Named in 1916 by M.P. Bridgland, (see Mount Bridgland) this mountain, which is 2065 m in altitude, has a rounded peak. "Greennoch" means "the sunny knoll" in Gaelic, so perhaps the mountain was climbed on a sunny day. Greenock is also a town in Scotland.

Gregg Lake (lake)

83 F/12 - Gregg Lake
32-52-26-W5
53°32'N 117°48'W
Approximately 21 km north-west of Hinton.

Mr. John James Gregg (1840-1941) lived in the district for 18 years and was the first person to tell the surveyor Saint-Cyn, who named the lake in 1908, about its location. Gregg was a prospector and trapper who travelled as far north as Lesser Slave Lake. He started a small trading store in 1894 along the Jasper Trail.

Gregg, Mount (mountain)

83 F/3 - Cadomin
5-47-24-W5
53°02'N 117°28'W
Approximately 93 km south-west of Edson.

This mountain stands approximately 2530 m in altitude. (see Gregg Lake)

Grey Owl Creek (creek)

83 B/12 - Harlech
35-43-13-W5
52°44'N 115°47'W
Flows north-west into Nordegg River, approximately 36 km north-east of Nordegg.

The name Grey Owl Creek was adopted officially by the Canadian Permanent Committee on Geographical Names 7 March 1957. A local hunter named Frank Lind suggested the name because while hunting, he and another member of his hunting party saw their first Great White Owl along this creek.

Griesbach, Mount (mountain)

83 E/1 - Snaring
8-47-3-W6
53°03'N 118°24'W
Approximately 30 km north-west of Jasper.

Alberta's 50th Jubilee emphasized memories of the "pioneers" of Alberta and to mark the occasion eleven mountains and one lake were named by the Geographic Board of Alberta after some of these pioneers. This 2682 m mountain is one of these commemorative mountains. It was named for the Honourable William Antrobus Griesbach, C.B., C.M.G., D.S.O., V.D. (Volunteer Officers' Decoration), K.C. (1878-1945), who served in the Boer War, was called to the Bar in 1901, and served as alderman and mayor of Edmonton, 1905-1906 and 1907, respectively. In the First World War, Griesbach served as a major, 2nd in command of the 1st Canadian Division, Cavalry Squadron, was promoted to Lieutenant-Colonel of the 49th Battalion (Infantry) of the Canadian Expeditionary Force in France and later to Brigadier General of the First Canadian Infantry Brigade, C.E.F. He was made a Senate member in September 1921 and was promoted to Major-General in November of that same year. He was Inspector General for Western Canada in 1940 and retired from that position 31 March 1943.

Grisette Mountain (mountain)

83 C/13 - Medicine Lake
12-46-28-W5
52°56′N 117°57′W
Approximately 12 km north-east of Jasper.

"Gris" is the French word for "grey." This mountain was named by M.P. Bridgland (see Mount Bridgland) in 1916 due to its composition of greyish limestone.

Grizzly Creek (creek)

83 C/6 - Sunwapta
14-39-24-W5
52°21′N 117°21′W
Flows west into Sunwapta River, approximately 77 km south south-east of Jasper.

The precise origin of the locally well-established name of this creek is not known. The Canadian Permanent Committee on Geographical Names approved the name for this stream 18 October 1976.

Grizzly Creek (creek)

82 G/8 - Beaver Mines
27-4-3-W5
49°19′N 114°20′W
Approximately 30 km south of Bellevue.

The origin of this locally well-established name is unknown, but the creek may possibly be named for the grizzly bears which frequent this area.

Grizzly Creek (creek)

82 J/14 Spray Lakes Reservoir
11-21-9-W5
50°46′N 115°09′W
Flows west into Kananaskis River, approximately 75 km west south-west of Calgary.

The origin of the name is unknown.

Grizzly Creek (creek)

83 L/3 - Copton Creek
30-59-9-W6
54°08′N 119°21′W
Flows north-east into Copton Creek, approximately 33 km north-west of Grande Cache.

This creek was officially named in 1951. A number of grizzly bears inhabit the area.

Grotto Mountain (mountain)

82 O/3 - Canmore
36-24-10-W5
51°05′N 115°15′W
Approximately 7 km east of Canmore.

This 2706 m mountain contains within it a large cave with a high arched roof which is narrow at the mouth. The descriptive name was applied by Eugene Bourgeau (see Mount Bourgeau) in 1858.

Grotto Mountain Pond (man-made pond)

82 O/3 - Canmore
14-21-24-9-W
51°04′00″N 115°12′00″W
Approximately 12 km east south-east of Canmore.

This pond was officially named 7 February 1983. (see Grotto Mountain)

Grouse Lake (lake)

82 O/12 - Barrier Mountain
28-30-13-W5
51°36′N 115°47′W
Approximately 50 km north-west of Banff.

Formerly called "Fagan Lake," the local name for this lake was approved by the Canadian Permanent Committee on Geographical Names in 1966. The name was likely suggested because there are Blue and Franklin Grouse in this area.

Gunderson Creek (creek)

83 L/5 - Two Lakes
29-63-11-W6
54°28′N 119°37′W
Flows north-east into Nose Creek, approximately 72 km north north-west of Grande Cache.

A.M. Floyd, a field topographer, suggested this name, which was officially adopted in 1947. The name commemorates a local hermit (according to local residents) named Gunderson, who lived in the area until circa 1940. He had many cabins on this creek. He chose to speak very little and was only seen in public a few times a year. Gunderson decided to leave the area for Fort Nelson, as he felt it was getting too crowded and he wanted more privacy.

Gunnery Grade (old road grade)

82 J/7 - Mount Head
15-34-16-5-W5
50°23′30″N 115°36′10″W
Approximately 78 km south south-west of Calgary, on secondary highway 541.

The date of construction for this old road grade would seem to be between 1900 and 1914. It is named after Billy Gunnery, a foreman for the Lineham Lumber Company. This tote road was used to get men and supplies to the timber limits.

Gustavs Creek (creek)

83 E/14 - Grande Cache
18-57-8-W6
53°56′N 119°10′W
Flows east into Smoky River, approximately 6 km north north-west of Grande Cache.

No information regarding the origin of the name is known.

Gustavs Flats (flats)

83 E/14 - Grande Cache
8-57-8-W6
53°55′N 119°10′W

Approximately 4 km north-west of Grande Cache.

No information regarding the origin of the name is known.

Gwendolyn Glacier (glacier)

82 O/5 - Castle Mountain
29-14-W5
51°26'45"N 115°53'25"W
Approximately 37 km north-west of Banff.

(see Gwendolyn Lake)

Gwendolyn Lake (lake)

82 O/5 - Castle Mountain
4-29-14-W5
51°27'N 115°55'W
Approximately 39 km north-west of Banff.

This lake, originally named Lake Gwendolyn, was named by Walter D. Wilcox (see Wilcox Pass) in 1921. One of Wilcox's companions, A.L. Castle, had three children: Alfred, Donald and Gwendolyn. This feature is named after Gwendolyn. (see Alfred Lake and Donald Lake)

Gypsum Creek (creek)

82 J/11 - Kananaskis Lakes
11-20-9-W5
50°40'50"N 115°08'25"W
Flows east into Lower Kananaskis Lake, approximately 60 km west of Turner Valley.

The east face of Mount Invincible was once mined for gypsum. Thus the name Gypsum Creek was suggested and officially approved 8 November 1978 for this mountain stream that arises near the old gypsum mines.

Habel Creek (creek)

83 C/5 - Fortress Lake
9-40-28-W5
52°15′N 117°30′W
Flows west into Athabasca River,
approximately 80 km south-east of Jasper.

(see Mount Habel)

Habel, Mount (mountain)

82 N/10 - Blaeberry River
51°39′32″N 116°33′47″W
Approximately 90 km north-west of Banff,
on the Alberta-B.C. boundary.

The name for this 3073 m mountain peak
was officially adopted 29 July 1986. It was
proposed by Graeme Pole to commemorate the
German-born explorer Jean Habel, who in
1897 made the first visit to the region along
the north fork of the Wapta River. Habel
was the first to cross the Yoho Pass. He also
discovered Takakkaw (a Cree name meaning
"it is magnificent") Waterfall, Yoho Valley,
and the glacier at its source. Habel was born
circa 1840 and died in 1902. His first
account of the area was published in the
journal *Appalachia* in 1898, and a mountain
therein referred to by him as "Hidden
Mountain" was subsequently given the name
Mount Habel by that same journal. The
name "Mount Habel" was changed to "Mt.
des Poilus" in 1919 by the Geographic
Board of Canada in appreciation of the
services of French foot soldiers in the First
World War but was later returned to Mount
Habel.

Haddo Glacier (glacier)

82 N/8 - Lake Louise
51°23′06″N 116°13′42″W
Approximately 55 km west north-west of
Banff.

The name was officially approved 21 January
1985. (see Haddo Peak)

Haddo Peak (peak)

82 N/8 - Lake Louise
7-28-16-W5
51°23′N 116°14′W
Approximately 55 km west north-west of
Banff.

This peak of Mount Aberdeen is 3070 m in
altitude. It is named after George, Lord
Haddo, eldest son of the Marquess of
Aberdeen and Temair.

Haglund Creek (creek)

83 L/12 - Nose Creek
16-66-13-W6
54°42′N 119°54′W
Flows west into Narraway River,
approximately 85 km south-west of Grande
Prairie.

This creek was officially named in 1946 after
a trapper whose trap line included this
stream.

Haiduk Creek (creek)

82 O/4 - Banff
25-25-15-W5
51°09′N 115°58′W
Flows north into Shadow Lake,
approximately 28 km south-west of Banff.

(see Haiduk Peak)

Haiduk Lake (lake)

82 O/4 - Banff
7-25-14-W5
51°06′N 115°57′W
Approximately 28 km west south-west of
Banff.

(see Haiduk Peak)

Haiduk Peak (peak)

82 O/4 - Banff
25-14-W5
51°06′N 115°57′W

H~I

Approximately 28 km south-west of Banff.

This 2920 m mountain peak has a name that
is well-established in the area. It may be
named after Haiduk district, Hungary, or
Haiduk village in Romania.

Haig Glacier (glacier)

82 J/11 - Kananaskis Lakes
26-20-10-W5
50°43′N 115°17′W
Approximately 75 km west north-west of
Turner Valley.

Sir Douglas Haig (1861-1928) was a British
Field Marshal and Commander-in-Chief of
the British Forces in France during most of
the First World War. He was given an Earldom
in 1919. After the war, Haig devoted himself
to the welfare of ex-servicemen and united
various organizations into the British Legion.
This glacier was named after him.

Haig, Mount (mountain)

82 G/8 - Beaver Mines
4-4-W5
49°17′N 114°27′W
Approximately 40 km south-west of Pincher
Creek.

Captain R.W. Haig, R.A. was the Chief
Astronomer and Senior Military Officer for the
British Boundary Commission of 1858-1862
from the Pacific to the Rockies. It was after
this man that Mount Haig, 2610 m in altitude,
was officially named 6 December 1916.

Haight Creek (creek)

83 E/14 - Grande Cache
31-55-10-W6
53°47′N 119°29′W

Flows north-east into Muddywater River, approximately 26 km west south-west of Grande Cache.

This creek is likely named after Alex Haight, an early resident of the area. No other information is available.

Hailstone Butte (butte)

82 J/1 - Langford reek
14-4-W5
50°12′N 114°27′W
Approximately 65 km north-east of Claresholm.

The name Hailstone Butte was officially applied to this feature 30 June 1915. The butte is 2373 m in altitude.

Halpenny Creek (creek)

83 F/2 - Foothills
19-48-19-W5
50°10′N 116°46′W
Flows north into Erith River, approximately 52 km south south-west of Edson.

This creek was named in 1927 for R.M. Halpenny of the Sterling Collieries, near where the creek rises.

Halstead Pass (pass)

82 0/5 - Castle Mountain
10-29-14-W5
51°28′N 115°53′W
Approximately 39 km north north-west of Banff.

This pass was named circa 1920 by Walter D. Wilcox (see Wilcox Pass) after one of his assistants, Albert Halstead Jr., who accompanied him through the pass.

Hamell, Mount (mountain)

83 E/14 - Grande Cache
36-57-9-W6
53°58′N 119°12′W
Approximately 11 km north north-west of Grande Cache.

This 2129 m mountain was named in 1953. A geologist who did some field work in this area said that the name Hamell was locally well-established for this mountain and that apparently it was named after a local hunter.

Hanlan Creek (creek)

83 F/2 - Foothills
15-49-18-W5
53°13′N 116°34′W
Flows north-west into Erith River, approximately 110 km east north-east of Jasper.

The origin of the name is unknown.

Hansen Creek (creek)

83 C/9 - Wapiabi Creek
12-43-17-W5
52°41′N 116°20′W
Flows west into Blackstone River, approximately 29 km north-west of Nordegg.

The origin of the name is unknown.

Hanson Creek (creek)

83 C/15 - Cardinal River
14-46-20-W5
52°58′N 116°47′W
Flows east into Pembina River, approximately 88 km east north-east of Jasper.

This creek is named after a local trapper and prospector.

Happy Creek (creek)

83 F/5 - Entrance
15-51-25-W5
50°24′N 117°36′W
Flows north into Athabasca River, approximately 1 km north of Hinton.

This creek was previously known as Coal Creek. Several bachelors had shacks along the creek. It was common practice at the time to run a still, and the area became known as "Happy Hollow" and the creek Happy Creek.

Happy Valley (valley)

82 J/1 - Langford Creek
13-2-W5
50°06′35″N 114°10′50″W
Approximately 43 km west north-west of Claresholm.

During the 1880s ranching began in this valley and in 1907-1908, many homesteaders arrived. Animosity developed between the ranchers and grew into open hostility which continued to 1916 when most of the homesteaders left the area. One of the ranchers, Denis Westropp, humorously christened the area Happy Valley. The name is well-established among local residents and was made official 29 July 1986.

Hardisty Creek (creek)

83 F/5 - Entrance
23-51-25-W5
53°25′N 117°34′W
Flows north-west into Athabasca River, approximately 1 km north of Hinton.

The origin of the name is unknown.

Hardisty Creek (creek)

83 C/12 - Athabasca Falls
20-43-27-W5
52°43′N 117°53′W
Flows south-west into Athabasca River, approximately 23 km south-east of Jasper.

(see Mount Hardisty)

Hardisty, Mount (mountain)

83 C/12 - Athabasca Falls
15-43-27-W5
52°42′N 117°50′W
Approximately 26 km south-east of Jasper.

This mountain, which is 2700 m in altitude, was named by Sir James Hector (see Mount Hector) in 1859 after Richard Hardisty (1831-1889). Hardisty was a Chief Trader with the Hudson's Bay Company, in charge of Fort Charleton, Saskatchewan in 1857-1858. He was Chief Factor in charge of the Edmonton district for many years after that. He was called to the Senate of Canada 23 February 1888, and died in Winnipeg the following year.

Hardscrabble Creek (creek)

83 E/11 - Hardscrabble Creek
30-55-9-W6
53°36'N 119°19'W
Flows north-west into Smoky River, approximately 33 km south-west of Grande Cache.

The origin of the name is unknown.

Harlequin Creek (creek)

83 C/14 - Mountain Park
29-46-24-W5
52°59'N 117°27'W
Flows north-east into Whitehorse Creek, approximately 44 km east south-east of Jasper.

No information regarding the origin of the name is known.

Harris Creek (creek)

82 O/2 - Jumpingpound Creek
23-4-W5
51°00'N 114°30'W
Flows south-east into Elbow River, approximately 24 km west south-west of Calgary.

The name for this creek was officially approved 12 December 1939. but its origin remains unknown.

Harris, Mount (mountain)

82 N/16 - Siffleur River
33-32-16-W5
51°47'N 116°13'W

Approximately 80 km north-west of Banff.

This mountain, which is 3299 m in altitude, was named after L.E. Harris of the Dominion Land Survey. In 1919 he was the first person to climb the mountain.

Harrison Flats (flat)

82 O/13 - Scalp Creek
25-33-14-W5
51°52'N 115°52'W
Approximately 80 km north north-west of Banff.

(see Harrison Lake)

Harrison Lake (lake)

82 O/12 - Barrier Mountain
30-13-W5
51°33'15"N 115°48'35"W
Approximately 45 km north-west of Banff.

This lake was officially named after George Harrison 14 January 1980. Harrison was born at Sibley, Iowa in 1887 and came to Canada from Montana in 1900. He first worked on a ranch in the Cochrane district, but in 1902, he joined James Brewster, working as a guide at the Yaha Tinda Ranch. He later became Brewster's chief guide at Lake Louise and then at Glacier House. In 1918 he started his own outfitting and guiding business at Glacier, which lasted until 1945.

Harvey Lake (lake)

83 E/1 - Snaring
24-48-4-W6
53°09'N 118°27'W
Approximately 40 km north-west of Jasper.

This lake was named in 1954 after Horace Harvey (1863-1949), Chief Justice of Alberta, 1910-1949. In 1940, he was Chairman of the Mobilization Board for National War Services. (see Mount Griesbach)

Harvey, Mount (mountain)

83 E/9 - Moberly Creek
24-53-4-W6
53°35'N 118°27'W
Approximately 83 km north north-west of Jasper.

This 2438 m mountain was named in 1949 in honour of Lieutenant F.M.W. Harvey, V.C., M.C., of Fort Macleod, Alberta. Harvey was killed in the First World War.

Harvey Pass (pass)

82 O/4 - Banff
18-25-13-W5
51°08'N 115°48'W
Approximately 17 km west south-west of Banff.

This pass, found in the Bourgeau Lake region, is named after Ralph L. Harvey, who was the first person to cross over it in the winter. It was officially named in 1959.

Harvie Heights (locality)

82 O/3 - Canmore
18-25-10-W5
51°07'N 115°23'W
Approximately 5 km north-west of Canmore.

The origin of the name is unknown.

Hastings Ridge (ridge)

82 G/8 - Beaver Mines
6-4-W5
49°30'N 114°25'W
Approximately 8 km south of Blairmore.

This ridge was named by J.B. Tyrrell after Tom Hastings, a member of the Geological Survey Party in 1884. The Cree name for Hastings Lake is "a-ka-ka-kwa- tikh," meaning "the lake that does not freeze." Hastings Creek is "kak-si-chi-wukh," which in Cree means "swift current." The lake and creek are not within this study area.

Hat Creek (creek)

83 L/4 - Kakwa Falls
30-60-11-W6
54°13′N 119°39′W
Flows south-west into Lynx Creek,
approximately 50 km north-west of Grande
Cache.

This creek runs off Hat Mountain. (see Hat
Mountain)

Hat Mountain (mountain)

83 L/5 - Two Lakes
24-61-12-W6
54°18′N 119°41′W
Approximately 55 km north-west of Grande
Cache.

The descriptive name for this mountain has
been in use since at least 1915. One of the
higher peaks on this 1800 m mountain
resembles a stetson, or cowboy hat.

Haultain, Mount (mountain)

83 E/1 - Snaring
31-48-2-W6
53°11′N 118°18′W
Approximately 37 km north north-west of
Jasper.

This 2621 m mountain was named in 1954
after Sir Frederick William Alpin Gordon
Haultain, K.C., LL.B. (1857-1942). Born in
England and educated in Canada, Haultain
was called to the Ontario Bar in 1882. He
came west to Fort Macleod in 1884 and from
1888 to 1905, was a member for Fort Macleod
in the Territorial Legislature. He was the
first editor of the *Lethbridge News*. Haultain
was knighted in 1916 and was Chancellor of
the University of Saskatchewan from
1917-1938. (see Mount Griesbach)

Haven Creek (creek)

83 C/8 - Nordegg
27-39-16-W5
52°23′N 116°13′W

Flows south into North Saskatchewan River,
approximately 13 km south-west of
Nordegg.

There was a cabin on this creek dating back
to 1910, owned by a rancher who had a
grazing lease in this valley. The man's last
name was Haven, but little else is known
about him.

Hawk Mountain (mountain)

83 E/1 - Snaring
26-46-1-W6
53°01′N 118°01′W
Approximately 16 km north of Jasper.

A hawk was seen flying about the summit of
this 2553 m mountain at the time of its
naming in 1916. The name was applied by
M.P. Bridgland. (see Mount Bridgland)

Hawke Island (island)

82 J/11 - Kananaskis Lakes
19-9-W5
50°37′N 115°09′W
Approximately 65 km west south-west of
Turner Valley, in Upper Kananaskis Lake.

The origin of the name is not known. (see
Hogue Island)

Hawkeye Creek (creek)

82 J/1 - Langford Creek
35-12-2-W5
50°03′N 114°10′W
Flows east into Nelson Creek,
approximately 40 km west of Claresholm.

The name was approved by the Canadian
Permanent Committee on Geographical
Names 22 April 1970. The origin of the
name is unknown.

Hawkins, Mount (mountain)

82 G/1 - Sage Creek
1,2-1-W5
49°05′N 114°05′W

Approximately 12 km west north-west of
Waterton Park.

Lieutenant-Colonel John Summerfields
Hawkins, R.E. was the Commissioner for
the British Boundary Commission of
1858-1862, from the Pacific to the Rockies.
This mountain is named after him.

Hay Creek (creek)

83 F/2 - Foothills
14-49-21-W5
53°14′N 116°57′W
Flows north-west into Embarras River,
approximately 54 km south-west of Edson.

No information regarding the origin of the
name is known.

Hay Creek (creek)

82 J/8 - Stimson Creek
11-16-3-W5
50°20′N 114°18′W
Flows north-east into Sheppard Creek,
approximately 40 km south of Turner Valley.

The name for this creek was officially
approved 12 December 1939. The origin of
the name is unknown.

Hayden Ridge (ridge)

83 E/14 - Grande Cache
29-55-7-W6
53°47′N 119°00′W
Approximately 14 km south south-east of
Grande Cache.

The origin of the name is unknown.

Hazell (hamlet)

82 G/10 - Crowsnest
7-8-5-W5
49°38′N 114°40′W
Approximately 11 km west of Coleman,
between Crowsnest and Island lakes.

This hamlet may possibly be named after Mr. E.G. Hazell, who owned and opened Summit Line Works Ltd. in 1903.

Head Creek (creek)

82 J/7 - Mount Head
17-5-W5
50°26′N 114°39′W
Flows north-east into Flat Creek, approximately 31 km south-west of Turner Valley.

The name was officially approved 8 December 1943. (see Mount Head)

Head, Mount (mountain)

82 J/7 - Mount Head
17-5-W5
50°26′N 114°39′W
Approximately 36 km south-west of Turner Valley.

This 2782 m mountain was named by Captain Palliser during his expedition (see Palliser Range). Sir Edmund Walker Head (1805-1868) was a scholar and statesman, and was Governor-in-Chief of Canada from 1854-1861. He played an important role in the early development of the concept of Confederation. After leaving Canada, he became Governor of the Hudson's Bay Company, 1863-1868.

Headwall Creek (creek)

82 J/14 - Spray Lakes Reservoir
11-21-10-W5
50°46′20″N 115°16′55″W
Flows south-west into Smith-Dorrien Creek, approximately 85 km west south-west of Calgary.

The name Headwall was officially applied to this stream 8 November 1978. The creek had been known as "Ranger Creek" since before the 1950s, when a forest ranger cabin was situated at the junction of the Spray Lakes Road and this feature. There are four official Ranger Creeks in Alberta; the closest is 48 km to the east in the Bragg Creek area. Another name was selected for the creek to avoid confusion, and because of the spectacular vertical rock headwall near the headwaters of the creek.

Headwall Lakes (lakes)

82 J/14 - Spray Lakes Reservoir
30-21-9-W5
50°48′15″N 115°14′45″W
Approximately 80 km west south-west of Calgary.

These two small alpine lakes are located at the headwaters of Headwall Creek. (see Headwall Creek)

Healy Creek (creek)

82 O/4 - Banff
18-25-12-W5
51°08′N 115°42′W
Flows east into Bow River, approximately 6 km south-west of Banff.

(see Healy Pass Lakes)

Healy Pass (pass)

82 O/4 - Banff
24-14-W5
51°06′N 115°52′W
Approximately 23 km south-west of Banff.

(see Healy Pass Lakes)

Healy Pass Lakes (lakes)

82 O/4 - Banff
51°05′30″N 115°51′45″W
Approximately 22 km west south-west of Banff.

Healy Pass Lakes were named for their proximity to Healy Pass and Healy Creek. Both were named in 1884 by G.M. Dawson (see Tombstone Mountain) after Captain John Gerome Healy. During the mid and late 19th century, he travelled extensively through the western half of North America and Mexico and was employed as a hunter, trapper, soldier, prospector, whiskey trader, editor, guide, Indian scout and sheriff. Eventually, he headed north to the Klondike and established the North American Trading and Transportation Company, an outfitting firm in Dawson City. The N.A.T. and T. Co. was the Alaska Commercial Company's main rival, building numerous supply posts and a fleet of steamboats and starting a price war in the Yukon. Healy and some of his associates laid claim to several copper finds in the vicinity of what is now known as Healy Creek. The N.A.T. and T. Co. no longer exists, but not due to misfortune, for John Healy died a wealthy man.

Heart Mountain (mountain)

82 O/3 - Canmore
12-24-9-W5
51°02′N 115°08′W
Approximately 17 km east south-east of Canmore.

This name was officially approved 4 April 1957, and, according to L.M. Clark, there is a heart-shaped configuration of strata at the crest of this 2042 m mountain.

Heath Creek (creek)

82 G/16 - Maycroft
32-9-1-W5
49°46′N 114°06′W
Flows west into Oldman River, approximately 30 km north-east of Blairmore.

Located across the North Fork, this creek is named after William H. Heath who, with his wife and family, lived on a ranch nearby. They were the first settlers on this creek, but did not live here very long.

Hector Creek (creek)

82 N/9 - Hector Lake
27-30-17-W5
51°36′N 116°19′W

Flows west into Bow River, approximately 70 km north-west of Banff.

(see Mount Hector)

Hector Glacier (glacier)

82 N/9 - Hector Lake
30-17-W5
51°35′N 116°16′W
Approximately 65 km north-west of Banff.

(see Mount Hector)

Hector Lake (lake)

82 N/9 - Hector Lake
20-30-17,18-W5
51°35′N 116°22′W
Approximately 70 km north-west of Banff.

(see Mount Hector)

Hector, Mount (mountain)

82 N/9 - Hector Lake
13-30-17-W5
51°35′N 116°15′W
Approximately 65 km north-west of Banff.

Sir James Hector, M.D., C.M.G., K.C.M.G., (1834-1907) was appointed Surgeon and Geologist to the Palliser Expedition 1857-1860 and was sent to explore the western parts of British North America (see Introduction). Hector made many important observations regarding the geology and ethnology of the Canadian West and Rocky Mountains. He was appointed geologist to the Provincial Government of Otago, New Zealand in 1861 and Director of the Geological Survey of New Zealand from 1865-1903. He returned to Canada in 1904 to visit some of his previous exploration grounds. This mountain, 3394 m in altitude, was named in 1884 by Dr. G.M. Dawson of the Geological Survey of Canada, after Sir James Hector.

Heel Creek (creek)

82 J/9 - Turner Valley
26-19-4-W5
50°39′N 114°27′W
Approximately 12 km west south-west of Turner Valley.

This creek was named for W.J. Heel, who emigrated from England in the late 1880s. The name was approved by the Geographic Board of Canada in 1928.

Heifer Lake (lake)

82 O/11 - Burnt Timber Creek
36-30-8-W5
51°37′N 115°00′W
Approximately 60 km north-east of Banff.

The origin of the name is unknown.

Helen Creek (creek)

82 N/9 - Hector Lake
17-31-17-W5
51°39′N 116°23′W
Flows south into Bow River, approximately 80 km north-west of Banff.

(see Helen Lake)

Helen Lake (lake)

82 N/9 - Hector Lake
30-31-17-W5
51°41′N 115°24′W
Approximately 80 km north-west of Banff.

This lake was named "Lake Ethel" by members of the Appalachian Mountain Club in 1899. "Ethel" was a duplicate name and was changed to Helen Lake, after a daughter of A.M.C. member, Reverend H.P. Nichols, (see Alice Lake) who climbed in the area at the turn of the century.

Helena Ridge (ridge)

82 O/5 - Castle Mountain
21-27-14-W5
51°19′N 115°54′W

Approximately 27 km north-west of Banff.

The locally well-established name for this ridge was proposed by C.D. Walcott circa 1910, after his first wife, Helena B. Stevens, and was officially approved 7 May 1959.

Hell-Roaring Creek (creek)

82 H/4 - Waterton Lakes
1-30-W4
49°02′N 113°54′W
Approximately 50 km south of Pincher Creek.

This turbulent creek empties west into Upper Waterton Lake. Originally, Hell-Roaring was the name of the canyon but has since been transferred to the creek and falls. It is one of the wildest and most impressive places in Waterton Lakes National Park. Its name is most descriptive.

Hell-Roaring Falls (falls)

82 H/4 - Waterton Lakes
12-1-30-W4
49°01′N 113°53′W
Approximately 50 km south of Pincher Creek.

(see Hell-Roaring Creek)

Hells Creek (creek)

83 E/14 - Grande Cache
29-57-8-W6
53°57′N 119°09′W
Flows south-east into Smoky River, approximately 41 km north-west of Grande Cache.

The origin of the name is unknown.

Helmer, Mount (mountain)

82 N/10 - Blaeberry River
31-20-W5
51°42′N 116°50′W
Approximately 105 km north-west of Banff, on the Alberta-B.C. boundary.

This mountain, which is 3030 m in altitude, was named after Brigadier-General Richard Alexis Helmer, C.M.G. (1864-1920). Helmer was a musketry expert and chemist whose life interest was in the Canadian Militia. He was an alderman for ten years and mayor twice for the city of Hull. He was one of the most capable members of the artillery of his time in Canada.

Helmet Creek (creek)

83 C/13 - Medicine Lake
35-43-25-W5
51°45′N 117°30′W
Flows north into Alpland Creek, approximately 38 km east south-east of Jasper.

(see Helmet Mountain)

Helmet Mountain (mountain)

83 C/13 - Medicine Lake
12-44-25-W5
52°46′N 117°29′W
Approximately 41 km east south-east of Jasper.

The locally well-established name for this 2612 m mountain was used by the Park Wardens at least as early as 1925. The name is likely descriptive of the mountain's shape, and was suggested by Mr. J. Henry Scattergood, an early alpinist in the area who wrote an article for *Appalachia*, April 1901 entitled "The Beaverfoot Valley and Mount Mollison." The name for the mountain was officially approved 1 March 1904.

Henday, Mount (mountain)

83 E/1 - Snaring
32-48-3-W6
53°11′N 118°24′W
Approximately 25 km north of Jasper.

This 2682 m mountain was named in 1954 after Anthony Henday, first European to set foot in what is now Alberta. (see Introduction)

Very little is known of the life of this fur trader who grew up in the Isle of Wight in England. He entered the service of the Hudson's Bay Company in 1750. Henday was the first white man to see the "Shining Mountains." He returned to England in 1763.

Hendrickson Creek (creek)

83 E/16 - Donald Flats
22-55-3-W6
53°46′N 118°22′W
Flows south-east into Cabin Creek, approximately 53 km east south-east of Grande Cache.

Frederick H. Hendrickson (1869-1962) was a local trapper who served the Alberta Forest Service for many years. He lived at Entrance beginning in 1927 and manned the Athabasca Lookout Tower for thirty years. This creek was named in 1947 after Mr. Hendrickson.

Henrietta Creek (creek)

83 E/14 - Grande Cache
34-56-9-W6
53°48′N 119°14′W
Flows south-east into Smoky River, approximately 13 km south-west of Grande Cache.

This creek was named prior to 1916 after a young coal engineer who worked with the Canadian Pacific Railway during its coal explorations. The name was recommended by a Mr. Elsworth. The correct name is believed to be Henraddy. Soon after leaving school, Mr. Henraddy and his sister, due to the misuse of their name, decided to adopt the name Henretta. C.M. "Henretta" occasionally spelled his name Henrietta.

Henry, Mount (mountain)

83 D/16 - Jasper
5-46-2-W6
52°56′N 118°15′W
Approximately 13 km north-west of Jasper.

Mount Henry is 2629 m in altitude. (see Henry House)

Henry House (locality)

83 D/16 - Jasper
22-46-1-W6
52°59′N 118°04′W
Approximately 12 km north of Jasper.

In 1811, while David Thompson was making his way over the Athabasca Pass, William Henry, the eldest son of Alexander Henry, provided support on the eastern side of the mountains. He built a post on the Athabasca River near the mouth of the Miette River, where the trail from the passes reached the head of navigation. At this point travellers coming over the mountains transferred from horses to canoes for the journey down river. William Henry (1783?-1846?) was a fur trader in the service of the North West Company, and it was after him that this trading post and later locality was named in 1912.

Henry MacLeod, Mount (mountain)

83 C/11 - Southesk Lake
15-41-24-W5
52°32′N 117°21′W
Approximately 86 km west of Nordegg.

This mountain, which is 3288 m in altitude, was named by Howard Palmer in 1923 after H.A.F. MacLeod, an engineer with the Canadian Pacific Railway. He made a reconnaissance survey up the Maligne Valley in 1875.

Herbert Lake (lake)

82 N/8 - Lake Louise
5-29-16-W5
52°28′N 116°13′W
Approximately 55 km north-west of Banff.

The origin of the name is unknown.

Hermit Creek (creek)

83 L/5 - Two Lakes
6-63-11-W6
54°25′N 119°38′W
Flows north into Gunderson Creek,
approximately 67 km north-west of Grande
Cache.

This creek was named in 1947 and is located in
close proximity to what is locally known as
the "hermit's cabin." (see Gunderson Creek)

Hibernia Lake (lake)

83 D/16 - Jasper
7-45-1-W6
52°52′N 118°08′W
Approximately 3 km west of Jasper.

Hibernia is the Latin name for Ireland. This
lake was named in 1914 by H. Matheson, of
the Dominion Land Survey. Its waters are
reputed to have a paddy-green colour.

Hidden Creek (creek)

82 G/16 - Maycroft
9-12-4-W5
49°59′N 114°29′W
Flows east into Oldman River,
approximately 40 km north of Blairmore.

This locally well-established name is likely
descriptive.

Hidden Lake (lake)

82 N/8 - Lake Louise
28-15-W5
51°29′N 116°06′W
Approximately 55 km north-west of Banff.

The name for this lake is likely descriptive.

Hidden Lake (lake)

82 J/11 - Kananaskis Lakes
16-19-9-W5
50°36′N 115°11′W
Approximately 65 km west south-west of
Turner Valley.

In 1901, W.D. Wilcox (see Wilcox Pass)
discovered that the waters of this lake reach
Upper Kananaskis Lake by a subterranean
stream. The name is therefore descriptive of
the "hidden" way in which the lake empties.

Hidden Lakes, Valley of the (valley)

83 L/4 - Kakwa Falls
5-15-60-13-W6
From 54°09′30″N 119°54′00″W
to 54°11′30″N 119°50′00″W
Approximately 130 km south south-west of
Grande Prairie.

The descriptive name for this valley is well-
established in local usage.

Hidden Lakes, Valley of the (valley)

82 O/5 - Castle Mountain
17-29-14-W5
51°29′N 115°55′W
Approximately 40 km north-west of Banff.

The mystery of the whereabouts of these
lakes lies in the fact that they may only be
seen from a certain vantage point. Prior to
1917, when Walter D. Wilcox (see Wilcox
Pass) finally discovered the location of the
lakes, they were reported only to be
somewhere in the valley. The official name
was given to this valley in 1965.

Hidden Valley Creek (creek)

83 L/4 - Kakwa Falls
8-20-60-13-W6
From 54°11′40″N 119°51′20″W
to 54°10′00″N 119°54′20″W
Flows north-west into Stinking Creek,
approximately 128 km south south-west of
Grande Prairie.

(see Hidden Valley)

High Lakes (lake)

83 D/16 - Jasper
45-2-W6
52°53′N 118°12′W

Approximately 8 km west of Jasper.

The origin of the name is unknown.

High Divide Ridge (ridge)

83 F/5 - Entrance
50-25-W5
53°18′N 117°31′W
Approximately 12 km south of Hinton.

The locally well-established name of this
ridge refers to the high divide crossed by the
Bighorn Trail between Hinton and the
McLeod River. The name was officially
approved in 1945.

High Rock Range (range)

82 J/2 - Fording River
14-6-W5
50°10′N 114°43′W
Approximately 70 km south-west of Turner
Valley.

This range was named in 1915. Its origin
comes from the precipitous, rocky character
of its summit compared to the hills near its
base.

Hightower Creek (creek)

83 F/13 - Hightower Creek
32-55-26-W5
53°47′N 117°50′W
Flows north-east into Pinto Creek,
approximately 45 km north north-west of
Hinton.

The name Hightower Creek was approved
for this feature 7 October 1948. It
commemorates Pilot Officer C.E.
Hightower, D.F.C. of Beverly, who was
killed in the Second World War.

Highwood Pass (pass)

82 J/11 - Kananaskis Lakes
19-8-W5
50°35′N 115°00′W

Approximately 50 km west south-west of Turner Valley.

(see Highwood River)

Highwood Range (range)

82 J/7 - Mount Head
17-5-W5
50°29′N 114°42′W
Approximately 36 km south-west of Turner Valley.

(see Highwood River)

Highwood River (river)

82 I/13 - Gleichen
26-21-28-W4
50°49′N 113°47′W
Flows north into Bow River, approximately 12 km south-west of Calgary.

This river has had a variety of names. It was "High-wood river" ("Ispasquehow") on the Palliser map; "Spitchee" on the David Thompson map, 1814; "Spitchi" or "Ispasquehow" on the Arrowsmith map, 1859; and was called "High Woods River" by Blakiston. The name appears to be a translation of an Indian name, "spitzee," and was so called because the river is on nearly the same level as the prairie instead of in a "bottom." The belt of timber along the stream is therefore higher than usual.

Hilda Creek (creek)

83 C/3 - Columbia Icefield
52°11′12″N 117°05′35″W
Approximately 105 km south-east of Jasper.

The origin of the name is unknown. The name was officially approved 21 January 1985.

Hilda Peak (peak)

83 C/3 - Columbia Icefield
52°11′25″N 117°10′35″W
Approximately 105 km south-east of Jasper.

The name was officially approved 21 January 1985. No other origin information is known.

Hillcrest (locality)

82 G/9 - Blairmore
20-7-3-W5
49°35′N 114°23′W
Approximately 45 km north-west of Pincher Creek.

Hillcrest was officially named 7 March 1905 after Charles P. Hill who was the Managing Director of Hillcrest Coal and Coke Company. He organized Hillcrest Mines Limited, and has a town in Idaho, called Post Hill, also named after him. The hamlet is located at the top of a hill, so the name may also be partially descriptive.

Hillcrest Mines (hamlet)

82 G/9 - Blairmore
17,20-7-3-W5
49°34′N 114°23′W
Approximately 41 km west north-west of Pincher Creek.

This hamlet, named after Charles Plummer Hill, came into being in 1905. It was the scene of one of the worst mining disasters in Canada when on 19 June 1914, an explosion took 189 lives. (see Hillcrest)

Hillcrest Mountain (mountain)

82 G/9 - Blairmore
1,12-7-4-W5
49°32′N 114°25′W
Approximately 40 km west north-west of Pincher Creek.

This mountain is 2164 in altitude. (see Hillcrest)

Hinton (town)

83 F/5 - Entrance
14-51-25-W5
53°25′N 117°34′W

Approximately 70 km north north-east of Jasper.

This town was named after William Pitman Hinton, President and General Manager of the Grand Trunk Pacific Coast Steamship Company. He received this appointment in August 1917, after several other positions with different railway companies throughout Canada.

Hoff Range (range)

83 E/9 - Moberly Creek
30-53-3-W6
53°36′N 118°25′W
Approximately 75 km north north-west of Jasper.

The name "Hoff Ridge" was adopted 1 September 1955, since it was thought the feature was part of the Berland Range. The name was initially submitted by A.H. Lang, a geologist for the Geological Survey, in 1947. The name "Hoff Ridge" was changed to Hoff Range 19 February 1959. It is named after Ludwig Hoff (1884-1968), a trapper and fur trader of the area. He worked on the construction of the Grand Trunk Pacific Railway and worked for the Forestry Department until he retired in 1955.

Hoffman, Mount (mountain)

82 J/10 - Mount Rae
18-19-5-W5
50°36′N 114°37′W
Approximately 28 km west of Turner Valley.

This mountain, which is 1829 m in altitude, was named by A.O. Wheeler (see Wheeler Flats) in 1896 after a member of his survey party. Mr. Hoffman subsequently became a hotel proprietor in Olds. The name was made official 6 September 1951.

Hogarth Lakes (lake)

82 J/14 - Spray Lakes Reservoir
21-21-10-W5
50°48′N 115°19′W

Approximately 85 km west south-west of Calgary.

These lakes were named after Jack Hogarth, a native of Scotland who was Warden at Spray Lakes prior to the First World War. He later managed the Lesson and Scott Store at Morley and was on staff for the Big Horn Trading Company Ltd., at Nordegg in 1924. The name was approved in 1977.

*Hogue Island (island)

82 J/11 - Kananaskis Lakes
23-19-9-W5
50°39′N 115°10′W
Approximately 60 km west of Turner Valley, in Upper Kananaskis Lake.

Before the Upper Kananaskis Lake was flooded, five of the larger islands were named; Ressy, Hogue, Pegasus, Hawke and Aboukir. The Hogue was one of the British naval vessels sunk by torpedoes 22 September 1914. The original Hogue Island is now under water, but its name, although rescinded, survives as an historical name of record.

Holcroft, Mount (mountain)

82 J/2 - Fording River
15-6-W5
50°14′N 114°46′W
Approximately 60 km south-west of Turner Valley.

This 2713 m mountain was officially named 5 February 1918, in memory of H.S. Holcroft, D.L.S., of the Surveyor General's staff, who died in active service in the First World War.

Hollebeke Mountain (mountain)

82 G/7 - Upper Flathead
11-5-5-W5
49°23′N 114°34′W

Approximately 30 km south south-west of the Coleman, on the Alberta-B.C. boundary.

This mountain, which was named in 1917, commemorates a village south-east of Ypres, Belgium, where Canadian troops fought during the First World War.

Holmes, Mount (mountain)

83 E/5 - Holmes River
36-50-12-W6
53°21′N 119°38′W
Approximately 117 km north-west of Jasper, on the Alberta-B.C. boundary.

It is not certain after whom this 2505 m mountain was named. It may commemorate William Holmes (d. 1792), a native of Ireland, who came to Canada a few years after the British conquest. He was a fur trader who was an original member of the North West Company. The name was approved in 1925.

Holy Cross Mountain (mountain)

82 J/7 - Mount Head
9-17-W5
50°25′20″N 114°37′50″W
Approximately 36 km south-west of Turner Valley.

The name for this 2650 m mountain comes from the presence in the spring of a cross of snow that remains on the side of the mountain. It was named by George Pocaterra (see Pocaterra Creek) circa 1900 and was officially approved 28 February 1980.

Honeymoon Lake (lake)

83 C/12 - Athabasca Falls
28-41-26-W5
52°33′N 117°40′W
Approximately 45 km south south-east of Jasper.

The origin of the name is unknown.

Hood Creek (creek)

82 J/11 - Kananaskis Lakes
35-20-9-W5
50°44′20″N 115°08′00″W
Flows west into Kananaskis River, approximately 60 km west north-west of Turner Valley.

This creek originates on the west slope of Mount Hood. The name was officially approved 2 February 1978. (see Mount Hood)

Hood, Mount (mountain)

82 J/11 - Kananaskis Lakes
6-21-8-W5
50°45′N 115°05′W
Approximately 55 km west north-west of Turner Valley.

Officially named in 1922, this 2903 m mountain is part of the Opal Range. The Honourable Horace Hood was a Rear Admiral with the British Navy. He participated in the Battle of Jutland in 1916 and went down with the battle cruiser H.M.S. Invincible.

Hoodoo Creek (creek)

83 D/16 - Jasper
52°57′15″N 118°01′45″W
Flows south-west into Athabasca River, approximately 9 km north-east of Jasper.

There are hoodoos at the outcrop of this creek.

Hooge, Mount (mountain)

83 C/3 - Columbia Icefield
35-22-W5
52°00′N 117°01′W
Approximately 120 km south-east of Jasper.

This mountain, which is 3216 m in altitude, was named in 1920, after the village located near Ypres, in the Ypres salient where the Canadian groups regained ground in the First World War.

Hooker Icefield (icefield)

83 D/8 - Athabasca Pass
40-1-W6
52°25′N 118°05′W
Approximately 47 km south of Jasper.

(see Mount Hooker)

Hooker, Mount (mountain)

83 D/8 - Athabasca Pass
4-40-1-W6
52°24′N 118°06′W
Approximately 53 km south of Jasper.

This 3286 m mountain was named in 1827 by David Douglas (see Mount Douglas) in honour of Sir William Jackson Hooker (1785-1865), an English botanist who was the first Director of the Royal Botanical Gardens at Kew, near London. He was also a professor of Botany at the University of Glasgow.

Horn Creek (creek)

83 E/14 - Grande Cache
26-57-10-W6
53°57′N 119°23′W
Flows south-east into Sheep Creek, approximately 18 km north-west of Grande Cache.

The origin of the name is unknown.

Horn Ridge (ridge)

83 L/4 - Kakwa Falls
60-13-W6
54°12′N 119°59′N
Approximately 64 km north-west of Grande Cache, on the Alberta-B.C. boundary.

The name for this ridge, which was officially approved in 1959, is descriptive of the appearance of the ridge.

Horns, Lake of the (lake)

82 J/7 - Mount Head
11-17-7-W5
50°24′50″N 114°51′15″W

Approximately 48 km south-west of Turner Valley, the source of McPhail Creek.

This permanent alpine lake was officially named 7 February 1983. During the late 1920s or early 1930s, Raymond Patterson found Horn Coral fossils scattered around this lake.

Horse Creek (creek)

83 F/13 - Hightower Creek
31-57-26-W5
53°57′N 117°51′W
Flows north-east into Berland River, approximately 64 km north north-west of Hinton.

The origin for the locally well-established name of this creek is unknown.

Horse Creek (creek)

82 O/2 - Jumpingpound Creek
8-26-4-W5
51°12′N 114°32′W
Flows south into Bow River, approximately 30 km north-west of Calgary.

The name of this creek was officially approved 12 December 1939. It is allegedly named Horse Creek because a lost horse was found drowned in the creek.

Horseshoe Basin (valley)

82 H/4 - Waterton Lakes
24-2-30-W4
49°08′N 113°55′W
Approximately 40 km south of Pincher Creek.

The name Horseshoe is descriptive of the shape of the basin. The name was made official 8 June 1971.

Horseshoe Falls (falls)

82 O/3 - Canmore
11-25-8-W5
51°07′N 115°02′W

Approximately 24 km east north-east of Canmore.

The name for these falls is likely descriptive of their appearance.

Horseshoe Glacier (glacier)

82 N/8 - Lake Louise
25-27-17-W5
51°21′N 116°15′W
Approximately 55 km west north-west of Banff.

The name for this horseshoe-shaped glacier is descriptive.

Horseshoe Lake (lake)

83 C/12 - Athabasca Falls
16-43-27-W5
52°42′N 117°40′W
Approximately 25 km south south-east of Jasper.

(see Horseshoe Glacier)

Horseshoe Ridge (ridge)

82 J/1 - Langford Creek
16-13-3-W5
50°05′N 114°20′W
Approximately 55 km west of Claresholm.

Named by M.P. Bridgland (see Mount Bridgland) 30 June 1915, this ridge measures 2131 m in altitude. Its name is descriptive of the sweeping curve in the ridge which faces north-east.

Howard Creek (creek)

82 J/15 - Bragg Creek
24-21-6-W5
50°48′N 114°43′W
Flows east into Quirk Creek, approximately 40 km west south-west of Calgary.

(see Mount Howard)

Howard, Mount (mountain)

82 J/15 - Bragg Creek
12-22-8-W5
50°51'N 114°59'W
Approximately 60 km west of Calgary.

Named in 1939 and officially approved 6
October 1949, this 2777 m mountain was
named after Terrence "Ted" Howard (1882-
1962), a homesteader (1902-1917), a First
World War veteran, (82nd Battalion) and the
local Forest Ranger from 1917 to 1937.

Howard Douglas Creek (creek)

82 O/4 - Banff
6-25-12-W5
51°06'N 115°40'W
Flows north-east into Brewster Creek,
approximately 11 km south-west of Banff.

This creek was named in 1904 after Howard
Douglas (1850-1929). In 1897, Douglas was
the Dominion Superintendent of Parks. He
was the first Commissioner of National
Parks and served in this capacity from 1908
to 1911.

Howard Douglas Lake (lake)

82 O/4 - Banff
51°02'15"N 115°44'50"W
Approximately 20 km south-west of Banff.

This 250 m by 100 m lake is at an elevation
of 2285 m. The name is taken from the fact
that it is the headwater branch of Howard
Douglas Creek. (see Mount Howard
Douglas) The lake was once known as
"Sundown Lake."

Howard Douglas, Mount (mountain)

82 O/4 - Banff
34-24-13-W5
51°05'N 115°44'W
Approximately 15 km south-west of Banff.

This 2820 m mountain was officially named
in 1958. (see Howard Douglas Creek)

Howse Pass (pass)

82 N/15 - Mistaya Lake
32-20-W5
51°48'N 116°46'W
Approximately 105 km north-west of Banff,
on the Alberta-B.C. boundary.

This pass was named after Joseph Howse
(1773-1852), a native of England who entered
the service of the Hudson's Bay Company in
1795. For much of the decade, from
1799-1809, he was in charge of Charleton
House (on the Saskatchewan River). In 1910,
he crossed the Rockies by this pass and
travelled southward to Montana, where he
built a post. During this trip, he explored
parts of the Columbia and Kootenay rivers
and wintered with the Flathead Indians.
Howse was appointed a Councillor of
Rupert's Land in 1815 but left the
appointment later that year to retire in
England. He published *A Grammar of the
Cree Language* (1844; second edition, 1865).

Howse Peak (peak)

82 N/15 - Mistaya Lake
7-33-20-W5
51°49'N 116°41'W
Approximately 105 km north-west of Banff,
on the Alberta-B.C. boundary.

This peak is 3290 m in altitude. (see Howse
Pass)

Howse River (river)

82 N/15 - Mistaya Lake
34-34-20-W5
51°57'N 116°45'W
Flows north into North Saskatchewan River,
approximately 105 km north-west of Banff.

(see Howse Pass)

Hudson Bay Lake (lake)

82 G/9 - Blairmore
8-8-2-W5
49°37'N 114°15'W

Approximately 27 km north-west of Pincher
Creek.

No information regarding the origin of the
name is known.

Huestis, Mount (mountain)

82 N/16 - Siffleur River
22-33-16-W5
51°50'N 116°11'W
Approximately 85 km north-west of Banff.

In recognition of his contribution to forestry
management in Alberta, this 3063 m
mountain was officially named by the
Canadian Permanent Committee on
Geographical Names 2 December 1967, after
the former Deputy Minister of Lands and
Forest, Mr. Eric Stephen Huestis (1900-1988).
He was a forest ranger, a timber inspector,
fish and game commissioner, and director of
forestry, before his position as Deputy
Minister of Forest and Wildlife. He retired in
1966 and began valuable work with Meals
on Wheels in Edmonton and at the Elves
Memorial Child Development Centre. Much
of his earlier employment was within the
Alberta mountain and foothills area. He did
much to help preserve Alberta lakeshores for
public use.

Hummingbird Creek (creek)

83 B/4 - Cripple Creek
9-36-14-W5
52°05'N 115°56'W
Flows east into Ram River, approximately 46
km south-east of Nordegg.

The origin for the locally well-established
name of this stream is unknown. The name
was officially approved 21 March 1942.

Hunt Creek (creek)

83 F/6 - Pedley
23-52-24-W5
53°30'N 117°25'W

Flows north into Athabasca River, approximately 65 km west of Edson.

This creek was likely named after Chauncey Murray Hunt (1885-1968), an early homesteader of the Pedley area. He and his sons trapped, logged, supplied firewood, and owned and operated a service station, which included a small grocery store and a restaurant.

Hunter Creek (creek)

82 J/1 - Langford Creek
8-13-2-W5
50°04′N 114°14′W
Flows east into Chaffin Creek, approximately 45 km west north-west of Claresholm.

The name was officially applied to this feature 28 October 1959. The origin of this locally well-established name is unknown.

Hunter, Mount (mountain)

83 E/9 - Moberly Creek
18-53-3-W6
53°34′N 118°26′W
Approximately 81 km north north-west of Jasper.

Formerly "Tip Top Mountain," this 2603 m mountain peak was named in 1948 after Flight Lieutenant R.H. Hunter, D.F.C. (1912-1945). Hunter was born in Botha, Alberta, and enlisted in the Royal Canadian Air Force in 1939 as an Aero-Engine Mechanic. He later was reassigned to aircrew in 1942, where he trained to become a pilot. Shortly afterward, he was awarded his Pilot's Flight Badge and appointed to a commission. He proceeded to the Far East and was listed as missing 19 February 1945.

Hunter Valley (valley)

82 O/11 - Burnt Timber Creek
13-31-9-W5
51°39′N 115°16′W

Approximately 55 km north north-east of Banff.

The origin of the name is unknown.

Huntington Creek (creek)

83 C/2 - Cline River
31-36-20-W5
52°09′N 116°50′W
Flows east into Cline River, approximately 115 km south-east of Jasper.

The name for this creek was officially approved 5 March 1935. Ellsworth Huntington (1876-1947) was an American cultural geographer, author and a professor at Yale from 1909 until his death. He imaginatively interpreted the effects of climate on human life and mapped out degrees of civilization.

Huntington Glacier (glacier)

83 C/2 - Cline River
17-37-21-W5
52°09′N 116°38′W
Approximately 110 km south-east of Jasper.

This glacier is the source of Huntington Creek. (see Huntington Creek)

Ice Lake (lake)

82 O/12 - Barrier Mountain
31-12-W5
51°39′20″N 115°35′00″W
Approximately 100 km north of Banff.

The name apparently stems from the fact that the lake remains covered with ice well into the summer due to its high altitude. The name was officially approved 7 September 1980.

Ice Water Creek (creek)

83 F/12 - Gregg Lake
19-52-27-W5
53°30′N 117°59′W

Flows north-west into Wildhay River, approximately 28 km west north-west Hinton.

The name may be descriptive of the water's temperature.

Icefall Mountain (mountain)

82 N/16 - Siffleur River
21-33-16-W5
51°50′N 116°13′W
Approximately 85 km north-west Banff.

This impressive, pinnacled mountain, which is 3221 m in altitude, has an isolated position and a hanging separate arm of Ram River Glacier is located on it. The name describes the feature.

Idalene Lake (lake)

83 E/2 - Resplendent Lake
53°09′45″N 118°42′00″W
Approximately 50 km north-west of Jasper.

The Idalene group of features was originally named by Jack Hargraves, an outfitter, and a group of tourists, in 1912. The significance of the name Idalene is not known.

Idlewilde Creek (creek)

83 B/4 - Cripple Creek
29-35-11-W5
52°02′N 115°29′W
Flows south into Clearwater River, approximately 56 km south-west of Rocky Mountain House.

Officially approved in 1941, the name for this creek is taken from the nearby mountain. The origin of the name is unknown.

Idlewilde Mountain (mountain)

83 B/3 - Tay River
27-35-11-W5
52°02′N 115°29′W
Approximately 51 km south-west of Caroline.

This well-established local name was officially adopted in 1941. The origin of the name is unknown.

Indefatigable, Mount (mountain)

82 J/11 - Kananaskis Lakes
19-9-W5
50°39′N 115°10′W
Approximately 65 km west of Turner Valley.

This mountain, which is 2670 m in altitude, was named in 1922 after a battle cruiser. The Indefatigable, along with The New Zealand, made up the 2nd Battle Cruiser Squadron at the Battle of Jutland. It was struck by five shells from the German battleship Vonder Tann and exploded, leaving only two survivors from a crew of 957.

Indian Creek (creek)

82 G/16 - Maycroft
6-10-1-W5
49°48′N 114°07′W
Flows south into Oldman River, approximately 30 km north-east of Blairmore.

The origin of the name is unknown.

Indian Creek (creek)

82 J/14 - Spray Lakes Reservoir
8-32-23-8-W5
50°59′55″N 115°04′25″W
Approximately 65 km west of Calgary.

This very small spring-fed creek is widely known for the quality of its water. Local legend has it that Indians came to this stream for drinking water. The name was officially approved 5 December 1984.

Indian Flat (alluvial flood plain)

82 O/3 - Canmore
10-27-24-10-W5
51°04′45″N 115°18′50″W
Approximately 3 km south-east of Canmore.

Indian Flat is an historical name for the place where Stoney Indians would camp during trips into and out of the mountains. This locally well-established name was officially approved 29 July 1986.

Indian Lookout (mountain)

82 N/16 - Siffleur River
35-33-15-W5
51°52′N 116°00′W
Approximately 85 km north north-west of Banff.

The name for this mountain, which is 2591 m in altitude, is likely descriptive.

Indian Pass (pass)

83 D/16 - Jasper
23-44-2-W6
52°48′N 118°11′W
Approximately 10 km south-west of Jasper.

(see Indian Ridge)

Indian Ridge (ridge)

83 D/16 - Jasper
44-2-W6
52°48′N 118°10′W
Approximately 9 km south-west of Jasper.

The name for this 2752 m ridge was suggested to M.P. Bridgland (see Mount Bridgland) in 1916 due to the red rock seen on the ridge.

Indian Spring (lake)

82 H/4 - Waterton Lakes
18-2-29-W4
49°08′N 113°53′W
Approximately 40 km south south-east of Pincher Creek.

There are a number of small springs flowing from the north of the north-west corner of the buffalo paddock that are locally called "Indian Springs." Evidence of an old Indian encampment near these springs makes the

name appropriate. The name was officially approved 8 June 1971.

Indian Springs Ridge (ridge)

82 H/4 - Waterton Lakes
2-29-W4
49°10′N 113°52′W
Approximately 40 km south south-east of Pincher Creek.

The name Indian Springs Ridge was officially applied to this feature 25 May 1943. (see Indian Springs)

Indianhead Creek (creek)

82 N/16 - Siffleur River
24-33-15-W5
51°52′N 116°00′W
Flows south-east into Clearwater River, approximately 80 km north north-west of Banff.

The origin of the name is unknown.

Inflexible, Mount (mountain)

82 J/14 - Spray Lakes Reservoir
17-21-9-W5
50°47′N 115°13′W
Approximately 80 km west south-west of Calgary.

This 3000 m mountain was named in 1922 after a warship engaged in the Battle of Jutland.

Inglismaldie, Mount (mountain)

82 O/3 - Canmore
23-26-11-W5
51°14′N 115°24′W
Approximately 13 km north-east of Banff.

This mountain, which is 2964 m in altitude, was named in 1886 or 1887 by George A. Stewart, Superintendent of Rocky Mountains Park, after Inglismaldie Castle, Kincardineshire, Scotland, seat of the Earl of Kintore, who visited Banff at that time.

Ings Creek (creek)

82 J/9 - Turner Valley
34-18-3-W5
50°34′N 114°20′W
Flows east into Highwood River,
approximately 13 km south of Turner Valley.

The name for this creek was officially
approved 8 December 1943. The Ings
brothers ranched near this creek until 1918.

Ink Pots (springs)

82 O/5 - Castle Mountain
6-27-13-W5
51°17′N 115°49′W
Approximately 20 km north-west of Banff.

This series of very interesting deep springs of
varying colour was officially named Ink Pots
in 1965. The colour of the springs varies
from shades of blue to slightly green.

Interpass Ridge (ridge)

83 E/12 - Pauline Creek
29-53-13-W6
53°37′N 119°57′W
Approximately 63 km west south-west of
Grande Cache, on the Alberta-B.C.
boundary.

The name for this ridge is likely descriptive.
It has been well-established in local usage
since at least 1924.

Intersection Mountain (mountain)

83 E/13 - Dry Canyon
3-56-14-W6
53°48′N 120°00′W
Approximately 58 km west south-west of
Grande Cache, on the Alberta-B.C. boundary.

The name for this 2452 m mountain was
chosen in preference to "Mount Haig" to
avoid duplication. This mountain is located
at the intersection of the Continental Divide
with the 120th Meridian. Intersection
Mountain was the name suggested by R.W.

Cautley of the Boundary Survey and this
name was officially applied to the feature in
1925.

Invincible Lake (lake)

82 J/11 - Kananaskis Lakes
3-5-20-9-W5
50°40′N 115°13′W
Approximately 65 km west of Turner Valley.

(see Mount Invincible)

Invincible, Mount (mountain)

82 J/11 - Kananaskis Lakes
20-9-W5
50°40′N 115°12′W
Approximately 64 km west of Turner Valley.

The Invincible, along with The Indomitable
and The Inflexible, made up the 3rd Battle
Cruiser Squadron at the Battle of Jutland
(see Jutland Mountain). This cruiser went
down after heavy German fire, leaving six
survivors from a crew of 1,034. This
mountain, 2670 m in altitude, was named in
1922 after this cruiser.

Iris Lake (lake)

83 D/16 - Jasper
16-45-2-W6
52°53′N 118°13′W
Approximately 9 km west of Jasper.

This lake was named in 1914. The origin of
the name is unknown.

Iron Creek (creek)

82 J/1 - Langford Creek
9-15-3-W5
50°15′N 114°23′W
Flows north-east into Willow Creek,
approximately 86 km south of Calgary.

No information regarding the origin of the
name is known.

Iron Creek (creek)

82 J/15 - Bragg Creek
50°57′N 114°36′W
Flows east into Elbow River, approximately
30 km west of Calgary.

This creek was officially named 12 December
1939. The origin of the name is unknown.

Iron Lakes (lakes)

82 J/1 - Langford Creek
11,14-1-15-4-W5
50°15′N 114°26′W
Approximately 87 km south south-west of
Calgary.

These lakes are named for their proximity as
the drainage for Iron Creek. (see Iron Creek)

Iron Ridge (ridge)

82 G/10 - Crowsnest
18-8-4-W5
49°39′N 114°32′W
Approximately 2 km north-west of
Coleman.

There is no origin information available for
this locally well-established name.

Isaac Creek (creek)

83 C/10 - George Creek
25-41-21-W5
52°33′N 116°54′W
Flows east into Brazeau River, approximately
55 km west of Nordegg.

This creek was officially named in 1947 after
Isaac Plante, who had a trap line near the
stream. The name was locally well-
established prior to its receiving official
status, but no other information is available.

Isaac Creek (creek)

83 E/15 - Pierre Greys Lakes
4-56-5-W6
53°48′N 118°40′W

Flows west into Mahon Creek, approximately 30 km east south-east of Grande Cache.

Isaac Plante, a local trapper, had his trap line near this creek. The name was officially approved in 1947.

Isaac, Mount (mountain)

83 C/10 - George Creek
28-41-21-W5
52°33′N 116°57′W
Approximately 61 km west of Nordegg.

This mountain, which is 2632 m in altitude, was officially named 5 March 1935. (see Isaac Creek)

Isabella Lake (lake)

82 N/16 - Siffleur River
1-32-17,18-W5
51°48′N 116°25′W
Approximately 90 km north-west of Banff.

This lake was named in 1898 by C.S. Thompson (see Thompson Pass) after his sister.

Isabelle Peak (peak)

82 N/1 - Mount Goodsir
15-25-15-W5
51°08′N 116°01′W
Approximately 35 km west south-west of Banff.

This peak is 2926 m in altitude. The origin of the name is unknown.

Ishbel, Mount (mountain)

82 O/5 - Castle Mountain
32-26-13-W5
51°16′N 115°46′W
Approximately 16 km north-west of Banff.

This mountain, which is 2908 m in altitude was named after Ishbel MacDonald, the daughter of Prime Minister Ramsay MacDonald (1866-1937) of Great Britain.

Island Creek (creek)

82 G/10 - Crowsnest
12-8-6-W5
49°38′N 114°41′W
Flows into Island Lake, approximately 13 km west of Coleman.

The name for this feature was sent to the Geographic Board for approval by Dr. R. Bell, acting Director of Geographical Survey, 19 March 1902. Island Creek is a descriptive name, and quite appropriate.

Island Lake (lake)

82 N/8 - Lake Louise
13-28-16-W5
51°23′N 116°07′W
Approximately 45 km north-west of Banff.

The locally well-established name for this feature was officially approved in 1958. There is an island located near the centre of the lake.

■ Island Lake (lake)

82 G/10 - Crowsnest
8-6-W5
49°37′N 114°40′W
Approximately 12 km west of Coleman.

This name has been in local usage since at least 1922. The origin of the name is unknown.

Island Muskeg (marsh)

82 O/11 - Burnt Timber Creek
9-32-8-W5
51°45′N 115°06′W
Approximately 70 km north-east of Banff.

No information regarding the origin of the name is known.

Island Ridge (ridge)

82 G/10 - Crowsnest
7,8-5,6-W5
49°37′N 114°41′W

Approximately 13 km west south-west of Coleman.

The origin of the name is unknown.

Isola Peak (peak)

82 J/1 - Langford Creek
13-4-W5
50°08′N 114°29′W
Approximately 65 km west north-west of Claresholm

This isolated mountain measures 2494 m in altitude. Its name is descriptive of its position.

Jack-knife Pass (pass)

83 E/10 - Adams Lookout
16-53-6-W6
53°34′30″N 118°49′30″W
Approximately 38 km south-east of Grande Cache.

Emil Moberly, a native hunter and trapper, said that this pass was so named because an early trapper found a jack-knife in the pass over 60 years ago. The name was officially approved 1 January 1979.

Jackfish Creek (creek)

83 B/5 - Saunders
17-40-11-W5
52°27′N 115°33′W
Flows south into North Saskatchewan River, approximately 36 km east of Nordegg.

The locally well-established name for this creek was officially approved in 1941. The origin of the name is unknown.

Jackfish Lake (lake)

83 B/12 - Harlech
6-41-11-W5
52°30′N 115°34′W
Approximately 34 km east of Nordegg.

The locally well-established name for this lake was made official in 1964. Presumably, the lake has a profusion of jackfish and this resulted in its name.

Jackknife Coulee (coulee)

82 G/16 - Maycroft
25-11-3-W5
49°47′N 114°17′W
Approximately 40 km north north-east of Blairmore.

The origin of this locally well-established name is unknown.

Jackpine Mountain (mountain)

83 E/5 - Holmes River
3-51-11-W6
53°22′N 119°34′W
Approximately 113 km north-west of Jasper.

This 2560 m mountain was named after the jackpine trees found on its slopes.

Jackpine Pass (pass)

83 E/6 - Twintree Lake
50-10-W6
53°22′N 119°26′W
Approximately 105 km west north-west of Jasper, on the Alberta-B.C. boundary.

The locally well-established name for this 2040 m pass was approved in 1925. It is the crossing watershed that allows access to the head of the Jackpine River. (see Jackpine Mountain)

Jackpine River (river)

83 E/11 - Hardscrabble Creek
28-54-10-W6
53°41′N 119°25′W
Flows north into Smoky River, approximately 29 km south-west of Grande Cache.

(see Jackpine Mountain)

Jackson Creek (creek)

82 G/8 - Beaver Mines
23-6-3-W5
49°29′N 114°17′W
Approximately 7 km south of Burmis.

There is no origin information available for this locally well-established name.

Jackson Creek (creek)

83 E/8 - Rock Lake
8-52-2-W6
53°28′N 118°15′W

Flows south-east into Wildhay River, approximately 68 km north north-west of Jasper.

This creek was named in 1948 in honour of Pilot Officer H.N. Jackson, D.F.C., of Millet, Alberta. Jackson was killed in the Second World War. The creek was formerly known as "Beirnes Creek."

Jackson Creek (creek)

83 F/3 - Cadomin
33-48-21-W5
53°11′N 117°00′W
Flows south-east into Embarras River, approximately 57 km south-west of Edson.

The origin of the name is unknown.

Jacob Creek (creek)

82 O/2 - Jumpingpound Creek
6-26-6-W5
51°12′N 114°49′W
Flows south-east into Bow River, approximately 45 km west north-west of Calgary.

The name of this stream is locally well-established and was made official 12 December 1939. The name was submitted by D.A. Nichols in 1929; it was the name of a Stoney Indian Chief who signed the Indian Treaty No. 7 in 1877.

Jacques Lake (lake)

83 C/13 - Medicine Lake
31-45-26-W5
52°55′N 117°45′W
Approximately 21 km east of Jasper.

(see Roche Jacques)

Jacques Pass (pass)

83 F/4 - Miette
1-46-28-W5
53°00′N 117°56′W
Approximately 52 km south south-west of Hinton.

(see Roche Jacques)

Jacques Range (range)

83 F/4 - Miette
47-27-W5
53°01′N 117°55′W
Approximately 49 km south-west of Hinton.

(see Roche Jacques)

Jacques, Roche (mountain)

83 F/4 - Miette
12-47-28-W5
53°02′N 117°57′W
Approximately 50 km south-west of Hinton.

"Old Jacques Cardinal" was a North West Company employee who was in charge of horse guard near Snaring River. He would provide horses to fur traders, and it is likely that this 2603 m mountain is named after him. The other features in the vicinity take their names from this mountain. (see Moberly Flats)

James Lake (lake)

82 O/11 - Burnt Timber Creek
12-32-11-W5
51°44′N 115°26′W
Approximately 60 km north of Banff.

(see James Pass)

James Pass (pass)

82 O/11 - Burnt Timber Creek
12-32-11-W5
51°44′N 115°27′W
Approximately 60 km north of Banff.

James Dickson, a celebrated Stoney Chief, was a councillor who signed Indian Treaty No. 7 in 1877. This pass, the nearby lake, and river are all named after this chief, whose name was "Ji-mis," in Cree.

James River (river)

82 O/14 - Limestone Mountain
13-34-5-W5
51°54′N 115°34′W
Approximately 85 km north of Banff.

The name for this river was officially approved 2 December 1941. (see James Pass)

James Walker Creek (creek)

82 J/11 - Kananaskis Lakes
31-20-9-W5
50°45 N 115°15′W
Flows south-west into Smith-Dorrien Creek, approximately 70 km west north-west of Turner Valley.

This creek was named for its proximity to Mount James Walker and the name was officially approved in November 1977. (see Mount James Walker)

James Walker, Mount (mountain)

82 J/14 - Spray Lakes Reservoir
29-21-9-W5
50°48′N 115°13′W
Approximately 80 km west south-west of Calgary.

Colonel James Walker (1846-1936) had several timber berths and a sawmill in the Kananaskis area. This ex-Mountie became the manager of the Cochrane Ranche in 1881. He was a soldier and businessman and followed his brother as postmaster of the first Kananaskis Post Office. This 3035 m mountain was officially named 16 June 1977.

Jarvis Creek (creek)

83 F/12 - Gregg Lake
1-54-26-W5
53°38′N 117°42′W
Flows north into Wildhay River, approximately 27 km north north-west of Hinton.

Edward William Jarvis and C.F. Hanington conducted a study of the Canadian Pacific Railway survey route during the winter of 1874-1875. This engineer joined the North West Mounted Police as superintendent after his long, arduous survey of the area from Fort George to the Athabasca River. The name for this creek was suggested by a local surveyor, Saint Cyr, in 1908, after E.W. Jarvis.

Jarvis Lake (lake)

83 F/5 - Entrance
5-52-26-W5
53°17′N 117°47′W
Approximately 14 km west north-west of Hinton.

(see Jarvis Creek)

Jasper (town)

83 D/16 - Jasper
7-16-45-1-W6
52°53′N 118°05′W
Approximately 66 km south-west of Hinton.

This town is named after the fur brigade post in the area. It was first mentioned in Franchère's *Relation d'un voyage à la Côte du Nord-Ouest de l'Amérique Septentrionale* (1820) (see Franchère Peak). The North West Company post was established in 1813 on Brule Lake as a "provision depot with the view of facilitating the passage of the mountains through Athabasca Pass." When Franchère was there (1814), the post was called "Rocky Mountain House," and was managed by Francois Décoigne (see Yellowhead Pass). This blonde-haired trapper used to store his furs near Tête Jaune Cache, in British Columbia; his nickname was "tête jaune" (yellow head). In 1817, the position was filled by Jasper Hawse, whose name was adopted to distinguish the post from the new Rocky Mountain House established on the North Saskatchewan River. In 1821, the Hudson's Bay Company and the North West Company amalgamated and by 1824, Michael Clyne (see Mount Cline) was in charge of the post. In 1829, Clyne built a new post at the junction of the Athabasca and the Snake Indian rivers. From 1835 to 1849, Colin Fraser (see Mount Colin) ran the post. In the early 1850s, it was closed as it was losing money. It was then reopened by Henry John Moberly (see Moberly Bend) and then closed in the late 1850s. The Jasper Forest Reserve, later Jasper Forest Park was named for the original fur brigade post. The confluence of the Miette and Athabasca rivers was selected by the Grand Trunk Pacific Railway as the location for their divisional point. This crew-changing station was given the name "Fitzhugh" in 1911, after a prominent Grand Trunk Pacific official. The following years, the name was changed to assume that of the new national park in which it was situated.

Jasper House (historic site)

83 D/16 - Jasper
23-48-28-W5
53°09′N 117°59′W
Approximately 25 km north north-east of Jasper.

To aid in provisioning the fur trading, a North West Company post was established at Brule Lake around 1813. Later relocated to a site near Jasper Lake, the post, which originally was known as Rocky Mountain House, was renamed Jasper House after Jasper Hawse, the trader placed in charge. The Hudson's Bay Company discontinued operations from Jasper House in 1884. (see Jasper)

Jasper Lake (lake)

83 E/1 - Snaring
35-47-1-W6
53°05′N 118°01′W
Approximately 23 km north of Jasper.

(see Jasper)

Jasper National Park (park)

83 C/13 - Medicine Lake
45-26-W5
52°53′N 117°45′W

Jasper National Park was established 14 September 1907 as a Forest Park. (see Jasper)

Jasper Park Lodge (locality)

83 D/16 - Jasper
15-45-1-W6
52°53′N 118°03′W
Approximately 2 km north-east of Jasper.

Jasper Park Lodge is unique in many respects. The main building, the largest single storey log structure of its kind in the world, comprises a spacious lounge, dining room, ballroom, as well as a limited number of bedrooms. The majority of guests who patronize this luxury hotel near Jasper find their sleeping accommodations in bungalow cabins, avenues of which encircle the shore of Lac Beauvert. These cabins vary in size from four to sixteen bedrooms (taken from *Jasper Park Lodge in the Canadian Rockies*, 1935. C.N.R.).

Jeffery Creek (creek)

83 C/13 - Medicine Lake
52°47′35″N 117°41′50″W
Flows north-east into Maligne River, approximately 25 km south-east of Jasper.

This creek was officially named 5 May 1987, after the Jeffery family of Jasper. Jeffery was a merchant in Jasper circa 1920. His sons and their partners are involved with Joe Weiss Mountain Climbers.

Jellicoe, Mount (mountain)

82 J/11 - Kananaskis Lakes
26-20-10-W5
50°43′N 115°17′W
Approximately 70 km west of Turner Valley.

Admiral Sir John R. Jellicoe (1859-1935) commanded the British Grand Fleet, 1914-1916, including the Battle of Jutland (see Jutland Mountain). This 3246 m mountain was officially named after him in 1918.

Jerram, Mount (mountain)

82 J/11 - Kananaskis Lakes
16-20-8-W5
50°42′N 115°03′W
Approximately 55 km west of Turner Valley.

This 2996 m mountain was named in 1922 after Admiral Sir Thomas Jerram, who commanded the Second Battle Squadron, 1915-1916, and led it in the Battle of Jutland.

Jessie, Mount (mountain)

83 E/6 - Twintree Lake
12-50-10-W6
53°18′N 119°20′W
Approximately 95 km west north-west of Jasper.

This mountain, which is 2652 m in altitude, was named in 1925 after Miss Jessie Campbell. She was the sister of A.J. Campbell, of the Dominion Land Survey.

Jewell Pass (pass)

82 O/3 - Canmore
5&8-17&18-24-8-W5
51°03′N 115°06′W
Approximately 20 km east south-east of Canmore.

This mountain pass, which is approximately 1600 m in elevation, is named after Mr. Bud Jewell, a gentleman who had a lease to log the Douglas Fir trees which remained after a fire had occurred in the area. The name Jewell Pass is locally well-established, and was officially proposed in 1986 by two groups of Calgary women known as the "Tuesday Hikers."

Jim Creek (creek)

82 J/9 - Turner Valley
19-18-3-W5
50°31′45″N 114°24′20″W
Flows south into Sullivan Creek, approximately 18 km south-west of Turner Valley.

The name Jim Creek has been used locally for this creek since 1905. It was named for Jim Bews, who started ranching at the junction of Sullivan Creek and the Highwood River in 1905. The name was officially approved 9 September 1979.

Jim Ridge (ridge)

82 J/9 - Turner Valley
18-4-W5
50°41'N 114°27'W
Approximately 16 km south-west of Turner Valley.

(see Jim Creek)

Jimmy Simpson, Mount (mountain)

82 N/10 - Blaeberry River
31-18-W5
51°41'N 116°30'W
Approximately 85 km north-west of Banff.

This mountain was named in 1973 in honour of James "Jimmy" Simpson (1877-1972). Originally from England, he came to Canada in 1896, working many jobs in Winnipeg, the United States, and finally, Banff. He started as a cook for one of the outfitting camps and eventually became a guide and outfitter in his own right. He built a tourist lodge at Bow Lake which is still in operation.

Joachim Creek (creek)

83 E/14 - Grande Cache
4-58-8-W6
53°59'N 119°08'W
Flows west into Smoky River, approximately 11 km north of Grande Cache.

This creek was named after Adam Joachim (1875-1959), a well-respected trapper, guide, prospector and spiritual leader. Joachim was a quiet and modest man who became the unofficial leader of his people. He was conversant in Cree, English, Latin and French. He was the grandson of Colin

Fraser (see Mount Colin). Adam's first wife was the daughter of Ewan Moberly (see Evan's Creek). His second wife gave birth to Marie, who married John Moberly, Ewan's brother. They, in turn, gave birth to eight children. One of his children, Caroline, married Felix Plante. (see Felix Creek and Mount Adam Joachim)

Joachim Lakes (lakes)

83 E/16 - Donald Flats
7-57-3-W6
53°55'N 118°27'W
Approximately 47 km east of Grande Cache.

(see Joachim Creek)

Job Creek (creek)

83 C/7 - Job Creek
6-41-20-W5
52°30'N 116°52'W
Flows north-west into Brazeau River, approximately 90 km east south-east of Jasper.

This creek and a nearby lake were named for a Stoney Indian, Job Beaver, an enterprising guide and explorer who marked out trails in this area around the late nineteenth century. A.P. Coleman (see Coleman Peak) was so impressed with the efforts of this man that he named the two features after Job, without ever having met the man.

Job Lake (lake)

83 C/7 - Job Creek
20-39-20-W5
52°22'N 116°51'W
Approximately 100 km south-east of Jasper.

(see Job Creek)

Job Pass (pass)

83 C/7 - Job Creek
24-39-20-W5
52°21'N 116°45'W

Approximately 105 km south-east of Jasper.

(see Job Creek)

Jock Creek (creek)

83 B/5 - Saunders
35-39-14-W5
52°20'N 115°56'W
Flows north-east into North Saskatchewan River, approximately 14 km south-east of Nordegg.

This creek was officially named in 1941 after Jock Richardson, who had a cabin at the mouth of this stream. No other information is available.

Jock Lake (lake)

83 B/5 - Sanders
32-38-14-W5
52°19'N 115°58'W
Approximately 19 km south-east of Nordegg.

(see Jock Creek)

Joffre, Mount (mountain)

82 J/11 - Kananaskis Lakes
20-18-9-W5
50°32'N 115°12'W
Approximately 70 km west south-west of Turner Valley.

This mountain, which is 3450 m in altitude, was officially named in 1918 after a French General. Marshal Joseph Jaques Cesaire Joffre (1852-1931) was appointed Commander-in-Chief of all French armies 2 December 1915 until December 1916. He was appointed to the French High Commission in 1917 and was made a member of the French Academy in December 1918.

John Ware Ridge (ridge)

82 J/9 - Turner Valley
3-20-4-W4
50°41'N 114°27'W

Approximately 11 km west of Turner Valley.

The name John Ware Ridge was officially applied to this feature 8 December 1943. John Ware was a negro pioneer rancher. He was born in the southern United States and came to Alberta on a cattle drive in 1882. He was one of the best known cattlemen of the ranching era. Formerly known as "Nigger John Ridge," this feature was renamed in 1970 to avoid use of discriminatory names. (see Mount Ware)

John-John Creek (creek)

83 C/6 - Sunwapta
35-39-22-W5
52°24′N 117°04′W
Flows north-east into Brazeau Lake, approximately 86 km south-east of Jasper.

This creek gets its name from John John Harrington who was the father of Mrs. Charlie Mathieson, the wife of a Park Warden living at Brazeau Lake Warden Station in 1931. The name was approved 18 October 1976.

Johnson Creek (creek)

82 J/1 - Langford Creek
9-15-3-W5
50°15′N 114°21′W
Flows north-east into Willow Creek, approximately 60 km north-west of Claresholm.

"Link" Johnson was one of the first men to cut timber along this creek for the Du Rocher sawmill. It is for Mr. Johnson that this creek is named.

Johnson Creek (creek)

82 O/6 - Lake Minnewanka
17-28-8-W5
51°23′N 115°05′W
Flows north-east into Waiparous Creek, approximately 38 km north north-east of Canmore.

Although it is not entirely clear who "Johnson" was, the name is locally well-established. The name was officially approved for this stream 5 May 1987.

Johnson Lake (lake)

82 O/3 - Canmore
4-26-11-W5
51°12′N 115°29′W
Approximately 6 km north-east of Banff.

Submitted by J.R.B. Coleman, this locally well-established name was made official in 1959. The origin of the name is unknown.

Johnston Canyon (canyon)

82 O/4 - Banff
23-26-14-W5
51°15′N 115°51′W
Approximately 21 km west north-west of Banff.

The prospector after whom this canyon was named in 1958 resided in this area circa 1882.

Johnston Creek (creek)

82 O/4 - Banff
23-26-14-W5
51°14′30″N 115°51′10″W
Flows south-west into Bow River, approximately 21 km west north-west of Banff.

(see Johnston Canyon)

Jonas Creek (creek)

83 C/6 - Sunwapta
4-40-24-W5
52°25′N 117°27′W
Flows west into Sunwapta River, approximately 68 km south south-east of Jasper.

Named for Jonas Good Stoney, chief of the Wesley band of Stoney Indians at Morley, who were located on the north side of the Bow River. During this time, the Reverend John McDougall established a mission at Morley and invited the Stoney to make their headquarters at that place. Jonas was one of the chiefs who signed Treaty Seven in 1877 and also provided A.P. Coleman (see Mount Coleman) with information regarding trails from the North Saskatchewan to the Athabasca Rivers.

Jonas Pass (pass)

83 C/6 - Sunwapta
1-39-23-W5
52°20′N 117°10′W
Approximately 87 km south-east of Jasper.

(see Jonas Creek)

Jonas Shoulder (ridge)

83 C/6 - Sunwapta
23-39-23-W5
52°22′N 117°13′W
Approximately 83 km south-east of Jasper.

(see Jonas Creek)

Jones Pass (pass)

83 E/12 - Pauline Creek
35-52-13-W6
53°32′N 119°48′W
Approximately 59 km south-west of Grande Cache.

This pass is named on the Jobe-Phillips map (see Avalanche Pass). It was presumably named after R.W. Jones, a Grand Trunk Pacific Railway engineer who made a reconnaissance survey on the Alberta side of the boundary in approximately 1906.

Joshua Creek (creek)

82 O/2 - Jumpingpound Creek
35-25-7-W5
51°10′N 114°53′W
Flows south-east into Bow River, approximately 50 km west north-west of Calgary.

The name for this stream is locally well-established and was made official 12

December 1939. The name was submitted by D.A. Nichols in 1929 after years of common local usage.

Joyce River (river)

83 C/8 - Nordegg
24-38-15-W5
52°17′N 116°00′W
Flows south-east into North Ram River, approximately 72 km west of Rocky Mountain House.

This river, formerly known as "Smallpox Creek," was officially renamed 16 January 1950 after Squadron Leader R.G. Joyce (1920-1946). He was killed in the Second World War as a result of a flying accident while on coastal and transport command duties. During his tour of duty, he was awarded a Mention in Dispatches and the Air Force Cross.

Jumpingpound Creek (creek)

82 0/2 - Jumpingpound Creek
4-26-4-W5
51°11′N 114°30′W
Approximately 53 km west north-west of Calgary.

(see Jumpingpound Mountain)

Jumpingpound Mountain (mountain)

82 J/15 - Bragg Creek
16-23-7-W5
50°57′N 114°55′W
Approximately 55 km west of Calgary.

This 2225 m mountain was officially named 6 October 1949. The only origin information for the name of the nearby creek, is as follows: the name in Blackfoot is "ninapiskan," meaning "men's pound," and in Stoney, it is "to-ko-jap-tab-wapta." There is a high, steep bank near the mouth of the creek where buffalo were driven over and killed.

Junction Creek (creek)

82 J/10 - Mount Rae
14-19-6-W5
50°36′N 114°44′W
Flows north into Sheep River, approximately 32 km west of Turner Valley.

The name was officially approved 6 September 1951. (see Junction Mountain)

Junction Mountain (mountain)

82 J/10 - Mount Rae
25-18-6-W5
50°35′N 114°43′W
Approximately 32 km west of Turner Valley.

The name for this 2682 m mountain is descriptive since it is at this point where two forks of the Sheep River join. The name "Junction Mountain" has been in local usage since 1895-1896 and was made official 6 September 1951.

Jura Creek (creek)

82 0/3 - Canmore
23-24-9-W5
51°04′N 115°09′W
Flows south-east into Bow River, approximately 15 km east south-east of Canmore.

The name for this creek was approved by the Canadian Permanent Committee on Geographical Names 17 March 1967. The origin of the name is unknown.

Jutland Brook (brook)

82 G/1 - Sage Creek
3-2-W5
49°13′N 114°14′W
Approximately 30 km north-west of Waterton Park.

(see Jutland Mountain)

Jutland Mountain (mountain)

82 G/1 - Sage Creek
3-3-W5
49°12′N 114°14′W
Approximately 30 km north-west of Waterton Park.

This 2408 m mountain is named for the Battle of Jutland. This battle was a major naval engagement fought between the British and German fleets during the First World War. The action took place about 122 km off the Danish coast of Jutland on 31 May and 1 June 1916. Although the material and human losses of the British were the greater, the German fleet made no further move toward breaking the Allied blockade. The end result was seen as a draw.

Kakwa Falls (falls)

83 L/4 - Kakwa Falls
19-59-13-W6
54°07′N 119°56′W
Approximately 60 km west north-west of Grande Cache.

This is a beautiful waterfall located on the Kakwa River. (see Kakwa Mountain)

Kakwa Mountain (mountain)

83 L/4 - Kakwa Falls
59-14-W6
54°05′N 119°59′W
Approximately 65 km west north-west of Grande Cache.

This mountain, which is 2259 m in altitude, was officially named 20 November 1925. The name Kakwa is the Cree Indian word for "porcupine," and was suggested by S. Prescott Fay (see Mount Fay), perhaps due to the shape of the mountain.

Kananaskis (hamlet)

82 0/3 - Canmore
31-24-8-W5
51°04′N 115°08′W

Approximately 17 km east of Canmore.

(see Kananaskis Range)

Kananaskis Falls (falls)

82 0/3 - Canmore
4-25-8-W5
51°06'N 115°04'W
Approximately 21 km east of Canmore.

(see Kananaskis Range)

Kananaskis Falls, 1889

Kananaskis Lakes (lakes)

82 J/11 - Kananaskis Lakes
19, 20-9-W5
50°38'N 115°08'W
Approximately 90 km south-west of Calgary.

(see Upper and Lower Kananaskis lakes)

Kananaskis Range (range)

82 J/14 - Spray Lakes Reservoir
21-10-W5
50°50'N 115°15'W
Approximately 80 km west of Calgary.

This range was formerly known as "Ship Mountains" due to the number of mountains named after ships engaged in the Battle of Jutland. The name Kananaskis is a corruption of "kin-e-ah-kis," the name of a Cree Indian about whom there is a legend giving an account of his recovery from a blow from an axe which had stunned but failed to kill him. Palliser (see Palliser Range) named the river and the pass after this extraordinary individual. The lakes take their names from the pass as well.

Kananaskis River (river)

82 O/3 - Canmore
24-8-W5
51°05'N 115°03'W
Flows north into Bow River, approximately 21 km east of Canmore.

(see Kananaskis Range)

Kananaskis River in valley, 1925

Kananaskis Village (small alpine village)

82 J/14 - Spray Lakes Reservoir
35-22-9-W5
50°55'56"N 115°09'10"W
Approximately 25 km south-east of Canmore.

This unincorporated resort facility received an official postal code in January 1987. (see Kananaskis Range)

Kane Glacier (glacier)

83 D/8 - Athabasca Pass
40-2-W6
52°24'N 118°09'W
Approximately 50 km south of Jasper.

(see Mount Kane)

Kane, Mount (mountain)

83 D/8 - Athabasca Pass
12-40-2-W6
52°26'N 118°08'W
Approximately 49 km south of Jasper.

This 3090 m mountain was named in 1921 by A.O. Wheeler (see Wheeler Flats) after Paul Kane (1810-1871). Kane was the most famous of all Canadian artist-explorers. After living in Toronto and having visited an exhibition of George Catlin's American Indian paintings in London, England, Kane was determined to paint a similar series of Canadian Indians. He recorded the impressions of his journey with George Simpson (see Simpson Pass) in *Wanderings of an Artist Among the Indians of North America* (1859).

Kangienos Lake (lake)

82 0/2 - Jumpingpound Creek
20-26-7-W5
51°14'N 114°33'W
Approximately 55 km west north-west of Calgary.

Officially approved 12 December 1939, the origin of the name for this 3 km long narrow lake, is unknown.

Karst Springs (springs)

82 J/14 - Spray Lakes Reservoir
2-22-11-W5
50°51'40"N 115°25'35"W
Approximately 82 km west of Calgary.

Karst is a geological term referring to pure limestone rock which is slowly eroded by water, forming incredible shapes and canyons. Karst topography includes springs, sinkholes, and underground streams. This spring is located in such a limestone formation. The water flows to the surface with considerable force and volume in approximately a 3 m wide area. Moss growth covers much of the surrounding area. The descriptive name for this spring was officially approved 12 December 1984.

Kataka Mountain (mountain)

83 D/16 - Jasper
32-44-3-W6
52°50′N 118°23′W
Approximately 22 km west south-west of Jasper.

The name for this 2621 m mountain is the Indian word for "fort." The flat-topped mountain resembles a fort and was given its descriptive name in 1916 by M.P. Bridgland. (see Mount Bridgland)

Katherine Lake (lake)

82 N/9 - Hector Lake
30-31-17-W5
51°41′N 116°23′W
Approximately 80 km north-west of Banff.

This lake, nearby Helen Lake and Margaret Lake were named for the daughters of Reverend H.P. Nichols. (see Lake Alice)

Katrine Lake (lake)

83 D/16 - Jasper
33-45-1-W6
52°55′N 118°04′W
Approximately 5 km north of Jasper.

H. Matheson of the Dominion Land Survey named this lake in 1914. The origin of the name is unknown.

Kaufmann Creek (creek)

82 N/15 - Mistaya Lake
12-34-20-W5
51°54′N 116°43′W

Flows north-east into Mistaya River, approximately 115 km north-west of Banff.

The Canadian Permanent Committee on Geographical Names approved this name in September 1971. (see Kaufmann Peaks)

Kaufmann Lake (lake)

82 N/15 - Mistaya Lake
2-34-20-W5
51°54′N 116°44′W
Approximately 115 km north-west of Banff.

The Canadian Permanent Committee on Geographical Names approved this name in September 1971. (see Kaufmann Peaks)

Kaufmann Peaks (peaks)

82 N/15 - Mistaya Lake
34-33-20-W5
51°53′N 116°45′W
Approximately 115 km north-west of Banff.

The name Kaufmann Peaks for these two mountain peaks, which are 3109 m and 3094 m in altitude, was suggested by Sir J. Outram (see Mount Outram), an alpinist who was the first to climb this mountain. With him was a Swiss guide, Christian Kaufmann (see Christian Peak). Outram thought the peaks should be named after them.

Kelsey, Mount (mountain)

83 E/7 - Blue Creek
25-50-5-W6
53°21′N 118°34′W
Approximately 64 km north-west of Jasper.

Officially named 1 May 1934, this mountain measures 2482 m in altitude. It is named after an early explorer, Henry Kelsey, who died in 1729. He was apprenticed to the Hudson's Bay Company circa 1688 and became the Governor of York Factory in 1717.

Kenny Creek (creek)

83 L/6 - Chicken Creek
20-62-9-W6
54°23′N 119°19′W
Flows north-east into Redrock Creek, approximately 45 km north north-west of Grande Cache.

This creek was named after Adam Kenny, a well-known native trapper and guide of the area. He married Miss Annie Sask in Grande Prairie in February 1939.

Kent Creek (creek)

82 J/11 - Kananaskis Lakes
14-20-9-W5
50°42′N 115°08′W
Flows south into Lower Kananaskis Lake, approximately 60 km west of Turner Valley.

Officially approved in November 1977, the name Kent Creek is locally well-established. (see Mount Kent)

Kent, Mount (mountain)

82 J/11 - Kananaskis Lakes
33-20-9-W5
50°43′N 115°11′W
Approximately 65 km west north-west of Turner Valley.

The origin for the name of this mountain, which is 2635 m in altitude, is unknown.

Kentigern, Mount (mountain)

82 N/16 - Siffleur River
32-17-W5
51°47′N 116°20′W
Approximately 85 km north-west of Banff.

This 3176 m mountain was named by R.W. Cautley of the Boundary Commission (see Mount Cautley) in 1928 after St. Kentigern, who lived in the sixth century. This mountain peak marks the Park Boundary at a point dividing the Clearwater and Siffleur rivers. The name Kentigern was thought by Cautley to be "euphonious and striking."

Kephala Creek (creek)

83 E/8 - Rock Lake
10-50-1-W6
53°18′N 118°03′W
Flows south-west into Moosehorn Creek,
approximately 47 km north of Jasper.

(see Mount Kephala)

Kephala, Mount (mountain)

83 E/8 - Rock Lake
35-50-1-W6
53°22′N 118°02′W
Approximately 54 km north of Jasper.

Kephala is the Greek word meaning
"head." This 2429 m mountain is the highest
peak in its range. The creek takes its name
from the mountain.

Kerkeslin, Mount (mountain)

83 C/12 - Athabasca Falls
28-42-27-W5
52°39′N 117°50′W
Approximately 32 km south south-east of
Jasper.

This 2984 m mountain was named by Sir
James Hector (see Mount Hector) in 1859.
The origin of the name is unknown.

Kerr, Mount (mountain)

83 D/16 - Jasper
4-46-2-W6
52°56′N 118°13′W
Approximately 11 km north-west of Jasper.

This mountain, which is 2560 m in altitude,
was named in 1951 after John Chipman
Kerr, V.C. (1887-1918), of the 49th Battalion
of the Canadian Expeditionary Force. He
was killed in action in the First World War.

Kesler Lake (lake)

82 H/4 - Waterton Lakes
2-3-30-W4
40°11′N 113°56′W

Approximately 35 km south of Pincher
Creek.

The origin of the name is unknown.

Kew Ridge (ridge)

82 J/9 - Turner Valley
9-20-3-W5
50°41′N 114°21′W
Approximately 4 km west of Turner Valley.

Officially named 8 December 1943, this
1447 m ridge is named after John and Kate
Quirk's cattle brand, "Q," and their ranch,
"Kew." (see Mount Quirk)

Kia Nea Lake (lake)

83 F/5 - Entrance
32-49-26-W5
53°16′N 117°47′W
Approximately 18 km south-west of Hinton.

The name was officially approved in 1979,
but its origin remains unknown.

Kicking Horse Pass (pass)

82 N/8 - Lake Louise
2-29-17-W5
51°27′N 116°17′W
Approximately 60 km north-west of Banff,
on the Alberta-B.C. boundary.

This pass commemorates an incident
involving Sir James Hector (see Mount
Hector) and his horse. The Palliser
Expedition was sponsored by Britain and
was to be a scientific exploration of the
Rocky Mountains and foothills (see
Introduction). During the exploration, one
of the pack horses, attempting to dodge
falling timber, plunged into the nearby
stream. While the men of the expedition
attempted to rescue the animal from the
water, Hector's horse wandered off. Hector
chased after his runaway horse, but in the
course of the pursuit, he was kicked in the
chest and was winded. Fortunately, the
geologist and surgeon was not seriously hurt

and was able to witness the naming of the
pass by his companions in memory of this
moment. The Kicking Horse River was
named for the same accident.

Kidd Creek (creek)

83 C/8 - Nordegg
5-39-16-W5
52°20′N 116°16′W
Flows north into North Saskatchewan River,
approximately 20 km south-west of Nordegg.

(see Mount Kidd)

Kidd, Mount (mountain)

82 J/14 - Spray Lakes Reservoir
28-22-9-W5
50°54′N 115°11′W
Approximately 75 km west of Calgary.

Stuart Kidd (1883-1956) came to Alberta in
1903. He was a homesteader, later manager
of Scott and Lesson's Trading Post at Morley,
and then manager of Brazeau Trading Co. at
Nordegg. He could speak fluent Stoney, and
in 1927, Mr. Kidd was made Honorary
Indian Chief of Stoneys, Chief "Tah-Osa"
(meaning "moose killer"). This mountain
was named after Kidd in 1907 by the geologist,
Dr. D.B. Dowling (see Dowling Ford).

King Creek (creek)

82 J/11 - Kananaskis Lakes
24-20-9-W5
50°43′N 115°08′W
Flows west into Kananaskis River,
approximately 60 km west of Turner Valley.

This creek is named after William Henry
"Willie" King (1875-1941), who came to the
Kananaskis Lakes area in 1910 with others to
look for coal, to hunt for big game, and to fish.

King Albert, Mount (mountain)

82 J/11 - Kananaskis Lakes
23-20-11-W5
50°40′N 115°25′W

Approximately 80 km west of Turner Valley, on the Alberta-B.C. boundary.

This mountain, 2987 m in altitude, was officially named in 1918 after King Albert of Belgium (1875-1934). This popular and progressive monarch was devoted to the welfare of his people and development of his country. Albert was killed in a mountain climbing accident 17 February 1934.

King Creek Ridge (ridge)

82 J/11 - Kananaskis Lakes
31-20-8-W5
50°44'15"N 115°06'15"W
Approximately 60 km west north-west of Turner Valley.

Officially approved 8 November 1978, this 2408 m ridge was named because of its proximity to King Creek. (see King Creek)

King Edward Glacier (glacier)

83 C/4 - Clemenceau Icefield
52°09'00"N 117°32'30"W
Approximately 94 km south-east of Jasper.

King Edward Glacier is named because it is located on the western slope of Mount King Edward. The name was first used in the *Alpine Journal* in 1925. The name was officially approved 7 November 1984. (see Mount King Edward)

King Edward, Mount (mountain)

83 C/4 - Clemenceau Icefield
9-37-25-W5
52°09'N 117°31'W
Approximately 90 km south-east of Jasper on the Alberta-B.C. boundary.

This 3490 m mountain was named by Mrs. Mary T.S. Schäffer, author of *A Hunter of Peace: Mary T.S. Schäffer's Old Indian Trails of The Canadian Rockies*, introduced and edited by E.J. Hart, 1980 (originally published 1911). The name King Edward commemorated Edward VII (1841-1910),

who succeeded to the throne of England in 1901.

Kingfisher Lake (lake)

82 N/8 - Lake Louise
32-28-16-W5
51°25'N 116°10'W
Approximately 50 km north-west of Banff.

Formerly "Betty Lake," the locally well-established name for this feature was officially approved in 1958.

Kinky Lake (lake)

83 F/5 - Entrance
6-50-26-W5
53°17'N 117°47'W
Approximately 53 km north north-east of Jasper.

The origin of the name is unknown.

Kinross, Mount (mountain)

83 D/16 - Jasper
11-46-2-W6
52°57'N 118°11'W
Approximately 24 km north-west of Jasper.

This 2560 m mountain was named in 1951. Private Cecil John Kinross, V.C. (1895-1957), was born in England but moved to Canada and enlisted in the Canadian Army 21 October 1915. He went to England 28 December 1915 and to France 16 March 1916 with the Canadian Expeditionary Force. He won the Victoria Cross 11 January 1918 while serving with the 49th Battalion. Private Kinross charged an enemy machine gun over open ground in broad daylight and killed six of the crew. He was severely wounded later in the day.

Kishinena Peak (peak)

82 G/1 - Sage Creek
2-2-W5
49°07'N 114°09'W

Approximately 18 km north-west of Waterton Park, on the Alberta-B.C. boundary.

The name Kishinena may be a corruption of the Indian word, "ish-nee-nee," which means "there it is." Evidently, the survey party was looking for water, and the Kootenays who crossed the summit with them answered saying "ish-nee-nee," and because they speak with a guttural sound, the survey party did not catch the proper pronunciation. There is another possibility for the origin of the name of this 2436 m peak: an Indian tribe, supposedly a branch of the Panther tribe, is known as the Kishinena tribe. The name was officially approved 5 March 1959.

Kiska Creek (creek)

83 C/8 - Nordegg
15-38-15-W5
52°16'N 116°03'W
Flows east into North Ram River, approximately 22 km south of Nordegg.

Named in 1932, this creek bears an abbreviated Indian name for "mountain sheep."

Kiskiu Creek (creek)

83 E/16 - Donald Flats
8-58-2-W6
54°00'N 118°16'W
Flows north into Little Smoky River, approximately 60 km north-east of Grande Cache.

The Indian name for this creek describes the flats where the creek enters the Little Smoky River. The name means "bob-tail" and was officially approved in 1947.

Kitchener, Mount (mountain)

83 C/3 - Columbia Icefield
26-35-24-W5
52°13'N 117°20'W

Approximately 90 km south-east of Jasper.

The name for this 3505 m mountain was officially approved 7 November 1916 over the name "Mount Douglas" to avoid duplication. Horatio Herbert Kitchener, Viscount Kitchener (1850-1916), was a British Field Marshal and Proconsul who was the Secretary of State for War (1914-1916) and who organized British armies at the outset of the First World War.

Klein Lake (lake)

82 0/11 - Burnt Timber Creek
33-30-10-W5
51°37'N 115°21'W
Approximately 50 km north north-east of Banff.

The Canadian Permanent Committee on Geographical Names officially approved the name of this lake in June 1968. Its origin is uncertain, but it may refer to Michael Cline. (see Mount Cline)

Knife Mountain (mountain)

83 E/14 - Grande Cache
28-55-8-W6
53°46'N 119°08'W
Approximately 12 km south of Grande Cache.

Knife Mountain is 2057 m in altitude. The origin of the name is unknown.

Knight, Mount (mountain)

83 E/1 - Snaring
36-47-4-W6
53°06'N 118°27'W
Approximately 35 km north-west of Jasper.

This was one of the 12 features named in 1954 for Alberta's 50th Jubilee (see Mount Griesbach). This 2845 m mountain was

named after Richard M. Knight (1876-1931). He was born in Ontario and came west in 1904, joining with Mr. A. Driscoll to form the company of Driscoll and Knight, Land Surveyors. He was actively involved with the Land Surveyors Association throughout his life and succeeded Maynard Rogers as the Superintendent of Jasper National Park, a postion in which he served until his death.

■ Kootenay Plains (plain)

83 C/1 - Whiterabbit Creek
20-36-17-W5
52°06'N 116°25'W
Approximately 120 km north-west of Banff.

In 1858, Sir James Hector (see Mount Hector) noted in his journal, "This plain, which is 12 km long and 4 km wide, is called the Kootanie Plain, as at the time that the Kootanie Indians exchanged their furs with the traders of the Saskatchewan forts, before there was any communication with them from the Pacific coast, an annual mart was held at this place, to which the Kootanie Indians crossed the mountains, while the traders came from the Mountain House." (see South Kootenay Pass)

Kotch Gap (gap)

82 J/8 - Stimson Creek
13-17-4-W5
50°26'N 114°25'W
Approximately 30 km south-west of Turner Valley.

Officially approved 12 December 1939, the name for this gap is presumably taken from A.J.H. Koch, who was a stepson of the Pflughaupts, ranchers in this area. Mr. Koch's name has obviously been corrupted.

*Kovach (former locality)

82 J/14 - Spray Lakes Reservoir
2-23-9-W5
50°56'N 115°12'W
Approximately 70 km west of Calgary.

This locality is likely named after Joe Kovach, a one-time resident in the Crowsnest Pass and former Canmore District Forest Ranger (1940-1953).

Kovach Pond (man-made pond)

82 J/14 - Spray Lakes Reservoir
14-1-23-9-W5
50°56'N 115°09'W
Approximately 70 km west of Calgary.

This man-made pond was once a gravel pit. A name was required because the feature was to be stocked with trout. The name is taken from the nearby locality of Kovach, a former mining community which took the name from Joe Kovach, a one-time resident.

Kvass Creek (creek)

83 E/11 - Hardscrabble Creek
24-54-8-W6
53°41'N 119°02'W
Flows north-east into Sulphur River, approximately 15 km south south-east of Grande Cache.

Fred Kvass was a pioneer trapper in this district. It is after this man that the stream and nearby flat are named.

Kvass Flats (flats)

83 E/14 - Grande Cache
55-9-W6
53°48'N 119°16'W
Approximately 10 km south-west of Grande Cache.

(see Kvass Creek)

---

*denotes rescinded name or former locality.

La Coulotte Peak (peak)

82 G/1 - Sage Creek
3-3-W5
49°12′N 114°19′W
Approximately 35 km north-west of
Waterton Park, on the Alberta-B.C.
boundary.

This peak was originally named in 1917,
after a place on the outskirts of Lens, France.

La Coulotte Ridge (ridge)

82 G/1 - Sage Creek
2-2-W5
49°11′N 114°17′W
Approximately 32 km north-west of
Waterton Park.

(see La Coulotte Peak)

La Crèche Mountain (mountain)

83 E/13 - Dry Canyon
2-58-19-W6
53°59′N 119°58′W
Approximately 57 km west north-west of
Grande Cache.

This name was suggested for this 2314 m
mountain because it had been adopted as a
cradle or nursery by mountain goats when
R.W. Cautley (see Mount Cautley) was
surveying in the area in 1925. Its precipitous
easterly face provided an excellent setting for
climbing practice for young goats.

La Grace, Mount (mountain)

83 C/14 - Mountain Park
7-44-22-W5
52°47′N 117°11′W
Approximately 60 km east of Jasper.

This 2822 m mountain is named after a
Métis hunter who accompanied Lord
Southesk (see Southesk River) on his trip to
the Rockies in 1859.

LaForce Creek (creek)

83 L/3 - Copton Creek
28-59-9-W6
54°08′N 119°18′W
Flows north-east into Copton Creek,
approximately 33 km north-west of Grande
Cache.

This creek was named in 1951 after Private
Alphonse Joseph LaForce, M.M., of Legal,
Alberta, who was a casualty of the Second
World War.

Labyrinth Mountain (mountain)

82 O/11 - Burnt Timber Creek
34-31-11-W5
51°41′40″N 115°28′25″W
Approximately 70 km north of Canmore.

The name Labyrinth Mountain was
proposed for this feature by Dr. D.M.
Cruden from the Department of Geology,
University of Alberta, 8 January 1981. The
name is descriptive, being a special feature of
the mountain itself. The name was officially
approved 5 May 1987.

Lagoon Lake (lake)

82 N/15 - Mistaya Lake
21-33-20-W5
51°50′N 116°47′W
Approximately 115 km north-west of Banff.

The origin of the name is unknown.

Lake Louise (hamlet)

82 N/8 - Lake Louise
28-28-16-W5
51°26′N 116°11′W
Approximately 60 km north-west of Banff.

Formerly Holt City (in 1883), and then
Laggan (1901-1914) which was suggested by
Lord Strathcona after Laggan, a hamlet in
Inverness, Scotland. The post office name,
Laggan was changed to Lake Louise 1 March

1914 in honour of Princes Louise Caroline
Alberta (1848-1939). (see Lake Louise)

Lakeview Ridge (ridge)

82 H/4 - Waterton Lakes
2-30-W4
49°04′N 113°54′W
Approximately 35 km south of Pincher
Creek.

The name is descriptive of the view from the
ridge and was officially approved 8 July 1954.

Lambe, Mount (mountain)

82 N/10 - Blaeberry River
18-32-20-W5
51°44′N 116°49′W
Approximately 110 km north-west of Banff,
on the Alberta-B.C. boundary.

This 3182 m mountain was named in 1920
after Lawrence Morris Lambe, C.M.G., a
vertebrate palaeontologist to the Geological
Survey of Canada, and an author of several
valuable scientific reports and papers.

Lancaster Creek (creek)

83 E/10 - Adams Lookout
9-55-6-W6
53°44′N 118°50′W
Flows east into Muskeg River, approximately
24 km south-east of Grande Cache.

The origin of the name is unknown.

Landslide Lake (lake)

83 C/2 - Cline River
7-36-18-W5
52°05′N 116°33′W

Approximately 115 km west south-west of Rocky Mountain House.

This scenic, 1.6 km long lake was created by a large rock slide. The lake has underground drainage for 0.8 km because of the slide. Landslide Lake was the name officially approved for this feature 30 October 1975.

Lanthier Creek (creek)

82 J/1 - Langford Creek
2-13-2-W5
50°03′N 114°10′W
Flows west into Nelson Creek, approximately 40 km west of Claresholm.

Officially approved by the Canadian Permanent Committee on Geographical Names 22 April 1970, this creek is named after Herman Lanthier who homesteaded by Nelson Creek. He was a Frenchman from Montreal and did a lot of fencing with Harry Beegle for Biley and Thompson. He returned to Quebec in 1917 after selling his farm.

Lapensée, Mount (mountain)

83 C/12 - Athabasca Falls
10-41-28-W5
52°31′N 117°57′W
Approximately 41 km south of Jasper.

This mountain, which is 3106 m in altitude, was named in 1920 by D.B. Dowling (see Dowling Ford). It commemorates Oliver Roy Lapensée. (see Mount Belanger and Franchère Peak)

Larch Valley (valley)

82 N/8 - Lake Louise
20-27-16-W5
51°20′N 116°13′W
Approximately 50 km west north-west of Banff.

This valley was named in May 1918. There is a profusion of the bright-foliaged deciduous

coniferous tree of the genus Larix (larch) in the vicinity.

Larkspur Creek (creek)

83 E/14 - Grande Cache
22-57-10-W6
53°56′N 119°24′W
Flows south-east into Sheep Creek, approximately 20 km west north-west of Grande Cache.

This creek was named in 1953 after the wildflower, the Larkspur, abundantly found throughout the foothills and forested regions during June and July. The flowers have an irregular shape with a long spur.

Lassie Creek (creek)

82 G/16 - Maycroft
11-12-2-W5
49°59′N 114°10′W
Flows west into Callum Creek, approximately 45 km north-east of Blairmore.

The name was officially applied to this feature 3 December 1941. The precise origin of this locally well-established name is unknown.

Last Break (hill)

82 J/15 - Bragg Creek
26-23-5-W5
50°59′N 114°35′W
Approximately 30 km west of Calgary.

This hill is the last one in the chain of hills extending northward. The descriptive name was officially approved 6 October 1949.

Laurie (Îyâmnathka), Mount (mountain)

82 O/3 - Canmore
7-25-8-W5
51°07′25″N 115°07′00″W
Approximately 17 km east north-east of Canmore.

The name Mount Laurie was originally made official 5 April 1961, in honour of Dr. John Laurie (1901-1959) who was a teacher among the Indians of southern Alberta and an advocate of aboriginal rights. This feature had been locally known by the name "Yamnuska" which is an approximation of the Stoney Indian name Îyâmnathka, meaning "flat faced mountain." In recognition of the widespread use of the local descriptive name, the form with the proper Stoney name in parentheses was officially approved 5 December 1984.

Laut, Mount (mountain)

83 E/9 - Moberly Creek
13-53-4-W6
53°34′N 118°27′W
Approximately 82 km north north-west of Jasper.

Wing Commander A. Laut, M.I.D. (1915-1943) was born in Crossfield, Alberta. He attended high school in Calgary, and then went to Mount Royal College, the University of Alberta and the University of Saskatchewan where he obtained a Bachelor of Science Degree in mechanical engineering. He was appointed to a commission in the Royal Canadian Air Force 4 July 1938 and served with the Coastal Command Squadrons in Canada. He was killed in a flying accident 3 October 1943. This 2408 m mountain was named in 1948 to commemorate him.

Lawrence Creek (creek)

83 E/14 - Grande Cache
34-55-9-W6
53°48′N 119°15′W
Flows south-east into Smoky River, approximately 14 km south-west of Grande Cache.

The locally well-established name for this creek was officially approved 4 June 1953. There are several Lawrences from the area after whom the creek may be named;

however, it is not certain after which one this creek was intended to be named.

Lawson Lake (lake)

82 J/11 - Kananaskis Lakes
12-20-10-W5
50°40'N 115°15'W
Approximately 70 km west of Turner Valley.

This lake was named in June 1902 by Walter D. Wilcox. (see Wilcox Pass and Mount Lawson)

Lawson, Mount (mountain)

82 J/14 - Spray Lakes Reservoir
9-21-9-W5
50°46'N 115°11'W
Approximately 75 km west south-west of Calgary.

The name for this 2795 m mountain was suggested by H.G. Bryant (see Bryant Creek) and W.D. Wilcox (see Wilcox Pass), after Major W.E. Lawson of the Geological Survey. Lawson was killed in France. The mountain was officially named in 1922.

Le Grande Brazeau (range)

83 C/6 - Sunwapta
40-23-W5
52°26'N 117°15'W
Approximately 67 km south-east of Jasper.

(see Brazeau Range)

Leach Lake (lake)

83 C/12 - Athabasca Falls
18-43-27-W5
52°42'N 117°54'W
Approximately 25 km south south-east of Jasper.

The name for this lake was officially approved 2 August 1956. The origin of the name is unknown.

Leah Lake (lake)

83 C/7 - Job Creek
7-39-20-W5
52°20'15"N 116°50'00"W
Approximately 100 km south-east of Jasper.

The name for this lake was officially approved 14 April 1980. Fred A. Kidd, the M.L.A. for Banff Constituency (1975-1979), proposed that this and another lake in the area could be named for two well-known Stoney Indian sheep hunters of the Wesley band, namely Samson and Leah Beaver. Shamosin (Samson) Beaver is the son of Job Beaver. (see Job Creek)

Leah Peak (peak)

83 C/12 - Athabasca Falls
21-43-25-W5
52°43'N 117°34'W
Approximately 39 km south-east of Jasper.

This mountain peak, which is 2801 m in altitude, bears the name of the wife of Samson Beaver (see Samson Peak). The peak is similar in shape to, but smaller than Samson Peak. It was named by Mrs. Mary T.S. Schäffer and was officially approved 23 November 1911.

Leather Peak (peak)

83 D/15 - Rainbow
18-45-5-W6
52°52'N 118°36'W
Approximately 35 km west of Jasper.

The name for this 2286 m mountain peak commemorates a disused name of the Yellowhead Pass. It was applied by A.O. Wheeler (see Wheeler Flats) in 1918, and refers to the supplies for trading posts, such as moose or caribou skins, carried by fur traders through the nearby pass.

Lectern Peak (peak)

83 D/9 - Amethyst Lakes
32-43-1-W6
52°45'N 118°07'W

Approximately 15 km south of Jasper.

This 2772 m mountain resembles the shape of a church lectern. Its descriptive name was suggested by G.E. Howard in 1914, and was made official in March 1917. No other information is known.

Lee Lake (lake)

82 G/9 - Blairmore
8-7-2-W5
49°32'N 114°15'W
Approximately 14 km west of Pincher Creek.

William Samuel Lee was born in England in 1830. In 1849, he went to the United States and joined the California Gold Rush. His next residence was Ohio until he moved to Canada in 1867. The creek upon which he established a trading post near Cardston became known as Lee's Creek. In 1871, Mr. Lee established a ranch near the Crowsnest Pass on the lake which now bears his name. This respected early settler discovered the hot springs in the Crowsnest Pass. Previous to 15 July 1977, when the name Lee Lake was officially adopted, the lake erroneously carried the name "Lees Lake" on maps and township plans.

Lefroy Glacier (glacier)

82 N/8 - Lake Louise
2-28-17-W5
51°22'N 116°15'W
Approximately 55 km west north-west of Banff.

(see Mount Lefroy)

Lefroy, Mount (mountain)

82 N/8 - Lake Louise
1-28-17-W5
51°22'N 116°16'W
Approximately 55 km west north-west of Banff.

This mountain, which is 3423 m in altitude, and the nearby glacier, were named in 1858 by Sir James Hector of the Palliser Expedition (see Mount Hector). General Sir John Henry Lefroy, R.A., C.B., K.C.M.G., F.R.S. (1817-1890) was a noted astronomer who measured the magnetic declination at a number of points in Canada as well as the Cape of Good Hope and St. Helena. He was head of the Toronto Observatory from 1842-1853.

Legace Lake (lake)

82 N/9 - Hector Lake
51°32'23"N 116°01'10"W
Approximately 22 km north-east of Lake Louise.

Legace Lake (pronounced legacy) was named for Ray Legace, an outfitter and guide who worked for the Brewsters (see Brewster Creek) in the Banff region during the early 1900s. The name Legace Lake has been in use by local residents for approximately 50 years, but was not officially approved until 20 October 1983.

Leman Lake (lake)

82 J/11 - Kananaskis Lakes
2-21-11-W5
50°45'N 115°25'W
Approximately 80 km west north-west of Turner Valley.

(see Mount Leman)

Leman, Mount (mountain)

82 J/11 - Kananaskis Lakes
35-20-11-W5
50°44'N 115°24'W
Approximately 80 km west north-west of Turner Valley.

This mountain, 2730 m in altitude, was named in 1918 after General G. Leman, who led the defence of Liege, Belgium, during the First World War.

Lendrum Creek (creek)

83 F/2 - Foothills
4-49-18-W5
53°12'N 116°35'W
Flows north-east into Erith River, approximately 43 km south of Edson.

This creek was named for an assistant forest ranger of the district. No other information is available.

Lesueur Creek (creek)

82 O/7 - Wildcat Hills
31-26-7-W5
51°16'N 114°58'W
Flows south-east into Ghost River, approximately 36 km west north-west of Cochrane.

This creek was named after the Payn Le Sueur family, who homesteaded in the area and were the first owners of the Bar C Ranch.

Leval Creek (creek)

82 J/14 - Spray Lakes Reservoir
11-21-11-W5
50°46'N 115°25'W
Flows north-east into Spray River, approximately 90 km west south-west of Calgary.

(see Mount Leval)

Leval, Mount (mountain)

82 J/14 - Spray Lakes Reservoir
3-21-11-W5
50°45'N 115°26'W
Approximately 85 km west south-west of Calgary.

Named in 1918, this mountain, which is 2713 m in altitude, commemorates Gaston de Leval, the famous Belgian lawyer who defended Edith Cavell. (see Mount Edith Cavell)

Levi Creek (creek)

83 F/5 - Entrance
19-51-27-W5
53°24'N 117°58'W
Flows east into Solomon Creek, approximately 24 km west of Hinton.

The origin of the name is unknown.

Lewis Creek (creek)

83 B/5 - Saunders
3-40-11-W5
52°24'N 115°30'W
Flows east into North Saskatchewan River, approximately 42 km east of Nordegg.

This creek was officially named in 1941 after J. Lewis of the Forestry Service.

Leyland (station)

83 F/3 - Cadomin
8-47-23-W5
53°02'N 117°19'W
Approximately 84 km south-west of Edson.

This Canadian National Railway station was opened in 1913, and was named after F.W. Leyland, an organizer of and an officer on the first executive of the Mountain Park Coal Company. (see Mountain Park)

Leyland Mountain (mountain)

83 F/3 - Cadomin
1-47-24-W5
53°01'N 117°22'W
Approximately 89 km south-west of Edson.

This name first appears on a sketch map drawn by W.H. Miller for the Geological Survey of Canada in 1922. The name was probably given to the feature, 2545 m in altitude, by geologists working in the area. (see Leyland)

Lick Creek (creek)

83 C/12 - Athabasca Falls
4-42-27-W5
52°36′N 117°49′W
Flows north-west into Athabasca River,
approximately 37 km south south-east of
Jasper.

(see Lick Peak)

Lick Creek (creek)

83 L/5 - Two Lakes
10-61-13-W6
54°15′N 119°52′W
Flows north-west into Torrens River,
approximately 56 km north-west of Grande
Cache.

The origin of the name is unknown.

Lick Peak (peak)

83 C/5 - Fortress Lake
14-40-28-W5
52°27′N 117°55′W
Approximately 50 km south south-east of
Jasper, on the Alberta-B.C. boundary.

This 2877 m peak was named in 1921 by
A.O. Wheeler (see Wheeler Flats). There was
a salt deposit, known as a "lick" near the
nearby creek where wild animals gathered to
obtain salt.

Lillian Lake (lake)

82 J/14 - Spray Lakes Reservoir
13-22-10-W5
50°52′N 115°15′W
Approximately 80 km west of Calgary.

The origin for this locally well-established
name, which was officially approved in 1964,
is unknown.

Limestone Creek (creek)

82 O/14 - Limestone Mountain
7-35-10-W5
51°59′N 115°25′W

Flows north-east into Clearwater River,
approximately 90 km north of Banff.

The name for this creek was officially
approved 2 November 1956. (see Limestone
Mountain)

Limestone Mountain (mountain)

82 J/14 - Spray Lakes Reservoir
10-22-9-W5
50°52′N 115°09′W
Approximately 70 km west of Calgary.

The name for this mountain is likely
descriptive of its composition.

Limestone Mountain (mountain)

82 O/14 - Limestone Mountain
24-34-11-W5
51°56′N 115°26′W
Approximately 88 km north of Banff.

Officially named 17 December 1941; likely
descriptive of the mountain's composition.

Lindsay, Mount (mountain)

83 C/14 - Mountain Park
11-45-24-W5
52°52′N 117°22′W
Approximately 47 km east of Jasper.

This 2743 m mountain was named by the
Earl of Southesk (see Southesk River) after
his friend, Sir Coutts Lindsay, of Balcarres,
Fifeshire, Scotland.

Lineham Creek (creek)

82 G/1 - Sage Creek
25-1-1-W5
49°04′N 114°00′W
Flows east into Cameron Creek,
approximately 45 km south of Pincher
Creek.

(see Lineham Creek)

Lineham Creek (creek)

82 J/9 - Turner Valley
2-20-3-W5
50°40′N 114°19′W
Flows east into Sheep River, approximately 3
km south-west of Turner Valley.

Lineham Creek and Mount Lineham are
named after one of the most notable
pioneers in the Highwood district. John
Lineham (1857-1913) came to Alberta in the
early 1880s. Originally from England, he
worked his way from Manitoba to Calgary
as a contractor with the Canadian Pacific
Railway. John Lineham was multitalented,
working for the lumber industry as well as
having interests in oil exploration, cattle and
horses. He became a member of the
Territorial Legislative Assembly in 1888. The
creek was officially named 8 December
1943.

Lineham Lakes (lakes)

82 G/1 - Sage Creek
33-1-1-W5
49°05′N 114°04′W
Approximately 13 km west of Waterton
Park.

Lineham Lakes consist of five individual
bodies of water. The five are identified
locally, but not officially, as Water Cugel,
Hourglass, Ptarmigan, Channel and Larch.
The name Lineham Lakes was officially
applied to this group of lakes 2 April 1959.
(see Lineham Creek)

Lineham, Mount (mountain)

82 G/1 - Sage Creek
27-1-1-W5
49°04′N 114°03′W
Approximately 10 km west of Waterton
Park.

This 2438 m mountain is listed in the
Canadian Gazette of 30 October 1915. Its
approval date is believed to predate this
listing. (see Lineham Creek)

Lingrell Creek (creek)

83 L/13 - Calahoo Creek
1-67-13-W6
54°47′N 119°51′W
Flows north-west into Wapiti River, approximately 75 km south-west of Grande Prairie.

The name for this creek has been in use since at least 1954, and is taken after a Mr. Charles Lingrell, an early trapper who had his trap line located nearby.

Lingrell Lake (lake)

83 L/12 - Nose Creek
24-66-13-W6
54°44′N 119°49′W
Approximately 81 km south-west of Grande Prairie.

(see Lingrell Creek)

Link Creek (creek)

82 J/10 - Mount Rae
26-20-5-W5
50°43′N 114°37′W
Flows north-east into Ware Creek, approximately 23 km west of Turner Valley.

The precise origin for this name, which was officially approved 6 September 1951, is unknown. It may be due to its proximity to Missinglink Mountain.

Linnet Lake (lake)

82 H/4 - Waterton Lakes
1-30-W4
49°04′N 113°54′W
Approximately 45 km south of Pincher Creek.

The origin of the name is unknown.

Lion Peak (peak)

83 C/2 - Cline River
36-19-W5
52°04′N 116°40′W

Approximately 125 km west south-west of Rocky Mountain House.

Resolute Mountain has two summits rising out of glaciers to the east of Mount Cline. The two summits appeared to Mr. Eric Hopkins, the man who suggested the names Lion and Lioness Peak, as lion's heads. (see Resolute Mountain)

Lioness Peak (peak)

83 C/2 - Cline River
36-19-W5
52°04′N 116°40′W
Approximately 125 km west south-west of Rocky Mountain House.

(see Lion Peak)

Lipalian Mountain (mountain)

82 N/8 - Lake Louise
30-28-15-W5
51°25′N 116°06′W
Approximately 50 km north-west of Banff.

The name for this mountain, which is 2710 m in altitude, was proposed in November 1958 by William C. Gussow, a geologist engineer from Calgary. This mountain is composed entirely of the youngest known Precambrian formations in North America. Lipalian, a geological term, refers to the interval of time represented by a widespread "unconformity" separating Precambrian and Cambrian rock layers. A new geological system, the Lipalian System, was proposed for these strata, and hence the name for the mountain where the best strata are exposed. Lipalian is from the Greek and means "lost interval." The term was coined by C.D. Walcott (see Chetang Ridge) in 1910.

Lipsett, Mount (mountain)

82 J/10 - Mount Rae
35-18-8-W5
50°34′N 114°59′W
Approximately 55 km west south-west of Turner Valley.

Officially named in February 1918, this 2560 m mountain commemorates Major-General L.J. Lipsett, C.M.G., of the Canadian Expeditionary Force.

Listening Mountain (mountain)

83 C/4 - Clemenceau Icefield
12-38-27-W5
52°15′N 117°45′W
Approximately 75 km south-east of Jasper.

This 3149 m mountain culminates in a wedge-shaped crest at the corners of which two points project above the snow like ears. The descriptive name was officially approved in 1921.

Little, Mount (mountain)

82 N/8 - Lake Louise
27-16-W5
51°18′N 116°11′W
Approximately 52 km west north-west of Banff.

This 3088 m mountain was named about 1916 after George F. Little of Boudoin College, Brunswick, Maine, a member of the group which was the first to ascend this mountain.

Little Alberta (mountain)

83 C/6 - Sunwapta
52°16′25″N 117°26′10″W
Approximately 76 km south south-east of Jasper.

The name for this mountain was officially approved 7 November 1984. It was named for its proximity to Mount Alberta.

Little Baker Lake (lake)

82 N/8 - Lake Louise
29-15-W5
51°29′N 116°01′W
Approximately 50 km north-west of Banff.

(see Baker Lake)

Little Berland River (river)

83 E/9 - Moberly Creek
9-55-2-W6
53°44′N 118°14′W
Flows north into Berland River, approximately 62 km east south-east of Grande Cache.

(see Berland Range)

Little Elbow River (river)

82 J/15 - Bragg Creek
24-21-7-W5
50°48′N 114°50′W
Flows east into Elbow River, approximately 50 km west south-west of Calgary.

Officially named 6 October 1949, this river is a tributary to the Elbow River. (see Elbow River)

Little Grayling Creek (creek)

83 L/12 - Nose Creek
17-65-11-W6
54°37′N 119°38′W
Flows north into Grayling Creek, approximately 80 km south south-west of Grande Prairie.

This creek was named as a tributary to Grayling Creek. (see Grayling Creek)

Little Highwood Pass (pass)

82 J/11 - Kananaskis Lakes
10-19-8-W5
50°35′55″N 115°01′20″W
Approximately 55 km west south-west of Turner Valley.

This pass, which is 2530 m in altitude, was locally referred to as Little Highwood Pass because of its size and its proximity to Highwood Pass. The name was made official 8 November 1978.

Little Jumpingpound Creek (creek)

82 O/2 - Jumpingpound Creek
19-24-4-W5
51°04′N 114°33′W
Flows east into Jumpingpound Creek, approximately 25 km west of Calgary.

(see Jumpingpound Creek)

Little MacKenzie Creek (creek)

83 F/3 - Cadomin
8-47-22-W5
53°02′N 117°10′W
Flows north-east into MacKenzie Creek, approximately 77 km south-west of Edson.

The locally well-established name for this creek was made official in 1944. It is a tributary of MacKenzie Creek. Originally spelled "Mackenzie," the spelling was changed to MacKenzie in 1961.

Little Muddy Creek (creek)

83 L/12 - Nose Creek
24-66-11-W6
54°44′N 119°32′W
Flows north into Muddy Creek, approximately 67 km south-west of Grande Prairie.

This creek was named for its proximity as a tributary to Muddy Creek. (see Muddy Creek)

Little Pipestone Creek (creek)

82 N/9 - Hector Lake
14-30-16-W5
51°34′N 116°09′W
Flows west into Pipestone River, approximately 60 km north-west of Banff.

(see Pipestone River)

Little Redrock Creek (creek)

83 L/6 - Chicken Creek
130-61-8-W6
54°19′N 119°11′W

Flows south-west into Redrock Creek, approximately 45 km north north-west of Grande Cache.

This creek was named for its proximity as a tributary to Redrock Creek. (see Redrock Creek)

Little Shovel Pass (pass)

83 C/12 - Athabasca Falls
52°44′00″N 117°45′30″W
Approximately 25 km south-east of Jasper.

(see Big Shovel Pass)

Little Snowbowl (basin)

83 C/12 - Athabasca Falls
52°44′10″N 117°46′20″W
Approximately 27 km east south-east of Jasper.

This descriptive name is used to differentiate this section of the basin from the rest of The Snowbowl. (see The Snowbowl)

Littlehorn Creek (creek)

83 C/8 - Nordegg
26-39-18-W5
52°23′N 116°29′W
Flows east into Bighorn River, approximately 112 km south-east of Jasper.

This creek is a tributary of the Bighorn River. Its name is therefore associated with the larger stream.

Livingstone Creek (creek)

82 O/2 - Jumpingpound Creek
28-24-5-W5
51°04′N 114°38′W
Flows south-east into Jumpingpound Creek, approximately 30 km west of Calgary.

The name for this feature was officially approved 12 December 1939. (see Livingstone Ridge)

Livingstone Falls (falls)

82 J/1 - Langford Creek
23-13-24-W5
50°06′05″N 114°26′30″W
Approximately 60 km west north-west of
Claresholm.

This waterfall has been known province
wide for many years as a major attraction on
the Forestry Trunk Road. The name was
officially approved 25 June 1979. (see
Livingstone River and Range)

Livingstone, Mount (mountain)

82 J/1 - Langford Creek
14-3-W5
50°08′N 114°24′W
Approximately 60 km west north-west of
Claresholm.

Officially approved 30 June 1915, this 3090
m mountain was named by Blakiston. (see
Mount Blakiston and Livingstone River)

Livingstone Range (range)

82 G/16 - Maycroft
10-3-W5
49°45′N 114°21′W
Runs in a north-south axis from Bellevue,
approximately 5 km east of Blairmore.

This mountain range was officially named 29
April 1941. Captain T. Blakiston (see Mount
Blakiston) called it the "Livingston"
(without the "e") Range, in 1858 after David
Livingstone (1813-1873), a Scottish born
doctor, missionary and noted African
explorer. Livingstone is considered one of
the greatest modern African explorers and
one of the pioneers in the abolition of the
slave trade.

Livingstone Ridge (ridge)

82 O/2 - Jumpingpound Creek
25-5-W5
51°06′N 114°41′W
Approximately 35 km west of Calgary.

Officially approved 12 December 1939, the
name for this ridge was suggested by D.A.
Nichols, after a local rancher, Mr. Sam
Livingstone (1830-1897).

Livingstone River (river)

82 G/16 - Maycroft
30-11-3-W5
49°57′N 114°23′W
Flows south into Oldman River, approximately
40 km north of Blairmore.

(see Livingstone Range)

Llama Creek (creek)

83 E/13 - Dry Canyon
35-55-11-W6
53°48′N 119°31′W
Flows south into Muddywater River,
approximately 27 km south-west of Grande
Cache.

The origin of the name is unknown.

Llama Flats (flats)

83 E/14 - Grande Cache
12-56-11-W6
53°49′N 119°28′W
Approximately 26 km west south-west of
Grande Cache.

No information regarding the origin of the
name is known.

Llama Mountain (mountain)

83 E/14 - Grande Cache
6-56-10-W6
53°48′N 119°28′W
Approximately 25 km west south-west of
Grande Cache.

This mountain is 2301 m in altitude. The
origin of the name is unknown.

Llysfran Peak (peak)

83 C/11 - Southesk Lake
35-41-25-W5
52°35′N 117°29′W

Approximately 58 km south-east of Jasper.

This mountain peak, which is 3141 m in
altitude, was named in 1911 by Mrs. Schäffer,
author of *Old Indian Trails* (1911). Llysfran
is the family name of a companion of hers,
Mary Vaux (see Mount Mary Vaux), after
whom a mountain of whose configuration
this peak forms, is named.

Loaf Mountain (mountain)

82 G/1 - Sage Creek
17-3-1-W5
49°13′N 114°06′W
Approximately 25 km north-west of
Waterton Park.

No origin information is available for this
2639 m mountain. The name may be
descriptive of the mountain's appearance.

Loder Peak (peak)

82 O/3 - Canmore
35-24-9-W5
51°05′N 115°08′W
Approximately 16 km east of Canmore.

Mr. Edwin Loder, after whom this 2088 m
peak is named, first settled in the Kananaskis
area circa 1880. He erected some lime-
burning kilns around the railway tracks
when the trans-continental railway went
through in 1884. In 1905, Edwin and his
brother, Richard, incorporated under the
name "Loder Lime Co. Ltd." Edwin Loder
was the first postmaster of the Kananaskis
district in 1900, a position which he held
until his death in 1935. The name was
approved by the Canadian Permanent
Committee on Geographical Names 3
October 1967.

Logan Creek (creek)

82 O/11 - Burnt Timber Creek
15-31-9-W5
51°39′N 115°12′W

Flows south into Red Deer River, approximately 55 km north-east of Banff.

This creek is likely named after Bill Logan, who owned a ranch in the area.

Logan Ridge (ridge)

82 O/2 - Jumpingpound Creek
24-5-W5
51°00′N 114°39′W
Approximately 30 km west of Calgary.

One of the oldest settlers in the area whose name was Logan, was remembered by D.A. Nichols, who suggested the name Logan Ridge for this feature. The name was officially approved 12 December 1939.

Lone Creek (creek)

82 G/1 - Sage Creek
9-2-1-W5
49°07′N 114°05′W
Approximately 15 km north-west of Waterton Park.

This creek empties into Blakiston Creek and is named for its proximity to Lone Mountain. (see Lone Mountain)

Lone Lake (lake)

82 G/1 - Sage Creek
6-2-1-W5
49°05′N 114°08′W
Approximately 15 km west north-west of Waterton Park.

(see Lone Mountain)

Lone Mountain (mountain)

82 G/1 - Sage Creek
7-2-1-W5
49°06′N 114°07′W
Approximately 15 km north-west of Waterton Park.

This isolated peak is located near the western boundary of Waterton Lakes National Park. It rises to a height of 2423 m.

Lone Teepee Creek (creek)

83 E/15 - Pierre Greys Lakes
15-57-5-W6
53°55′N 118°39′W
Flows north-west into Muskeg River, approximately 31 km east of Grande Cache.

Formerly "Teepee Creek," the name Lone Teepee Creek was adopted in 1947 to avoid duplication. No other origin information is known.

Lonesome Lake (lake)

82 H/4 - Waterton Lakes
36-1-30-W4
49°04′N 113°53′W
Approximately 45 km south of Pincher Creek.

The origin of the name is unknown.

Long Prairie Creek (creek)

82 J/9 - Turner Valley
19-4-W5
50°38′N 114°28′W
Flows east into Sheep River, approximately 55 km south south-west of Calgary.

There is a long meadow near this creek. The name was officially approved 8 December 1943.

Longview (village)

82 J/9 - Turner Valley
20-18-2-W5
50°32′N 114°14′W
Approximately 17 km south south-east of Turner Valley.

Thomas Long came west from Ontario to the Okotoks region in 1895. He taught school, then with his brother Oliver, who came later, took a homestead at Big Hill. The name Longview is derived from Mr. Long's name and the fact that there was a good view from the post office. It was officially approved 8 December 1943. The

words long view are "ba-ha-a-humbi" in Stoney, and describes a hill north-east of the village which was once used as a lookout. (Dempsey)

■ Longview Hill (hill)

82 J/9 - Turner Valley
21-18-2-W5
50°33′N 114°13′W
Approximately 20 km south-west of Turner Valley.

The name for this hill was officially approved 8 December 1943. It is sometimes referred to as "Big Hill." (see Longview)

Longview Mountain (mountain)

83 C/7 - Job Creek
22-39-21-W5
52°22′N 116°56′W
Approximately 95 km south-east of Jasper.

The origin for the name of this mountain is unknown. The name was officially approved 5 March 1935.

Lookout Creek (creek)

83 C/9 - Wapiabi Creek
13-42-17-W5
52°38′N 116°19′W
Flows north into Blackstone River, approximately 23 km north-west of Nordegg.

This creek was named after the locally known "Lookout Hill" nearby, which has a good vantage point.

Loomis Creek (creek)

82 J/7 - Mount Head
29-17-6-W5
50°28′N 114°48′W
Flows east into Highwood River, approximately 41 km south-west of Turner Valley.

(see Mount Loomis)

Loomis, Lake (lake)

82 J/7 - Mount Head
21-17-7-W5
52°27'15"N 114°54'15"W
Approximately 50 km south-west of Turner Valley.

This small alpine lake is called Lake Loomis because of its proximity to Mount Loomis. It forms part of the headwaters of Loomis Creek adjacent to the Alberta-B.C. boundary. It was officially approved 7 February 1983. (see Mount Loomis)

Loomis, Mount (mountain)

82 J/7 - Mount Head
29-17-7-W5
50°27'N 114°55'W
Approximately 50 km south-west of Turner Valley.

Officially approved in 1918, this 2822 m mountain was named in honour of Brigadier-General F.V.W. Loomis, C.M.G., D.S.O. (1870-1914), who commanded the 13th Battalion, C.E.F., in 1914.

Loop Ridge (ridge)

82 G/10 - Crowsnest
3-8-6-W5
49°37'N 114°44'W
Approximately 18 km west south-west of Coleman.

Loop Ridge, which stands approximately 2003 m in altitude, was named by the Alberta-British Columbia Boundary Commission in 1915, and suggested by Coleman (see Mount Coleman). The Crowsnest branch of the Canadian Pacific Railway forms a rather extensive loop at the western base of this ridge and it is for this turn that the ridge is named. Part of the ridge forms the Alberta-B.C. boundary.

Loren Pass (pass)

83 E/5 - Holmes River
23-52-13-W6
53°30'N 119°49'W
Approximately 135 km north-west of Jasper, on the Alberta-B.C. boundary.

The altitude of this pass measures 1542 m. There is a "Loren Lake" in British Columbia after which this pass takes its name, but its precise origin is unknown.

Lorette Creek (creek)

82 J/14 - Spray Lakes Reservoir
13-23-9-W5
50°58'N 115°07'W
Flows south-east into Kananaskis River, approximately 70 km west of Calgary.

The name for this creek was approved in 1948. (see Mount Lorette)

Lorette, Mount (mountain)

82 J/14 - Spray Lakes Reservoir
23-9-W5
50°59'N 115°08'W
Approximately 70 km west of Calgary.

This 2469 m mountain was named in 1922 after Lorett Ridge in France.

Lorraine Lake (lake)

83 C/12 - Athabasca Falls
26-43-26-W5
52°44'N 117°39'W
Approximately 32 km south-east of Jasper.

The origin for this locally well-established name is unknown.

Lost Creek (creek)

82 G/8 - Beaver Mines
35-5-4-W5
49°26'N 114°26'W
Approximately 18 km west south-west of Beaver Mines.

This creek, a tributary of Carbondale River, has been known as Lost Creek since at least 1917. Its precise origin, however, is unknown.

Lost Creek (creek)

82 J/7 - Mount Head
18-15-5-W5
50°16'N 114°40'W
Flows north into Cataract Creek, approximately 52 km south-west of Turner Valley.

The name for this creek was officially approved 8 December 1943, but its origin is unknown.

Lost Creek (creek)

83 F/3 - Cadomin
21-48-21-W5
53°10'N 117°00'W
Flows east into Embarras River, approximately 40 km south-west of Edson.

The origin of the name is unknown.

Lost Mountain (mountain)

82 G/1 - Sage Creek
16-2-1-W5
49°07'N 114°05'W
Approximately 10 km west of Waterton Park.

This 2512 m mountain appears lost in Waterton Lakes National Park because of its isolated position. Its descriptive name is well-established in local usage.

Lost Guide Creek (creek)

82 O/13 - Scalp Creek
29-33-13-W5
51°52'N 115°48'45"W
Flows south-east into Clearwater River, approximately 80 km north north-west of Banff.

In the early 1920s, Jack Browning, veteran packer and guide, took a party of three on a hunting trip to the headwaters of Ranger Creek. Crossing the divide between Ranger Creek and Lost Guide Creek, on their return trip, they split up. Two continued down the south side of the creek along the familiar trail, while Browning and the other member of the party went exploring along the ridge with the intention of coming down a trail on the north side of this stream. They were lost in a sudden fog, and, much to Browning's chagrin, other guides in the area dubbed this creek Lost Guide Creek. The name was officially approved 8 November 1978.

Lost Guide Lake (lake)

82 O/13 - Scalp Creek
9-34-14-W5
51°54′20″N 115°56′00″W
Approximately 85 km north north-west of Banff.

(see Lost Guide Creek)

Lost Horse Creek (creek)

82 O/4 - Banff
26-14-W5
51°12′N 115°52′W
Flows north into Redearth Creek, approximately 22 km west of Banff.

The name was officially approved 7 May 1959. It is rumoured that F.O. "Pat" Brewster lost his pack horse at Hillsdale.

Lost Horse Creek (creek)

82 H/4 - Waterton Lakes
18-2-30-W4
49°07′N 114°00′W
Flows south-west into Blakiston Creek, approximately 40 km south of Pincher Creek.

At the mouth of this stream, along the banks of Pass Creek, the North West Mounted Police established a summer training camp. During one of their exercises,

a horse became lost in the upstream area of this creek. Thus it became known as Lost Horse Creek. The name was approved 7 September 1969 by the Canadian Permanent Committee on Geographical Names.

Lost Knife Creek (creek)

82 O/7 - Wildcat Hills
6-28-7-W5
51°22′N 114°59′W
Flows north-east into Waiparous Creek, approximately 42 km north-west of Cochrane.

The origin of the name is unknown.

Lost Lake (lake)

82 N/8 - Lake Louise
5-29-17-W5
51°28′N 116°16′W
Approximately 60 km north-west of Banff.

This descriptive name refers to the lake's remoteness and the surrounding nondescript mountain ridges.

Lost Lake (lake)

82 G/1 - Sage Creek
20-2-2-W5
49°09′N 114°09′W
Approximately 20 km north-west of Waterton Park.

(see Lost Mountain)

Loudon Creek (creek)

83 C/1 - Whiterabbit Creek
24-35-18-W5
52°01′N 116°26′W
Flows north into North Saskatchewan River, approximately 110 km north-west of Banff.

This creek runs off Mount Loudon. (see Mount Loudon)

Loudon, Mount (mountain)

82 N/16 - Siffleur River
34-18-W5
51°55′N 116°26′W
Approximately 100 km north-west of Banff.

Mount Loudon is 3221 m in altitude. The origin of the name is unknown.

Lougheed, Mount (mountain)

82 J/14 - Spray Lakes Reservoir
13-23-10-W5
50°58′N 115°16′W
Approximately 80 km west of Calgary.

The Honourable Sir James Alexander Lougheed, K.C.M.G. (1854-1925) was a lawyer and politician who moved to Calgary in 1882 from Toronto. He was called to the Canadian Senate in 1889. In 1920, Lougheed assumed the appointment of federal Minister of the Interior, Superintendent General of Indian Affairs and Minister of Mines, which he held until the election of the Mackenzie King Government the following year. The name for this 3105 m mountain was officially approved 5 January 1926.

Louie, Mount (mountain)

83 E/14 - Grande Cache
32-56-7-W6
53°53′N 119°00′W
Approximately 7 km west of Grande Cache.

Mount Louie is 1844 m in altitude. This mountain may be named after Louie Delorme. (see Delorme Pass)

Louis, Mount (mountain)

82 O/4 - Banff
12-26-13-W5
51°13′N 115°41′W
Approximately 9 km south-west of Banff.

Louis B. Stewart D.L.S., D.T.S. (1861-1937), after whom this 2682 m mountain was named, was a professor of surveying at the

University of Toronto. He surveyed Banff National Park with his father, G.A. Stewart, the first Park Superintendent, in 1904.

Louise Creek (creek)

82 N/8 - Lake Louise
27-28-16-W5
51°25′N 116°11′W
Flows east into Bow River, approximately 55 km north-west of Banff.

(see Lake Louise)

Lake Louise and Mount Whyte, 1890

Louise, Lake (lake)

82 N/8 - Lake Louise
19-28-16-W5
51°25′N 116°14′W
Approximately 55 km north-west of Banff.

This lake, once known as "Emerald Lake" to Tom Wilson (see Mount Wilson), and as "Lake of the Little Fishes" to the Stoneys, was named in 1884 after Her Royal Highness Princess Louise Caroline Alberta (1848-1939)

*denotes rescinded name or former locality.

the fourth daughter of Queen Victoria. She was married to the Marquis of Lorne, later Governor General of Canada (1878-1883). Hindman, an assistant to Major Rogers, an engineer on the construction of the Canadian Pacific Railway, named both Lake Louise and Lorne Lake, which is now officially Mirror Lake. (see Mirror Lake) According to Harriet Hartley Thomas (*Barnacle to Banff*), Dr. G.M. Dawson and Lord Temple (see Mount Temple) named the lake. Lord Temple named the lake after his daughter, but then it was renamed after the Princess.

Lovett River (river)

83 C/15 - Cardinal River
23-46-19-W5
52°59′N 116°39′W
Approximately 100 km east of Jasper.

Formerly "Little Pembina River," the name was changed to avoid confusion. As the river flows through Lovettville, the name Lovett River was chosen in 1927 to replace the old name. (see Lovettville)

***Lovettville** (former locality)

83 F/2 - Foothills
3-47-19-W5
53°02′N 116°41′W
Approximately 105 km east north-east of Jasper.

The locality's centre, a C.N.R. station, was established in 1913 to service the coal mines in the area. It was named after H.A. Lovett, President of the North American collieries.

Low, Mount (mountain)

82 N/10 - Blaeberry River
32-20-W5
51°43′N 116°48′W
Approximately 90 km north-west of Banff.

This 2722 m mountain was named in 1920 after A.P. Low (1861-1942). The former

Arctic explorer (1903-1904) and director of the Dominion Survey (1906-1908) was, at the time of naming, Deputy Minister of the Department of Mines.

Lower Colefair Lake (lake)

83 C/12 - Athabasca Falls
52°43′45″N 117°52′10″W
Approximately 21 km south-east of Jasper.

The precise origin for this locally well-established name is unknown.

Lower Kakwa Falls (chutes)

83 L/4 - Kakwa Falls
NW-11-59-12-W6
54°05′30″N 119°40′30″W
Approximately 130 km south south-west of Grande Prairie.

Several features in this area share this name. (see Kakwa Mountain)

Lower Kananaskis Lake (lake)

82 J/11 - Kananaskis Lakes
35-19-9-W5
50°39′N 115°08′W
Approximately 60 km west of Turner Valley.

(see Kananaskis Range)

Lower Longview Lake (lake)

83 C/7 - Job Creek
26-39-21-W5
52°23′05″N 116°55′25″W
Approximately 95 km south-east of Jasper.

The name was officially approved 20 March 1979. Both this lake and Upper Longview Lake were named for their proximity to Longview Mountain.

Lower Rowe Lake (lake)

82 G/1 - Sage Creek
22-1-1-W5
49°03′N 114°03′W

Approximately 10 km west of Waterton Park.

The lake may be found on the north face of Mount Rowe, which was named after Lieutenant Rowe, a surveyor on the British Boundary Commission of 1872-1876. (see Mount Rowe)

Lower Spray Falls (falls)

82 O/3 - Canmore
7-24-9-W5
51°02′N 115°15′W
Approximately 10 km south-east of Canmore.

(see Spray River)

Lower Waterton Lake (lake)

82 H/4 - Waterton Lakes
6-2-29-W4
49°06′N 113°51′W
Approximately 45 km south south-east of Pincher Creek.

The name Lower Waterton Lake is used for specific identification, as each lake is a clearly defined entity in its own right. The name was approved 7 September 1969 by the Canadian Permanent Committee on Geographical Names. (see Waterton Lakes)

Lucerne Peak (peak)

83 D/15 - Rainbow
18-45-5-W6
52°52′N 118°34′W
Approximately 34 km west of Jasper.

This mountain peak overlooks Lucerne Railway Station which was named after Lucerne Lake and Canton, in Switzerland. It was named in 1918 by A.O. Wheeler. (see Wheeler Flats)

Lucifer, Mount (mountain)

83 E/5 - Holmes River
27-51-11-W6
53°26′N 119°32′W

Approximately 115 km north-west of Jasper.

The name Mount Lucifer was officially adopted for this 3060 m mountain 1 May 1934. Its precise origin is not known.

Luellen Lake (lake)

82 O/5 - Castle Mountain
28-27-14-W5
51°20′N 115°55′W
Approximately 30 km north-west of Banff.

This lake was named in 1959 after the daughter of Mr. Jack Martin, who was superintendent of the fish hatchery.

Lund Creek (creek)

83 C/15 - Cardinal River
7-46-18-W5
53°00′N 115°34′W
Flows east into Pembina River, approximately 99 km east of Jasper.

This creek may have been named after Einar Lund, who was warden on the Lovett Branch in early days.

Lundbreck (hamlet)

82 G/9 - Blairmore
26-7-2-W5
49°35′N 114°12′W
Approximately 6 km west north-west of Cowley.

The Canadian Pacific Railway came through this hamlet in 1904. The name Lundbreck is derived from a combination of Lund and Breckenridge (Breckenridge & Lund Coal Co.) who operated collieries and sawmills at the site. The mine was opened in 1906, when the population of the town was about 600.

■ Lundbreck Falls (falls)

82 G/9 - Blairmore
28-7-2-W5
49°35′N 114°12′W

Approximately 3 km west of Lundbreck.

(see Lundbreck)

Lundine Creek (creek)

83 B/5 - Saunders
34-39-12-W5
52°24′N 115°39′W
Flows north-east into Trout Creek, approximately 31 km south-east of Nordegg.

This creek was named after an old Norwegian trapper who was a permanent resident of the area, the proper spelling of the name, which was approved in 1941, should be "Lundin."

Luscar (station)

83 F/3 - Cadomin
23-47-24-W5
53°04′N 117°24′W
Approximately 86 km south-west of Edson.

The village of Luscar, now a ghost town, was named in 1922 after Luscar, Dunfermline, Fifeshire, Scotland. The Mitchell family, associated with the Mountain Park Coal Mine, were originally from Luscar.

Luscar Creek (creek)

83 F/3 - Cadomin
16-47-24-W5
53°03′N 117°19′W
Flows east into MacLeod River, approximately 84 km south-west of Edson.

(see Luscar)

Luscar Mountain (mountain)

83 F/3 - Cadomin
3-47-24-W5
53°02′N 117°26′W
Approximately 91 km south-west of Edson.

(see Luscar)

Lusk Creek (creek)

82 O/3 - Canmore
15-24-8-W5
51°02′N 115°02′W
Flows north-west into Kananaskis River,
approximately 24 km east south-east of
Canmore.

Originally from Texas, Thomas "Coffee"
Lusk made his way north-west to Alberta,
and in 1895 he arrived in Banff where he
served as a trail guide. His heavy drinking
concerned some of his employers, but he
was always sober by the morning. He took
up homestead in 1900, but his interest in
following new trails was still very much
alive. In 1907 when Martin Nordegg (see
Nordegg) and Donaldson Dowling (see
Dowling Ford) set out to seek large coal
deposits, they engaged Tom Lusk as their
head packer. Eventually, Tom made his way
back to Texas, was turned away by his son,
and died on his way back to Alberta.

Lyall Creek (creek)

82 J/2 - Fording River
21-13-5-W5
50°06′N 114°38′W
Flows north-east into Oldman River,
approximately 65 km south-west of Turner
Valley.

(see Mount Lyall)

Lyall, Mount (mountain)

82 J/2 - Fording River
13-6-W5
50°05′N 114°42′W
Approximately 70 km south-west of Turner
Valley.

Dr. David Lyall, M.D. was the surgeon and
naturalist on the British Boundary
Commission of 1858-1862, from the Pacific
to the Rockies. The name for this 2950 m
mountain was officially approved in 1917.

Lyautey Glacier (glacier)

82 J/11 - Kananaskis Lakes
18-19-9-W5
50°37′N 115°13′W
Approximately 65 km west south-west of
Turner Valley.

(see Mount Lyautey)

Lyautey, Mount (mountain)

82 J/11 - Kananaskis Lakes
17-19-9-W5
50°36′N 115°13′W
Approximately 70 km west south-west of
Turner Valley.

This 3082 m mountain was named officially
in 1918 after General Louis Hubert
Gonzalve Lyautey (1854-1934). This member
of the French Academy was the Minister of
War in 1916.

Lychnis Mountain (mountain)

82 O/5 - Castle Mountain
29-14-W5
51°29′N 115°59′W
Approximately 44 km north-west of Banff.

This 2819 m mountain was named in 1911
by J.F. Porter of Chicago (see Merlin Lake),
after the wild flower (Lychnis), commonly
known as alpine campion or catchfly.

Lyell Icefield (icefield)

82 N/14 - Rostrum Peak
29-34-22-W5
51°56′N 117°05′W
Approximately 140 km north-west of Banff,
on the Alberta-B.C. boundary.

The name "Lyell Snowfield" was approved 6
February 1912 for this feature, but was
officially changed to Lyell Icefield by the
Canadian Permanent Committee on
Geographical Names 6 July 1964. (see
Mount Lyell)

Lyell, Mount (mountain)

82 N/14 - Rostrum Peak
31-34-22-W5
51°57′N 117°06′W
Approximately 140 km north-west of Banff.

This mountain, 3504 m in altitude, was
named in 1858 by Sir James Hector (see
Mount Hector) after a noted British
geologist and member of the British Survey
Commission, Sir Charles Lyell (1795-1875).
Lyell made observations on geological
formations and their relation to fossils in
Britain, Europe and North America. The
nearby glacier takes its name from this
mountain. (see Christian Peak)

Lynch Creek (creek)

83 B/5 - Saunders
17-38-14-W5
52°17′N 115°58′W
Flows south into North Ram River,
approximately 22 km south-east of Nordegg.

The name for this creek is printed in large
letters on a tree at the junction of the Ram
River Trail and this stream. Officially
approved in 1941, the creek takes its name
from Morris Lynch, a packer in the district.

Lynx Creek (creek)

82 G/8 - Beaver Mines
12-6-4-W5
49°27′N 114°24′W
Approximately 15 km west of Beaver Mines.

The name for this creek was officially
applied 4 August 1960. The creek has been
known locally as Lynx Creek for several
years, possibly to differentiate it from
another nearby creek known as Link Creek.

Lynx Creek (creek)

83 B/4 - Cripple Creek
26-36-13-W5
52°07′N 115°45′W

Flows east into Ram River, approximately 46 km south-east of Nordegg.

The origin of the name is unknown.

Lynx Creek (creek)

83 C/5 - Fortress Lake
30-38-25-W5
52°18′N 117°33′W
Flows west into Athabasca River,
approximately 75 km south-east of Jasper.

(see Lynx Mountain)

Lynx Creek (creek)

83 L/4 - Kakwa Falls
35-60-11-W6
54°14′N 119°32′W
Flows east into Kakwa River, approximately 48 km north-west of Grande Cache.

The origin of the name is unknown.

Lynx Mountain (mountain)

83 E/3 - Mount Robson
11-48-8-W6
53°07′N 119°04′W
Approximately 70 km west north-west of Jasper.

This mountain, 3192 m in altitude, was named in 1911 when a dead lynx was found on the ice floe below it.

■ Lyons Creek (creek)

82 G/9 - Blairmore
35-7-4-W5
49°36′N 114°26′W
Flows north into Crowsnest River, at Blairmore.

Lieutenant-Colonel Harry E. Lyon was an early settler in this region. He was the first mayor of Blairmore. He also served as a postmaster and raised an infantry battalion in the Crowsnest Pass area during the First World War.

Ma Butte (mountain)

82 G/10 - Crowsnest
18-9-4-W5
49°44′N 114°31′W
Approximately 10 km north of Coleman.

This mountain is part of the Crowsnest volcanics. It is approximately 2331 m in altitude. No origin information about the name is known.

Mac Creek (creek)

82 J/15 - Bragg Creek
27-21-6-W5
50°48′N 114°45′W
Flows north-east into Quirk Creek, approximately 45 km west south-west of Calgary.

This creek appears as Mac Creek on a Land Survey map of 1921. The name was officially approved 6 October 1949, but its precise origin is not known.

Macabee Creek (creek)

82 J/9 - Turner Valley
29-19-3-W5
50°38′N 114°23′W
Flows east into Sheep River, approximately 7 km south-west of Turner Valley.

First appearing on a map in 1894, the name Macabee Creek was officially approved 8 December 1943. It is named after three brothers who migrated to the area around 1887. The brothers, whose name is said to have been spelled McAbee, became pioneer ranchers and later importers of race horses.

Macabee Gap (gap)

82 J/9 - Turner Valley
11-20-4-W5
50°40′50″N 114°27′00″W
Approximately 15 km west south-west of Turner Valley.

The name for this gap is taken from its proximity to Macabee Creek. (see Macabee Creek)

Maccarib Creek (creek)

83 D/9 - Amethyst Lakes
25-43-2-W6
52°43′N 118°12′W
Flows north-west into Moat Creek, approximately 22 km south-west of Jasper.

(see Mount Maccarib)

Maccarib, Mount (mountain)

83 D/9 - Amethyst Lakes
43-2-W6
52°42′N 118°12′W
Approximately 21 km south-west of Jasper.

This mountain, which is 2749 m in altitude, was officially named in 1916 by M.P. Bridgland (see Mount Bridgland). Caribou were seen below the peak and since maccarib is the Quinnipiac Indian word for caribou, the mountain received this name.

Maccarib Pass (pass)

83 D/9 - Amethyst Lakes
15-43-3-W6
52°43′N 118°12′W
Approximately 19 km south-west of Jasper.

(see Mount Maccarib)

MacDonald Creek (creek)

82 G/7 - Upper Flathead
19-5-4-W5
49°24′N 114°31′W
Flows north into Carbondale River, approximately 25 km south of Coleman.

Major MacDonald, an experienced Canadian militia officer, helped organize the International Boundary Commission of 1872-1876, from Lake of the Woods to the Rockies (see Cameron Lake). The creek may be named after this man.

M~N~O

Machray, Mount (mountain)

83 E/2 - Resplendent Creek
11-47-6-W6
53°03′N 118°47′W
Approximately 50 km west north-west of Jasper.

This 2749 m mountain was officially named in 1923 after the Most Reverend Robert Machray, D.D., LL.D., D.C.L. (1831-1904). He was appointed Bishop of Rupert's Land in 1865, and became Archbishop of Rupert's Land and Primate of all Canada in 1893. He was the first Chancellor of the University of Manitoba, a position he held from 1877 until his death in 1904.

MacKenzie Creek (creek)

83 F/3 - Cadomin
29-47-22-W5
53°05′N 117°10′W
Flows north into McLeod River, approximately 74 km south-west of Edson.

Named in March 1935, the creek's original spelling was "Mackenzie." The origin of the name is unknown.

Mackenzie, Mount (mountain)

83 C/14 - Mountain Park
2-45-23-W5
52°51′N 117°14′W
Approximately 56 km east of Jasper.

This 2764 m mountain was named after D. M. Mackenzie, a Forest Supervisor on the Brazeau Reserve. No other information is available for this man.

MacLaren, Mount (mountain)

82 J/7 - Mount Head
16-6-W5
50°21'N 114°49'W
Approximately 49 km south-west of Turner Valley.

Brigadier-General Charles H. MacLaren, C.M.G., D.S.O. (1878-1962) was born in Wakefield, Quebec, and received his Bachelor of Arts from Queen's University in 1902 and Bachelor of Laws from Osgoode Hall in 1905. He commanded a brigade of Canadian artillery during the First World War. It is after this man that Mount MacLaren was officially named in 1918.

Magog, Mount (mountain)

82 J/13 - Mount Assiniboine
22-12-W5
50°52'N 115°38'W
Approximately 105 km west of Calgary, on the Alberta-B.C. boundary.

The names Og, Gog and Magog are mentioned in the Old Testament of the Bible. Og was a giant King of Bashan, conquered by Moses. Gog was a prince, and Magog was the land he came from (Ezekiel: 38,39). Gog and Magog are also mentioned in The Book of Revelation: chapter 20. Robert Young, author of *Analytical Concordance to the Bible* (1893) gives a meaning of Gog as "high, mountain." Magog was the second son of Japheth (Genesis: chapter 10). An English myth relates that Gog and Magog were the last of a giant race and were captured and brought to England. Early Celtic mythology also mentions these giants. It is possible that this 3095 m mountain was named from one of these origins. (see Og Pass and Og Mountain)

Magog Pass (pass)

82 J/13 - Mount Assiniboine
22-12-W5
50°52'N 115°38'W
Approximately 100 km west of Calgary, on the Alberta-B.C. boundary.

This pass was named for its proximity to Mount Magog. (See Mount Magog)

Mahon Creek (creek)

83 E/15 - Pierre Greys Lakes
17-56-5-W6
53°50'N 118°41'W
Flows north into Muskeg River, approximately 28 km east south-east of Grande Cache.

The locally well-established name of this creek was officially approved in 1947. It takes its name from Sam Mahon who was studying the coal deposits in the area with his brother-in-law, John McVicar, on a September day when a vicious storm erupted. Sam, without a coat and the means for lighting a fire, left the trail. The members of his party searched the area until dusk and kept a night vigil until the morning, when they went out to search again, only to find that Sam had died of exposure in the night.

Mahood, Mount (mountain)

83 E/2 - Resplendent Creek
1-47-5-W6
53°02'N 118°35'W
Approximately 37 km west north-west of Jasper.

This 2896 m mountain was named in 1924 after a Canadian Pacific Railway engineer and explorer. No other information is available.

Majestic Mountain (mountain)

83 D/16 - Jasper
3-44-2-W6
52°45'N 118°13'W
Approximately 16 km south-west of Jasper.

This mountain, which stands 3086 m in altitude, is the highest peak in its range. The descriptive name was applied in 1916 by M.P. Bridgland. (see Mount Bridgland)

Makwa Creek (creek)

83 B/5 - Saunders
27-38-11-W5
52°18'N 115°30'W
Flows south-east into Ram River, approximately 45 km south-east of Nordegg.

The term "makwa" is the Cree Indian word for a loon. Presumably there are a number of loons in the area.

Makwa Creek (creek)

83 F/4 - Miette
5-48-27-W5
53°06'N 117°54'W
Flows north-west into Rocky River, approximately 42 km south-west of Hinton.

(see Makwa Ridge)

Makwa Ridge (ridge)

83 F/4 - Miette
47-27-W5
53°03'N 117°47'W
Approximately 45 km south south-west of Hinton.

The intended name for this ridge was "muskwa," the Cree Indian word for a bear. However, it became "makwa," the Cree Indian word for "loon." It was named by M.P. Bridgland (see Mount Bridgland) in 1916.

Malcolm Creek (creek)

83 E/14 - Grande Cache
8-57-8-W6
53°54'N 119°09'W
Flows north-east into Smoky River, approximately 3 km north north-west of Grande Cache.

This creek was officially named in 1953 after Malcolm Moberly, who owned land at its mouth.

Maligne Canyon (canyon)

83 D/16 - Jasper
36-45-1-W6
52°55'N 118°00'W
Approximately 7 km north-east of Jasper.

(see Maligne River)

Maligne Lake (lake)

83 C/12 - Athabasca Falls
8-43-25-W5
52°40'N 117°35'W
Approximately 34 km south-east of Jasper.

The earliest known visit to the lake was
made by the railway surveyor Henry
MacLeod in the summer of 1875. Searching
for a possible route for the Canadian Pacific
Railway, MacLeod and his party ascended
the Maligne River from the Athabasca
Valley, and after what was obviously a
difficult journey, named the lake there
"Sore-foot Lake." The lake's name was
subsequently changed to Maligne Lake. (see
Maligne River)

Maligne Lake, 1911

Maligne Mountain (mountain)

83 C/11 - Southesk Lake
28-42-24-W5
52°39'N 117°24'W
Approximately 50 km south of Jasper.

This 3150 m mountain was named in 1911
by Mary T.S. Schäffer who wrote that she
had named the mountain for: "no reason at
all except that a fine double-headed peak is
seen 3 or 4 miles up that valley and one
mountain should surely bear the name of
the Lake." (see Maligne Lake)

Maligne Pass (pass)

83 C/11 - Southesk Lake
1-41-25-W5
52°30'N 117°27'W
Approximately 60 km south-east of Jasper.

The name for this pass was officially
approved 5 March 1935. (see Maligne Lake)

Maligne Range (range)

83 C/13 - Medicine Lake
44-27-W5
52°48'N 117°55'W
Approximately 7 km east of Jasper.

The name was officially approved 5 March
1935. (see Maligne River)

Maligne River (river)

83 D/16 - Jasper
2-46-1-W5
52°56'N 118°02'W
Flows west into Athabasca River,
approximately 6 km north north-east of
Jasper.

Maligne is a French word for something
which is wicked or malignant. This is a
reference probably made by an early fur
trade voyageur, to the current of the river
near its intersection with the Athabasca
River. Father de Smet (see Roche de Smet)
used this name for the river in 1846.

Mallard Peak (peak)

83 D/8 - Athabasca Pass
22-40-2-W6
52°27'N 118°12'W

Approximately 47 km south south-west of
Jasper.

This mountain peak, which is 2835 m in
altitude, was given its descriptive name in
1921 by A.O. Wheeler. The shape of the
rock resembles a mallard duck.

Malloch Creek (creek)

82 N/16 - Siffleur River
15-33-15-W5
51°50'N 116°03'W
Flows south-east into Clearwater River,
approximately 80 km north north-west of
Banff.

(see Mount Malloch)

Malloch, Mount (mountain)

82 N/16 - Siffleur River
13-33-16-W5
51°50'N 116°08'W
Approximately 80 km north-west of Banff.

This 3068 m mountain was named in 1920
after George Malloch, a Canadian geologist
who mapped the area overlooked by this
peak. Malloch was also one of the first to
climb the mountain. He died in Siberia in
1914.

Mangin Glacier (glacier)

82 J/11 - Kananaskis Lakes
20-18-9-W5
50°33'N 115°13'W
Approximately 70 km west south-west of
Turner Valley.

(see Mount Mangin)

Mangin, Mount (mountain)

82 J/11 - Kananaskis Lakes
20-18-9-W5
50°32'N 115°13'W
Approximately 70 km west south-west of
Turner Valley.

Charles Emmanuel Mangin (1866-1925) was a French soldier who spent the best part of twenty-six years in different parts of Africa, striving to develop the resources of France's colonies. General Mangin, during the First World War, was a successful commander and later held the Grand Cross of the Legion of Honour. He was a member of the Supreme War Council, was Inspector General of the Colonial Forces, and presided over the Supreme Council for National Defence. This 3057 m mountain was named in 1918, after this French hero.

Manx Peak (peak)

83 D/16 - Jasper
11-44-2-W6
52°47′N 118°12′W
Approximately 13 km south-west of Jasper.

The shape of the contours of this 3044 m mountain peak resembles the coat of arms of the Isle of Man. The name was applied in 1916 by M.P. Bridgland. (see Mount Bridgland)

Manystick Creek (creek)

82 G/16 - Maycroft
3-12-4-W5
49°58′N 114°28′W
Flows south-west into Oldman River, approximately 40 km north of Blairmore.

The origin of the name is unknown.

Marble Mountain (mountain)

82 O/14 - Limestone Mountain
34-9-W5
51°54′N 115°13′W
Approximately 85 km north north-east of Banff.

The name for this mountain was officially approved 25 November 1941, and is likely descriptive.

Marble Mountain (mountain)

83 C/6 - Sunwapta
14-39-22-W5
52°21′N 117°03′W
Approximately 82 km south-east of Jasper.

The precise origin for the name of this mountain, which was officially approved by the Geographic Board of Canada 5 March 1935, is unknown but it is likely descriptive.

March Creek (creek)

82 J/10 - Mount Rae
25-19-5-W5
50°38′N 114°33′W
Flows east into Sheep River, approximately 20 km west of Turner Valley.

The origin of the name which was officially approved 6 September 1951, is unknown.

Margaret Lake (lake)

82 N/9 - Hector Lake
19-30-17-W5
51°34′N 116°23′W
Approximately 75 km north-west of Banff.

This lake was named by C.S. Thompson (see Thompson Pass) before 1898 after the eldest daughter of Reverend H.P. Nichols. (see Lake Alice)

Margaret Lake (lake)

82 O/6 - Lake Minnewanka
22-28-9-W5
51°23′N 115°11′W
Approximately 35 km east north-east of Banff.

This feature is named after Margaret Hill, the late wife of Ray Hill who was the Chief Ranger at the Canmore Ranger Station. The name was officially approved 5 May 1987.

Maria Creek (creek)

83 F/12 - Gregg Lake
36-54-27-W5
53°42′N 117°51′W
Flows north-west into Pinto Creek, approximately 39 km north north-west of Hinton.

The locally well-established name of this creek was officially approved in 1945. Its precise origin is unknown.

Marjorie Lake (lake)

83 D/16 - Jasper
7-45-1-W6
52°52′N 118°07′W
Approximately 2 km west of Jasper.

The name for this lake was applied by H. Matheson, D.L.S. in 1914. The origin of the name is unknown.

Marl Creek (creek)

83 F/6 - Pedley
15-52-24-W5
53°29′N 117°28′W
Flows north-east into Athabasca River, approximately 69 km west of Edson.

The name for this creek, which was adopted in 1944, may be descriptive of the area in which the creek flows. Marl is a soil containing calcium carbonate, used as cement or as fertilizer. (see Marl Lake)

Marl Lake (lake)

82 J/11 - Kananaskis Lakes
31-19-8-W5
50°39′25″N 115°06′25″W
Approximately 60 km west of Turner Valley.

The bottom of this lake is covered with a limy marl deposit (a valuable fertilizer), which gives the lake a distinctive colour. The name for this lake was officially approved 8 November 1978.

Marlborough, Mount (mountain)

82 J/11 - Kananaskis Lakes
32-18-9-W5
50°34′N 115°12′W
Approximately 65 km west south-west of
Turner Valley.

"The Marlborough" was a battleship
engaged in the Battle of Jutland in May,
1916. This 2973 m mountain was officially
named in 1918 after this ship. (see Jutland
Mountain)

Marmot Creek (creek)

82 J/14 - Spray Lakes Reservoir
12-23-9-W5
50°57′N 115°08′W
Flows east into Kananaskis River,
approximately 70 km west of Calgary.

The marmot or siffleur (French meaning
"whistler") is a large diurnal burrowing
rodent of the squirrel family. Four of the
thirteen species occur in Canada
(groundhogs are also a species of marmot).
These mammals inhabit grassy areas and
rocky slopes of mountains and lowlands.
The marmot is well known for its shrill call
and yellow grey body. This creek may be
named for the alpine animal.

Marmot Mountain (mountain)

82 N/16 - Siffleur River
23-32-18-W5
51°45′N 116°27′W
Approximately 90 km north-west of Banff.

Marmot Mountain is 2606 m in altitude. (see
Siffleur Mountain)

Marmot Mountain (mountain)

83 D/16 - Jasper
18-44-1-W6
52°47′N 118°07′W
Approximately 10 km south south-west of
Jasper.

This mountain, which is 2608 m in altitude,
was named in 1916 after the marmots found
on its slope. (see Marmot Creek)

Marmot Pass (pass)

83 D/16 - Jasper
7-44-1-W6
52°47′N 118°08′W
Approximately 11 km south south-west of
Jasper.

(see Marmot Mountain)

Marna Lake (lake)

82 G/8 - Beaver Mines
32-5-1-W5
49°26′00″N 114°05′45″W
Approximately 8 km east south-east of
Beaver Mines.

The name Marna Lake was officially
approved 5 July 1961, but was used prior to
1890 by settlers of the area. It is not known
for whom the lake is named.

Marsh Creek (creek)

82 O/3 - Canmore
13-24-10-W5
51°02′N 115°16′W
Flows north into Bow River, approximately
9 km south-east of Canmore.

This name is likely descriptive of the area
through which the creek flows.

Marsh Creek (creek)

83 F/11 - Dalehurst
12-55-24-W5
53°45′N 117°25′W
Flows north into Oldman Creek,
approximately 67 km west north-west of
Edson.

The descriptive name for this creek and the
nearby lake was suggested by J.A.
McDonald, one of the first topographers on

a survey of the area. It was officially
approved in 1946. The creek is surrounded
by a marsh.

Marsh Lake (lake)

83 F/11 - Dalehurst
36-54-24-W5
53°42′N 117°24′W
Approximately 65 km west north-west of
Edson.

(see Marsh Creek)

Marshybank Creek (creek)

83 C/15 - Cardinal River
29-44-19-W5
52°50′N 116°43′W
Flows north into Brazeau River,
approximately 91 km east south-east of
Jasper.

The name for this creek is likely descriptive.

Marshybank Lake (lake)

83 C/15 - Cardinal River
17-44-19-W5
52°47′N 116°44′W
Approximately 92 km east south-east of
Jasper.

(see Marshybank Creek)

Marston Creek (creek)

82 J/7 - Mount Head
5-17-4-W5
50°24′N 114°30′W
Flows south-east into Highwood River,
approximately 33 km south-west of Turner
Valley.

The name for this feature was officially
approved 8 December 1943. The creek flows
through Mr. E. Marston's ranch, and it is
after him that it is named.

Martin Creek (creek)

82 N/16 - Siffleur River
36-32-16-W5
51°48′N 116°07′W
Flows east into Martin Lake, approximately
80 km north-west of Banff.

The origin of the name is unknown.

Martin Creek, 1916

Martin Lake (lake)

82 N/16 - Siffleur River
36-32-16-W5
51°47′N 116°08′W
Approximately 80 km north-west of Banff.

No information regarding the origin of the
name is known.

Marvel Creek (creek)

82 J/13 - Mount Assiniboine
22-12-W5
50°53′N 115°31′W
Flows east into Bryant Creek, approximately
95 km west of Calgary.

(see Marvel Peak)

Marvel Lake (lake)

82 J/13 - Mount Assiniboine
23-22-12-W5
50°52′N 115°33′W
Approximately 95 km west of Calgary.

(see Marvel Peak)

Marvel Pass (pass)

82 J/13 - Mount Assiniboine
22-12-W5
50°50′30″N 115°34′45″W
Approximately 100 km west of Calgary, on
the Alberta-B.C. boundary.

(see Marvel Peak)

Marvel Peak (peak)

82 J/13 - Mount Assiniboine
22-12-W5
50°51′N 115°33′W
Approximately 105 km west of Calgary.

This 2713 m peak was named in 1917. Its
name is descriptive of its beauty: one
marvels at it.

Mary Gregg Creek (creek)

83 F/3 - Cadomin
28-48-23-W5
53°09′N 117°15′W
Flows north-east into McLeod River,
approximately 73 km south-west of Edson.

Mary was the daughter of Chief Cardinal, a
Stoney Indian. This well-respected woman
became the wife of John Gregg (see Gregg
Lake), and helped him find important coal
deposits at Mountain Park, near Lovettville.
(see Lovettville)

Mary Gregg Lake (lake)

83 F/3 - Cadomin
4-48-24-W5
53°07′N 117°27′W

Approximately 84 km south-west of Edson.

(see Mary Gregg Creek)

Mary Vaux, Mount (mountain)

83 C/11 - Southesk Lake
30-41-25-W5
52°33′N 117°27′W
Approximately 55 km south south-east of
Jasper.

This 3200 m mountain was named in 1911
by Mary T.S. Schäffer, after Miss Mary Vaux
(pronounced "Vox"), who had taken great
interest in the Canadian Rockies.

Maskinonge Lake (lake)

82 H/4 - Waterton Lakes
8-2-29-W4
49°06′N 113°50′W
Approximately 216 km south of Calgary.

This feature is the smallest of the four
Waterton Lakes. It was officially named 19
May 1943.

Maskuta Creek (creek)

83 F/5 - Entrance
5-51-25-W5
53°23′N 117°39′W
Flows north-east into Athabasca River,
approximately 3 km south-west of Hinton.

The name for this creek, previously known
as "Prairie Creek," was officially approved 1
May 1945. "Maskuta" is the Cree Indian
word meaning "meadows" or "prairie."

Mason Creek (creek)

83 E/15 - Pierre Greys Lakes
16-57-6-W6
53°56′N 118°50′W
Flows north into Muskeg River, approximately
20 km east north-east of Grande Cache.

This creek, which was officially named in
1947, may take its name after Crawford or

Cyrus Mason. The Masons were local homesteaders who arrived in the area in 1928.

Massive (locality)

82 O/4 - Banff
17-26-13-W5
51°13′N 115°47′W
Approximately 16 km west north-west of Banff.

(see Massive Mountain)

Massive Mountain (mountain)

82 O/4 - Banff
32-25-13-W5
51°11′N 115°47′W
Approximately 16 km west of Banff.

This 2435 m mountain was officially given its descriptive name 5 February 1918. (see Massive Range)

Massive Range (range)

82 O/4 - Banff
25-13-W5
51°09′N 115°49′W
Approximately 17 km west south-west of Banff.

This well-established local name was officially given to this range, which includes Pilot Mountain, Mount Brett, Mount Bourgeau and Massive Mountain, by the Canadian Permanent Committee on Geographical Names 17 May 1976.

Mastodon Glacier (glacier)

83 D/9 - Amethyst Lakes
42-3-W6
52°37′N 118°20′W
Approximately 34 km south-west of Jasper.

(see Mastodon Mountain)

Mastodon Mountain (mountain)

83 D/9 - Amethyst Lakes
11-42-3-W6
52°36′N 118°20′W
Approximately 35 km south-west of Jasper.

This mountain, which is 2987 m in altitude, has a fancied resemblance in shape to the extinct mammal which was a type of elephant. It was given its descriptive name in 1922 by A.O. Wheeler. (see Wheeler Flats)

Matkin, Mount (mountain)

82 G/1 - Sage Creek
9-3-2-W5
49°12′N 114°13′W
Approximately 28 km north-west of Waterton Park.

Henry Matkin was born 11 April 1875 at Hyde Park, Cache County, Utah. At 12 years of age, he moved to Cardston, Alberta with his father. He made a life for himself and his family in southern Alberta. His son, Philip K. Matkin, R.C.A.F. was killed in the Second World War. It is for this young sergeant that this 2418 m mountain was named.

Maude Brook (brook)

82 J/11 - Kananaskis Lakes
12-20-10-W5
50°41′N 115°15′W
Flows east into Upper Kananaskis Lake, approximately 70 km west of Turner Valley.

(see Mount Maude)

Maude Lake (lake)

82 J/11 - Kananaskis Lakes
14-20-10-W5
50°41′N 115°17′W
Approximately 70 km west of Turner Valley.

(see Mount Maude)

Maude, Mount (mountain)

82 J/11 - Kananaskis Lakes
15-20-10-W5
50°42′N 115°18′W
Approximately 70 km west of Turner Valley.

This 3042 m mountain was officially named in 1918. It commemorates Major General Sir Frederick Stanley Maude, K.C.B. (1864-1917), a British soldier, captor of Baghdad, and Military Secretary to the Governor General of Canada, 1901-1904.

Maverick Hill (hill)

82 G/8 - Beaver Mines
17-6-3-W5
49°28′N 114°22′W
Approximately 13 km west north-west of Beaver Mines.

There is no origin information available for this locally well-established name.

Mawdsley, Mount (mountain)

83 E/14 - Grande Cache
33-55-8-W6
53°48′N 119°08′W
Approximately 10 km south of Grande Cache.

This 2134 m mountain was named in 1953. A geologist who did some field work in the area said that the name was locally well-established. Mawdsley was an Indian boy who lived in this area.

Maxwell Lake (lake)

83 F/5 - Entrance
11-51-25-W5
53°23′N 117°35′W
Approximately 1 km south of Hinton.

The locally well-established name for this lake was officially adopted in 1944. Its precise origin is unknown.

May, Mount (mountain)

83 L/4 - Kakwa Falls
58-13-W6
54°03′N 119°55′W
Approximately 62 km west north-west of
Grande Cache.

(see Francis Peak and George Peak)

*Maycroft (former locality)

82 G/16 - Maycroft
17-10-2-W5
49°49′N 114°14′W
Approximately 30 km north-east of Blairmore.

This former locality was officially named in
1910 after Mrs. May Raper, wife of the area's
first postmaster, A.C. Raper. The post office
in this locality was closed in September,
1964.

McBeath, Mount (mountain)

83 C/14 - Mountain Park
5-44-22-W5
52°46′N 117°07′W
Approximately 64 km east south-east of
Jasper.

This 2845 m mountain was officially named
in 1925 after a member of Lord Southesk's
party of 1859 (see Southesk River). Morrison
McBeath was a tall Scottish man with a dark
moustache and beard who was frequently
seen in the area on his coal black horse.

McCardell Creek (creek)

83 F/3 - Cadomin
20-49-23-W5
53°14′N 117°19′W
Flows north-west into McLeod River,
approximately 70 km south-west of Edson.

This creek was named for a member of the
local forest staff, prior to 1927. No other
information is available for the name.

McCarty, Mount (mountain)

82 G/8 - Beaver Mines
17-5-4-W5
49°23′N 114°30′W
Approximately 23 km west south-west of
Beaver Mines.

M.P. Bridgland (see Mount Bridgland)
named this mountain in 1915, but it is not
known for whom it was named.

■ **McConnel Falls** (falls)

82 J/8 - Stimson Creek
12-14-16-4-W5
50°20′50″N 114°27′20″W
Approximately 35 km south-east of Turner
Valley.

These falls were officially named 5 May
1983, after Gordon McConnel, who bought
the quarter section on which the feature is
located in the early 1890s. Mr. McConnel
died in the area in 1900.

McConnell Creek (creek)

82 O/12 - Barrier Mountain
15-31-14-W5
51°39′N 115°54′W
Flows south-east into Red Deer River,
approximately 60 km north-west of Banff.

(see Mount McConnell)

McConnell, Mount (mountain)

82 N/9 - Hector Lake
31-15-W5
51°39′N 116°00′W
Approximately 60 km north-west of Banff.

This 3109 m mountain was named in 1884
by Dr. G.M. Dawson (see Tombstone
Mountain), of the Geological Survey of
Canada, after Richard George McConnell,
F.R.S.C. (1857-1942), who was his assistant in
1882. McConnell joined the Geological
Survey in 1880 and carried out many
important explorations in western Canada.

He became Director of the Geological
Survey in 1914 and later became Deputy
Minister of the Department of Mines.

McConnell Ridge (ridge)

82 O/3 - Canmore
17,20-24-8-W5
From 51°03′N 115°05′W
to 51°04′N 115°07′W
Approximately 20 km east south-east of
Canmore.

This highly visible landmark is
approximately 4.5 km in length and has a
lookout on its high point. The name is
derived from the McConnell Fault which
underlies the area.

McCord, Mount (mountain)

83 E/2 - Resplendent Creek
9-47-5-W6
53°02′N 118°39′W
Approximately 44 km west north-west of
Jasper.

William C. McCord was the head of the
Canadian Pacific Railway trail-making party
of 1872. This 2511 m mountain was named
after Mr. McCord in 1923 by A.O. Wheeler.
(see Wheeler Flats)

McCormick Creek (creek)

83 C/16 - Brown Creek
6-44-17-W5
52°46′N 116°27′W
Flows south-east into Brown Creek,
approximately 41 km north-west of
Nordegg.

The origin for the name of this stream is
unknown.

McCue Creek (creek)

82 O/11 - Burnt Timber Creek
16-31-9-W5
51°38′N 115°14′W

Flows north-east into Red Deer River, approximately 60 km north north-east of Banff.

The origin of the name is unknown.

McDonald Creek (creek)

83 C/2 - Cline River
7-37-19-W5
52°10'N 116°42'W
Flows south-east into Cline River, approximately 120 km south-east of Jasper.

The name for this creek was officially approved in 1959, for the feature's proximity to Mount McDonald. (see Mount McDonald)

McDonald Flats (flats)

83 E/15 - Pierre Greys Lakes
23-57-7-W6
53°56'N 118°56'W
Approximately 14 km north-east of Grande Cache.

Officially approved in 1947, the name for these flats may be taken from Donald McDonald, a well-known guide in the area. He was a packer and guide who also worked doing cable splicing and maintaining steam engines. He homesteaded near the old Grand Trunk Pacific Railway grade.

McDonald, Mount (mountain)

83 C/7 - Job Creek
10-39-21-W5
52°21'N 116°57'W
Approximately 100 km south-east of Jasper.

The name was officially approved 5 March 1935. The origin of the name is unknown.

McDonell Peak (peak)

83 D/9 - Amethyst Lakes
25-42-3-W6
52°39'N 118°19'W

Approximately 29 km south-west of Jasper.

This 3261 m mountain peak is one of the peaks of Mount Fraser, and was named in 1922 after Simon Fraser's wife (see Mount Fraser). She was the daughter of Colonel Allan McDonell, of Ontario.

McDougall, Mount (mountain)

82 J/14 - Spray Lakes Reservoir
28-22-8-W5
50°54'N 115°04'W
Approximately 65 km west of Calgary.

This mountain, which is 2591 m in altitude, was named by G.M. Dawson (see Tombstone Mountain) in 1884, after Reverend George M. McDougall (1821-1876) and his sons David and John. The Reverend George McDougall laboured for many years among the Stoney Indians, work which was continued by his son, Reverend John McDougall, after his death. He went to Edmonton in 1871, and soon after built the city's first Protestant church.

McGillivray Creek (creek)

82 G/10 - Crowsnest
8-8-4-W5
49°38'N 114°31'W
Flows south into Crowsnest River, approximately 1 km west of Coleman.

(see McGillivray Ridge)

McGillivray, Mount (mountain)

82 O/3 - Canmore
3-24-9-W5
51°02'N 115°10'W
Approximately 16 km east south-east of Canmore.

This 2454 m mountain was officially named 4 April 1957, after Duncan McGillivray (1770-1808), a wintering partner of the North West Company. This early fur trader established himself with the North West Company prior to 1793. He established

Rocky Mountain House trading post in 1799 and was the first white man to use the Athabasca Pass.

McGillivray Pond (mountain lake)

82 O/3 - Canmore
2-21-24-9-W5
51°03'30"N 115°11'50"W
Approximately 18 km east south-east of Canmore.

This low mountain lake is named for its proximity to Mount McGillivray. (see Mount McGillivray)

McGillivray Ridge (ridge)

83 D/8 - Athabasca Pass
35-39-2-W6
52°23'N 118°11'W
Approximately 57 km south of Jasper.

According to *Place Names of Alberta*, (1928): "Franchère says that it was named McGillivray's Rock by J. Henry" in his diary 10 October 1923, John Work says, "after William McGillivray;" William McGillivray (1764-1825), elder brother of Simon McGillivray and uncle of Duncan McGillivray (see Mount McGillivray), was one of the leading members of the North West Company; he is also commemorated by Fort William city, Ontario; he was a member of the House of Assembly of Lower Canada, 1808-1809 for Montreal West and of the Legislative Council of Lower Canada 1814-25; he died in 1825 in London, England.

McGillivray Ridge (ridge)

82 G/10 - Crowsnest
8-4-5
49°41'N 114°31'W
Approximately 7 km north of Coleman.

The name McGillivray Ridge was submitted by A.O. Wheeler (see Wheeler Flats) in October of 1921 and it was officially approved 6 December 1921. This ridge is part of the Crowsnest volcanics, and was named after a contractor in the area.

McGladrey, Mount (mountain)

82 G/10 - Crowsnest
34-7-5-W5
49°35′N 114°34′W
Approximately 15 km south south-west of Coleman, forming part of the Alberta-B.C. boundary.

The origin of the name is unknown.

McGregor Lake (lake)

83 B/12 - Harlech
17-42-13-W5
52°37′N 115°34′W
Approximately 22 km north-east of Nordegg.

The locally well-established name for this lake was officially approved 20 January 1955. Its precise origin, however, is unknown.

McGuire, Mount (mountain)

83 C/6 - Sunwapta
52°22′N 117°27′W
Approximately 70 km south south-east of Jasper.

Fenton John Alexander (''Mickey'') McGuire (1911-1973) moved to Edson from his birth place in Kamsack, Saskatchewan in 1922, and to Jasper in 1924. He served the National Parks Service in Jasper National Park for 34 years and one of his duties included that of Chief Park Warden. He was considered a loyal and dedicated Park Warden and was highly respected by his peers and associates. This 3030 m mountain and the nearby valley were both officially named for Mr. McGuire 14 August 1975.

McGuire Valley (valley)

83 C/6 - Sunwapta
52°24′N 117°25′W

Approximately 68 km south south-east of Jasper.

(see Mount McGuire)

McHarg, Mount (mountain)

82 J/11 - Kananaskis Lakes
22-19-10-W5
50°37′N 115°18′W
Approximately 70 km west south-west of Turner Valley.

Mount McHarg was named after Lieutenant-Colonel Hart McHarg, who was killed in action in 1915 leading the 7th Battalion, B.C. Regiment. This 2888 m mountain was officially named in 1918.

McKean, Mount (mountain)

83 D/16 - Jasper
13-46-3-W6
52°57′N 118°18′W
Approximately 17 km north-west of Jasper.

This 2743 m mountain was officially named 7 March 1951 in honour of Captain George Burdon McKean, V.C., M.C., M.M. (1888-1926) of the 14th Battalion of the Canadian Expeditionary Force, in the First World War.

McLaren, Mount (mountain)

82 G/10 - Crowsnest
26-7-5-W5
49°35′N 114°34′W
Approximately 7 km south-west of Coleman.

This 2286 m mountain was named as a tribute to Senator Peter McLaren, one of the lumber kings along Ontario's Clyde River, who brought his expertise in the lumber industry to the Crowsnest Pass area in 1881. Beginning in 1882, the Peter McLaren Lumber Company business set up operations in the village of Mountain Mill. It is in honour of Senator McLaren's contribution to this first industry in the Crowsnest Pass

area that the mountain was officially named 31 December 1962. (see Mill Creek)

McLean Creek (creek)

82 J/15 - Bragg Creek
29-22-5-W5
50°54′N 114°40′W
Flows north into Elbow River, approximately 35 km west of Calgary.

(see McLean Hill)

McLean Hill (hill)

82 J/15 - Bragg Creek
16-22-5-W5
50°53′N 114°38′W
Approximately 30 km west of Calgary.

This feature was named for Jack McLean, an early resident who established an independent logging operation in the area. He is said to be the first man to float his logs down the Elbow River to Calgary and other destinations. He left the area by 1907. Both the creek and the hill were officially named 6 October 1949.

McLean Pond (reservoir)

82 J/15 - Bragg Creek
20-22-5-W5
50°53′25″N 114°40′00″W
Approximately 35 km west of Calgary.

Named for its proximity to McLean Hill, this small man-made reservoir is located on McLean Creek. It was officially named 7 February 1983.

McLeod River (locality)

83 F/7 - Erith
7-52-18-W5
53°28′N 116°38′W
Approximately 18 km south-west of Edson.

The Canadian National Railway station opened here in 1912. The station took its name from the McLeod River which flows nearby.

McLeod River (river)

83 J/4 - Whitecourt
35-59-12-W5
54°09'N 115°42'W
Flows north-east into Athabasca River, approximately 250 km north-east of Jasper.

The name has been in use since at least 1814, when it was marked "McLeods River" on C.S. Thompson's map. (see Thompson Pass)

McMurtry Ridge (ridge)

82 O/14 - Limestone Mountain
16-32-8-W5
51°46'N 115°07'W
Approximately 70 km north north-east of Banff.

This ridge was officially named 25 November 1941 after David A. McMurtry of Blaeberry, who hunted on this ridge for many years.

McNab, Mount (mountain)

82 J/10 - Mount Rae
25-19-5-W5
50°38'N 114°34'W
Approximately 20 km west south-west of Turner Valley.

This mountain, which is 1690 m in altitude, was officially named 6 September 1951 after an old-time resident of the Turner Valley/Black Diamond area, named Sandy McNab. There is some question of the correct spelling of the name: McNab or McNabb.

McNair Pond (pond)

82 N/8 - Lake Louise
51°24'20"N 116°09'15"W
Approximately 50 km north-west of Banff.

The name was officially approved 21 January 1985. The origin of the name is unknown.

McNeill Creek (creek)

83 F/7 - Erith
13-50-21-W5
53°19'N 116°56'W
Flows south-east into Prest Creek, approximately 45 km south-west of Edson.

The locally well-established name for this creek was made official 15 December 1944. It is named for a timber operator who located on the stream for several years.

McPhail Creek (creek)

82 J/7 - Mount Head
17-6-W5
50°26'N 114°45'W
Flows east into Highwood River, approximately 44 km south-west of Turner Valley.

(see McPhail Mountain)

McPhail, Mount (mountain)

82 J/7 - Mount Head
17-7-W5
50°24'N 114°51'W
Approximately 50 km south-west of Turner Valley.

This 2883 m mountain was officially named in 1918, after N.R. McPhail, of the Surveyor General's Staff. He was killed in active service in November 1917, during the First World War.

McPherson Creek (creek)

83 F/6 - Pedley
36-50-23-W5
53°22'N 117°13'W
Flows east into McLeod River, approximately 57 km south-west of Edson.

The locally well-established name of this creek was officially applied 21 December 1944. The McPherson and Quigley Lumber Company operated on the stream.

McQueen, Mount (mountain)

83 E/11 - Hardscrabble Creek
17-55-8-W6
53°45'N 119°10'W
Approximately 13 km south south-west of Grande Cache.

This 2286 m mountain was named in 1953 after Dr. David George McQueen (1854-1930). A native of Ontario, McQueen served as the Minister for the First Presbyterian Church of Edmonton from 1887-1930.

Mead Creek (creek)

82 G/16 - Maycroft
19-10-1-W5
49°50'N 114°07'W
Flows west into Callum Creek, approximately 35 km north-east of Blairmore.

Dr. H.R. Mead, was Pincher Creek's first doctor. He died in 1898 and it may be for him that this creek was named. The name is well-established in local usage.

Meadow Creek (creek)

82 O/6 - Lake Minnewanka
12-28-8-W5
51°22'N 115°00'W
Flows east into Waiparous Creek, approximately 40 km north-east of Canmore.

The precise origin of the name is unknown.

Meadow Creek (creek)

83 D/16 - Jasper
8-45-2-W6
52°52'N 118°15'W
Flows north into Miette River, approximately 11 km west of Jasper.

This creek rises at Meadow Glacier and has a descriptive name which was approved prior to 1918.

Meadow Glacier (glacier)

83 D/16 - Jasper
44-3-W6
52°46′N 118°23′W
Approximately 24 km south-west of Jasper.

(see Meadow Creek)

Meadowland Creek (creek)

83 E/12 - Pauline Creek
28-52-12-W6
53°31′N 119°42′W
Flows north-east into Jackpine River,
approximately 55 km south-west of Grande
Cache.

The descriptive name for this creek has been
in use since 1924.

Meadows Creek (creek)

83 B/5 - Saunders
18-39-11-W5
52°22′N 115°34′W
Flows north-east into Rough Creek,
approximately 37 km south-east of Nordegg.

The name for this creek was suggested by the
Topographical Survey due to its proximity to
the Meadows Fire Ranger Station.

Meadows Lake (lake)

83 C/15 - Cardinal River
32-45-18-W5
52°55′N 116°35′W
Approximately 103 km east of Jasper.

The origin for the name of this lake is
unknown.

Meda, Mount (mountain)

83 C/14 - Mountain Park
4-44-23-W5
52°46′N 117°17′W
Approximately 57 km south-east of Jasper.

This 2858 m mountain was named in 1925,
after the Indian heroine of the Earl of

Southesk's (see Southesk River) poem, "The
Meda Maiden."

Medallion Lakes (lake)

83 L/5 - Two Lakes
14-62-14-W6
54-80- 22′N 119°59′W
Approximately 65 km north-west of Grande
Cache.

The descriptive name for these two small
lakes was officially approved in 1922.

Medicine Lake (lake)

83 C/13 - Medicine Lake
6-45-26-W5
52°51′N 117°45′W
Approximately 17 km east of Jasper.

The name for this lake is recorded as early as
1846 by Father De Smet (see De Smet
Range). Its precise origin is not known. In
Cree, it is known as "muskiki" and
"nipagwasinow sipi;" in Stoney as "to-go-
wap-ta." (Tyrrell)

Medicine Lake Slabs (ridge)

83 C/13 - Medicine Lake
From 52°51′30″N 117°44′10″W
to 52°52′10″N 117°48′00″W
Approximately 16 km east of Jasper, along
the north-east bank of Medicine Lake.

This ridge has been given a name that
describes the rock formation above the
treeline which is likened to a series of slabs
or dominoes. (see Medicine Lake)

Medicine Lodge (locality)

83 F/11 - Dalehurst
34-52-21-W5
53°34′N 117°00′W
Approximately 38 km west of Edson.

The Canadian National Railway opened a
station here in 1911. No other origin
information is available.

Medicine Tent River (river)

83 C/14 - Mountain Park
34-44-24-W5
52°50′N 117°23′W
Flows south-west into Rocky River,
approximately 46 km east of Jasper.

The name for this river is a translation of an
Indian word indicating magic and mystery. It
dates back to 1875, where it may be found
on Southesk's map. (see Southesk River)

Meinsinger Creek (creek)

82 J/8 - Stimson Creek
32-15-2-W5
50°18′N 114°14′W
Flows north into Sheppard Creek,
approximately 30 km south south-east of
Turner Valley.

Henry Minesinger arrived in the area around
1885. He later settled with his family near
the lake, the source of this creek, which are
now both named after him. The spelling has
been altered over time, but this version of
the name was officially approved 12
December 1939.

Meinsinger Lake (lake)

82 J/8 - Stimson Creek
20-15-2-W5
50°16′N 114°15′W
Approximately 45 km south of Turner
Valley.

(see Meinsinger Creek)

Mercer Creek (creek)

82 J/13 - Mount Assiniboine
35-22-12-W5
50°55′N 115°33′W
Flows south-west into Bryant Creek,
approximately 100 km west of Calgary.

(see Mount Mercer)

Mercer, Mount (mountain)

82 J/13 - Mount Assiniboine
22-11-W5
50°55′N 115°30′W
Approximately 95 km west of Calgary.

This 2970 m peak was officially named in 1918 after Major-General M.S. Mercer, C.B., C.E.F., who was killed in action near Zillebeke, Flanders 2 June 1916.

Mercoal (locality)

83 F/3 - Cadomin
24-48-22-W5
53°10′N 117°05′W
Approximately 64 km south-west of Edson.

The name for this Canadian National Railway station which opened in 1913 is a compound of the name, "McLeod River Hard Coal Company Limited."

Mercoal Creek (creek)

83 F/3 - Cadomin
29-48-22-W5
53°10′N 117°11′W
Flows west into McLeod River, approximately 67 km south-west of Edson.

This stream, named prior to 1927, flows through Mercoal, after which it is named. (see Mercoal)

Merlin Castle (mountain)

82 N/9 - Hector Lake
6-30-15-W5
51°32′N 116°07′W
Approximately 55 km north-west of Banff.

This cluster of tower-like rocks was named by James F. Porter. (see Merlin Lake)

Merlin Creek (creek)

83 F/4 - Miette
11-47-27-W5
53°03′N 117°51′W

Flows north into Rocky River, approximately 46 km south south-west of Hinton.

(see Mount Merlin)

Merlin Lake (lake)

82 N/9 - Hector Lake
25-29-16-W5
51°30′N 116°07′W
Approximately 55 km north-west of Banff.

This lake was named after Merlin the Magician in 1911 by James F. Porter, a surveyor and alpinist from Chicago. It is located at the foot of a mountain bearing a fancied resemblance to Merlin's Castle.

Merlin, Mount (mountain)

83 C/13 - Medicine Lake
27-46-27-W5
52°59′N 117°50′W
Approximately 20 km north-east of Jasper.

This 2711 m mountain was named after Merlin, from Alfred Lord Tennyson's *Idylls of the King* (1859?).

Merlin Pass (pass)

83 C/13 - Medicine Lake
21-46-27-W5
52°58′N 117°52′W
Approximately 19 km north-east of Jasper.

(see Mount Merlin)

Merlin Ridge (ridge)

82 N/9 - Hector Lake
26-29-16-W5
51°31′N 116°08′W
Approximately 55 km north-west of Banff.

(see Merlin Lake)

Merlin Ridge (ridge)

83 F/4 - Miette
33-46-27-W5
53°01′N 117°51′W

Approximately 50 km south south-west of Hinton.

(see Mount Merlin)

Merlin Valley (valley)

82 N/9 - Hector Lake
25-29-16-W5
51°31′N 116°07′W
Approximately 55 km north-west of Banff.

(see Merlin Lake)

Mesa Butte (butte)

82 J/15 - Bragg Creek
13-21-5-W5
50°47′N 114°34′W
Approximately 35 km west of Calgary.

A "mesa" is the geological term for a type of landform produced by differential erosion where the rock strata are horizontal. The descriptive name for this butte was officially approved 6 October 1949.

Mesa Creek (creek)

82 J/16 - Priddis
9-21-4-W5
50°46′N 114°29′W
Flows east into Fisher Creek, approximately 34 km south-west of Calgary.

The name Mesa Creek was officially approved 6 October 1949. (see Mesa Butte)

Messines, Mount (mountain)

82 N/15 - Mistaya Lake
13-33-22-W5
51°50′N 116°59′W
Approximately 122 km north-west of Banff, on the Alberta-B.C. boundary.

This 3100 m mountain was named in 1920 in commemoration of the fighting of Canadian troops in Messines, West Flanders, south of Ypres, in June 1917 and April 1918.

■ **Michener, Mount** (mountain)

83 C/1 - Whiterabbit Creek
29-37-17-W5
52°12′N 116°23′W
Approximately 120 km north-west of Banff.

Mount Michener, a 2545 m peak, forms the north end of the Ram Range, a portion of the Front Ranges of the Rocky Mountains. Officially named 26 August 1979, this mountain honours the Right Honourable Daniel Roland Michener P.C., C.C., C.M.M., C.D., Q.C., B.A.(Alberta), M.A. and B.C.L.(Oxon.), LL.D., D.C.L. He was the Governor General of Canada from 1967-1974.

Middle Kootenay Pass (pass)

82 G/8 - Beaver Mines
36-3-4-W5
49°16′N 114°24′W
Approximately 26 km south-west of Beaver Mines.

(see South Kootenay Pass)

Middle Waterton Lake (lake)

82 H/4 - Waterton Lakes
24-1-30-W4
49°03′N 113°53′W
Approximately 50 km south of Pincher Creek.

The name is used for specific identification, as each lake is clearly a defined entity in its own right. The name was approved 7 September 1969 by the Canadian Permanent Committee on Geographical Names. (see Waterton Lakes)

Middle Whirlpool River (river)

83 D/9 - Amethyst Lakes
22-41-1-W6
52°33′N 118°04′W
Flows north-east into Whirlpool River, approximately 37 km south of Jasper.

(see Whirlpool River)

Midway Peak (peak)

82 N/15 - Mistaya Lake
32-19-W5
51°48′N 116°39′W
Approximately 100 km north-west of Banff, on the Alberta-B.C. boundary.

The descriptive name for this 2871 m mountain peak was officially approved 2 July 1918. A.O. Wheeler noticed it was located "midway" between Stairway Peak and Mount Synge.

Miette (locality)

83 F/4 - Miette
7-49-27-W5
53°13′N 117°57′W
Approximately 34 km south-west of Hinton.

(see Roche Miette)

Miette Hill (hill)

83 D/16 - Jasper
4-45-3-W6
52°51′N 118°23′W
Approximately 20 km west of Jasper.

(see Roche Miette)

Miette Hotsprings (hotsprings)

83 F/4 - Miette
8-48-26-W5
53°07′N 117°47′W
Approximately 35 km south south-west of Hinton.

These natural hotsprings were discovered by Ewan Moberly (see Evan's Creek) in the early 1870s. (see Roche Miette)

Miette Hotsprings (locality)

83 F/4 - Miette
8-48-26-W5
53°08′N 117°47′W
Approximately 35 km south south-west of Hinton.

(see Roche Miette)

Miette Lake (lake)

83 E/2 - Resplendent Creek
2-47-5-W6
53°01′N 118°38′W
Approximately 40 km west north-west of Jasper.

(see Roche Miette)

Miette Pass (pass)

83 E/2 - Resplendent Creek
46-5-W6
53°01′N 118°40′W
Approximately 40 km west north-west of Jasper.

(see Roche Miette and South Pass)

Miette Range (range)

83 F/4 - Miette
48-27-W5
53°08′N 117°58′W
Approximately 36 km north north-east of Jasper.

(see Roche Miette)

Miette River (river)

83 D/16 - Jasper
10-45-1-W6
52°52′N 118°04′W
Flows east into Athabasca River, approximately 1 km south-east of Jasper.

(see Roche Miette)

Miette, Roche (mountain)

83 F/4 - Miette
29-48-27-W5
53°10′N 117°55′W
Approximately 37 km south-west of Hinton.

This 2316 m mountain is named after a voyageur by the name of Miette who climbed its south side, according to G. Franchère (see Franchère Peak). Another suggestion for the origin of the name states that Miette is from the Cree Indian word, "my-a-tick," meaning "mountain sheep" — a word perhaps corrupted throughout the years to eventually become Miette. The other features take their names from this mountain.

Roche Miette, 1910

Mildred Lake (lake)

83 D/16 - Jasper
22-45-1-W6
52°53'N 118°03'W
Approximately 2 km north-east of Jasper.

The origin of the name is unknown.

Miles Coulee (coulee)

82 G/16 - Maycroft
11-11-3-W5
49°54'N 114°18'W
Approximately 35 km north of Blairmore.

Mr. Thomas Miles settled on the north fork of Livingstone Range around 1889, on the ranch where Mr. Thomas Heap, now lives. This coulee may be named after Mr. Miles.

Mill Creek (creek)

82 G/8 - Beaver Mines
19-6-1-W5
49°29'N 114°07'W
Approximately 12 km west of Pincher Creek.

This creek was named for a sawmill that flourished for several years. The mill started in 1880 as a government project to encourage development of natives, but was not a success. Two years later, the project was sold to Senator Peter McLaren (see Mount McLaren), and he gradually made the mill profitable. By 1900, work on a modern sawmill was started and the mill was fully operational by 1902. The village of Mountain Mill grew up around this industry, but within thirty years, the mill and the town had shut down.

Millarville (locality)

82 J/16 - Priddis
3-21-3-W5
50°45'N 114°19'W
Approximately 27 km south-west of Calgary.

This locality, officially named in January 1940, takes its name from Malcolm T. Millar (1860-1937). He was the first settler and postmaster of the area. Mr. Millar came to Canada from his native Scotland in 1879. After five years with the North West Mounted Police, he began homesteading on Sheep Creek.

Miller Creek (creek)

82 J/8 - Stimson Creek
24-16-4-W5
50°22'N 114°27'W
Flows north-east into Pekisko Creek, approximately 35 km south-west of Turner Valley.

Herbert William Millar (1860-1955) came north from Illinois in 1882, and started his career as a bronco buster on Tom Lynch's ranch. Herb Millar worked at the Bar U

Ranch for over 50 years, for various owners. The name "Miller Creek" was officially approved 12 December 1939, the spelling slightly different from the name of the man for whom the creek was named.

Miller Gap (gap)

82 J/8 - Stimson Creek
30-17-3-W5
50°28'N 114°24'W
Approximately 40 km south south-west of Turner Valley.

The name was officially approved 12 December 1939. (see Miller Creek)

Mina Lake (lake)

83 D/16 - Jasper
17-45-1-W6
52°53'N 118°07'W
Approximately 2 km north-west of Jasper.

This lake was named in 1914 by H. Matheson, of the Dominion Land Survey. The origin of the name is unknown.

Minaga Creek (creek)

83 D/16 - Jasper
45-2-W6
52°52'N 118°15'W
Flows south-east into Miette River, approximately 10 km west of Jasper.

"Minaga" is the Cree Indian word for "spruce." This creek has a name descriptive of its surroundings and the name is well-established in local usage.

Minewakun Lake (lake)

82 N/8 - Lake Louise
25-28-17-W5
51°25'N 116°16'W
Approximately 55 km north-west of Banff.

The name for this lake was officially approved in 1958. "Minewakun" is the

Stoney word for "cascade." This feature was named by S.E.S. Allen. (see Mount Allen)

Minnewanka, Lake (lake)

82 O/3 - Canmore
22-26-10-W5
51°14'N 115°19'W
Approximately 9 km north-east of Banff.

The name Lake Minnewanka was applied by the Department of the Interior in 1888, replacing all former names, including: Devil's Lake; "m'ne-sto" or Cannibal Lake (Stoney); "ki'noo-ki'mow" or Long Lake (Cree); and Peechee Lake (G.M. Dawson). Minnewanka is translated as "lake of the water-spirit," and has a legend attached to it. One of the first Indians who saw the lake, viewed it from the summit of one of the highest mountains which surround it. From this vantage point, he noticed a fish in the water which appeared to be as long as the lake, leading him to call the lake "The lake of the Evil Water-Spirit."

Lake Minnewanka, ca. 1890

Minnow Lake (lake)

83 D/16 - Jasper
21-45-2-W6
52°53'N 118°13'W
Approximately 9 km west north-west of Jasper.

The name for this lake is likely descriptive; its precise origin, however, is unknown.

Mirror Lake (lake)

82 N/8 - Lake Louise
19-28-16-W5
51°25'N 116°14'W
Approximately 55 km north-west of Banff.

This lake was named in 1894 by W.D. Wilcox (see Wilcox Pass) from the reflection in the lake when seen from the height above. It had previously been named "Lorne Lake" by Hindman, who assisted Major Rogers in the exploration of the Rockies for a railroad route, after the Marquis of Lorne, Governor General of Canada from 1878-1883.

Minster Mountain (mountain)

83 C/2 - Cline River
35-36-20-W5
52°08'N 116°45'W
Approximately 120 km south-east of Jasper.

This mountain was named by A.P. Coleman (see Mount Coleman) as it presents "the imposing walls of cathedral-shaped mountains." (see Cloister Mountain)

Missinglink Mountain (mountain)

82 J/10 - Mount Rae
8-20-5-W5
50°40'N 114°39'W
Approximately 26 km west of Turner Valley.

Officially approved 6 September 1951, the origin for this locally well-established name is unknown.

Mist Creek (creek)

82 J/10 - Mount Rae
13-18-7-W5
50°31'N 114°50'W
Flows south-east into Storm Creek, approximately 43 km south-west of Turner Valley.

(see Mist Mountain)

Mist Mountain (mountain)

82 J/10 - Mount Rae
28-18-7-W5
50°33'N 114°55'W
Approximately 46 km west of Turner Valley.

At the time of its naming in 1884 by Dr. G.M. Dawson of the Geological Survey of Canada (see Tombstone Mountain), this 3057 m mountain had a gathering of clouds surrounding it, giving it a misty appearance.

Mistanusk Creek (creek)

83 L/12 - Nose Creek
35-64-14-W6
54°35'N 120°00'W
Flows north into Red Deer Creek, in British Columbia, approximately 95 km west south-west of Grande Prairie.

The name for this creek was proposed by R.W. Cautley (a Commissioner for the Provincial Boundary Survey) in 1922, and is the Cree word for "badger."

Mistaya Canyon (canyon)

82 N/15 - Mistaya Lake
51°56'50"N 116°42'40"W
Approximately 115 km north-west of Banff.

The name for this canyon was officially approved 20 October 1983. (see Mistaya River and Mistaya Mountain)

Mistaya Lake (lake)

82 N/15 - Mistaya Lake
2-32-19-W5
51°47'N 116°35'W
Approximately 118 km north-west of Banff.

This lake is named for its proximity to Mistaya River. It was named from the Indian word for bear and has been in use since at least 1916. Previously, it had been known as Bear River or Little Fork, but was changed to avoid duplication.

Mistaya Mountain (mountain)

82 N/10 - Blaeberry River
6-32-18-W5
51°43′N 116°35′W
Approximately 90 km north-west of Banff, on the Alberta-B.C. boundary.

This mountain, which is 3078 m in altitude, takes its name from the nearby river. Mistaya Mountain was the name officially applied to this feature 2 July 1918. (see Mistaya River)

Mistaya River (river)

82 N/15 - Mistaya Lake
35-34-20-W5
51°58′N 116°43′W
Flows north-west into North Saskatchewan River, approximately 120 km north-west of Banff.

This river carries the Indian word for "bear." It has been in use since at least 1916. Previously, it had been known as Bear River or Little Fork, but was changed to avoid duplication.

Misty Range (range)

82 J/10 - Mount Rae
18-7-W5
50°35′N 114°57′W
Approximately 47 km west of Turner Valley.

In 1884 when this range was named, clouds covered the summits of the mountains, inspiring Dr. G.M. Dawson (see Tombstone Mountain), of the Geological Survey of Canada, to give it this descriptive name.

Mitayimin Creek (creek)

83 E/14 - Grande Cache
8-57-7-W6
53°55′N 119°00′W
Flows south-east into Peavine Creek, approximately 9 km north-east of Grande Cache.

In order to avoid duplication of the name "Strawberry Creek," the Cree Indian word for strawberry, "mitayimin," was adopted for this stream in 1952.

Mitchell Creek (creek)

83 F/7 - Erith
14-50-20-W5
53°19′N 116°49′W
Flows north into Embarras River, approximately 39 km south-west of Edson.

This creek, formerly known as "Middle Embarras" was renamed Mitchell Creek after an old trapper who lived near its mouth. The change was adopted in 1927 to avoid confusion with Embarras River.

Mitchell Group (peaks)

83 C/5 - Fortress Lake
33-39-25-W5
52°24′N 117°30′W
Approximately 65 km south south-east of Jasper.

(see Mount Mitchell)

Mitchell Lake (lake)

83 B/3 - Tay River
36-37-8-W5
52°12′N 115°00′W
Approximately 20 km south south-west of Rocky Mountain House.

The locally well-established name for this lake was officially approved 5 May 1987. It was named after Jack Mitchell, a homesteader of the area.

Mitchell, Mount (mountain)

83 C/6 - Sunwapta
13-39-25-W5
52°24′N 117°30′W
Approximately 65 km south south-east of Jasper.

Mr. J.H. Mitchell was the Senior Assistant Engineer for the Department of the Interior in charge of construction of the Banff-Jasper Highway from Jasper to the Banff-Jasper boundary. Through his efforts, drive and initiative, the road continued through to a point approximately 16 km beyond his target. This work was done during the Great Depression with mainly men and horse equipment. Mr. Mitchell was certainly the key figure in the construction and location of the northern portion of the now famous highway. He also pioneered the first water and sewer system in the Jasper townsite, the first Athabasca River Bridge at Mile 12, and the Snaring River Bridge on Highway 16E.

Mitre, The (mountain)

82 N/8 - Lake Louise
1-28-17-W5
51°22′N 116°15′W
Approximately 55 km west north-west of Banff.

This 2886 m mountain was named by S.E.S. Allen (see Mount Allen) circa 1893. Its shape resembles that of a Bishop's mitre.

Mitre Glacier (glacier)

82 N/8 - Lake Louise
51°21′20″N 116°16′05″W
Approximately 55 km west north-west of Banff.

The name was officially approved 21 January 1985. (see The Mitre)

Moab Lake (lake)

83 C/12 - Athabasca Falls
34-42-28-W5
52°40′N 117°57′W
Approximately 26 km south of Jasper.

The name for this lake was officially approved 1 May 1934. The origin of the name is unknown.

Moat Creek (creek)

83 D/9 - Amethyst Lakes
36-44-2-W6
52°45′N 118°18′W
Flows north into Meadow Creek,
approximately 22 km south-west of Jasper.

Formerly "Meadow Creek," the name for
this creek is likely taken from the lake out of
which it flows. The change in name was
made in 1978. (see Moat Lake)

Moat Creek (creek)

83 D/16 - Jasper
20-44-2-W6
52°45′10″N 118°18′10″W
Flows north into Meadow Creek,
approximately 20 km south-west of Jasper.

The Canadian Permanent Committee on
Geographical Names approved the name for
this creek 4 October 1978. (see Moat Lake)

Moat Lake (lake)

83 D/9 - Amethyst Lakes
23-43-3-W6
52°43′N 118°20′W
Approximately 24 km south-west of Jasper.

The descriptive name for this lake was
applied in 1927. The lake resembles a ditch
or moat located in front of The Ramparts.

Moat Pass (pass)

83 D/9 - Amethyst Lakes
43-3-W6
52°43′N 118°20′W
Approximately 25 km south-west of Jasper.

Moat Pass is 2115 m in altitude. (see Moat
Lake)

Moberly Bend (bend)

83 C/12 - Athabasca Falls
52°34′55″N 117°44′30″W
Approximately 40 km south south-east of
Jasper.

This feature was named after John Moberly
in the early 1950s. The name is well-
established locally.

Moberly Creek (creek)

83 F/12 - Gregg Lake
9-53-27-W5
53°33′N 117°56′W
Flows east into Wildhay River,
approximately 29 km north-west of Hinton.

(see Moberly Flats)

Moberly Flats (flats)

83 E/1 - Snaring
47-1-W6
53°03′N 118°07′W
Approximately 19 km north of Jasper.

Henry John Moberly (1835-1931), after
whom these flats were named in 1859 by Sir
James Hector (see Mount Hector), was born
in Penetanguishene, Upper Canada. He
entered the service of the Hudson's Bay
Company in 1854 as a clerk. He served as
Chief Factor of the once-abandoned Jasper
House post from 1855-1861. He married
Suzanne Cardinal (?-1905), a descendant of
Jacques Cardinal (see Jacques Roche)
Together, they had two children, Ewan and
John (see Evan's Creek). He served the
Company until 1894 when, as a factor, he
retired and settled in Saskatchewan. Moberly
had intervals when he was a free trader. His
biography, entitled *When Fur Was King*
(1929) is an amalgamation of several articles
he wrote for *The Beaver* about his life in the
fur trade. (see Pyramid Mountain)

Moberly Hill (hill)

83 F/5 - Entrance
10-52-27-W5
53°28′N 117°54′W
Approximately 22 km west north-west of
Hinton.

(see Moberly Flats)

Molar Creek (creek)

82 N/9 - Hector Lake
11-30-16-W5
51°31′N 116°09′W
Flows south-east into Pipestone River,
approximately 60 km north-west of Banff.

(see Molar Mountain)

Molar Glacier (glacier)

82 N/9 - Hector Lake
36-30-17-W5
51°37′N 116°16′W
Approximately 70 km north-west of Banff.

(see Molar Mountain)

Molar Mountain (mountain)

82 N/9 - Hector Lake
32-30-16-W5
51°37′N 116°12′W
Approximately 65 km north-west of Banff.

The descriptive name for this tooth-shaped
mountain, 3022 m in altitude, was given by
Sir James Hector (see Mount Hector) in 1859.

Molar Pass (pass)

82 N/9 - Hector Lake
51°38′35″N 116°15′20″W
Approximately 70 km north-west of Banff.

The name for this pass was officially approved
21 January 1985. (see Molar Mountain)

Mona Lake (lake)

83 C/12 - Athabasca Falls
26-43-26-W5
52°44′N 117°40′W
Approximately 31 km south-east of Jasper.

The origin of the name is unknown.

Monarch Mountain (mountain)

83 D/16 - Jasper
25-46-4-W6
53°00′N 118°27′W

Approximately 27 km north-west of Jasper.

The origin of the name is unknown.

Monchy Mountain (mountain)

83 C/3 - Columbia Icefield
35-22-W5
52°01′N 117°01′W
Approximately 120 km south-east of Jasper.

This 3210 m mountain was named in 1920 after the village in France which British soldiers attacked and took, 26 August 1918.

Monkhead Mountain (mountain)

83 C/11 - Southesk Lake
4-42-24-W5
52°35′N 117°23′W
Approximately 90 km west of Nordegg.

The name for this bold noticeable formation on the north corner of Mount Warren, is descriptive. The mountain measures 3211 m in altitude. It was named in 1911 by Mrs. Mary Schäffer.

Monoghan Creek (creek)

83 E/10 - Adams Lookout
23-53-7-W6
53°35′N 118°56′W
Flows north-east into Sulphur River, approximately 36 km south south-east of Grande Cache.

This creek was likely named after Tom Monoghan, a licensed guide, trapper and outfitter of the Entrance region. The earliest reference to the name Monoghan Creek was in 1962.

Mons Creek (creek)

83 C/10 - George Creek
9-42-19-W5
52°36′N 116°40′W

Flows north into Blackstone River, approximately 43 km west north-west of Nordegg.

(see Mons Peak)

Mons Peak (peak)

82 N/14 - Rostrum Peak
27-33-22-W5
51°52′N 117°02′W
Approximately 125 km north-west of Banff, on the Alberta-B.C. boundary.

This 3083 m mountain peak was named in 1920 after the Belgian town which saw the first battle of the British Expeditionary Force in the First World War, 23 August 1914. The town was recaptured and entered by Canadian troops immediately before the armistice, 11 November 1918.

Monte Cristo Mountain (mountain)

83 E/7 - Blue Creek
27-50-7-W6
53°20′N 118°57′W
Approximately 78 km north-west of Jasper.

The name for this mountain was officially approved 1 May 1934. The origin of the name is unknown.

Moon Creek (creek)

83 E/9 - Moberly Creek
15-53-3-W6
53°45′N 118°21′W
Flows north-east into Berland River, approximately 99 km north north-west of Jasper.

No information regarding the origin of the name is known.

Moose Basin (lake)

83 B/12 - Harlech
3-18-43-12-W5
52°42′00″N 115°43′45″W

Approximately 35 km north-east of Nordegg.

The name for this small lake is descriptive of fauna — moose at one time came here to drink. The name was locally well-established prior to its official approval 7 February 1983.

Moose Creek (creek)

82 O/2 - Jumpingpound Creek
8-24-6-W5
51°02′N 114°48′W
Flows north into Jumpingpound Creek, approximately 40 km west of Calgary.

The name was officially approved 12 December 1939. (see Moose Mountain)

Moose Creek (creek)

83 B/3 - Tay River
29-37-8-W5
52°12′15″N 115°06′30″W
Flows north-east into Vetch Creek, approximately 20 km south-west of Rocky Mountain House.

The name of this creek was officially adopted 7 February 1983. This locally well-established name, applied to the feature since at least 1906, describes the local wildlife.

Moose Lake (lake)

83 C/12 - Athabasca Falls
52°42′00″N 117°38′20″W
Approximately 28 km east south-east of Jasper.

The name for this lake is locally well-established. The origin of the name is unknown.

Moose Lake (lake)

83 C/15 - Cardinal River
14-46-19-W5
52°58′N 116°35′W
Approximately 97 km east north-east of Jasper.

The origin of the locally well-established name of this lake is unknown.

Moose Mountain (mountain)

82 J/15 - Bragg Creek
7-23-6-W5
50°56'N 114°50'W
Approximately 50 km west of Calgary.

The name for this 2437 m mountain was officially approved 6 October 1949. The name has been in local usage since 1895-1896, and is likely named after the moose, a large ungulate commonly found in this area.

Moose Pass (pass)

83 E/3 - Mount Robson
13-49-8-W6
53°13'N 119°02'W
Approximately 75 km north of Jasper.

(see Moose Creek)

Moose Dome Creek (creek)

82 J/15 - Bragg Creek
29-22-6-W5
50°54'N 114°48'W
Flows south into Canyon Creek, approximately 45 km west of Calgary.

This creek empties into the Elbow River near the Moose Dome oil well. The name Moose Dome Creek was officially approved 6 October 1949, but the local name is "Sulphur Creek" or "Sulphur Spring Creek," due to the location of a sulphur spring near the stream.

Moose Mountain Creek (creek)

82 J/15 - Bragg Creek
35-22-7-W5
50°55'N 114°52'W
Flows south off Moose Mountain into Canyon Creek, approximately 50 km west of Calgary.

(see Moose Mountain)

Moosehorn Creek (creek)

83 F/4 - Miette
17-49-27-W5
53°13'N 117°53'W
Flows south into Athabasca River, approximately 33 km south-west of Hinton.

The name for this creek was formerly "Moose Creek," but was changed in 1917 to avoid duplication. No other information is available.

Moosehorn Lake (lake)

83 E/8 - Rock Lake
31-50-1-W6
53°21'50"N 118°08'10"W
Approximately 55 km north of Jasper.

The lake is named after Moosehorn Creek, the origin of which is unknown. The name was officially approved 20 October 1983.

Moosehound Creek (creek)

83 C/15 - Cardinal River
8-45-18-W5
52°52'N 116°34'W
Flows north into Brazeau River, approximately 102 km east of Jasper.

The origin for the name of this creek is unknown.

Moraine Creek (creek)

82 N/8 - Lake Louise
6-28-15-W5
51°22'N 116°06'W
Flows north-east into Bow River, approximately 45 km west north-west of Banff.

(see Moraine Lake)

Moraine Lake (lake)

82 N/8 - Lake Louise
21-27-16-W5
51°19'N 116°11'W

Approximately 45 km west north-west of Banff.

A moraine is the term used to describe an area of debris carried down and deposited by a glacier. At the lower end of this lake there is a ridge of glacial formation measuring 2.5 km in length. This formation inspired W.D. Wilcox (see Wilcox Pass) to descriptively name the feature Moraine Lake in 1899. Geologists once considered the deposit to be a moraine, but now believe it is material deposited by a slide.

Morden Long, Mount (mountain)

83 C/5 - Fortress Lake
9-40-25-W5
52°25'N 117°32'W
Approximately 60 km south south-east of Jasper.

This mountain, which is 3040 m in altitude, was officially named by the Canadian Permanent Committee on Geographical Names 30 August 1966 after Professor Morden H. Long of the University of Alberta. He was Chairman of the Geographic Board of Alberta from its inception to his death, and was recognized as an outstanding Canadian historian who did much to make the history of Canada meaningful.

Moren, Mount (mountain)

83 E/2 - Resplendent Creek
36-46-4-W6
53°00'N 118°35'W
Approximately 34 km west north-west of Jasper.

This 2562 m mountain was named in 1923 after Arthur Moren, M.D., a member of the Sandford Fleming Party of 1872 (see Introduction). No other information is available.

Morin Creek (creek)

82 G/9 - Blairmore
18-8-3-W5
49°39'N 114°24'W

Flows south-west into Gold Creek, approximately 5 km north of Frank.

The name Morin Creek has been in local usage since about 1908. The origin of the name is unknown.

Morkill, Mount (mountain)

83 E/12 - Pauline Creek
27-54-13-W6
53°42′N 119°50′W
Approximately 52 km west south-west of Grande Cache.

This 2286 m mountain was officially named 7 May 1965. (see Morkill Pass)

Morkill Pass (pass)

83 E/12 - Pauline Creek
19-54-12-W6
53°40′N 119°45′W
Approximately 49 km west south-west of Grande Cache.

This 1656 m pass was named in 1923 after D.B. Morkill, a land surveyor from British Columbia. The nearby mountain takes its name from this pass.

Morley (locality)

82 O/2 - Jumpingpound Creek
25-25-7-W5
51°09′N 114°52′W
Approximately 60 km north-west of Calgary.

This area is named after the ranch of Reverend George McDougall (see Mount McDougall), which McDougall had named after prominent Methodist preacher Reverend Morley Punshon. In 1872, McDougall secured approval from Punshon to establish a mission at the foot of the Rockies. John McDougall (George's son) also lived at the mission established in 1873, which gave rise to the Blackfoot name "Tsawn-occowoy" (where John lives), "tsawn" being Blackfoot for John.

Morley Hill (hill)

83 L/3 - Copton Creek
28-60-8-W6
54°13′N 119°08′W
Approximately 136 km south of Grande Prairie.

The origin of the name is unknown.

Morleyville Settlement (settlement)

82 O/2 - Jumpingpound Creek
7,14,16-26-6-W5
51°12′N 114°46′W
Approximately 55 km north-west of Calgary.

(see Morley)

Morris Creek (creek)

83 F/4 - Miette
35-48-27-W5
53°11′N 117°50′W
Flows north-east into Fiddle River, approximately 34 km south-west of Hinton.

This creek was named after a representative of the Geological Survey of 1911.

Morrison, Mount (mountain)

82 J/14 - Spray Lakes Reservoir
5-22-1-W5
50°50′N 115°29′W
Approximately 95 km west south-west of Calgary.

This 2896 m mountain was officially named in 1918 after Major-General Sir Edward Whipple Bancroft Morrison, D.S.O., C.B., K.C.M.G. (1867-1925). He was the Editor-in-Chief of the *Ottawa Citizen* newspaper. Morrison was the officer commanding the 1st Artillery Brigade for the Canadian Expeditionary Force of 1914-1915, and was the General Officer commanding the Canadian Corps Artillery from 1916-1919.

Morro Creek (creek)

83 E/1 - Snaring
4-47-1-W6
53°01′N 118°05′W
Flows west into Athabasca River, approximately 16 km north of Jasper.

(see Morro Peak)

Morro Peak (peak)

83 E/1 - Snaring
10-47-1-W6
53°02′N 118°04′W
Approximately 17 km north of Jasper.

The descriptive name for this 1678 m peak is Spanish for "round hill." The name was applied to the feature in 1916 by M.P. Bridgland (see Mount Bridgland). The creek takes its name from the peak.

Mosquito Creek (creek)

82 N/9 - Hector Lake
3-31-17-W5
51°38′N 116°20′W
Flows south into Bow River, approximately 75 km north-west of Banff.

The locally well-established name for this creek was officially approved 7 May 1959. Its precise origin is unknown.

Mosquito Hill (hill)

82 J/7 - Mount Head
11-30-15-5-W5
50°18′N 114°26′W
Approximately 90 km south-west of Calgary.

This hill was named by the Boundary Survey in 1915. The name seems to be descriptive.

Mount Brown Icefield (icefield)

83 D/8 - Athabasca Pass
22-39-2-W6
52°22′N 118°13′W

Approximately 57 km south south-west of Jasper.

(see Mount Brown)

Mount Lorette Ponds (man-made ponds)

82 J/14 - Spray Lakes Reservoir
4-19-23-8-W5
50°58′00″N 115°06′55″W
Approximately 68 km west of Calgary.

These ponds were developed as fishing ponds for handicapped people; there is a day use site nearby with the same name. The name was officially approved 20 October 1983. (see Mount Lorette)

Mountain Creek (creek)

83 F/4 - Miette
7-49-27-W5
53°12′N 117°55′W
Flows north-east into Athabasca River, approximately 33 km south-west of Hinton.

The locally well-established name of this creek is descriptive of the area through which it flows.

*Mountain Park (former locality)

83 C/14 - Mountain Park
33-45-23-W5
52°55′N 117°16′W
Approximately 56 km east of Jasper.

Robert Thornton (see Cheviot Mountain) named this feature in 1910. According to an anonymous document quoted in *Oh! The Coal Branch*, "Thornton, in awe, looked at tall straight spruce trees, wild mountain flowers and the whole basin surrounded by snow-capped mountains. 'If we do start a mine, I have chosen a name for the town that will be built. This is like a park in the old country. We will call it Mountain Park.'" The town and mine were established in 1950.

—————————
*denotes rescinded name or former locality.

Mouse Cache Creek (creek)

83 L/4 - Kakwa Falls
21-59-13-W6
54°07′N 119°52′W
Flows south-east into Kakwa River, approximately 47 km north-west of Grande Cache.

A forest superintendent for Grande Prairie, Mr. R. Smuland, proposed this name which has been in use since 1927 or 1928. It was officially approved 4 December 1958. The precise origin of the name is unknown.

Mowitch Creek (creek)

83 E/8 - Rock Lake
22-51-4-W6
53°25′N 118°29′W
Flows north-east into Rock Creek, approximately 66 km north north-west of Jasper.

The origin of the name is unknown.

Mud Lake (lake)

82 N/8 - Lake Louise
4-29-16-W5
51°27′N 116°12′W
Approximately 55 km north-west of Banff.

The name for this lake is descriptive.

Mud Lake (lake)

82 J/14 - Spray Lakes Reservoir
22-21-10-W5
50°47′N 115°19′W
Approximately 85 km west south-west of Calgary.

The name for this muddy-watered lake was officially approved in 1965 and is probably descriptive.

Muddy Creek (creek)

83 L/12 - Nose Creek
27-66-11-W6
54°45′N 119°34′W

Flows north into Nose Creek, approximately 69 km south-west of Grande Prairie.

The name is likely descriptive.

Muddywater River (river)

83 E/14 - Grande Cache
25-55-10-W6
53°47′N 119°21′W
Flows east into Smoky River, approximately 19 km south-west of Grande Cache.

The name for this watercourse is likely descriptive.

Mudsprings Creek (creek)

82 J/9 - Turner Valley
12-22-20-4-W5
50°42′40″N 114°29′00″W
Flows north into Ware Creek, approximately 14 km west of Turner Valley.

The locally well-established name for this creek was officially approved 5 May 1983. The wet and muddy areas near the headwaters of the creek explain its descriptive name.

Muhigan Creek (creek)

83 D/16 - Jasper
8-45-2-W6
52°52′N 118°12′W
Flows north-east into Miette River, approximately 6 km west of Jasper.

(see Muhigan Mountain)

Muhigan Mountain (mountain)

83 D/16 - Jasper
28-44-2-W6
52°49′N 118°13′W
Approximately 11 km south-west of Jasper.

"Muhigan" is an Indian word for "wolf." This 2609 m mountain and the nearby creek were officially named 1 May 1934. The name was established in 1916 by M.P. Bridgland. (see Mount Bridgland)

Muir Creek (creek)

82 J/7 - Mount Head
17-6-W5
50°25′N 114°47′W
Approximately 49 km south-west of Turner Valley.

(see Mount Muir)

Muir, Mount (mountain)

82 J/7 - Mount Head
16-6-W5
50°24′N 114°49′W
Approximately 48 km south-west of Turner Valley.

This 2758 m mountain was officially named in 1918, after Alexander Muir (1830-1906), the author of *The Maple Leaf*, written in 1867.

***Muirhead** (former locality)

82 J/8 - Stimson Creek
15-16-1-W5
50°21′N 114°03′W
Approximately 19 km south south-east of Turner Valley.

Peter Muirhead (circa 1853-1917) was the original owner of the townsite after whom this locality was named. Officially approved 12 December 1939, the name was rescinded 12 February 1970 when the population dropped to nil.

Mule Shoe Lake (lake)

82 O/4 - Banff
26-25-13-W5
51°09′N 115°43′W
Approximately 11 km west of Banff.

This locally well-established name is descriptive of the outline of the lake. Previously "Muleshoe Lake," the spelling

was officially changed to Mule Shoe Lake in 1958.

Mumm Creek (creek)

83 E/8 - Rock Lake
10-52-2-W6
53°28′N 118°12′W
Flows south-east into Wildhay River, approximately 67 km north of Jasper.

(see Mumm Peak)

Mumm Peak (peak)

83 E/3 - Mount Robson
35-48-8-W6
53°11′N 119°10′W
Approximately 79 km north-west of Jasper.

This mountain peak is 2962 m in altitude and was ascended first by Arnold L. Mumm, F.R.G.S. (1859-1928) after whom it is named. Mumm was a lifetime member of the Alpine Club of Canada and played a prominent role in the early explorations of the Canadian Rockies.

Mummery Group (range)

82 N/10 - Blaeberry River
36-31-21-W5
51°42′N 116°51′W
Approximately 105 km north-west of Banff.

Albert Frederick Mummery (1855-1895) was a British political economist and alpinist. He is famous for his extensive mountain climbing excursions in the Himalayas. He published *My Climbs in the Alps and Caucasus* in 1895, a work of great merit. He died in 1895, in a lone attempt to reach the summit of the Nanga Parbat group of the Kashmir Himalayas. This range, and a mountain in British Columbia, are named for A.F. Mummery.

Mummy Lake (lake)

82 O/4 - Banff
25-14-W5
51°05′N 115°55′W

Approximately 26 km south-west of Banff.

The locally well-established name for this lake, which lies south of the Pharoah Peaks, has been in use for at least seventy years. It is part of an area known locally as the "Egyptian Connection."

Municipality of Crowsnest Pass

(municipality - town)
82 G/8 & 9
8-5-W5
49°38′N 114°41′W
Along Highway #3, in the Crowsnest Pass.

The Crowsnest Pass Municipal Unification Act, consolidated the towns of Blairmore, Coleman, the villages of Bellevue, Frank and Hillcrest Mines, and several other small hamlets in the Crowsnest Pass, under one municipal government. The area included is about 150 square kilometres, with a population of 7500. The Act provided for municipal government by nine aldermen (three wards) and a mayor elected by the entire municipality. The towns and villages are now mainly residential areas and the names are still used. The Act came into effect 1 January 1979.
(see Crowsnest Mountain)

Murchison Creek (creek)

82 N/15 - Mistaya Lake
6-35-19-W5
51°59′N 116°41′W
Flows north-west into North Saskatchewan River, approximately 127 km north-west of Banff.

(see Mount Murchison)

Murchison Icefield (icefield)

82 N/15 - Mistaya Lake
34-19-W5
51°55′N 116°38′W
Approximately 122 km north-west of Banff.

(see Mount Murchison)

*denotes rescinded name or former locality.

Murchison, Mount (mountain)

82 N/15 - Mistaya Lake
34-19-W5
51°55′N 116°39′W
Approximately 125 km north-west of Banff.

This 3390 m mountain was named by Dr. James Hector (see Mount Hector) in 1859 after Sir Roderick Impey Murchison (1792-1871), a Scottish geologist. Murchison was the Director General of the Geological Survey of Great Britain and it was on his recommendation that Hector was chosen for the post of surgeon and geologist to the Palliser Expedition. (see Introduction)

Murray Creek (creek)

82 J/11 - Kananaskis Lakes
36-20-10-W5
50°44′45″N 115°15′15″W
Flows east into Smith-Dorrien Creek, approximately 70 km west north-west of Turner Valley.

Officially approved 8 November 1978, the name is taken from the stream which flows from the south side of Mount Murray. (see Mount Murray)

Murray, Mount (mountain)

82 J/11 - Kananaskis Lakes
3-21-10-W5
50°45′N 115°17′W
Approximately 85 km west south-west of Calgary.

This 3024 m mountain was officially named in 1918. General Sir A.J. Murray was the Chief of the Imperial General Staff in 1915. He was the general officer commanding the Forces in Egypt in 1916-1917.

Mushroom Peak (peak)

83 C/6 - Sunwapta
38-24-W5
52°19′N 117°24′W

Approximately 77 km south south-east of Jasper.

The appropriately descriptive name for this 3210 m mountain peak was approved in 1971.

Muskeg Creek (creek)

82 J/15 - Bragg Creek
8-21-5-W5
50°46′N 114°40′W
Flows south-east into Threepoint Creek, approximately 40 km south-west of Calgary.

This creek runs through an area locally known as "The Muskeg." Muskeg corresponds to the Ojibway Indian term for a type of bog. The descriptive name for this creek was officially approved 6 October 1949.

Muskeg River (locality)

83 E/15 - Pierre Greys Lakes
11-57-5-W6
53°55′N 118°39′W
Approximately 25 km north north-east of Jasper.

This settlement was named after the nearby river. (see Muskeg River)

Muskeg River (river)

83 L/3 - Copton Creek
13-58-8-W6
54°01′N 119°03′W
Approximately 16 km north north-east of Grande Cache.

The locally well-established name of this creek appears to have been jokingly applied to this stream, which is teeming with Dolly Varden Trout, since the water is as clear as crystal.

Muskiki Creek (creek)

83 C/15 - Cardinal River
10-45-20-W5
52°52′N 116°49′W

Flows north-east into Cardinal River, approximately 84 km east of Jasper.

The name for this lake is the translation of the Cree Indian word for "medicine."

Muskiki Lake (lake)

83 C/15 - Cardinal River
5-45-20-W5
52°51′N 116°52′W
Approximately 82 km east of Jasper.

(see Muskiki Creek)

Myosotis Lake (lake)

82 N/9 - Hector Lake
20-29-15-W5
51°30′N 116°05′W
Approximately 50 km north-west of Banff.

The name for this lake was suggested by James F. Porter, an American who made an extensive exploration of the area in 1911. The blue colour of the lake resembles the colour of the Forget-me-not flower, whose genus name is Myosotis, and which grows in the area. (see Forgetmenot Mountain)

Mystery Lake (lake)

83 F/4 - Miette
13-48-26-W5
53°09′N 117°40′W
Approximately 32 km south south-west of Hinton.

The precise origin of the name is unknown.

Mystic Lake (lake)

82 O/5 - Castle Mountain
27-13-W5
51°17′N 115°45′W
Approximately 17 km north-west of Banff.

No information regarding the origin of the name is known.

Mystic Pass (pass)

82 O/5 - Castle Mountain
51°18′20″N 115°45′45″W
Approximately 20 km north-west of Banff.

Mystic Pass was officially named 21 January 1985, for its proximity to Mystic Lake.

Nanga Parbat Mountain (mountain)

82 N/10 - Blaeberry River
2-31-21-W5
51°42′N 116°52′W
Approximately 110 km north-west of Banff, on the Alberta-B.C. boundary.

The name for this mountain, which is 3240 m in altitude, was adopted 3 February 1920. The feature was named after a feature in the Himalaya Mountains. Nanga Parbat is descriptive; "nanga" means "bare" and "parbat" refers to a hill or a mountain.

Narraway River (river)

83 L/12 - Nose Creek
20-66-13-W6
54°44′N 119°55′W
Flows north into Wapiti River, approximately 74 km west south-west of Grande Prairie.

This river was officially named in 1923 after A.M. Narraway, D.L.S., who was the Controller of Surveys, in Ottawa. His duties took him to the 120th Meridian and to this river in 1922.

Narrow Canyon (canyon)

83 C/6 - Sunwapta
7-38-23-W5
52°15′N 117°19′W
Approximately 85 km south south-east of Jasper.

The name for this canyon is likely descriptive.

Narrow Creek (creek)

83 E/13 - Dry Canyon
57-13-W6
53°58′N 119°49′W
Flows north-east into Trench Creek, approximately 46 km west north-west of Grande Cache.

The name for this creek was submitted for approval by Mr. R. Eben, a local outfitter and guide. The descriptive name was officially approved in 1958.

Nashan Creek (creek)

83 F/4 - Miette
15-47-27-W5
53°03′N 117°52′W
Flows north-east into Rocky River, approximately 45 km south south-west of Hinton.

The name of this creek was officially approved in 1927 and is the Stoney Indian word for "wolverine."

Nasswald Peak (peak)

82 J/13 - Mount Assiniboine
23-12-W5
51°00′N 115°39′W
Approximately 20 km south of Banff, on the Alberta-B.C. boundary.

This 2995 m mountain peak was named in 1913, after Nasswald, Austria. This was the birthplace of Conrad Kain, a member of a survey party who climbed the peak in that year.

Natural Arch, The (cliff)

83 E/7 - Blue Creek
30-51-6-W6
53°26′N 118°51′W
Approximately 81 km north-west of Jasper.

The name for this cliff is likely descriptive of its formation and appearance. The cliff is part of The Ancient Wall. (see The Ancient Wall)

Needle Peak (peak)

83 D/9 - Amethyst Lakes
26-41-2-W6
52°33′N 118°12′W
Approximately 38 km south south-west of Jasper.

The name for this 2850 m mountain peak, given by A.O. Wheeler (see Wheeler Flats) in 1922, is descriptive of the outline of the summit.

Neighbor Creek (creek)

83 E/14 - Grande Cache
16-57-10-W6
53°56′N 119°25′W
Flows north into Sheep Creek, approximately 16 km west of Grande Cache.

This creek may be named after Rufus Neighbor, a local homesteader.

Neilson Creek (creek)

83 C/10 - George Creek
8-43-20-W5
52°41′N 116°52′W
Flows east into Southesk River, approximately 58 km west north-west of Nordegg.

The creek was named after a local forest ranger. No other information is available.

Nelson Creek (creek)

82 J/1 - Langford Creek
23-13-2-W5
50°07′N 114°10′W
Flows north into Chaffin Creek, approximately 45 km west north-west of Claresholm.

The origin of the name is unknown.

Nelson Flat (flat)

83 C/9 - Wapiabi Creek
30-42-16-W5
52°38′N 116°18′W

Approximately 23 km north-west of Nordegg.

The origin of the name is unknown.

Nelson, Mount (mountain)

83 C/6 - Sunwapta
12-39-25-W5
52°20′N 117°28′W
Approximately 78 km south south-east of Jasper.

This name was recommended by J. Monroe Thorington in 1952. According to Thorington, this 3150 m mountain was first ascended by a climbing family called the Ebersbachers. It seems they had a particular Englishman in mind, but it is not clear to whom the name refers. The locally established name was made official 3 May 1961.

Neptuak Mountain (mountain)

82 N/10 - Blaeberry River
27-17-W5
51°18′N 116°15′W
Approximately 50 km west north-west of Banff.

This 3233 m mountain was named by S.E.S. Allen (see Mount Allen). "Neptuak" is the Stoney Indian term for "nine." As this mountain is ninth of the "Ten Peaks," its name is numerically descriptive.

Nestor, Mount (mountain)

82 J/14 - Spray Lakes Reservoir
5-23-10-W5
50°56′N 115°22′W
Approximately 85 km west of Calgary.

The origin of the name is unknown.

Newman Peak (peak)

82 G/1 - Sage Creek
5-3-1-W5
49°10′N 114°05′W

Approximately 20 km north-west of Waterton Park.

This 2621 m mountain was named in 1858 by Thomas Blakiston (see Mount Blakiston). Edward Newman (1801-1876) was born of Quaker parents in Middlesex, England. He was a very famous naturalist in his own time and published several works on natural history, plant and bird life, and insects. His reputation as an enthusiastic and laborious naturalist was well-deserved. He remained in England for most of his life, but his reputation carried to Canada where Blakiston chose to commemorate him.

Nez Percé Creek (creek)

82 G/10 - Crowsnest
9-8-4-W5
49°38′N 114°29′W
Flows west into Blairmore Creek, approximately 4 km north of Blairmore.

The name for this creek was made official 8 April 1905. The creek is named after the Nez Percés, a name applied by the French to all Indian tribes whose members pierced the nose for the insertion of a shell ornament (Dentalium). The term is now applied only to the tribe of the Shahaptian family, now found in northern Idaho and Oregon.

Niblock, Mount (mountain)

82 N/8 - Lake Louise
25-28-17-W5
51°25′N 116°16′W
Approximately 60 km north-west of Banff.

This 2976 m mountain was named in 1904 after Superintendent John Niblock, who was attached to the Canadian Pacific Railway.

Nice Creek (creek)

83 B/5 - Saunders
16-38-13-W5
52°15′N 115°49′W

Flows south into North Ram River, approximately 28 km south-east of Nordegg.

This creek was officially named in 1941. "Nice" describes the topography surrounding the stream.

Nickerson Creek (creek)

83 E/14 - Grande Cache
16-32-57-9-W6
53°58′30″N 119°17′30″W
Flows south-east into Sheep Creek, approximately 15 km north-west of Grande Cache.

Nick Nickerson, a prominent figure of Hinton and area, was likely the person after whom this creek was named.

Nigel Creek (creek)

83 C/3 - Columbia Icefield
11-37-22-W5
52°10′N 117°02′W
Flows south-east into North Saskatchewan River, approximately 105 km south-east of Jasper.

(see Nigel Peak)

Nigel Pass (pass)

83 C/3 - Columbia Icefield
5-38-22-W5
52°15′N 117°06′W
Approximately 95 km south-east of Jasper.

(see Nigel Peak)

Nigel Peak (peak)

83 C/3 - Columbia Icefield
2-38-23-W5
52°15′N 117°10′W
Approximately 95 km south-east of Jasper.

This mountain peak and the other features located in close proximity are named for Nigel Vavasour. He was a packer who accompanied the first major climbing expedition to this area in 1898 with J.N. Collie (see Collie Creek) and H.E.M. Stutfield. (see Stutfield Peak)

Nihahi Creek (creek)

82 J/15 - Bragg Creek
22-21-7-W5
50°48′N 114°55′W
Flows south-east into Elbow River, approximately 55 km west south-west of Calgary.

(see Nihahi Ridge)

Nihahi Ridge (ridge)

82 J/15 - Bragg Creek
22-7-W5
50°50′N 114°55′W
Approximately 55 km west south-west of Calgary.

This feature was named "Nihahi Range" in 1922, after the Stoney Indian word for "rock." The descriptive name is appropriate, but the generic "range" was changed to "ridge" 12 December 1939.

Nikanassin Range (range)

83 F/4 - Miette
6-47-25-W5
53°04′N 117°34′W
Approximately 40 km south of Hinton.

This mountain range was officially named 1 December 1909. The name was suggested by D.B. Dowling of the Geological Survey (see Dowling Ford) and is the Cree Indian word combination for "first range" as it is the first or front range when approaching the Rockies from the east.

Niverville Glacier (glacier)

82 N/15 - Mistaya Lake
32-21-W5
51°47′N 116°57′W

Approximately 115 km north-west of Banff, on the Alberta-B.C. boundary.

(see Mount Niverville)

Niverville, Mount (mountain)

82 N/15 - Mistaya Lake
32-21-W5
51°48′N 116°57′W
Approximately 115 km north-west of Banff.

Joseph Claude Boucher de Niverville (1715-1804) was the soldier and explorer after whom this 2963 m mountain was named. He took part in the war with the English colonies in 1746 and 1747. In 1750 he accompanied Legardeur de St. Pierre's expedition to Upper Saskatchewan, where he built Fort de Jonquière in 1751. He took part in the American invasion in 1775, and from 1775-1795, he was superintendent of Indians in the district of Three Rivers. He was made a colonel of militia in 1790.

Noire, Roche (peak)

83 D/16 - Jasper
26-44-3-W6
52°49′N 118°19′W
Approximately 22 km south-west of Jasper.

The descriptive name for this 2878 m mountain peak refers to its black summit. Noire is French for black. M.P. Bridgland (see Mount Bridgland) applied the name to the peak in 1916.

Nomad Creek (creek)

83 C/14 - Mountain Park
8-45-21-W5
52°52′N 116°59′W
Flows east into Cardinal River, approximately 73 km east of Jasper.

The origin for the name of this stream is unknown.

Noonday Peak (peak)

83 E/7 - Blue Creek
15-51-6-W6
53°25′N 118°47′W
Approximately 76 km north-west of Jasper.

The origin of the name of this mountain peak, which was adopted 1 May 1934, is unknown.

Nordegg (hamlet)

83 C/8 - Nordegg
27-40-15-W5
52°28′N 116°05′W
Approximately 155 km north-west of Red Deer.

This C.N.R. station was named after Martin Nordegg, Manager of the Brazeau Collieries. (see Lusk Creek)

Nordic Ridge (ridge)

82 N/16 - Siffleur River
28-33-16-W5
51°51′N 116°12′W
Approximately 85 km north-west of Banff.

The name for this ridge was approved by the Canadian Permanent Committee on Geographical Names 2 December 1967. The ridge is steep and sharply mountainous and has twin summits distinct from the mountain to the north. Nordic Ridge retains its original name which had faint reference to the original leader of the Ram River Hydrological and Glaciological project and to the person who first climbed it, as they were both from a Nordic country.

Norman Creek (creek)

83 C/2 - Cline River
2-36-21-W5
52°04′N 116°54′W
Flows south into North Saskatchewan River, approximately 120 km south-east of Jasper.

The origin of the name is unknown.

Norman Lake (lake)

83 C/2 - Cline River
14-36-21-W5
52°06′N 116°54′W
Approximately 120 km south-east of Jasper.

The origin of this name, which is well-established in local usage, is unknown.

Norman Lake (lake)

82 O/2 - Jumpingpound Creek
22-24-5-W5
51°03′N 114°37′W
Approximately 30 km west of Calgary.

This lake was officially named 12 December 1939, after the Norman family which lived nearby.

Norquay, Mount (mountain)

82 O/4 - Banff
8-26-12-W5
51°12′N 115°38′W
Approximately 5 km north-west of Banff.

This 2522 m mountain was named in 1904 in honour of the Honourable John Norquay (1841-1889) who climbed the peak in 1887 or 1888. He was born in the Red River settlement, Assiniboia, and held several political positions in the Manitoba Government. He was Premier of Manitoba from 1878-1887, after which he became a Railway Commissioner. The correct pronunciation is "Nor-kway."

Norris Creek (creek)

83 L/2 - Bolton Creek
4-60-5-W6
54°09′N 118°42′W
Flows north into Smoky River, approximately 40 km north-east of Grande Cache.

The origin of the name is unknown.

North Creek (creek)

82 G/16 - Maycroft
19-11-1-W5
49°56′N 114°08′W

Flows west into Callum Creek, approximately 40 km north-east of Blairmore.

The origin of the name is unknown.

North Glacier (glacier)

82 N/15 - Mistaya Lake
31-34-22-W5
51°52′N 116°57′W
Approximately 125 km north-west of Banff.

(see East Glacier)

North Belly River (river)

82 H/4 - Waterton Lakes
16-1-28-W4
49°02′N 113°41′W
Flows north-east into the Belly River, approximately 55 km south-east of Pincher Creek.

The Belly River is named after the Atsina, a detached branch of the Arapaho Indian tribe, now on a reserve in Montana. This tribe is known by the other Arapaho people as Hituena, meaning "beggars or spongers." The tribal sign was, commonly but incorrectly, rendered "belly people," "big bellies," or "Gros Ventres" by French Canadian traders and explorers. The name North Belly River was officially approved 4 November 1942.

North Berland River (river)

83 E/10 - Adams Lookout
29-54-5-W6
53°42′N 118°42′W
Flows north-east into Berland River, approximately 33 km south-east of Grande Cache.

(see Berland Range)

North Burnt Timber Creek (creek)

82 O/11 - Burnt Timber Creek
11-30-9-W5
51°33′N 115°11′W

Flows east into Burnt Timber Creek, approximately 50 km north-east of Banff.

(see Burnt Timber Creek)

North Cascade River (river)

82 O/5 - Castle Mountain
21-28-13-W5
51°24′N 115°46′W
Approximately 29 km north north-west of Banff.

This river was officially named 10 December 1959. (see Cascade Mountain)

North Castle (mountain)

82 G/8 - Beaver Mines
20-4-2-W5
49°18′N 114°15′W
Approximately 30 km south-west of Pincher Creek.

(see Castle Peak)

North Coal Creek (creek)

82 J/10 - Mount Rae
24-19-5-W5
50°37′N 114°34′W
Flows north-east into Coal Creek, approximately 21 km west of Turner Valley.

The name for this feature was officially approved 6 September 1951. (see Coal Creek)

North Cutbank River (river)

83 L/6 - Chicken Creek
30-63-10-W6
54°29′N 119°30′W
Flows north-east into Cutbank River, approximately 70 km north north-west of Grande Cache.

Named for its proximity as a tributary to Cutbank River, which has a name descriptive of its flow pattern. (see also Flume Creek)

North Fork Pass (pass)

82 G/15 - Tornado Mountain
24-11-6-W5
49°55′N 114°41′W
Approximately 35 km north-west of
Coleman, on the Alberta-B.C. boundary.

This 1992 m pass may be found west of
Gould Dome at the head of what was
formerly designated the North Fork of
Oldman River.

North Isaac Creek (creek)

83 C/6 - Sunwapta
2-41-22-W5
52°30′N 117°03′W
Flows south-east into Isaac Creek,
approximately 81 km south-east of Jasper.

(see Isaac Creek)

North Kananaskis Pass (pass)

82 J/11 - Kananaskis Lakes
20-10-W5
50°42′N 115°18′W
Approximately 70 km west of Turner Valley.

(see Kananaskis Range)

North Kootenay Pass (pass)

82 G/7 - Upper Flathead
14-5-5-W5
49°24′N 114°34′W
Approximately 28 km south of Coleman, on
the Alberta-B.C. boundary.

(see South Kootenay Pass)

North Lost Creek (creek)

82 G/8 - Beaver Mines
4-6-4-W5
49°27′N 114°30′W
Approximately 2 km south south-west of
Beaver Mines.

(see Lost Creek)

North Pass (pass)

83 E/2 - Resplendent Creek
10-47-5-W6
53°02′N 118°38′W
Approximately 42 km west north-west of
Jasper.

(see Centre Pass)

North Racehorse Creek (creek)

82 G/15 - Tornado Mountain
17-10-4-W5
49°50′N 114°31′W
Flows east into Racehorse Creek,
approximately 20 km north of Coleman.

The name for this creek was officially
approved in 1960. (see Racehorse Creek)

North Ram River (river)

83 B/5 - Saunders
9-38-12-W5
52°15′N 115°38′W
Flows east into Ram River, approximately 38
km south-east of Nordegg.

(see Ram River)

North Ribbon Creek (creek)

82 J/14 - Spray Lakes Reservoir
3-23-9-W5
50°56′N 115°11′W
Flows south-east into Ribbon Creek,
approximately 75 km west of Calgary.

(see Ribbon Creek)

North Saskatchewan River (river)

83 C/3 - Columbia Icefield
52°09′N 117°07′W
At Source.

One of the Canadian Prairies' great rivers,
the North Saskatchewan is fed by the
Saskatchewan Glacier, an outlet of the
Columbia Icefield. The water of this river
furnishes the drinking water for several

settlements along its course, measuring some
1223 kilometres in length, eventually ending
up at Hudson Bay. (see Saskatchewan
Mountain).

North Sullivan Creek (creek)

82 J/9 - Turner Valley
13-18-4-W5
50°32′N 114°25′W
Flows south-east into Sullivan Creek,
approximately 19 km south-west of Turner
Valley.

Officially approved 8 December 1943, the
name of this creek was taken from its
proximity to Sullivan Creek. (see Sullivan
Creek)

North Twin Creek (creek)

82 J/1 - Langford Creek
22-14-4-W5
50°10′N 114°28′W
Flows south-west into Livingstone River,
approximately 65 km north-west of
Claresholm.

The precise origin of the name is unknown,
but it may be descriptive. (see South Twin
Creek)

North Twin Peak (peak)

83 C/3 - Columbia Icefield
37-25-W5
52°13′30″N 117°26′00″W
Approximately 85 km south-east of Jasper.

(see The Twins)

North Wolf Creek (creek)

83 L/6 - Chicken Creek
32-63-11-W6
54°28′N 119°12′W
Flows east into Wolf Creek, approximately
60 km north of Grande Cache.

This creek was named for its proximity as a
tributary to Wolf Creek. (see Wolf Creek)

North York Creek (creek)

82 G/10 - Crowsnest
29-7-4-W5
49°36′N 114°31′W
Flows north-east into York Creek,
approximately 4 km south of Coleman.

(see York Creek)

Northover, Mount (mountain)

82 J/11 - Kananaskis Lakes
7-19-9-W5
50°35′N 115°15′W
Approximately 70 km west south-west of
Turner Valley.

This 3048 m mountain was officially named
after Lieutenant A.W. Northover, V.C., 28th
Battalion of the Canadian Expeditionary
Force, in 1917. He was awarded the Victoria
Cross for his courage.

Nose, The (mountain)

83 L/12 - Nose Creek
29-64-11-W6
54°34′N 119°38′W
Approximately 86 km south-west of Grande
Prairie.

This prominent point at the western
extremity of Nose Mountain is 1456 m in
altitude. (see Nose Mountain)

Nose Creek (creek)

83 L/13 - Calahoo Creek
7-8-68-11-W6
54°52′N 119°38′W
Flows north into Wapiti River,
approximately 67 km west south-west of
Grande Prairie.

The name for this stream is taken from the
Cree Indian "os-kewun," and the Stoney,
"tap-o-ol wapta," which are translated to
mean nose. (see also Nose Mountain)

Nose Lake (lake)

83 L/12 - Nose Creek
16-64-11-W6
54°32′N 119°36′W
Approximately 86 km south south-west of
Grande Prairie.

(see Nose Mountain)

Nose Mountain (mountain)

83 L/12 - Nose Creek
64-11-W6
54°30′N 119°32′W
Approximately 90 km south south-west of
Grande Prairie.

The name for this mountain, which is 1500
m in altitude, is likely descriptive of the
mountain's appearance. The other features in
the vicinity take their names from this
mountain.

Noseeum Creek (creek)

82 N/9 - Hector Lake
33-30-17-W5
51°37′N 116°20′W
Flows south-west into Bow River,
approximately 75 km north-west of Banff.

The locally well-established name for this
creek was officially approved 7 May 1959.
The precise origin is unknown, but the
name may refer to presence of "no-see-um"
biting insects in the area.

Noyes Creek (creek)

82 N/15 - Mistaya Lake
4-22-33-19-W5
51°50′N 116°37′W
Flows south-west into Mistaya River,
approximately 105 km north-west of
Banff.

(see Mount Noyes)

Noyes, Mount (mountain)

82 N/15 - Mistaya Lake
33-19-W5
51°49′N 116°33′W
Approximately 105 km north-west of Banff.

This 3084 m mountain was named by
Stutfield (see Stutfield Peak) and Collie (see
Collie Creek) in 1902, after Rev. Charles
Lathrop Noyes (1851-1923), a member of the
Appalachian Mountain Club. Along with
making several first ascents in the Canadian
Rockies, Dr. Noyes performed the first
exploration of Peyto Glacier.

Nuisance Creek (creek)

82 O/6 - Lake Minnewanka
17-29-8-W5
51°29′N 115°05′W
Flows east into Fallentimber Creek,
approximately 50 km north-east of Banff.

The name was officially approved in 1960.
Its origin remains unknown.

O'Beirne Mountain (mountain)

83 D/15 - Rainbow
26-45-5-W6
52°55′N 118°37′W
Approximately 36 km west of Jasper.

This 2560 m mountain was named in 1918
by A.O. Wheeler (see Wheeler Flats). It is
named after "O'B," (Eugene Francis O'Beirne)
who travelled across western Canada with
the Overlanders and Reverend John McDougall
and was left at Fort Edmonton. In 1863 he
attached himself to the expedition of Lord
Milton and Dr. Walter B. Cheadle (see
Bingley Peak), co-authors of *The North West
Passage by Land* (1965). His overbearing and
freeloading nature added greatly to the
difficulties of their journey.

O'Brien Lake (lake)

82 N/8 - Lake Louise
6-27-15-W5
51°17′N 116°05′W

Approximately 40 km west north-west of Banff.

The name for this lake, also known as Larch Lake, may be taken from the mountain of the same name in British Columbia. The mountain was named after a section foreman who was associated with the Canadian Pacific Railway.

O'Hagan Creek (creek)

83 F/4 - Miette
20-48-26-W5
53°09′N 117°47′W
Flows south into Fiddle River, approximately 34 km south south-west of Hinton.

(see Mount O'Hagan)

O'Hagan, Mount (mountain)

83 F/4 - Miette
1-48-27-W5
53°07′N 117°49′W
Approximately 37 km south south-west of Hinton.

The name for this 2445 m mountain and the nearby creek was officially approved in 1960. Dr. Thomas O'Hagan (1879-1957) arrived in Jasper in 1924 and founded the Seton Hospital. He served as resident physician in the area for twenty-eight years and was highly respected.

O'Shaughnessey Falls (man-made waterfalls)

82 J/14 - Spray Lakes Reservoir
8-32-23-8-W5
51°59′55″N 115°04′15″W
Approximately 65 km west of Calgary.

When the nearby road was built, the creek was diverted to a culvert to flow under the road. The creek would not be diverted easily; some of it went through the culvert and some of it over the road, causing problems to travellers. John O'Shaugnessy was an engineer for the highway

construction crew and he created a new way of dealing with the problem by building a waterfall to control the stream's flow. He then did some landscaping and built a wishing well. These falls were named for John O'Shaugnessy, and the name was officially approved 5 December 1984.

Oates, Mount (mountain)

83 D/8 - Athabasca Pass
14-40-1-W6
52°26′N 118°02′W
Approximately 48 km south of Jasper.

This 3120 m mountain was named by G.E. Howards in 1914 after Captain Oates (1880-1912) of the Inniskilling Dragoons, a member of the British Antarctic Expedition under Captain R.F. Scott. (see Mount Scott)

Obed (locality)

83 F/11 - Dalehurst
6-53-22-W5
53°33′N 117°12′W
Approximately 51 km west of Edson.

The locality and nearby lake are named after Lieutenant-Colonel John Obed Smith (1864-1937), a Briton who came to Canada as a young man. He served the Manitoba Government until 1901 when he began serving on various commissions for emigration.

Observation Peak (peak)

82 N/16 - Siffleur River
15-32-18-W5
51°45′N 116°28′W
Approximately 90 km north-west of Banff.

This peak, which stands 3174 m in altitude, was named in 1898 by the Reverend C.L. Noyes (see Mount Noyes) because when climbed, it was the most satisfactory viewpoint he and his crew had reached in the Rockies.

Obstruction Lakes (lakes)

83 C/7 - Job Creek
6-40-20-W5
52°25′N 116°52′W
Approximately 95 km south-east of Jasper.

The name of these lakes was officially approved 20 March 1979.

Obstruction Mountain (mountain)

83 C/7 - Job Creek
25-39-21-W5
52°23′N 116°53′W
Approximately 95 km south-east of Jasper.

The descriptive name for this 3199 m mountain was officially approved 5 March 1935. The lakes take their name from the mountain.

Ochre Lake (lake)

83 D/16 - Jasper
26-45-1-W6
52°54′N 118°02′W
Approximately 4 km north-east of Jasper.

The name for this lake is descriptive of its colour. The name was officially approved in 1915.

Odlum Creek (creek)

82 J/7 - Mount Head
6-18-6-W5
50°29′N 114°48′W
Flows east into Highwood River, approximately 42 km south-west of Turner Valley.

(see Mount Odlum)

Odlum, Mount (mountain)

82 J/7 - Mount Head
17-7-W5
50°29′N 114°55′W
Approximately 50 km south-west of Turner Valley.

This 2670 m mountain was officially named in 1917. It commemorates Brigadier-General

V.W. Odlum, C.M.G., a member of the Canadian Expeditionary Force.

Og Mountain (mountain)

82 J/13 - Mount Assiniboine
23-12-W5
50°57′N 115°37′W
Approximately 100 km west of Calgary, on the Alberta-B.C. boundary.

This mountain is 2874 m in altitude. (see Mount Magog)

Og Pass (pass)

82 J/13 - Mount Assiniboine
16-23-12-W5
50°57′N 115°37′W
Approximately 100 km west of Calgary, on the Alberta-B.C. boundary.

Named for its proximity to Og Mountain. (see Mount Magog)

Ogre Canyon (canyon)

83 F/4 - Miette
49-27-W5
53°14′N 117°53′W
Approximately 29 km south-west of Hinton.

The name of this canyon is descriptive in that from one viewpoint, there appears an outine of a grotesque head apparently trying to take a bite out of a rock on the opposite side. A local legend says that an Indian went into the gorge and was never seen again.

Oil Basin (valley)

82 H/4 - Waterton Lakes
35-2-30-W4
49°10′N 113°55′W
Approximately 35 km south of Pincher Creek.

This basin is the headwaters of Cottonwood Creek. The name derives from oil explorations at this location before the area was included within the boundaries of Waterton Lakes National Park. Today, the only reminders of the oil explorations are the tops of well casing that can be found on the valley floor below the basin.

Okotoks Mountain (mountain)

82 J/9 - Turner Valley
12-19-4-W5
50°36′N 114°26′W
Approximately 13 km south-west of Turner Valley.

The name for this mountain, which is 1762 m in altitude, was officially applied 8 December 1943. "Okotoks" is a Blackfoot Indian word meaning "big rock" or "crossing by the big rock." This landmark is also known as "The Big Rock." Another Indian tribe, the Sarcees, call this place "chachosika," translating to "valley of the big rock." The Stoney Indian name, "ipabitun- gaingay," means "where the big rock is." The name is, therefore, appropriately descriptive.

Old Baldy Mountain (mountain)

82 O/14 - Limestone Mountain
21-33-8-W5
51°51′N 115°05′W
Approximately 80 km north-east of Banff.

Officially approved 25 November 1941, the name for this mountain is descriptive of its bare rounded top. (see Mount Baldy)

Old Entrance (locality)

83 F/5 - Entrance
2-51-26-W5
53°22′N 117°43′W
Approximately 8 km west south-west of Hinton.

(see Entrance)

Old Fort Creek (creek)

82 O/2 - Jumpingpound Creek
13-25-8-W5
51°08′N 114°59′W
Flows east into Bow River, approximately 55 km west north-west of Calgary.

There was once a Hudson's Bay Company fort, known as the "Old Bow Fort," near the mouth of this creek. It was established in 1832 but abandoned in 1834, and the creek was named for the ruins.

Old Fort Point (point)

83 D/16 - Jasper
10-45-1-W6
52°52′N 118°03′W
Approximately 2 km south-east of Jasper.

The hunting lodge of the North West Company, which was built by William Henry in 1812, and called "Old Fort," was located near this rocky eminence.

Oldhorn Mountain (mountain)

83 D/9 - Amethyst Lakes
11-43-2-W6
52°41′N 118°11′W
Approximately 23 km south-west of Jasper.

The name for this 3000 m mountain was applied in 1916 by W.P. Hinton (see Hinton-town). Oldhorn is descriptive of the shape of the mountain.

Oldhouse Creek (creek)

83 F/5 - Entrance
26-50-27-W5
53°21′N 117°51′W
Flows east into Solomon Creek, approximately 16 km west south-west of Hinton.

This creek is near the original site of Jasper House. The name has been in use since at least 1927.

Oldman Creek (creek)

83 F/11 - Dalehurst
14-55-22-W5
53°45′N 117°10′W

Flows east into Athabasca River, approximately 51 km west north-west of Edson.

The legend attached to the locally well-established name of this creek states that this was the playground of the "Old Man" of Cree Indian mythology.

Ole Buck Mountain (mountain)

82 O/2 - Jumpingpound Creek
24-7-W5
51°04′N 114°51′W
Approximately 45 km west of Calgary.

This 1905 m mountain was officially named 12 December 1939. It appears on A.O. Wheeler's (see Wheeler Flats) map, but the origin is unknown.

Olin Creek (creek)

82 G/9 - Blairmore
16-9-1-W5
49°44′N 114°05′W
Flows west into Oldman River, approximately 18 km north of Cowley.

Bill Olin, after whom this creek was named, claimed to be one of the original buffalo hunters of the plains. Early in the 1870s, some traders and buffalo hunters, including Mr. Olin, travelled north from Montana with the Indians. Bill Olin set up his ranch near this creek. He died in 1906, after a drinking binge.

Olive, Mount (mountain)

82 N/9 - Hector Lake
32-30-18-W5
51°37′N 116°29′W
Approximately 80 km north-west of Banff, on the Alberta-B.C. boundary.

This 3130 m mountain was named in 1899 by H.B. Dixon, of The Alpine Club (London), after his wife.

Oliver, Mount (mountain)

82 O/6 - Lake Minnewanka
28-11-W5
51°25′N 115°28′W
Approximately 28 km north north-east of Banff.

The origin of the name is unknown.

Oliver, Mount (mountain)

83 E/1 - Snaring
3-47-3-W6
53°01′N 118°21′W
Approximately 24 km north-west of Jasper.

One of twelve features named in 1954 for Alberta's 50th Jubilee, this 2865 m mountain was named after the Honourable Frank Oliver, P.C. (1853-1933), founder of the *Edmonton Bulletin*, 1880. He was a member of the North West Council in 1883. Oliver was elected to the Legislative Assembly, 1888-1896, and was elected to the House of Commons in 1896 as an Independent Liberal. He became Minister of the Interior in the Laurier Administration 8 April 1905 until 1911. (see Mount Griesbach)

Olson Creek (creek)

83 E/16 - Donald Flats
19-56-1-W6
53°51′N 118°07′W
Flows east into Berland River, approximately 70 km east of Grande Cache.

This creek was named in 1947 after Ben Olson, who trapped in the area near the stream for over forty years. The name is locally well-established.

Olympus, Mount (mountain)

83 C/6 - Sunwapta
25-40-22-W5
52°28′N 117°01′W
Approximately 84 km south-east of Jasper.

The name for this 3088 m mountain was approved 5 March 1935. Its precise origin is

unknown; however, there may be an allusion to the locally named "Greek Mountain": Olympus was the home of the gods of Greek mythology.

Omega Peak (peak)

83 C/4 - Clemenceau Icefield
35-36-26-W5
52°08′N 117°35′W
Approximately 90 km south-east of Jasper, on the Alberta-B.C. boundary.

The origin of the name is unknown, but likely descriptive of the peak's appearance.

Onion Creek (creek)

83 C/1 - Whiterabbit Creek
13-36-15-W5
52°06′N 116°01′W
Flows south-east into Hummingbird Creek, approximately 105 km north north-west of Banff.

This creek was named in 1960. The name's origin is unknown.

Onion Lake (lake)

83 C/1 - Whiterabbit Creek
26-36-16-W5
52°07′N 116°11′W
Approximately 120 km north-west of Banff.

(see Onion Creek)

Opabin Creek (creek)

83 C/10 - George Creek
28-42-20-W5
52°38′N 116°49′W
Flows north into Brazeau River, approximately 53 km west north-west of Nordegg.

The name for this creek was proposed by George Malloch (see Mount Malloch), and was officially approved 1 February 1910. "Opabin" is the Stoney Indian term for "stone" (W.A. Wixton). It officially replaced the former names, "Boulder" and "Rocky," both of which were descriptive of the area in which the creek flows.

Opal Creek (creek)

82 J/11 - Kananaskis Lakes
7-20-8-W5
50°40'45"N 115°05'45"W
Flows north-west into Pocaterra Creek, approximately 60 km west of Turner Valley.

The name for this creek was approved 8 November 1978. (see Opal Range)

Opal Falls (falls)

82 J/11 - Kananaskis Lakes
8-20-8-W5
50°40'20"N 115°04'55"W
Approximately 55 km west of Turner Valley.

This tall, conspicuous waterfall on Opal Creek is situated approximately 183 m east of the junction of Opal Creek and the Kananaskis Valley. The name was officially approved 8 November 1978. (see Opal Range)

Opal Hills (range of hills)

83 C/12 - Athabasca Falls
52°45'00"N 117°36'00"W
Approximately 40 km south-east of Jasper.

These hills, located at the north-east end of Maligne Lake, have a descriptive name. Mary Schäffer, who discovered Maligne Lake in 1908 and measured it in 1911, named the hills and the lake because of their beautiful colouring: blue in morning mists, green and yellow in the sunshine. The locally well-established name was officially applied to this feature 5 May 1987.

Opal Lake (lake)

83 C/13 - Medicine Lake
52°45'15"N 117°35'00"W

Approximately 36 km south-west of Jasper.

This lake is likely named for its proximity to Opal Hills. The name is well-established in local usage. (see Opal Hills)

Opal Peak (peak)

83 C/13 - Medicine Lake
52°45'10"N 117°35'40"W
Approximately 36 km south-west of Jasper.

(see Opal Hills)

Opal Range (range)

82 J/14 - Spray Lakes Reservoir
21-9-W5
50°47'N 115°08'W
Approximately 60 km west north-west of Turner Valley.

This range was named by G.M. Dawson, of the Geological Survey of Canada (see Tombstone Mountain), after the quartz crystals, coated with films of opal he found in small cavities there.

Oppy Mountain (mountain)

82 N/14 - Rostrum Peak
36-34-23-W5
51°58'N 117°09'W
Approximately 140 km north-west of Banff, on the Alberta-B.C. boundary.

This mountain, which is 3335 m in altitude, was named in 1920 after the village 10 km south-east of Lens, France. The name commemorates the fighting that took place there during the First World War.

Orchard Creek (creek)

83 F/5 - Entrance
3-51-26-W5
53°22'N 117°45'W
Flows south into Athabasca River, approximately 9 km west of Hinton.

The locally well-established name for this creek was officially approved in 1944. The origin of the name is unknown.

Orient Point (peak)

82 O/3 - Canmore
26-9-W5
51°15'N 115°10'W
Approximately 23 km north-east of Canmore.

The 2636 m peak is part of the Fairholme Range. The origin of this locally well-established name is unknown.

Osborne Creek (creek)

83 C/13 - Medicine Lake
52°56'10"N 117°44'30"W
Flows north-west into Jacques Creek, approximately 20 km east of Jasper.

According to the Canadian Park Warden Service, the name Osborne Creek predates the Park. The precise origin is unknown but the name may have been that of a local telephone line person during the 1920s. The locally well-established name was officially approved 5 May 1987.

Osprey Lake (lake)

83 C/12 - Athabasca Falls
27-33-25-W5
52°41'N 117°30'W
Approximately 44 km south-east of Jasper.

The origin of the name is unknown.

Otter Creek (creek)

83 B/4 - Cripple Creek
19-37-14-W5
52°11'N 115°59'W
Flows north into Cripple Creek, approximately 33 km south-east of Nordegg.

The locally well-established name for this creek was made official 29 September 1976. The origin of the name is unknown.

Otuskwan Peak (peak)

82 O/12 - Barrier Mountain
30-12-W5
51°32′N 115°32′W
Approximately 40 km north of Banff.

This 2530 m mountain peak resembles an elbow. The Cree word is "o-toos-kwa-na," and it is from this word the name appears to come.

Outpost Peak (peak)

83 D/9 - Amethyst Lakes
29-42-2-W6
52°39′N 118°15′W
Approximately 28 km south-west of Jasper.

The name for this 2774 m mountain peak is descriptive of its height and outlying position.

Outram Lake (lake)

82 N/15 - Mistaya Lake
29-33-20-W5
51°52′N 116°49′W
Approximately 117 km north-west of Banff.

(see Mount Outram)

■ Outram, Mount (mountain)

82 N/15 - Mistaya Lake
35-33-21-W5
51°53′N 116°52′W
Approximately 120 km north-west of Banff.

This 3240 m mountain was named in 1920 after Sir James Outram, Baronet (1864-1925), a noted mountain climber who made first ascents of many of the highest peaks in the Rockies. He was the author of *In the Heart of the Canadian Rockies* (1905), and a member of the Appalachian Mountain Club.

Overturn Mountain (mountain)

83 F/4 - Miette
14-47-26-W5
53°03′N 117°42′W

Approximately 43 km south south-west of Hinton.

This 2560 m mountain derives its name from the overturned rock strata and structure in it and along its ridges. The name was made official in 1960.

Owen Creek (creek)

83 C/2 - Cline River
35-19-W5
52°00′N 116°40′W
Flows south-east into North Saskatchewan River, approximately 125 km west south-west of Rocky Mountain House.

Officially named in 1960, the origin for the name of this feature is unknown.

Owl Creek (creek)

82 J/1 - Langford Creek
13-13-3-W5
50°05′N 114°17′W
Flows east into Hunter Creek, approximately 50 km west north-west of Claresholm.

No origin information is available for this locally well-established name.

Owl Creek (creek)

82 J/13 - Mount Assiniboine
19-22-11-W5
50°53′10″N 115°30′30″W
Flows north-east into Owl Lake, following an underground course, emerging to flow into Bryant Creek, approximately 95 km west of Calgary.

The origin of the name is unknown.

Owl Lake (lake)

82 J/13 - Mount Assiniboine
12-22-12-W5
50°51′N 115°31′W
Approximately 95 km west of Calgary.

The precise origin of this locally well-established name is unknown.

Oyster Creek (creek)

82 J/2 - Fording River
29-13-5-W5
50°07′N 114°39′W
Flows south into Oldman River, approximately 65 km south-west of Turner Valley.

There are large beds of fossil oysters (Ostrea) in the banks of this creek.

Oyster Lake (lake)

82 N/9 - Hector Lake
26-29-15-W5
51°30′N 116°00′W
Approximately 50 km north-west of Banff.

(see Oyster Creek)

Oyster Peak (peak)

82 N/9 - Hector Lake
34-29-15-W5
51°31′N 116°01′W
Approximately 50 km north-west of Banff.

This mountain peak measures 2777 m in altitude. (see Oyster Creek)

Ozada (locality)

82 O/2 - Jumpingpound Creek
13-25-8-W5
51°07′N 114°58′W
Approximately 55 km west of Calgary.

This locality is found at the junction of the Bow and Kananaskis Rivers. The name, "Ozada," is the Stoney Indian word for "the forks of the river." The name was officially approved 12 December 1939.

Pack Trail Coulee (coulee)

82 J/7 - Mount Head
35-16-5-W5
50°23′35″N 114°34′50″W
Approximately 37 km south-west of Turner Valley.

The locally well-established name was made official 28 February 1980. The name comes from a well-used pack trail that follows the 2 km long coulee.

Packenham, Mount (mountain)

82 J/14 - Spray Lakes Reservoir
6-21-8-W5
50°45′N 115°05′W
Approximately 70 km west south-west of Calgary.

This 3000 m mountain was named in 1922 after Rear Admiral W.C. Packenham, in command of the 2nd Battle Cruiser Squadron at the Battle of Jutland, 1916. (see Jutland Mountain)

Packrat Creek (creek)

83 E/16 - Donald Flats
10-57-1-W6
53°55′N 118°04′W
Flows east into Berland River, approximately 71 km east of Grande Cache.

A cabin known as "Packrat Cabin" was located on this creek. The creek was named in 1947, after the cabin.

Paddy's Flat (flat)

82 J/15 - Bragg Creek
24-22-6-W5
50°53′00″N 114°42′30″W
Approximately 38 km west south-west of Calgary.

This flat, officially named 5 December 1984, takes its name from Patrick (Paddy) McCarthy, who came to the area in the 1920s. In June 1929, he was employed as a

teamster by Herron Petroleum which was drilling a well on the flat. In December 1930, the well was closed down and Paddy was made caretaker until such time as the well was reopened. Paddy remained at the well for over ten years until 1942, when he returned to Winnipeg, where he later died.

Pakakos Mountain (mountain)

82 O/6 - Lake Minnewanka
16-27-10-W5
51°18′10″N 115°20′30″W
Approximately 12 km north-east of Banff.

The name for this isolated, distinct mountain, which is 2440 m in altitude, was the winning entry in the Geographical Naming Contest for Elementary Schoolchildren to commemorate the International Year of the Child in 1979. Officially approved 14 April 1980, the name "Pakakos" is the Peace Indian word for the spirit of the "flying skeleton."

Palisade, The (ridge)

83 D/16 - Jasper
46-1-W6
52°59′N 118°06′W
Approximately 10 km north of Jasper.

Cliffs below Pyramid Mountain comprise this ridge, which resembles a defensive enclosure, such as a palisade. The descriptive name for this feature was applied in 1924.

Palliser Pass (pass)

82 J/11 - Kananaskis Lakes
20-11-W5
50°42′N 115°23′W
Approximately 80 km west of Turner Valley.

(see Palliser Range)

Palliser Range (range)

82 O/5 - Castle Mountain
28-12-W5
51°23′N 115°34′W

P~Q~R

Approximately 15 km north of Banff.

This range is noted on the Palliser map of 1859. The range and nearby pass are named for Captain John Palliser (1817-1887) who commanded an expedition between 1857-1860, to explore and survey the country between the 49th parallel and the North Saskatchewan River, and between the Red River and the Rockies. He was further instructed to ascertain and report the feasibility for a railway route south of Athabasca Pass.

Palmer, Mount (mountain)

83 C/5 - Fortress Lake
52°21′10″N 117°31′10″W
Approximately 68 km south-east of Jasper.

Mount Palmer, 3150 m in altitude, was named for Howard Palmer, an early mountaineer in the Canadian Rockies. He conducted extensive explorations and took numerous photographs of the Canadian Rockies greatly enhancing the knowledge of this region of Canada. These photos are on file with the American Alpine Club in New York. Palmer was Secretary of the American Alpine Club from 1912-1915.

Palu Mountain (mountain)

83 E/3 - Mount Robson
33-49-9-W6
53°15′N 119°15′W
Approximately 87 km north-west of Jasper.

This 2929 m mountain was named in 1923 after a mountain in Switzerland which is similar in shape.

Pangman Glacier (glacier)

82 N/15 - Mistaya Lake
19-32-21-W5
51°46′N 116°57′W
Approximately 115 km north-west of Banff.

(see Pangman Peak)

Pangman Peak (peak)

82 N/15 - Mistaya Lake
19-32-21-W5
51°46′N 116°58′W
Approximately 115 km north-west of Banff,
on the Alberta-B.C. boundary.

This 3473 m mountain peak was named in
1920, after Peter Pangman (1774?-1819). This
early fur trader was born in New England of
German descent. He joined the firm of
Gregory, McLeod, and Co. in 1783. The
North West Company absorbed this firm in
1787, and Pangman became a partner at that
time. In 1790, Pangman carved his name on
a pine tree, 5 km upstream from what was
then Rocky Mountain House.

Panorama Ridge (ridge)

82 N/8 - Lake Louise
23-27-16-W5
51°19′N 116°08′W
Approximately 45 km west north-west of
Banff.

The name for this ridge describes the view
from its summit.

Panther Falls (falls)

83 C/3 - Columbia Icefield
15-37-22-W5
52°10′55″N 117°03′20″W
Approximately 105 km south-east of Jasper.

The locally well-established name for this
waterfall was officially approved 28 February
1980. A group of hikers in 1907 discovered
the falls and later found that a panther (or
wildcat) had followed in their footsteps for a
considerable distance along their trail (Mary
Schäffer in *Old Indian Trails*, 1911).

Panther Lake (lake)

82 O/5 - Castle Mountain
51°27′55″N 115°52′00″W
Approximately 37 km north-west of Banff,
east of the Valley of the Hidden Lakes.

This name is well-used locally and perhaps
stems from the fact the lake feeds Panther
River. (see Panther Mountain)

Panther Mountain (mountain)

82 O/12 - Barrier Mountain
30-29-12-W5
51°31′N 115°40′W
Aproximately 40 km north north-east of
Banff.

Applied by Dr. G.M. Dawson (see
Tombstone Mountain), the name for this
2943 m mountain, and the nearby river,
corresponds to the Indian name of the
stream which signifies "the river where the
mountain lion was killed." Officially
approved 16 January 1912, the Stoney
expression is "it-mos-tunga-moos-ta-ga-te-
wap-ta"; in Cree, it is "mis-si-pi-sioo-ka-nipa-
hiht-si-pi."

Panther River (river)

82 O/11 - Burnt Timber Creek
14-31-10-W5
51°39′N 115°19′W
Flows north-east into Red Deer Creek,
approximately 60 km north north-east of
Banff.

(see Panther Mountain)

Paradise Creek (creek)

82 N/8 - Lake Louise
14-28-16-W5
51°24′N 116°09′W
Flows north-east into Bow River,
approximately 50 km north-west of Banff.

(see Paradise Valley)

Paradise Creek (creek)

83 F/5 - Entrance
13-52-28-W5
53°29′N 117°59′W
Flows north-west into Ice Water Creek,
approximately 28 km west north-west of
Hinton.

The origin of the name is unknown.

Paradise Valley (valley)

82 N/8 - Lake Louise
5-2-8-16-W5
51°22′N 116°13′W
Approximately 50 km west north-west of
Banff.

The descriptive name for this sun-flooded
valley, full of many streams, open country
and charming contrasts, was given in 1894
by Walter D. Wilcox. (see Wilcox Pass)

Paragon Peak (peak)

83 D/9 - Amethyst Lakes
6-43-2-W6
52°41′N 118°18′W
Approximately 27 km south-west of Jasper.

This magnificent peak, which stands 3030 m
in altitude, was given its descriptive name in
1921 by Howard Palmer. (see Mount Palmer)

Parapet Glacier (glacier)

82 N/10 - Blaeberry River
16-32-19-W5
51°45′N 116°39′W
Approximately 100 km north-west of Banff,
on the Alberta-B.C. boundary.

The locally well-established name for this
feature is descriptive since the glacier is
surrounded by rock walls, similar in shape to
a parapet (a low wall).

Parker Creek (creek)

82 O/14 - Limestone Mountain
20-33-8-W5
51°50′N 115°06′W
Flows north-west into South James River, approximately 80 km north-east of Banff.

Tyson Parker (1891-1965) was born in England and came to Canada in 1909. He homesteaded in the Blaeberry district for a number of years, and was the Forest Ranger of the district for some time. This creek was officially named after Mr. Parker 25 November 1941.

Parker Ridge (ridge)

83 C/3 - Columbia Icefield
17-37-22-W5
52°11′N 117°06′W
Approximately 100 km south-east of Jasper.

The Canadian Permanent Committee on Geographical Names approved the name for this ridge 9 June 1978. It was likely named for Mr. Herchel C. Parker. No other information is available about this man.

Parker Ridge (ridge)

82 O/14 - Limestone Mountain
33-8-W5
51°47′N 115°05′W
Approximately 75 km north-east of Banff, running parallel to Parker Creek.

(see Parker Creek)

Parrish, Mount (mountain)

82 G/10 - Crowsnest
23-7-5-W5
49°35′N 114°35′W
Approximately 9 km south-west of Coleman.

This 2530 m mountain was named after Sherman Parrish, the first homesteader in the western portion of the Crowsnest Pass. The name was proposed by Dr. R.A. Price,

of the Geological Survey of Canada, 3 October 1961 and was officially applied to this feature 31 December 1962.

Pasque Creek (creek)

82 J/2 - Fording River
13-5-W5
50°06′N 114°37′W
Flows south into Oldman River, approximately 70 km south-west of Turner Valley.

(see Pasque Mountain)

Pasque Mountain (mountain)

82 J/2 - Fording River
14-5-W5
50°09′N 114°36′W
Approximately 60 km south-west of Turner Valley.

This 2541 m mountain was named by M.P. Bridgland (see Mount Bridgland) in 1914, due to the abundance of pasque flowers found near its summit.

Pasture Creek (creek)

83 E/16 - Donald Flats
1-56-2-W6
53°48′N 118°09′W
Flows north-east into Big Creek, approximately 66 km east south-east of Grande Cache.

The good pasture land found near this creek inspired its descriptive name, which was officially approved in 1947.

Patricia Lake (lake)

83 D/16 - Jasper
29-45-1-W6
52°54′N 118°06′W
Approximately 6 km north-east of Jasper.

This slim, narrow lake was named in honour of Her Royal Highness, Princess Patricia of Connaught, daughter of His Royal Highness,

the Duke of Connaught, Governor General of Canada, from 1911-1916. The name was applied in 1914 by Mr. H. Matheson of the Dominion Land Survey. (see Princess Lake)

Patterson, Mount (mountain)

82 N/10 - Blaeberry River
14-32-19-W5
51°45′N 116°35′W
Approximately 95 km north-west of Banff.

The name for this mountain, 3197 m in altitude, was officially approved 2 July 1918. John Duncan Patterson (1864-1940) was originally from Ontario. He was an active mountaineer and became a member of the Alpine Club of Canada at its inception, and was made President of the Club in 1914.

Pattison, Mount (mountain)

83 D/16 - Jasper
17-46-3-W6
52°58′N 118°24′W
Approximately 24 km north-west of Jasper.

This 2316 m mountain was named in 1951 after Private John George Pattison, V.C. (1875-1917) of the 50th Battalion, C.E.F., who was killed in action 3 June 1917. He was awarded a Victoria Cross for his bravery.

Pattison Pass (pass)

83 D/16 - Jasper
53°57′55″N 118°22′45″W
Approximately 23 km north-west of Jasper.

This feature was named for its proximity to Mount Pattison. (see Mount Pattison)

Paul, Mount (mountain)

83 C/11 - Southesk Lake
18-42-24-W5
52°37′N 117°26′W
Approximately 53 km south-east of Jasper.

This 2805 m mountain was named in 1911 by Mrs. Mary Schäffer, after Paul Sharples, the first white child (9 years old) to visit the Maligne Lake country. Paul navigated the waters and made all the climbs. He was favoured with having his name on this mountain while his behaviour was good. When he was bad, the name was removed. It has remained to this day, in his honour.

Pauline Creek (creek)

83 E/12 - Pauline Creek
20-53-11-W6
53°35'N 119°35'W
Flows north-east into Jackpine River, approximately 43 km south-west of Grande Cache.

(see Mount Pauline)

Pauline, Mount (mountain)

83 E/12 - Pauline Creek
31-52-13-W6
53°32'N 119°54'W
Approximately 65 km south-west of Grande Cache.

F.A. Pauline was the new Agent General for British Columbia at the time of this 2653 m mountain's naming in 1925. The mountain was formerly known as "Curly Mountain," after Donald "Curly" Phillips (see Chetang Ridge), a local guide, but the present name was chosen to avoid duplicaton.

Peach Lake (lake)

83 F/5 - Entrance
6-50-26-W5
53°17'N 117°47'W
Approximately 17 km south-west of Hinton.

The origin of the name is unknown.

Pearl Creek (creek)

83 E/14 - Grande Cache
6-56-7-W6
53°49'N 119°02'W

Flows north-west into Sulphur River, approximately 10 km south south-east of Grande Cache.

No information regarding the origin of the name is known.

Peavine Lake (lake)

83 E/14 - Grande Cache
8-57-7-W6
53°55'N 119°00'W
Approximately 8 km east north-east of Grande Cache.

The name for this lake is likely taken after the Purple Wild Vine, a native legume which is a perennial wildflower, and which is found throughout the wooded areas of Alberta, particularly in July. The name was officially approved in 1947.

Pedley (station)

83 F/16 - Pedley
3-52-24-W5
53°28'N 117°27'W
Approximately 78 km north north-west of Jasper.

The name for this station located on the Canadian National Railway line has been in use since at least 1916. Its precise origin is unknown.

Pedley Reservoir (reservoir)

83 F/6 - Pedley
2-52-24-W5
53°27'25"N 117°26'15"W
Approximately 77 km north north-east of Jasper.

This reservoir was built as a water supply for the Grand Trunk Pacific Railway (1910-1911) steam engines. The name is taken after the nearby Pedley station. The reservoir is a popular fishing spot and has contained Rainbow trout for many years.

Peechee, Mount (mountain)

82 O/3 - Canmore
7-26-10-W5
51°13'N 115°23'W
Approximately 14 km east north-east of Banff.

This 2935 m mountain was named by Dr. G.M. Dawson, of the Geological Survey of Canada (see Tombstone Mountain), in 1884. During Sir George Simpson's trip across the continent (see Simpson Pass), he had, as a guide, a Métis Indian, named Piché (later corrupted to Peechee). It is after this guide that this feature is named. (see Introduction)

*Pegasus Island (island)

82 J/11 - Kananaskis Lakes
23-19-9-W5
50°38'N 115°09'W
Approximately 60 km west south-west of Turner Valley, in Upper Kananaskis Lake.

(see Hogue Island)

Pekisko (locality)

82 J/8 - Stimson Creek
13-17-2-W5
50°26'N 114°10'W
Approximately 66 km south of Calgary.

"Pekisko" is the Blackfoot Indian word for "foothills" or "rolling hills." The name was suggested by Fred Stimson (see Stimson Creek) in 1896.

Pekisko Creek (creek)

82 J/8 - Stimson Creek
31-17-1-W5
50°29'N 114°08'W
Flows north into Highwood River, approximately 60 km south of Calgary.

This creek, known as "Bull Pond River" in Fidler's time (see Introduction), was officially named after the nearby locality 12 December 1939.

Pelletier Creek (creek)

82 G/9 - Blairmore
3-8-4-W5
49°37′N 114°28′W
Flows south into Crowsnest River,
approximately 1 km east of Coleman.

This creek was probably named after Henry
S. Pelletier who opened a brickyard in
Blairmore in 1912. Mr. Pelletier was also
instrumental in prospecting for coal along
the Frank seams in the area north of the
railway in the early 1900s. He worked
closely with S.W. Gebo and A.L. Frank. (see
Frank)

Pengelly, Mount (mountain)

82 G/10 - Crowsnest
27-6-5-W5
49°30′N 114°36′W
Approximately 16 km south south-west of
Coleman, on the Alberta-B.C. boundary.

Pengelly was the family name of the wife of
A.J. Campbell, who was assistant to A.O.
Wheeler (see Wheeler Flats) on the Alberta-
British Columbia Boundary Survey. The
name for this 2560 m mountain was
officially approved in January 1917.

Peppers Creek (creek)

83 B/4 - Cripple Creek
33-35-12-W5
52°03′N 115°39′W
Flows east into Clearwater River,
approximately 56 km south-east of Nordegg.

The origin for the locally well-established
name for this creek is unknown.

Peppers Creek (creek)

83 F/5 - Entrance
5-52-26-W5
53°28′N 117°47′W
Flows west into Jarvis Lake, approximately
16 km west north-west of Hinton.

The name for this creek was suggested by the
Topographical Survey and was officially
approved in 1944. The origin of the name is
unknown.

Peppers Lake (lake)

83 B/4 - Cripple Creek
30-35-12-W5
52°02′N 115°42′W
Approximately 54 km south-east of Nordegg.

This lake takes its name from Peppers Creek
which drains it. The name was officially
approved in July 1941.

Peppers Lake (lake)

83 F/5 - Entrance
35-51-26-W5
53°26′N 117°43′W
Approximately 9 km west north-west of
Hinton.

References to this lake date back to 1916.
Many of the names of features in this area
were given by early surveyors, and several are
names of surveyors or members of the
survey parties. The precise origin of the
name of this lake is not known.

Percé, Mount (mountain)

83 E/7 - Blue Creek
31-51-6-W6
53°27′N 118°51′W
Approximately 83 km north-west of Jasper.

The name of this 2754 m mountain was
officially approved 1 May 1934. Its precise
origin is unknown.

Perdrix, Roche à (peak)

83 F/4 - Miette
7-49-26-W5
53°13′N 117°47′W
Approximately 28 km south-west of Hinton.

The name Roche à Perdrix was first given to
this mountain, which stands 2134 m in

altitude, by Principal Grant (see Grant Pass).
The name is derived from the peculiar
folding of the strata, which when observed
from certain directions, resembles the tail of
a partridge. "Perdrix" is the French word for
"partridge."

Perren, Mount (mountain)

82 N/8 - Lake Louise
8-27-16-W5
51°18′N 116°12′W
Approximately 50 km west north-west of
Banff, on the Alberta-B.C. boundary.

The Canadian Permanent Committee on
Geographical Names approved the name for
this 3051 m mountain 12 February 1970. Walter
Perren, a mountain guide of international
repute, came to Canada from Switzerland in
1950. He was a climbing guide for the
Canadian Pacific Railway for five years. In
May 1955, at the termination of his contract,
he joined the National Parks Service and
remained in service until his death in
December 1967. He was responsible for
developing search and rescue training
techniques still used by the wardens.

Perserverence Mountain (mountain)

83 E/5 - Holmes River
30-51-12-W6
53°26′N 119°46′W
Approximately 128 km north-west of Jasper.

The name of this 2426 m mountain
describes the character quality required to
reach the summit. It was named in 1925.

Persimmon Creek (creek)

83 E/10 - Adams Lookout
15-53-5-W6
53°34′N 118°39′W
Flows east into South Berland River,
approximately 47 km south-east of Grande
Cache.

Persimmon is a corruption of an Algonquin
Indian word, and refers to any of several

tropical trees (genus Diospyrus) bearing edible round fruit. The fruit of these trees is typically orange-red in colour, and is sweet tasting when ripe. This creek is likely a red-orange colour; hence the name.

Persimmon Range (range)

83 E/10 - Adams Lookout
42-5-W6
53°32′N 118°42′W
Approximately 30 km south south-east of Grande Cache.

(see Persimmon Creek)

Peskett, Mount (mountain)

82 N/16 - Siffleur River
26-34-18-W5
51°57′N 116°27′W
Approximately 105 km north-west of Banff.

The name for this 3124 m mountain was approved by the Canadian Permanent Committee on Geographical Names 3 December 1968. The feature was named after Louis Peskett, a man that "in life and even in death was instrumental in seeing many teenagers come to a greater understanding of themselves their fellow man and their Creator," according to Len Siemens of the Frontier Lodge Youth Camp, who proposed the name. Peskett was director of an organization known as Youth for Christ. Peskett initiated the idea of a camp designed for delinquent teenagers. The camp was subsequently built.

Petain, Mount (mountain)

82 J/11 - Kananaskis Lakes
18-9-W5
50°32′N 115°11′W
Approximately 65 km west south-west of Turner Valley.

This 3183 m mountain was officially named in 1918 after a French soldier and politician. General Henri Philippe Benoni Omer Petain

(1856-1951) successfully led the defence of Verdun against crushing German attacks from February to May 1916. Petain was elected to the French Academy in 1929 and started his political career 1 June 1940, he succeeded Paul Reynaud as Premier and surrendered to German invaders 22 June 1940, establishing and heading the Nazi-controlled Vichy Government of France. A series of reforms earned him the title "Revolution Nationale."

Peters Creek (creek)

82 O/13 - Scalp Creek
28-33-14-W5
51°51′N 115°56′W
Flows north into Clearwater River, approximately 80 km north-west of Banff.

(see Mount Peters)

Peters, Mount (mountain)

82 O/13 - Scalp Creek
18-33-14-W5
51°49′N 116°00′W
Approximately 80 km north-west of Banff.

This 2850 m mountain was named in 1928 by Canadian surveyor R.W. Cautley (see Mount Cautley), after Frederick Hathaway Peters, O.B.E., D.L.S. Peters had a long and distinguished career as an engineer and surveyor. He was the Surveyor General of Canada from 1924-1948.

Petite Lake (lake)

83 F/12 - Gregg Lake
12-54-26-W5
53°39′00″N 117°47′30″W
Approximately 30 km north north-west of Hinton.

The name Petite Lake was given to the lake in 1962 by Don Jenkins who was the Forestry District Representative at the time. Mr. Jenkins thought the name was fitting because the lake is small and very

picturesque. The name was officially approved 2 February 1978.

Peveril Peak (peak)

83 D/16 - Jasper
36-43-2-W6
52°45′N 118°09′W
Approximately 14 km south south-west of Jasper.

The origin for the name of this 2650 m mountain peak, which was officially adopted 1 May 1934, is unknown.

Peyto Creek (creek)

82 N/10 - Blaeberry River
51°42′55″N 116°31′10″W
Flows north-east into Peyto Glacier, approximately 90 km north-west of Banff.

The name for this creek was officially approved 21 January 1985. (see Peyto Peak)

Peyto Glacier (glacier)

82 N/10 - Blaeberry River
30-31-18-W5
51°41′N 116°32′W
Approximately 90 km north-west of Banff.

(see Peyto Peak)

Peyto Peak and Peyto Glacier

Peyto Lake (lake)

82 N/10 - Blaeberry River
8-32-18-W5
51°43′N 116°31′W
Approximately 90 km north-west of Banff.

(see Peyto Peak)

Peyto Peak (peak)

82 N/10 - Blaeberry River
25-31-19-W5
51°41′N 116°34′W
Approximately 90 km north-west of Banff.

Ebenezer William "Bill" Peyto (1868-1943) became one of the best guides in the Canadian Rockies. He arrived in Canada from England in 1886. He worked for the railway, homesteaded near Cochrane, hunted, prospected and trapped, and finally signed up as an apprentice guide in 1893-1894 with Tom Wilson. He became friends with Walter D. Wilcox (see Wilcox Pass) and learned some mountain climbing techniques. This 2970 m mountain was named after Bill Peyto. The other features in the area are also named for him.

Phantom Crag (mountain)

82 O/6 - Lake Minnewanka
9-27-9-W5
51°18′N 115°11′W
Approximately 25 km north north-east of Canmore.

This 2332 m mountain was officially named 7 August 1958. Phantom (a vague, dim, shadowy image of the mind, without material substance) is a name concurrent with the "supernatural" idiom, long established in the area.

Pharoah Creek (creek)

82 O/4 - Banff
32-25-14-W5
51°11′N 115°55′W

Approximately 25 km west of Banff.

(see Pharoah Peaks)

Pharoah Lake (lake)

82 O/4 - Banff
9-25-14-W5
51°07′N 115°55′W
Approximately 25 km west south-west Banff.

The name for this lake was officially approved in 1966. (see Pharoah Peaks)

Pharoah Peaks (peak)

82 O/4 - Banff
25-14-W5
51°07′N 115°54′W
Approximately 25 km west south-west of Banff.

This prominent feature in the landscape of the area measures 2711 m in altitude. There is some resemblance to a row of Egyptian mummies to be seen in the peaks, resulting in this descriptive name.

Philip Creek (creek)

83 B/5 - Saunders
15-38-14-W5
52°16′N 115°55′W
Flows south into North Ram River, approximately 26 km south-east of Nordegg.

The creek is named after Philip House, an Indian who camped at its mouth. The name was officially approved in 1941.

Phillipps Lake (lake)

82 G/10 - Crowsnest
18-8-5-W5
49°38′30″N 114°39′40″W
Approximately 11 km west north-west of Coleman.

The lake was officially named 15 February 1978 due to its proximity to Phillipps Peak and Pass. (see Phillipps Pass)

Phillipps Pass (pass)

82 G/10 - Crowsnest
18-8-5-W5
49°38′N 114°39′W
Approximately 11 km west north-west of Coleman, on the Alberta-B.C. boundary.

This pass and nearby mountain were named by the Geographic Board of Canada in 1917 after Michael Phillipps. He was the Hudson's Bay Company clerk in charge of the post at Wild Horse River in British Columbia. In 1860, he discovered and subsequently explored this feature and was the first man to report the numerous coal deposits in the Crowsnest area. He died in 1916. The name Phillipps Pass was officially adopted 2 May 1959.

Phillipps Peak (peak)

82 G/10 - Crowsnest
19-8-5-W5
49°40′N 114°39′W
Approximately 11 km west north-west of Coleman.

This peak is on a 2500 m mountain which at one time had two names: "Mount Phillipps" and "Mount Tecumseh." The name Mount Tecumseh won the battle, and 4 August 1960 an application for Phillipps Peak (see Phillipps Lake) was confirmed for the west peak of Mount Tecumseh to avoid confusion.

Phillips, Mount (mountain)

83 E/3 - Mount Robson
28-48-9-W6
53°10′N 119°16′W
Approximately 86 km west north-west of Jasper.

This 3249 m mountain was named in 1923 after Donald "Curly" Phillips, a well-known guide associated with the region. (see Chetang Ridge)

Phroso Creek (creek)

83 E/10 - Adams Lookout
54-7-W6
53°38′N 118°59′W
Flows south-west into Sulphur River,
approximately 29 km south south-east of
Grande Cache.

(see Anthony Creek)

Picklejar Creek (creek)

82 J/10 - Mount Rae
7-18-6-W5
50°31′N 114°49′W
Flows south into Highwood River,
approximately 42 km west of Turner Valley.

(see Picklejar Lakes)

Picklejar Lakes (lakes)

82 J/10 - Mount Rae
16-8-6-W5
53°31′15″N 114°46′30″W
Source of Picklejar Creek, approximately 37
km west of Turner Valley.

These four small lakes at the head of
Picklejar Creek were officially named 5 May
1983. The fishing in these lakes is excellent:
so good, in fact, that it is said to be like
catching fish in a picklejar.

Pierre Creek (creek)

83 L/12 - Nose Creek
26-66-11-W6
54°44′N 119°33′W
Flows west into Muddy Creek,
approximately 68 km south-west of Grande
Prairie.

This creek was named by J. Haan, a local
forest ranger, after Pierre Shetler, a well-
known Indian trapper, who died in the
1940s. He apparently had a cabin located on
the nearby lake.

Pierre Lake (lake)

83 L/12 - Nose Creek
15-66-11-W6
54°43′N 119°34′W
Approximately 71 km south-west of Grande
Prairie.

(see Pierre Creek)

Pierre Greys Lakes (lakes)

83 E/15 - Pierre Greys Lakes
6-57-4-W6
53°54′N 118°35′W
Approximately 34 km east of Grande Cache.

The Canadian Permanent Committee on
Geographical Names approved the name for
this chain of small lakes 4 September 1947.
Pierre Grey was a trader and successful
entrepreneur who maintained a free trade in
furs. His trading post was located in the
Grande Cache area and his homestead was at
Lake Isle.

Pigeon Creek (creek)

82 O/3 - Canmore
13-24-10-W5
51°01′N 115°15′W
Flows north into Bow River, approximately
9 km south-east of Canmore.

(see Pigeon Mountain)

Pigeon Mountain (mountain)

82 O/3 - Canmore
24-9-W5
51°02′N 115°12′W
Approximately 13 km east south-east of
Canmore.

This 2394 m mountain was originally named
"Pic des Pigeons" by Eugene Bourgeau (see
Mount Bourgeau), of the Palliser Expedition.
At the time of its naming (1858), flocks of
wild pigeons were seen in the vicinity.

Pika Peak (peak)

82 N/8 - Lake Louise
19-29-15-W5
51°30′N 116°06′W
Approximately 55 km north-west of Banff.

Numerous Pika, or "rock rabbits," are found
in the rocks around the base of this 3053 m
mountain. There is also a curious rock
formation on the peak which resembles the
little animal.

Pilkington, Mount (mountain)

82 N/10 - Blaeberry River
4-32-21-W5
51°43′N 115°56′W
Approximately 115 km north-west of Banff,
on the Alberta-B.C. boundary.

This mountain whose peaks measure 3285 m
and 3210 m, was officially named 3 January
1911, after Charles Pilkington, then
President of The Alpine Club (London).

Pilot Mountain (mountain)

82 O/4 - Banff
1-26-14-W5
51°12′N 115°50′W
Approximately 18 km west of Banff.

This 2935 m square-topped tower was named
by Dr. G.M. Dawson (see Tombstone
Mountain) of the Geological Survey of
Canada in 1884, because it is visible for a
long distance down the valley. (see Massive
Range)

Pilot Pond (lake)

82 O/4 - Banff
19-26-13-W5
51°14′N 115°49′W
Approximately 18 km north-west of Banff.

Named for its proximity to Pilot Mountain,
this feature was formerly known as "Lizard
Lake" and "Osprey Lake." The official
decision of the Canadian Permanent

Committee on Geographical Names to use the name Pilot Pond was made in 1966.

Pincher Ridge (ridge)

82 G/8 - Beaver Mines
4-1-W5
49°17′N 114°07′W
Approximately 23 km south-west of Pincher Creek.

This ridge was named in 1917 after the nearby Pincher Creek, which in turn, has several possibilities for its specific origin, but all related to an incident involving a pair of horseshoe pincers (pinchers) lost near the stream.

Pineneedle Creek (creek)

82 O/14 - Limestone Mountain
6-36-34-10-W5
51°57′15″N 115°17′30″W
Flows north into Clearwater River, approximately 90 km north north-east of Banff.

The descriptive name applied to this creek stems from the abundance of pine needles scattered on the ground along the shores of the stream. The name was officially approved by the Canadian Permanent Committee on Geographical Names 16 April 1970.

Pinetop Hill (hill)

82 O/2 - Jumpingpound Creek
35-23-6-W5
51°00′N 114°44′W
Approximately 35 km west of Calgary.

This likely descriptive name was officially approved 12 December 1939.

Pinnacle Mountain (mountain)

82 N/8 - Lake Louise
29-27-16-W5
51°20′N 116°14′W
Approximately 50 km west north-west of Banff.

The descriptive name for this 3067 m mountain was given by W.D. Wilcox (see Wilcox Pass) and was officially approved 1 February 1904.

Pinto Creek (creek)

83 B/5 - Saunders
15-38-13-W5
52°16′N 115°47′W
Flows north into North Ram River, approximately 31 km south-east of Nordegg.

Officially approved in 1941, the name for this creek was taken from A.P. Coleman's (see Mount Coleman) most troublesome packhorse. (see Pinto Lake)

Pinto Creek (creek)

83 F/13 - Hightower Creek
24-56-25-W5
53°51′N 117°35′W
Flows north-east into Wildhay River, approximately 49 km north of Hinton.

The origin of the name is unknown.

Pinto Lake (lake)

83 C/2 - Cline River
25-36-21-W5
52°07′N 116°51′W
Approximately 115 km south-east of Jasper.

A.P. Coleman (see Mount Coleman) stated: "One morning . . . Pinto was missing. But our loads were now light and none of us was sorry to lose him, so we left him behind. Though he was more trouble as a packhorse than all the others put together, we immortalized him by giving his name to an exquisite lake near the head of Cataract River" (*The Canadian Rockies: New and Old Trails*) (1912).

Pipestone Mountain (mountain)

82 N/9 - Hector Lake
16-30-15-W5
51°34′N 116°02′W

Approximately 55 km north-west of Banff.

(see Pipestone River)

Pipestone Pass (pass)

82 N/9 - Hector Lake
36-31-27-W5
51°42′N 116°15′W
Approximately 75 km north-west of Banff.

This pass is 2448 m in altitude. (see Pipestone River)

Pipestone River (river)

82 N/8 - Lake Louise
28-28-16-W5
51°25′N 116°11′W
Flows south into Bow River, approximately 55 km north-west of Banff.

Sir James Hector (see Mount Hector) named this river from the existence on the river of soft, fine-grained, grey-blue argillite, which the Indians used in the manufacture of pipes. "Blue Pipestone River" is the translation of the Stoney Indian name for the river, "pa-hooh-to-hi-agoo-pi-wap-ta," and the Cree name, "moni-spaw-gun-na-nis-si-pi."

Pipit Lake (lake)

82 O/12 - Barrier Mountain
7-30-14-W5
51°37′N 115°52′W
Approximately 55 km north-west of Banff.

The name for this lake, officially approved by the Canadian Permanent Committee on Geographical Names in 1966, was taken from the water pipit. This is a favourite nesting area for these gregarious brownish ground birds which bear a general resemblance to the lark.

Planet Creek (creek)

83 E/9 - Moberly Creek
1-54-4-W6
53°38′N 118°28′W

Flows east into Moon Creek, approximately 89 km north north-west of Jasper.

The precise origin of the name is unknown.

Plante Creek (creek)

83 E/16 - Donald Flats
11-57-4-W6
53°55′N 118°29′W
Flows south into Lone Teepee Creek, approximately 43 km east of Grande Cache.

B.L. Anderson, a field topographer in the area stated that this creek was named after a local trapper 17 April 1947.

Plante Creek (creek)

83 F/11 - Dalehurst
31-53-22-W5
53°37′N 117°13′W
Flows south-east into Athabasca River, approximately 52 km west of Edson.

This creek was officially named in 1946 after Tommy Plante, a trapper in the district. The name was suggested by J.A. McDonald, a field topographer on a survey in the area.

Plateau Creek (creek)

82 J/7 - Mount Head
25-15-5-W5
50°17′N 114°34′W
Approximately 77 km north-west of Claresholm.

(see Plateau Mountain)

Plateau Mountain (mountain)

82 J/2 - Fording River
15-4-W5
50°13′N 114°31′W
Approximately 70 km west north-west of Claresholm.

The name is descriptive of the mountain's appearance and was officially approved in 1915. The mountain measures 2438 m in altitude.

Playle Creek (creek)

82 G/16 - Maycroft
25-11-2-W5
49°56′N 114°09′W
Flows west into Callum Creek, approximately 40 km north-east of Blairmore.

The origin of the name is unknown.

Poacher Basin (basin)

83 C/11 - Southesk Lake
52°31′40″N 117°05′58″W
Approximately 14 km south-east of Southesk Lake, 70 km south-east of Jasper.

This basin was named by the Jasper Park Wardens in the area because of the history of Bighorn Sheep being poached there.

Poachers Creek (creek)

83 F/4 - Miette
16-47-25-W5
53°03′N 117°35′W
Flows north-east into Fiddle River, approximately 43 km south of Hinton.

The name for this creek was suggested by E.W. Mountjoy of the Geographic Board after hunters who illegally entered Jasper Park via Fiddle Pass to poach game. The name was officially approved in 1960.

Poboktan Creek (creek)

83 C/6 - Sunwapta
2-41-22-W5
52°30′N 117°03′W
Flows south-east into Isaac Creek, approximately 81 km south-east of Jasper.

(see Poboktan Pass)

Poboktan Mountain (mountain)

83 C/6 - Sunwapta
14-40-23-W5
52°26′N 117°13′W

Approximately 76 km south-east of Jasper.

This mountain is 3320 m in altitude. (see Poboktan Pass)

Poboktan Pass (pass)

83 C/6 - Sunwapta
25-39-23-W5
52°23′N 117°10′W
Approximately 83 km south-east of Jasper.

This pass, the nearby creek and mountain were named by A.P. Coleman (see Mount Coleman) in 1892, from the big owls that blinked out at him and his party from the spruce trees. "Poboktan" is the Stoney Indian word for "owl."

Pocahontas (locality)

83 F/4 - Miette
6-49-27-W5
53°12′N 117°55′W
Approximately 34 km south-west of Hinton.

The town of Pocahontas opened in 1909 as a construction camp, and was an outlet for Jasper Park Collieries in 1910. This coal mining village was named by W.H. Morris, then Manager of the mine, after Pocahontas, a Virginian coal town.

Pocaterra Creek (creek)

82 J/11 - Kananaskis Lakes
13-20-9-W5
50°42′N 115°07′W
Flows north-west into Kananaskis River, approximately 60 km west of Turner Valley.

George Pocaterra (1882-1972) was an Italian pioneer rancher after whom this creek was officially named in 1914. He established "Buffalo Head" on Highwood River, which was one of Canada's first dude ranches. He was one of the first to prospect the Kananaskis district for coal. He went back to Italy in 1933 and returned to Canada in 1940. He and his wife, an opera singer, established a home and school for singers in Calgary in 1955. (see Spotted Wolf Creek)

Pocket Creek (creek)

82 G/16 - Maycroft
13-10-4-W5
49°49′N 114°25′W
Flows west into Daisy Creek, approximately 25 km north of Blairmore.

The origin of the name is unknown.

Police Flats (flat)

82 G/9 - Blairmore
14-7-3-W5
49°33′30″N 114°18′20″W
Approximately 32 km west north-west of Pincher Creek.

This spot hidden in the hills at the eastern end of the Crowsnest Pass was a favourite stopping place for cattle rustlers to rest their stolen cattle before they were transported to be sold in the United States. To combat rustling, the North West Mounted Police established a post on the flat in the late 1870s, shortly after the founding of Fort Macleod. Since that time, the name Police Flats has been attached to this feature. The name was officially approved 21 July 1982, and is still known locally as an historical name.

Policeman Creek (creek)

82 O/3 - Canmore
28-24-10-W5
51°04′N 115°20′W
Flows south-east into Bow River, approximately 2 km south-east of Canmore.

Charles Carey, an engineer, built his home on this creek, which soon became known as "Carey Creek." The North West Mounted Police erected a station building directly across from Carey's home, and the name for the spring-fed stream was gradually changed to Policeman Creek.

Polkington, Mount (mountain)

82 N/10 - Blaeberry River
4-32-21-W5
51°43′N 116°56′W
Approximately 115 km north-west of Banff, on the Alberta-B.C. boundary.

This mountain, which is 3301 m in altitude, was officially named 3 January 1911 after Charles Polkington, then President of The Alpine Club (London).

Ponoka Creek (creek)

83 F/11 - Dalehurst
10-53-23-W5
53°33′N 117°18′W
Flows north into Athabasca River, approximately 57 km west of Edson.

The origin of the name is unknown.

Poole Creek (creek)

82 G/16 - Maycroft
11-12-2-W5
49°59′N 114°10′W
Approximately 35 km north-east of Blairmore.

No information regarding the origin of the name is known.

Pope Creek (creek)

83 E/10 - Adams Lookout
26-53-5-W6
53°36′N 118°37′W
Flows north-west into South Berland River, approximately 46 km south-east of Grande Cache.

The exact origin of the name is not known.

Pope's Peak (peak)

82 N/8 - Lake Louise
23-28-17-W5
51°24′N 116°17′W

Approximately 60 km west north-west of Banff, on the Alberta-B.C. boundary.

Formerly known as "Boundary Peak," this 3163 m mountain had its name changed by a federal government Order-In-Council 4 April 1887 to commemorate John Henry Pope (1824-1889). He was federal Minister of Agriculture from 1871-1873 and 1878-1885, and Minister of Railways and Canals from 1885-1889.

Porcupine Creek (creek)

82 N/16 - Siffleur River
6-34-17-W5
51°53′N 116°24′W
Flows east into Siffleur River, approximately 100 km north-west of Banff.

The origin of the name is unknown.

Porcupine Creek (creek)

82 J/14 - Spray Lakes Reservoir
29-23-8-W5
50°59′15″N 115°05′25″W
Flows north-west into Kananaskis River, approximately 63 km west of Calgary.

The locally well-established name for this creek was officially approved 20 October 1983, and refers to the number of porcupines in the general area. Although the stream has a year-round flow, in dry years, it tends to run underground toward the end of the summer.

Porcupine Flats (flats)

83 L/3 - Copton Creek
NW-1-61-10-W6
54°14′N 119°23′W
Approximately 116 km south of Grande Prairie.

This feature gets its name from a translation of the Cree Indian word, "kakwa." These flats are located along the Kakwa River.

Porcupine Lake (lake)

82 N/16 - Siffleur River
34-33-18-W5
51°52′N 116°28′W
Approximately 100 km north-west of Banff.

This lake feeds Porcupine Creek. (see Porcupine Creek)

Portal, The (pass)

83 D/9 - Amethyst Lakes
25-43-1-W6
52°43′N 118°08′W
Approximately 15 km south south-west of Jasper.

This pass resembles a doorway or gateway. (see Portal Peak)

Portal Creek (creek)

83 D/16 - Jasper
31-44-1-W6
52°48′20″N 118°02′05″W
Flows east into Athabasca River, approximately 9 km south south-east of Jasper.

This creek was so named because of its proximity to Portal Peak. (see Portal Peak)

Portal Peak (peak)

82 N/10 - Blaeberry River
17-31-18-W5
51°39′N 116°31′W
Approximately 85 km north-west of Banff.

The descriptive name for this 2911 m mountain peak was applied to the feature in 1916 by C.S. Thompson, a member of the Appalachian Club, Boston (see Thompson Pass). The location of the peak, at the entrance to the valley, is likened to a portal (a doorway, or a point of entry).

Powder Creek (creek)

83 F/5 - Entrance
17-52-26-W5
53°29′N 117°48′W

Flows north-east into Cache Lake, approximately 17 km north-west of Hinton.

The origin of the name is unknown.

Powder Face Ridge (ridge)

82 J/15 - Bragg Creek
22-7-W5
50°51′N 114°51′W
Approximately 50 km west south-west of Calgary.

Tom Powderface and his family were Stoney Indians who lived in the Bragg Creek district. This ridge and the nearby creek were officially named for them 6 December 1949.

Powderface Creek (creek)

82 J/15 - Bragg Creek
17-22-6-W5
50°52′N 114°47′W
Flows west into Elbow River, approximately 45 km west south-west of Calgary.

(see Powder Face Ridge)

Prairie Bluff (mountain)

82 G/8 - Beaver Mines
29-4-1-W5
49°20′N 114°06′W
Approximately 20 km south-west of Pincher Creek.

The name Prairie Bluff was officially approved 14 August 1941. The name, however, was established locally some years earlier. This bluff has an altitude of 2254 m.

Prairie Creek (creek)

82 J/15 - Bragg Creek
17-22-6-W5
50°52′N 114°47′W
Flows east into Elbow River, approximately 45 km west of Calgary.

(see Prairie Mountain)

Prairie Creek (creek)

83 L/7 - Prairie Creek
24-61-6-W6
54°18′N 118°46′W
Flows north-west into Kakwa River, approximately 42 km north north-west of Grande Cache.

The origin of the name is unknown, but it is likely descriptive.

Prairie Mountain (mountain)

82 J/15 - Bragg Creek
20-22-6-W5
50°53′N 114°48′W
Approximately 45 km west of Calgary.

This 2210 m mountain receives its name from the prairie landscape which lies immediately east of the Rockies. The name has been on record since before 1928.

Prairie de la Vache (prairie)

83 D/16 - Jasper
44-1-W6
52°48′N 118°01′W
Approximately 10 km south south-east of Jasper.

The name for this flat stretch of land has been in use since at least 1827. Prairie de la Vache refers to the buffalo which formerly ranged in the mountains to this point, although its literal translation is "cow prairie."

Prest Creek (creek)

83 F/7 - Erith
17-50-20-W5
53°18′N 116°53′W
Flows east into Embarras River, approximately 44 km south-west of Edson.

The name for this creek was suggested by Dr. Rutherford of the Alberta Research Council Survey during the course of the railway survey of 1910-1911. It commemorates Benjamin J. Prest, born in England in 1884, who came to Canada in 1904. He was a member of the Grand Trunk Pacific Railway Survey Party of 1910, and in later years joined the staff of the Department of Lands and Mines. He moved to the Department of Public Works in 1938. Mr. Prest retired from the Civil Service in 1950.

Pretty Place Creek (creek)

82 O/6 - Lake Minnewanka
20-29-9-W5
51°30′00″N 115°14′00″W
Flows north into Burnt Timber Creek, approximately 44 km north north-east of Banff.

This stream was named by Ole Olson, a prominent old-timer from the Water Valley area who died in 1975. During frequent hunting trips to the area, Mr. Olson and other hunters from the Water Valley district used to camp at a "pretty place" along this creek. This was 50-60 years ago, and the creek has been called Pretty Place Creek since that time. The name was officially approved 8 November 1978.

Priddis Creek (creek)

82 J/16 - Priddis
22-22-3-W5
50°53′N 114°20′W
Flows east into Fish Creek, approximately 14 km west south- west of Calgary.

The post office in the area opened in 1894. The locality of Priddis is named after Charles Priddis, an early settler, surveyor and postmaster. This creek takes its name from the locality.

Princess Lake (lake)

83 E/1 - Snaring
34-48-1-W6
53°11′N 118°04′W
Approximately 35 km north of Jasper.

The locally well-established name for this lake is taken from Princess Patricia. (see Patricia Lake)

Princess Margaret Mountain (mountain)

82 O/3 - Canmore
29-25-10-W5
51°10′N 115°22′W
Approximately 8 km north of Canmore.

Officially approved 21 July 1958, this 2515 m mountain located in the Fairholme Range, was named in honour of Her Royal Highness Princess Margaret, who visited Banff, and overnighted in a spot from where this mountain may be clearly seen, in 1958.

Prine Creek (creek)

83 F/5 - Entrance
34-50-27-W5
53°21′N 117°51′W
Flows east into Solomon Creek, approximately 17 km west south-west of Hinton.

The origin of the well-established name for this creek is unknown.

Pringle, Mount (mountain)

82 O/2 - Jumpingpound Creek
22-26-7-W5
51°14′N 114°53′W
Approximately 55 km west north-west of Calgary.

This mountain, which is 1692 m in altitude, was officially named 9 April 1970, after the Indian agent of the same name who worked at Morley in the early to mid-1920s.

Prior Peak (peak)

82 N/10 - Blaeberry River
5-32-21-W5
51°43′N 116°56′W
Approximately 115 km north-west of Banff, on the Alberta-B.C. boundary.

The origin of the name is unknown.

Prisoner Point Lake (lake)

82 O/12 - Barrier Mountain
22-31-12-W5
51°40′N 115°36′W
Approximately 55 km north of Banff.

M.E. Kraft, the Regional Fisheries Biologist for Rocky Mountain House, Alberta, originally recommended the name "Barrier Lake" because of this lake's proximity to Barrier Mountain. Local residents, however, preferred the name Prisoner Point Lake, after the Prisoner of War camp that was located just near the lake during the Second World War. The name was officially approved 5 May 1987.

Prospect Creek (creek)

83 C/14 - Mountain Park
8-45-21-W5
52°52′N 116°59′W
Flows north-east into McLeod River, approximately 52 km east north-east of Jasper.

The locally well-established name for this creek was applied prior to 1922, when the nearby mountain was officially named after this stream.

Prospect Mountain (mountain)

83 C/14 - Mountain Park
28-45-24-W5
52′55°N 116°23′W
Approximately 46 km east of Jasper.

This mountain is 2819 m in altitude. (see Prospect Creek)

Protection Mountain (mountain)

82 N/8 - Lake Louise
30-28-15-W5
51°22′45″N 116°00′50″W
Approximately 45 km north-west of Banff.

The 2850 m mountain was named in 1911 by James F. Porter (see Merlin Lake) who explored the locality. The mountain itself separates an unusually beautiful valley from Baker Creek Valley. The application to have the name include the entire massif was made official 20 October 1983. Up to that date, the name was indicated for the north-west extremity only of the feature on maps.

Prow Mountain (mountain)

82 O/12 - Barrier Mountain
11-31-14-W5
51°38′N 115°53′W
Approximately 55 km north-west of Banff.

This 2858 m mountain has a descriptive name, since its shape resembles that of the prow of a ship.

Ptarmigan Lake (lake)

82 N/8 - Lake Louise
17-29-15-W5
51°29′N 116°04′W
Approximately 50 km north-west of Banff.

(see Ptarmigan Peak)

Ptarmigan Lake (lake)

83 E/12 - Pauline Creek
31-52-11-W6
53°32′N 119°36′W
Approximately 50 km south south-west of Grande Cache.

A ptarmigan is a bird of the grouse family which inhabits high altitudes. The name for this lake was suggested by W.D. Wilcox (see Wilcox Pass) and F. Bryant. (see Mount Bryant)

Ptarmigan Peak (peak)

82 N/8 - Lake Louise
13-29-16-W5
51°30′N 116°05′W
Approximately 50 km north-west of Banff.

This 3035 m mountain peak was named for the large quantity of ptarmigan found in the

alplands around the lake and at the base of the peak.

Ptolemy Creek (creek)

82 G/10 - Crowsnest
36-7-6-W5
49°36′N 114°41′W
Flows north into Crowsnest Creek, approximately 13 km west south-west of Coleman.

This creek was named for its proximity to Mount Ptolemy in 1915. (see Mount Ptolemy)

Ptolemy, Mount (mountain)

82 G/10 - Crowsnest
9-7-5-W5
49°33′N 114°37′W
Approximately 13 km south-west of Coleman.

Mount Ptolemy was named by A.O. Wheeler (see Wheeler Flats) owing to its resemblance in shape to a man sitting with his arms folded. The name was officially approved 2 November 1915. The mountain has had two other local names, "Mummy Mountain" and "The Sleeping Giant." Both have the same general meaning. It has also been suggested that this 2815 m mountain may have been named for the celebrated astronomer by the name of Ptolemy who lived in ancient times.

Ptolemy Pass (pass)

82 G/10 - Crowsnest
12-7-6-W5
49°33′N 114°41′W
Approximately 16 km west south-west of Coleman.

The pass was officially named Ptolemy Pass 2 November 1915. (see Mount Ptolemy)

Pulpit Peak (peak)

82 N/9 - Hector Lake
18-30-17-W5
51°34′N 116°22′W

Approximately 75 km north-west of Banff.

The descriptive name for this 2725 m mountain peak was given by C.S. Thompson (see Thompson Pass) in 1898.

Pulsatilla Mountain (mountain)

82 O/5 - Castle Mountain
28-14-W5
51°24′N 115°59′W
Approximately 39 km north-west of Banff.

This mountain, which is 3035 m in altitude, was named in 1911 by J.F. Porter of Chicago (see Merlin Lake), after the mountain plant of this name. "Pulsatilla" is a sub-generic name for one section of the genus of western Anemones which grow in the area.

Punchbowl Falls (falls)

83 F/4 - Miette
5-49-27-W5
53°12′N 117°55′W
Approximately 35 km south-west of Hinton.

The descriptive name for these falls has been in use since 1917. The solid rock of the area has been eroded so as to resemble a huge goblet, or punchbowl.

Purple Mountain (mountain)

83 C/2 - Cline River
15-36-19-W5
52°06′N 116°38′W
Approximately 120 km west south-west of Rocky Mountain House.

This mountain, which is 2591 m in altitude, was named in 1959. It is composed of a large amount of purple coloured rock and its name is thus descriptive.

Putnik, Mount (mountain)

82 J/11 - Kananaskis Lakes
25-19-10-W5
50°39′N 115°14′W
Approximately 70 km west of Turner Valley.

This mountain, 2940 m in altitude, was officially named in 1918 after Field Marshal Radomir Putnik (1847-1917), a Serbian Army Officer, War Minister and Chief of Staff.

Putzy Creek (creek)

83 L/4 - Kakwa Falls
13-58-13-W6
54°01′N 119°47′W
Flows north-west into South Kakwa River, approximately 35 km north-west of Grande Cache.

The origin of the name is unknown.

Pyramid Creek (creek)

83 D/16 - Jasper
34-45-1-W6
52°55′N 118°03′W
Flows south-east into Athabasca River, approximately 4 km north north-east of Jasper.

(see Pyramid Mountain)

Pyramid Lake (lake)

83 D/16 - Jasper
32-45-1-W6
52°55′N 118°06′W
Approximately 4 km north of Jasper.

(see Pyramid Mountain)

Pyramid Mountain (mountain)

83 D/16 - Jasper
12-46-2-W6
52°57′N 118°09′W
Approximately 9 km north-west of Jasper.

This 2766 m pyramid-shaped mountain was given its name in 1859 by Sir James Hector (see Mount Hector) while on an expedition through Athabasca Pass, accompanied by H.J. Moberly (see Moberly Flats) and their Iroquois guide, Tekarra (see Mount Tekarra). It is a well-known landmark near Jasper. The

other features take their names from this mountain.

Pyriform Mountain (mountain)

82 J/10 - Mount Rae
13-18-6-W5
50°32′N 114°42′W
Approximately 33 km west of Turner Valley.

This mountain, which was named in 1922, is 2621 m in altitude. The name was not officially applied until 6 September 1951. "Pyriform" is a descriptive name in that in scientific and technical use, the word means "pear-shaped."

Quadra Mountain (mountain)

82 N/8 - Lake Louise
27-16-W5
51°17′N 116°09′W
Approximately 45 km west north-west of Banff, on the Alberta-B.C. boundary.

There are four pinnacles at this 3173 m mountain's summit. The name Quadra Mountain was first adopted in the 18th Report of 1924 and appeared in this form in *Place Names of Alberta* (1928). The name was later changed to "Mount Quadra" (3 April 1952) because it was thought that the feature had been "named after a person, as Quadra Island." The name, however, is descriptive, and was therefore officially changed back to Quadra Mountain 6 November 1983.

Quartz Hill (hill)

82 O/4 - Banff
24-13-W5
51°02′N 115°46′W
Approximately 21 km south south-west of Banff.

This prominent hill, which is 2568 m in altitude, has a crest covered with broken blocks of quartz, falling very steeply on the south-west side. The name was very well-established in local usage before its official approval.

Queen Elizabeth, Mount (mountain)

82 J/11 - Kananaskis Lakes
24-20-11-W5
50°43′N 115°24′W
Approximately 80 km west of Turner Valley.

This mountain was named in 1918 after the former Duchess of Bavaria, Queen of Belgium, consort of King Albert I, Munich Elizabeth. It is 2850 m in altitude.

Queen Elizabeth Ranges (range)

83 C/13 - Medicine Lake
44-25-W5
52°45′N 117°17′W
Approximately 24 km east of Jasper.

Named in 1953 after Her Majesty Queen Elizabeth II to commemorate her coronation, these ranges practically encircle Maligne Lake.

Quigley Creek (creek)

83 F/6 - Pedley
26-50-23-W5
53°21′N 117°15′W
Flows east into McLeod River, approximately 61 km south-west of Edson.

The McPherson and Quigley Lumber Company operated a sawmill on this stream around 1925. The locally well-established name was officially adopted in 1924.

Quincy Creek (creek)

83 C/5 - Fortress Lake
5-39-25-W5
52°19′N 117°34′W
Flows north-east into Athabasca River, approximately 70 km south-east of Jasper.

(see Mount Quincy)

Quincy, Mount (mountain)

83 C/5 - Fortress Lake
3-39-26-W5
52°20′N 117°39′W

Approximately 65 km south-east of Jasper.

This 3150 m mountain was named in 1892 by A.P. Coleman (see Coleman Peak) after his brother, Lucius Quincy Coleman, a rancher at Morley. Quincy was a common family name. Their mother, one of the New England "Adams," was a relative of John Quincy Adams, the 6th President of the United States.

Quirk Creek (creek)

82 J/15 - Bragg Creek
4-22-6-W5
50°52'N 114°47'W
Flows north-west into Elbow River, approximately 45 km west south-west of Calgary.

(see Mount Quirk)

Quirk, Mount (mountain)

82 J/15 - Bragg Creek
30-21-5-W5
50°49'N 114°41'W
Approximately 40 km west south-west of Calgary.

This mountain, which is 1905 m in altitude, was officially named 6 October 1949 after becoming well-established in local usage. John "Johnny" Quirk, after whom this feature was named, was an early local rancher. (see Kew Ridge)

Quoin, The (peak)

83 E/7 - Blue Creek
30-51-4-W6
53°26'N 118°35'W
Approximately 70 km north-west of Jasper.

The name for this mountain peak, 2454 m in altitude, is descriptive since the peak forms a cornerstone of the Starlight Range. The name was officially adopted 1 May 1934.

Race Creek (creek)

83 C/10 - George Creek
43-20-W5
52°42'N 116°49'W
Flows north-west into Brazeau River, approximately 54 km west north-west of Nordegg.

This creek was named by Sam McEvoy of Brazeau Collieries. There was a race between their competitors and the Brazeau Company to stake the coal claim. The Brazeau Company won. The name Race Creek was used locally for many years after 1910, when the race ended.

Racehorse Creek (creek)

82 G/16 - Maycroft
32-10-3-W5
49°52'N 114°23'W
Flows north-east into Oldman River, approximately 30 km north of Blairmore.

The precise origin of this name is not known; however, the stream is very swift, and the name may reflect the speed of the water.

Racehorse Pass (pass)

82 G/15 - Tornado Mountain
29-9-5-W5
49°46'N 114°39'W
Approximately 15 km north-west of Coleman, on the Alberta-B.C. boundary.

(see Racehorse Creek)

Radiant Creek (creek)

83 B/4 - Cripple Creek
29-35-11-W5
52°02'N 115°32'W
Flows east into Idlewilde Creek, approximately 56 km south-west of Rocky Mountain House.

The name for this creek was suggested by the Topographical Survey and was officially

approved in July, 1941. The creek flows through a bright valley with gently sloping hills on each side, and its name is appropriately descriptive.

Rae Creek (creek)

82 J/10 - Mount Rae
33-19-7-W5
50°39'N 114°55'W
Flows north-east into Sheep Creek, approximately 45 km west of Turner Valley.

(see Mount Rae)

Rae, Lake (lake)

82 J/10 - Mount Rae
36-19-8-W5
50°39'25"N 114°58'25"W
Source of Rae Creek, approximately 48 km west of Turner Valley.

The lake is named for its proximity to Mount Rae. The name Lake Rae was officially approved 7 February 1983. (see Mount Rae)

Rae, Mount (mountain)

82 J/10 - Mount Rae
24-19-8-W5
50°37'N 114°59'W
Approximately 48 km west of Turner Valley.

Dr. John Rae, F.R.S., F.R.G.S., (1813-1893) was a Scottish explorer and surgeon to the Hudson's Bay ship in 1833. He became resident surgeon at Moose Fort in 1835. His career in exploration began in 1846 and continued until 1864, when he conducted the survey for a telegraph line from Red River Settlement to the Pacific Coast. This 3218 m mountain was named by Sir James Hector, M.D. (see Mount Hector) after Dr. John Rae.

Rainbow Creek (creek)

83 F/3 - Cadomin
25-47-22-W5
53°05'N 117°05'W

Flows north into Beaverdam Creek, approximately 70 km south-west of Edson.

The name for this creek was suggested by J.R. Matthews, a local resident. It was officially approved in 1944, but no other origin information is available.

Rainbow Lake (lake)

82 O/5 - Castle Mountain
51°21'20"N 115°43'50"W
Approximately 25 km north-west of Banff.

The name for this lake was officially approved 21 January 1985, but its origin is not known.

Rainy Creek (creek)

82 J/15 - Bragg Creek
8-22-6-W5
50°51'20"N 114°47'45"W
Flows north-east into Elbow River, approximately 45 km west south-west of Calgary.

The name was officially approved 7 February 1983. The origin of the name is unknown.

Rainy Ridge (ridge)

82 G/1 - Sage Creek
3-3-W5
49°15'N 114°23'W
Approximately 40 km north-west of Waterton Park, on the Alberta-B.C. boundary.

The origin for this locally well-established name is unknown.

Rajah, The (peak)

83 E/7 - Blue Creek
29-49-4-W6
53°16'N 118°33'W
Approximately 53 km north-west of Jasper.

The name of this 3018 m mountain peak is the word for "king" in Hindi, an official

language of India. The name was applied to the feature by R.W. Cautley of the Alberta Boundary Commission, and later of the National Parks Branch. (see Mount Cautley)

Ram Falls (falls)

83 B/4 - Cripple Creek
18-36-13-W5
52°05'20"N 115°51'20"W
Approximately 46 km south-east of Nordegg.

This high waterfall on the Ram River is known throughout Alberta as a major attraction on the Forestry Trunk Road, south of Nordegg. The name was officially approved 22 June 1979. (see Ram River)

Ram Falls

Ram Range (range)

83 B/4 - Cripple Creek
35-14-W5
52°04'N 115°54'W
Approximately 47 km south-east of Nordegg.

(see Ram River)

Ram River (river)

83 B/6 - Crimson Lake
20-39-10-W5
52°23'N 115°25'W
Flows north-east into North Saskatchewan River, approximately 32 km west of Rocky Mountain House.

The name of this river is "Ram Rivulet" on David Thompson's map of 1814, probably after the term for a male Rocky Mountain Sheep.

Ram River Glacier (glacier)

82 N/16 - Siffleur River
22-33-16-W5
51°51'N 116°12'W
Approximately 85 km north-west of Banff.

(see Ram River)

Rampart Creek (creek)

83 C/2 - Cline River
25-35-21-W5
52°02'N 116°52'W
Flows south into North Saskatchewan River, approximately 125 km south-east of Jasper.

This creek was officially named in 1960, after the name had been in local use for many years previously. There is a great cliff on Mount Wilson, after which this creek is named.

Ramparts, The (range)

83 D/9 - Amethyst Lakes
43-3-W6
52°42'N 118°18'W
Approximately 26 km south-west of Jasper.

The previously well-established name for this series of low, guardian-like black mountains was officially approved in 1916. A rampart is a defensive stone wall. (see Dungeon Peak)

Ranee, The (peak)

83 E/2 - Resplendent Creek
19-49-4-W6
53°15′N 118°35′W
Approximately 54 km north-west of Jasper.

"Ranee" is the word for "queen" in Hindi.
(see The Rajah)

Ranger Canyon (canyon)

82 O/4 - Banff
26-13-W5
51°14′N 115°46′W
Approximately 15 km north-west of Banff.

The name was officially approved in 1959,
the name for this canyon stems from a visit
by the Boy Scouts (also known as Rangers)
in the summer prior to its naming.

Ranger Creek (creek)

83 B/4 - Cripple Creek
22-35-14-W5
52°01′N 115°56′W
Flows north-west into Ram River,
approximately 45 km south-east of Nordegg.

The origin of the name is unknown.

Ranger Creek (creek)

82 O/4 - Banff
26-13-W5
51°13′N 115°46′W
Flows south into Bow River, approximately
15 km west north-west of Banff.

(see Ranger Canyon)

Ranger Creek (creek)

82 J/15 - Bragg Creek
30-22-5-W5
50°54′N 114°41′W
Flows south-east into Elbow River,
approximately 35 km west of Calgary.

Between 1910 and 1915, the first ranger
station was built on the site near this creek.
The name for the creek has been locally
known as Ranger Creek since that time and
was officially approved 6 October 1949.

Rapid Creek (creek)

83 B/12 - Harlech
28-43-13-W5
52°44′N 115°49′W
Flows north-west into Nordegg River,
approximately 33 km north-east of Nordegg.

The name for this creek is likely descriptive
of the speed of the stream's flow. No other
information is available.

Rat Creek (creek)

83 C/15 - Cardinal River
17-46-20-W5
52°58′N 116°52′W
Flows north-east into Pembina River,
approximately 86 km east north-east of
Jasper.

The precise origin of this name is not
known.

Rat Lake (lake)

83 B/3 - Tay River
11-38-9-W5
52°14′N 115°10′W
Approximately 23 km south-west of Rocky
Mountain House.

Don Macdonald, a resident of the Strachan
area for 77 years, named this small lake after
the many muskrats which used to inhabit
the lake. The name Rat Lake was officially
approved 5 May 1987.

Rathlin Lake (lake)

83 D/16 - Jasper
14-45-2-W6
52°52′N 118°11′W
Approximately 7 km west of Jasper.

The name for this lake was applied in 1914
by H. Matheson, D.L.S. The origin of the
name is unknown.

Raven Creek (creek)

83 F/7 - Erith
18-50-18-W5
53°19′N 116°38′W
Flows north-west into Erith River,
approximately 33 km south south-west of
Edson.

No information regarding the origin of the
name is known.

Ravine Creek (creek)

83 L/5 - Two Lakes
12-61-11-W6
54°15′N 119°31′W
Flows north-west into Torrens River,
approximately 56 km north-west of Grande
Cache.

A.M. Floyd, a field topographer of the area,
suggested the descriptive name for this very
steep and deep creek.

Rawson Creek (creek)

82 J/11 - Kananaskis Lakes
14-19-9-W5
50°35′N 115°08′W
Flows north into Upper Kananaskis Lake,
approximately 60 km west south-west of
Turner Valley.

Dr. D.S. Rawson (1905-1961) was a world-
famous limnologist who conducted biological
surveys throughout Canada. He was a member
of numerous boards and associations
throughout his lifetime, and at the time of
his death he was the Canadian Representative
on the Council of the International Association
for Theoretical and Applied Limnology.
This creek is the headwaters of Rawson
Lake, accessible only by foot (a five hour climb
from Upper Kananaskis Lake) or by helicopter.
The name was officially approved 16 June 1977.

Rawson Lake (lake)

82 J/11 - Kananaskis Lakes
11-19-9-W5
50°35′N 115°09′W
Approximately 60 km west south-west of Turner Valley.

(see Rawson Creek)

Razorback Mountain (mountain)

83 D/15 - Rainbow
28-46-5-W6
52°58′N 118°40′W
Approximately 41 km west north-west of Jasper.

The name for this mountain is likely descriptive of its appearance.

Recondite Peak (peak)

82 N/16 - Siffleur River
51°49′00″N 116°18′00″W
Approximately 85 km north-west of Banff.

The origin of the locally well-established name of this mountain peak, 3356 m in altitude, is unknown. The name was officially approved 21 January 1985.

Red Deer Lakes (lakes)

82 N/9 - Hector Lake
3-30-15-W5
51°32′N 116°02′W
Approximately 55 km north-west of Banff.

The name of these lakes was taken from the Red Deer River since the largest lake of this group forms its headwaters.

Red Man Mountain (mountain)

82 J/13 - Mount Assiniboine
21-12-W5
50°47′N 115°32′W
Approximately 100 km west south-west of Calgary, on the Alberta-B.C. boundary.

This mountain, which is 2905 m in altitude, was named in 1918 from the red colour of

the rock which is in contrast to White Man Mountain. (see White Man Mountain)

Red Man Pass (pass)

82 J/13 - Mount Assiniboine
29-21-11-W5
50°48′10″N 115°33′10″W
Approximately 40 km south south-east of Banff.

Place Names of Alberta (1928) suggested that the name "Red Man" was derived from the red colour of the rock, in contrast to the colour of White Man Mountain. Other sources indicate that the name Red Man Pass is well-established in local usage, but its precise origin is unknown. (see Red Man Mountain)

Red Rock Canyon (canyon)

82 G/1 - Sage Creek
23-2-1-W5
49°08′N 114°01′W
Approximately 12 km north-west of Waterton Park.

The name Red Rock Canyon was officially approved 27 April 1972. (see Red Rock Creek)

Red Rock Creek (creek)

82 G/1 - Sage Creek
14-2-1-W5
49°08′N 114°02′W
Approximately 12 km north-west of Waterton Park.

Red Rock Creek empties into Bauerman Creek and was officially named 7 April 1960. Iron compounds in the rocks give the canyon and the creek the colour for which they are named.

Redan Mountain (mountain)

83 E/1 - Snaring
11-48-2-W6
53°08′N 118°11′W

Approximately 27 km north north-west of Jasper.

This 2560 m mountain is situated at a bend in the ridge, with steep rock walls to the east. The descriptive name was applied in 1916 by M.P. Bridgland (see Mount Bridgland): a "redan" is a military field word describing two faces forming a salient angle.

Redcap Mountain (mountain)

83 C/14 - Mountain Park
2-46-22-W5
52°56′N 117°05′W
Approximately 67 km east of Jasper.

The name for this 2393 m mountain is descriptive.

Redearth Creek (creek)

82 O/4 - Banff
18-26-13-W5
51°13′N 115°49′W
Flows north-east into Bow River, approximately 18 km west north-west of Banff.

This creek, which was formerly called "Vermilion Creek," was named by Dr. G.M. Dawson (see Tombstone Mountain) of the Geological Survey of Canada to avoid duplication. It gets its name from the red ochre found in places on its banks.

Redearth Pass (pass)

82 O/4 - Banff
24-14-W5
51°05′N 115°52′W
Approximately 24 km south-west of Banff.

This pass is 2063 m in altitude. (see Redearth Creek)

Redfern Lake (lake)

82 G/9 - Blairmore
16-7-3-W5
49°33′40″N 114°20′25″W

Approximately 2 km south-east of Blairmore.

The locally well-established name for this lake is taken from Jim and Mary Alice Redfern, a family who lived in Passburg around 1907. The name was officially approved 21 July 1982.

Redoubt Lake (lake)

82 N/8 - Lake Louise
8-29-15-W5
51°28′N 116°04′W
Approximately 50 km north-west of Banff.

The name was officially approved 6 February 1912. (see Redoubt Mountain)

Redoubt Mountain (mountain)

82 N/8 - Lake Louise
8-29-15-W5
51°28′N 116°05′W
Approximately 50 km north-west of Banff.

This 2902 m mountain was named in 1908 by A.O. Wheeler (see Wheeler Flats) since the rock formation resembles a huge redoubt, or military outer defence lacking flanking defences. The lake takes its name from the mountain.

Redoubt Peak (peak)

83 D/9 - Amethyst Lakes
12-43-3-W6
52°41′N 118°17′W
Approximately 26 km south-west of Jasper.

(see Bastion and Dungeon Peaks)

Redrock Creek (creek)

83 L/6 - Chicken Creek
19-61-8-W6
54°17′N 119°11′W
Flows south into Kakwa River, approximately 45 km north of Grande Cache.

The name for this creek was officially approved in 1955. There is a wall of red coloured rock along the banks, caused either by an iron deposit or once heated coal deposits.

Reef Icefield (icefield)

83 E/3 - Mount Robson
48-8-W6
53°08′N 119°02′W
Approximately 68 km west north-west of Jasper.

Dr. A.P. Coleman (see Mount Coleman) named this icefield in 1911 on one of his explorations to Mount Robson. Several long, low rock ridges rise from the surface of the snowfields, and these rock reefs inspired the name.

Remus, Mount (mountain)

82 J/15 - Bragg Creek
24-21-8-W5
50°48′N 114°58′W
Approximately 60 km west south-west of Calgary.

Mount Remus is 2688 m in altitude. (see Mount Romulus)

Replica Peak (peak)

83 C/11 - Southesk Lake
7-41-24-W5
52°31′N 117°26′W
Approximately 60 km south-east of Jasper.

This mountain peak was named by Howard Palmer in 1923. No origin information is available; presumably it closely resembles another peak.

Resolute Mountain (mountain)

83 C/2 - Cline River
9-36-19-W5
52°04′N 116°39′W
Approximately 130 km south-east of Jasper.

Lion and Lioness Peaks are situated on this mountain. "Resolute" has a remote association with lions and is not duplicated anywhere in Alberta. The name was officially approved in 1959. (see Lion Peak)

Resthaven Icefield (icefield)

83 E/6 - Twintree Lake
51-10-W6
53°26′N 119°28′W
Approximately 110 km west north-west of Jasper.

(see Mount Resthaven)

Resthaven Mountain (mountain)

83 E/5 - Holmes River
35-51-11-W6
53°27′N 119°30′W
Approximately 114 km north-west of Jasper.

This 3125 m mountain is the highest peak in Willmore Wilderness Park. The origin for the name "Resthaven," approved in 1925, is unknown.

Restless River (river)

83 C/14 - Mountain Park
22-44-24-W5
52°48′N 117°23′W
Flows north into Rocky River, approximately 46 km east south-east of Jasper.

The name for this river describes its course: it changes its bed every high water. It was named by a W.H. Miller in 1925. Nothing else is known about Mr. Miller.

Reunion Peak (peak)

83 D/9 - Amethyst Lakes
10-42-3-W6
52°36′N 118°20′W
Approximately 36 km south-west of Jasper.

During the Alberta-British Columbia Boundary Survey, there was a station to link up with other stations located on this peak, i.e., a reunion point.

Revenant Mountain (mountain)

82 O/6 - Lake Minnewanka
4-28-11-W5
51°22'N 115°28'W
Approximately 21 km north north-east of Banff.

The name for this mountain, which is 3065 m in altitude, was suggested by T.W. Swaddle of Calgary in 1970. Swaddle maintained that "Revenant" (meaning "ghost-like" in both English and French) was concurrent with the "supernatural" idiom long established in the area. The Canadian Permanent Committee on Geographical Names approved the name 7 February 1971.

Rhondda, Mount (mountain)

82 N/10 - Blaeberry River
7-31-18-W5
51°39'N 116°33'W
Approximately 85 km north-west of Banff on the Alberta-B.C. boundary.

David Alfred Thomas, 1st Viscount Baron Rhondda (1856-1918), was a Welsh coal mining entrepreneur, a Liberal member in the British House of Commons for 22 years, and a government administrator who visited Canada in 1915. This 3055 m mountain was named in 1918 after this man who introduced an effective food rationing system for Britain during the First World War.

Ribbon Creek (creek)

82 J/14 - Spray Lakes Reservoir
1-23-9-W5
50°56'N 115°08'W
Flows north-east into Kananaskis River, approximately 70 km west of Calgary.

(see Ribbon Lake and North Ribbon Creek)

Ribbon Lake (lake)

82 J/14 - Spray Lakcs Reservoir
19-22-9-W5
50°53'00"N 115°14'35"W
Approximately 75 km west of Calgary.

This locally well-established name was officially approved 7 February 1983. Ribbon Lake forms part of the Ribbon Creek drainage system. The name is somewhat descriptive in that at this lake, the valley fans out into a number of high alpine cirques.

Rice Creek (creek)

82 J/1 - Langford Creek
26-14-2-W5
50°12'N 114°10'W
Flows north-east into Willow Creek, approximately 45 km north-west of Claresholm.

William Henry Rice (1870-1948) came from Ontario to homestead in 1892. He located on this creek which was officially named after him 3 December 1941.

Richards, Mount (mountain)

82 H/4 - Waterton Lakes
1-30-W4
49°01'N 113°56'W
Approximately 55 km south of Pincher Creek.

This 2377 m mountain was named after Captain (later, Admiral) G.R. Richards, R.N., the Second Commissioner of the British Boundary Commission from the Pacific to the Rockies, 1856-1863. Admiral Richards made hydrographic surveys of the British Columbia coast. This mountain is still known by many people as "Sleeping Indian Mountain," referring to the east ridge of the mountain which resembles the profile of an Indian face.

Richardson, Mount (mountain)

82 N/8 - Lake Louise
23-29-16-W5
51°30'N 116°08'W

Approximately 55 km north-west of Banff.

This mountain, which is 3086 m in altitude, was named by Sir James Hector (see Mount Hector) after Sir John Richardson, F.R.S., M.D. (1787-1865). He was appointed Surgeon and Naturalist to Franklin's Arctic expeditions of 1819 and 1825, and in 1840 he was appointed Inspector of Hospitals.

Ridge Creek (creek)

83 E/10 - Adams Lookout
27-54-6-W6
53°42'N 118°48'W
Flows north-west into Muskeg River, approximately 29 km south-east of Grande Cache.

The origin of the name is unknown.

Ridges Creek (creek)

83 C/3 - Columbia Icefield
28-35-22-W5
52°02'N 117°04'W
Flows north into Alexandra River, approximately 115 km south-east of Jasper.

The descriptive name for this creek, which flows between two striking mountain ridges, was officially approved by the Canadian Permanent Committee on Geographical Names 6 July 1964.

Riley Creek (creek)

82 J/1 - Langford Creek
17-13-2-W5
50°05'N 114°14'W
Flows east into Chaffin Creek, approximately 45 km west north-west of Claresholm.

Daniel E. Riley (later Senator) held a lease in Happy Valley with a fellow rancher named Thompson. Politics caused a commotion in 1917 when Dr. G.D. Stanley, a Conservative from High River was running for re-election against Dan Riley, a Liberal. Riley lost by 38 votes, 885 to 923. There was a conflict of interest debate with respect to the lease. By the next year, most of the troubles had resolved themselves.

Riley Lake (lake)

83 D/16 - Jasper
19-45-1-W6
52°54'N 118°07'W
Approximately 3 km north-west of Jasper.

The origin of the name is unknown.

Rim Ridge (ridge)

83 L/4 - Kakwa Falls
7-58-12-W6
54°00'N 119°45'W
Approximately 42 km west north-west of Grande Cache.

No information regarding the origin of the name is known.

Ringrose Peak (peak)

82 N/8 - Lake Louise
26-27-17-W5
51°21'N 116°17'W
Approximately 55 km west north-west of Banff, on the Alberta-B.C. boundary.

This mountain, which is 3278 m in altitude, was named in 1894 by S.E.S. Allen (see Mount Allen) after A.E.L. Ringrose of London, England. Ringrose travelled extensively in the Rocky Mountains.

Rink Brook (brook)

83 D/15 - Rainbow
32-45-4-W6
52°56'N 118°32'W
Flows north-east into Miette River, approximately 31 km west of Jasper.

(see Rink Lake)

Rink Lake (lake)

83 D/15 - Rainbow
25-45-5-W6
52°54'N 118°36'W

Approximately 34 km west of Jasper.

The name of this lake was applied in 1918 by A.O. Wheeler. (see Wheeler Flats)

Ripple Rock Creek (creek)

82 J/11 - Kananaskis Lakes
2-21-9-W5
50°44'45"N 115°08'30"W
Flows west into Kananaskis River, approximately 60 km west of Turner Valley.

The descriptive name of this stream was officially approved 5 December 1984. It was selected because rock ripples from an old ocean bed may be found in the creek.

Roaring Creek (creek)

82 N/16 - Siffleur River
31-32-15-W5
51°48'N 116°07'W
Flows north into Clearwater River, approximately 80 km north-west of Banff.

The name for this creek is descriptive.

Robb (hamlet)

83 F/2 - Foothills
15-49-21-W5
52°13'N 116°58'W
Approximately 54 km south-west of Edson.

The name of this hamlet was originally "Minehead," but was changed to Robb, in 1912. At that time, it became also known as "Mile 33" on the railroad. The Balkan Coal Company bought the mine located there in the 1910s. The hamlet was possibly named after Peter Addison "Baldy" Robb (1887-1955). (see Robb Station)

Robb Station (station)

83 F/2 - Foothills
10-49-21-W5
53°13'00"N 116°58'40"W
Approximately 55 km south-west of Edson.

This station, built in 1912, was probably named after local firefighter Peter Addison "Baldy" Robb (1887-1955), who freighted supplies for the Grand Trunk Pacific main line and the branch line to Luscar. Around 1915, he took up prospecting in the coal branch area, buying up the Balkan Coal Company property in the area in 1927, after a complicated legal struggle. (see Robb)

Robertson, Mount (mountain)

82 J/11 - Kananaskis Lakes
22-20-10-W5
50°43'N 115°19'W
Approximately 75 km west of Turner Valley.

Sir William Robert Robertson (1860-1933) was a British Field Marshal and Chief of the Imperial General Staff during the greater part of the First World War. This 3194 m mountain was officially named after this man in 1918.

Robertson Peak (peak)

82 G/9 - Blairmore
34-7-3-W5
49°36'N 114°19'W
Approximately 4 km north-west of Bellevue.

The precise origin of this name is unknown, but it may be named after Joseph Robertson, an early resident of the area.

Robinson Hill (hill)

82 J/15 - Bragg Creek
20-22-4-W5
50°53'N 114°31'W
Approximately 25 km west of Calgary.

This hill was officially named 6 October 1949 after Richard George Robinson, who arrived in the area in 1888. He operated the "Robinson Cow Camp" at the turn of the century.

Robson Pass (pass)

83 E/3 - Mount Robson
29-48-8-W6
53°09'N 119°08'W

Approximately 77 km west north-west of Jasper.

This 1569 m pass takes its name from Mount Robson (3954 m), the highest peak in the Canadian Rockies. The pass is located directly north of the mountain. It is uncertain after whom the impressive peak in British Columbia is named. Mount Robson was first mentioned in *The North West Passage by Land* (1865, Milton & Cheadle), possibly referring to Colin Robertson of both the Hudson's Bay Company and the North West Company. It is possible that 1814 Jasper House Manager, Francois Décoigne (see Yellowhead Pass) named the feature after Robertson, his superior officer.

Roche Miette Creek (creek)

83 F/4 - Miette
6-49-27-W5
53°12′N 117°56′ W
Flows north-west into Athabasca River, approximately 36 km south-west of Hinton.

(see Roche Miette)

Rock Creek (creek)

82 G/9 - Blairmore
20-7-2-W5
49°35′N 114°14′W
Flows south into Crowsnest River, approximately 5 km west of Lundbreck.

This locally well-established name is likely descriptive of the area through which the creek flows.

Rock Creek (creek)

82 J/10 - Mount Rae
23-20-6-W5
50°43′N 114°44′W
Flows east into Volcano Creek, approximately 32 km west of Turner Valley.

The name was officially approved 6 September 1951. The origin of the name is unknown but it is likely descriptive.

Rock Creek (creek)

83 E/8 - Rock Lake
6-52-2-W6
53°27′N 118°16′W
Flows east into Wildhay River, approximately 66 km north of Jasper.

The locally well-established name for this creek was approved in 1955. It is likely descriptive.

Rock Lake (lake)

83 E/8 - Rock Lake
5-52-2-W6
53°27′N 118°15′W
Approximately 65 km north north-west of Jasper.

The legend of the origin of the name of this lake is found in *Jungling in Jasper* by L.J. Burpee (Ottawa: 1929, pp. 45-50.) The legend was related as follows: "So named because of an incident that almost brought about war between the Stoney and the Dogrib. A young Dogrib maiden, the daughter of a chief, was betrothed to a warrior. One day a stranger, exhausted and half-starved, arrived at their village. He was a Stoney medicine man who had been banished from his tribe. The Dogrib took him in and cared for him, but he repaid their kindness by making away with the maiden in the form of a dog. The tribe pursued him, but not seeing the girl, allowed him to escape. He returned to his tribe and told them how he had "rescued" the maiden; he returned her to her human form but stole her power of speech. The tribe thought him very clever and took him back. One day a young man came to the village and made an offer to the chief. He told the chief that the medicine man had done a grave wrong to the Dogrib and challenged him to a test of power to show his guilt. The Dogrib warrior was willing to forfeit his life if he could not prove the medicine man's guilt: the test was to see which of them could return the maiden's power of speech. Realizing the trap that had been set, the medicine man had no

alternative but to return her speech, and she told the story of his evil trick. The chief ordered him killed. The young warrior and his bride were loaded with gifts and sent on their way. The evil medicine man was bound with thongs, weighted with a heavy stone and thrown into the middle of Rock Lake."

Rockbound Lake (lake)

82 O/5 - Castle Mountain
17-27-14-W5
51°19′N 115°56′W
Approximately 29 km west north-west of Banff.

The name for this lake is descriptive in that it has cliffs bordering it on three sides, and the north side is a solid rock shore. The name Rockbound Lake was officially applied to this feature in 1959.

Rockslide Creek (creek)

83 E/6 - Twintree Lake
16-52-9-W6
53°28′N 119°15′W
Flows north-west into Smoky River, approximately 102 km west north-west of Jasper.

The origin of the name is unknown.

Rockwall Fen (marsh)

82 J/11 - Kananaskis Lakes
13-20-9-W5
50°41′40″N 115°06′50″W
Approximately 60 km west of Turner Valley.

Officially approved 8 November 1978, the name comes from the distinctive natural rock wall on the west side of the fen.

Rockwall Lake (lake)

82 J/11 - Kananaskis Lakes
12-20-9-W5
50°41′10″N 115°06′35″W
Approximately 60 km west of Turner Valley.

(see Rockwall Fen)

Rocky Creek (creek)

82 J/14 - Spray Lakes Reservoir
3-22-9-W5
50°51′N 115°10′W
Flows north-west into Kananaskis Lake,
approximately 75 km west south-west of
Calgary.

There is no origin information for this
locally well-established name. It is likely
descriptive.

Rocky Creek (creek)

82 O/14 - Limestone Mountain
11-35-11-W5
51°59′N 115°28′W
Flows north-east into Clearwater River,
approximately 90 km north of Banff.

The descriptive name for this creek was
officially approved 2 November 1956.

Rocky Creek (creek)

83 E/16 - Donald Flats
6-58-2-W6
53°59′N 118°18′W
Flows north into Little Smoky River,
approximately 55 km east north-east of
Grande Cache.

The "very descriptive" name of this creek
was proposed 17 April 1947 by B.L.
Anderson, a field topographer of the area.

Rocky Flat (flat)

82 J/8 - Stimson Creek
33-16-3-W5
50°23′30″N 114°21′35″W
Approximately 30 km south south-west of
Turner Valley.

This is a large, open flat adjacent to Pekisko
Creek. Dr. Brian Burke suggested this locally
well-established name and stated that the flat
is called Rocky Flat because of the boulders
strewn across it. The name was officially
approved 25 September 1979.

Rocky Pass (pass)

83 E/10 - Adams Lookout
10-54-7-W6
53°39′N 118°56′W
Approximately 27 km south-east of Grande
Cache.

The locally well-established name of this
pass is likely descriptive.

Rocky River (river)

83 F/4 - Miette
14-48-28-W5
53°08′N 117°59′W
Flows north-west into Athabasca River,
approximately 43 km south-west of Hinton.

The name of this river dates back to 1859,
when Sir James Hector (see Mount Hector)
mentioned it in his journal.

Roddy Creek (creek)

83 E/14 - Grande Cache
18-57-8-W6
53°55′N 119°10′W
Flows north-east into Smoky River,
approximately 5 km north north-west of
Grande Cache.

The name of this creek may be taken from
Rod McCrimmon, a young packer and guide
who was a member of the survey party for
the final line of the Grand Trunk grade.

Rodney Creek (creek)

83 F/7 - Erith
19-51-18-W5
53°25′N 116°39′W
Flows north-west into Embarras River,
approximately 23 km south-west of Edson.

The creek was named in 1910 after a
member of the Grand Trunk Pacific Railway
Survey Party. The name was officially
approved 3 December 1957. No other
information is available.

Romulus, Mount (mountain)

82 J/15 - Bragg Creek
24-21-8-W5
50°47′N 114°59′W
Approximately 60 km west south-west of
Calgary.

This mountain rises 2832 m in altitude.
Remus and Romulus were the children of
Rhea Silvia and the god Mars in Roman
mythology. Romulus is said to be the
founder of the settlement of Rome, later the
capital of a great empire. There are a few
legends as to the outcome of the lives of
these twin brothers, each equally fascinating.
Some historians believe that the story of the
two brothers, whichever one is correct,
symbolizes a duality, either racial or
political, in the early history and
development of Rome.

Ronde Creek (creek)

83 F/4 - Miette
12-49-28-W5
53°13′N 117°58′W
Flows south-east into Athabasca River,
approximately 32 km south-west of Hinton.

(see Roche Ronde)

Ronde, Roche (mountain)

83 E/1 - Snaring
12-49-1-W6
53°14′N 118°00′W
Approximately 40 km north of Jasper.

The descriptive name for this 2138 m mountain
is the French word for "round." The mountain
is mentioned in G.M. Grant's (see Grant
Pass) *Ocean to Ocean* (1873).

Rooster Creek (creek)

83 F/11 - Dalehurst
7-23-53-23-W5
53°36′N 117°16′W
Flows north into Athabasca River,
approximately 55 km west of Edson.

The locally well-established name of this creek was made official in 1946. The origin of the name is unknown.

Ropeo Gap (pass)

82 J/1 - Langford Creek
10-13-2-W5
50°04′N 114°12′W
Approximately 45 km west of Claresholm.

The name "Ropeo" is a distortion of the name, "Rupio," an early settler in Happy Valley. He was killed instantly when a falling log pinned him against a stump. His unfinished cabin remained standing several years after his death in the Gap which now bears his name. The neighbours, careless with pronunciation, soon had Mr. Rupio's name transformed.

Rose, Mount (mountain)

82 J/10 - Mount Rae
18-20-6-W5
50°41′N 114°49′W
Approximately 38 km west of Turner Valley.

The origin of the name of this 2515 m mountain is unknown.

Ross Creek (creek)

82 G/9 - Blairmore
14-8-2-W5
49°39′N 114°10′W
Flows east into Cow Creek, approximately 8 km north of Lundbreck.

John Ross was an early settler in the area. In 1886, he brought cattle from Colorado and settled near the creek, which is now named after him.

Ross Lake (lake)

82 G/9 - Blairmore
21-8-2-W5
49°39′N 114°13′W

Approximately 10 km north-west of Lundbreck.

(see Ross Creek)

Ross Cox Creek (creek)

83 D/9 - Amethyst Lakes
4-41-1-W6
52°31′N 118°05′W
Flows north into Whirlpool River, approximately 42 km south of Jasper.

(see Mount Ross Cox)

Ross Cox, Mount (mountain)

83 D/8 - Athabasca Pass
24-40-1-W6
52°28′N 118°01′W
Approximately 46 km south of Jasper.

Ross Cox was an Irishman who joined Astor's Pacific Fur Company. In 1813, the North West Company bought out the Pacific Fur Company, and Cox joined the new firm. He travelled into the interior of Canada from the Columbia River in 1817. He crossed Athabasca Pass, went on to Henry House, Fort William and finally Montreal, where he remained.

Rostrum Hill (hill)

83 D/16 - Jasper
13-44-3-W6
52°48′N 118°18′W
Approximately 17 km south-west of Jasper.

This 2256 m hill is said to resemble a pulpit or platform: it is a low peak, standing by itself between the entrances to two small valleys. (see Curia Mountain)

Route Creek (creek)

83 L/6 - Chicken Creek
30-61-8-W6
54°18′N 119°05′W

Flows south-east into Kakwa River, approximately 40 km north of Grande Cache.

The origin of the name is unknown.

Rowan Lake (lake)

83 F/7 - Erith
18-52-18-W5
53°29′N 116°38′W
Approximately 17 km south-west of Edson.

This lake was officially named in 1958, but its origin remains unknown.

Rowand, Mount (mountain)

83 E/1 - Snaring
48-3-W6
53°11′N 118°20′W
Approximately 37 km north north-west of Jasper.

This is one of 12 features named in 1954 for Alberta's 50th Jubilee. John Rowand (1787-1854), after whom this 2682 m mountain was named, was a fur-trader born in Montreal. He entered the service of the North West Company in 1804 and in 1808 was involved in building a fort on the site of what is now the city of Edmonton. He remained in the area most of the rest of his life. He became a chief trader with the Hudson's Bay Company in 1821, just after the union with the North West Company. He was promoted to Chief Factor at Edmonton in 1826. He married an Indian girl who saved his life when he was thrown from his horse. (see Mount Griesbach)

Rowe Creek (creek)

82 G/1 - Sage Creek
24-1-1-W5
49°03′N 114°01′W
Approximately 5 km west of Waterton Park.

The name Rowe Creek was officially applied to this feature 7 April 1960. (see Mount Rowe)

Rowe, Mount (mountain)

82 G/1 - Sage Creek
22-1-1-W5
49°03'N 114°03'W
Approximately 9 km west of Waterton Park, on the Alberta-B.C. boundary.

The name Mount Rowe was officially approved 16 March 1917. The 2452 m mountain was named after Lieutenant V.F. Rowe, R.E., who was the Surveying Officer for the International Boundary Commission of 1872-1876. (see Lower Rowe Lake)

Royalties (town)

82 J/9 - Turner Valley
28-18-2-W5
50°33'N 114°14'W
Approximately 14 km south of Turner Valley.

This hamlet was once a booming centre called "Little Chicago." The unpleasant odour in the area, due in part to the oil lying below the surface of the land, inspired the title. Another explanation for the name "Little Chicago" suggests that one of the general storekeepers named Rex Warman earned the nickname "Little Al Capone." Al Capone was a gangster who operated in Chicago during the 1920s. The "twin cities," as they were called, of "Little Chicago" and "Little New York," which was located four km south, owed their existence to the Turner Valley Royalties Number 1 Well. In 1914, a commercial deposit of crude oil was first tapped by this well; one of the first of its kind in Western Canada. The community of "Little Chicago" exploded into existence in 1936. Boom towns such as this grew quickly in response to a demand for housing and services for the workers and their families. Mr. R.A. Brown, owner of Turner Valley Royalties Ltd., set the wheels in motion for the growth of "Little Chicago." Its first store was located on the shores of a slough which American drillers dubbed "Lake Michigan." The small community was eventually

renamed Royalties by the government in honour of the first well to produce crude oil. The post office in the area was officially closed 7 June 1969, leaving one service station in operation. This is all that was left of the service facilities and businesses. The name Royalties is still in local usage. By 1970, the all-night poker games, the 350-500 people, and the rowdy nature of an oil boom town were nothing but memories for this little community.

Ruby Creek (creek)

82 H/4 - Waterton Lakes
6-2-30-W4
49°06'N 113°59'W
Approximately 40 km south of Pincher Creek.

Ruby Creek is the drainage of Ruby Lake and flows north-east into Blakiston Creek. The name Ruby Creek was officially approved 8 June 1971. (see Ruby Ridge)

Ruby Creek (creek)

83 C/15 - Cardinal River
11-45-21-W5
52°52'N 116°57'W
Flows north-west into Cardinal River, approximately 76 km east of Jasper.

(see Ruby Mountain)

Ruby Lake (lake)

82 G/1 - Sage Creek
27-3-3-W5
49°14'N 114°19'W
Approximately 10 km north-west of Waterton Park, on the north side of Ruby Ridge.

The name Ruby Lake was suggested 7 August 1958 by Mr. J. Elliott, a local forest officer of Beaver Mines. The official location for the lake was changed and approved 31 October 1973. (see Ruby Ridge)

Ruby Mountain (mountain)

83 C/14 - Mountain Park
10-44-22-W5
52°46'N 117°06'W
Approximately 66 km east south-east of Jasper.

The descriptive name for this 2900 m mountain refers to the red strata in the mountain.

Ruby Ridge (ridge)

82 H/4 - Waterton Lakes
6-2-30-W4
49°06'N 114°00'W
Approximately 45 km south of Pincher Creek.

Near the summit of this 2436 m ridge, beds of bright red shale can be seen, which suggest ruby.

Rudolph Peak (peak)

82 N/14 - Rostrum Peak
34-22-W5
51°57'48"N 117°05'19"W
Approximately 135 km north-west of Banff.

This peak on Mount Lyell, which is 3507 m in altitude, was named after Rudolph Aemmer, a member of the "Swiss Guide's Group." (see Christian Peak)

Rufus Creek (creek)

83 E/14 - Grande Cache
1-57-11-W6
53°53'N 119°29'W
Flows north-west into Sheep Creek, approximately 24 km west of Grande Cache.

This creek may possibly be named after Rufus Neighbor, a local homesteader.

Rum Punch Creek (creek)

82 O/13 - Scalp Creek
25-33-14-W5
51°51'40"N 115°50'40"W

Approximately 75 km north north-west of Banff.

In the early 1930s, Jimmy Simpson, Sr. took a party of freshmen from Yale University into the mountain area for a month-long fishing and hunting trip. At the fording point of this creek, the banks are steep and formed with loose shale. A horse carrying the trip's supply of liquor was led along the bank, despite warnings from the guide, and fell over backwards into the creek, smashing every bottle of rum in the pack. Members of the party quickly drank as much of the water downstream as they could and christened the creek Rum Punch Creek.

Rummel Creek (creek)

82 J/14 - Spray Lakes Reservoir
1-8-22-10-W5
50°51′00″N 115°20′40″W
Flows west into Smuts Creek, approximately 83 km west of Calgary.

This creek and the nearby lake were officially named 5 December 1984 in commemoration of Baroness Elizabeth (Lizzie) von Rummel (1897-1980), who came with her family to a ranch in the Millarville area in 1911. She and her two sisters successfully operated the ranch. Lizzie later left ranching and worked as a hostess and guide at Mount Assiniboine, ran a ski lodge, and bought and operated the Sunburst Lodge at Mount Assiniboine. In 1980, she received the Order of Canada in recognition of her achievements.

Rummel Lake (lake)

82 J/14 - Spray Lakes Reservoir
1-3-22-10-W5
50°50′05″N 115°17′45″W
Approximately 80 km west of Calgary.

(see Rummel Creek)

Rundell Creek (creek)

83 C/16 - Brown Creek
27-44-15-W5
52°49′N 116°06′W

Flows north-east into Blackstone River, approximately 3 km north of Nordegg.

The origin for the name of this creek is unknown.

Rundle, Mount (mountain)

82 O/3 - Canmore
25-11-W5
51°07′N 115°28′W
Approximately 5 km south-east of Banff.

The Reverend Robert Terrill Rundle (1811-1896) was the first missionary to enter this part of what is now the Bow Valley, in Alberta. This Methodist missionary was actively interested in the language and lifestyle of the Cree Indians. He spoke Cree and wrote hymns in Cree. The 2846 m mountain was named in 1858 by James Hector (see Mount Hector), and the name was officially approved 11 April 1911.

Running Rain Lake (lake)

82 J/10 - Mount Rae
7-18-7-W5
50°31′N 114°57′W
Approximately 49 km south-west of Turner Valley.

Tom Kaquitts (Running Rain) was a Stoney Indian born at Lundbreck in the Porcupine Hills in southern Alberta in 1893. He survived an influenza epidemic in the early part of the 20th century to become a significant force in tribal politics. He was a councillor of the Chiniki band, a chief of the Wesley band as well as honorary chief in the commemorative 100th anniversary of the signing of Treaty #7 in 1977. Running Rain died in 1983. This creek was officially named 5 May 1987 after this long-time member of the Stoney Indian tribe at Morley.

Russell Creek (creek)

83 C/14 - Mountain Park
22-45-22-W5
52°53′N 117°06′W

Flows north into Cardinal River, approximately 65 km east of Jasper.

(see Mount Russell)

Russell, Mount (mountain)

83 E/11 - Hardscrabble Creek
11-55-8-W6
53°44′N 119°05′W
Approximately 15 km south of Grande Cache.

The origin of the name is unknown.

Russell, Mount (mountain)

83 C/14 - Mountain Park
29-44-22-W5
52°49′ N 117°10′W
Approximately 61 km east of Jasper.

This 2819 m mountain and the nearby creek were named in 1925 after Thomas Russell, a local pioneer prospector.

Rutherford, Mount (mountain)

83 E/1 - Snaring
30-48-3-W6
53°10′N 118°25′W
Approximately 40 km north-west of Jasper.

This 2847 m mountain was named in 1954 after Alexander Cameron Rutherford, B.A., LL.D. (1858-1941). Born in Ontario, Rutherford came west in 1895 to set up a law firm in Strathcona (now part of south Edmonton). He served as Secretary-Treasurer of the town of Strathcona and of Strathcona School District. With the formation of the Province of Alberta in 1905, Mr. Rutherford, a prominent member of the Liberal Party, was called upon to assume the Premiership of the Province and to set up a cabinet. In an election that year, he was elected as Premier. He resigned in 1910. In 1928 Rutherford became the first Chancellor of the University of Alberta, and remained in that capacity until his death in 1941. This is one of 12 features named in 1954 for Alberta's 50th Jubilee. (see Mount Griesbach)

Saddle Mountain (mountain)

82 N/8 - Lake Louise
18-28-16-W5
51°24′N 116°13′W
Approximately 55 km north-west of Banff.

The name for this 2433 m mountain was applied by S.E.S. Allen (see Mount Allen) in 1894, to describe its shape, which resembles that of a saddle.

Saddle Peak (peak)

82 O/3 - Canmore
20-26-9-W5
51°14′N 115°13′W
Approximately 19 km north-east of Canmore.

(see Saddle Mountain)

Sage Mountain (mountain)

82 G/1 - Sage Creek
2-3-2-W5
49°11′N 114°10′W
Approximately 25 km north-west of Waterton Park.

This 2368 m mountain has been known locally as Sage Mountain since at least 1916. The precise origin is not known, but it may be named after the sagebrush found here and in other arid districts of the western United States and Canada.

Sage Pass (pass)

82 G/1 - Sage Creek
23-2-W5
49°11′N 114°10′W
Approximately 20 km north-west of Waterton Park, on the Alberta-B.C. boundary.

(see Sage Mountain)

Saghali Mountain (mountain)

83 E/6 -Twintree Lake
19-51-7-W6
53°25′N 119°01′W

Approximately 88 km west north-west of Jasper.

The name was adopted 1 May 1934. No information regarding the origin of the name is known.

St. Bride, Mount (mountain)

82 O/12 - Barrier Mountain
30-29-14-W5
51°30′N 115°57′W
Approximately 45 km north-west of Banff.

This 3315 m mountain is located directly south of Mount Douglas. St. Bride is the patron saint of the Douglas family, and it is after this saint that the mountain was named.

St. Eloi Mountain (mountain)

82 G/8 - Beaver Mines
4-4-W5
49°20′N 114°28′W
Approximately 42 km south-west of Pincher Creek, on the Alberta-B.C. boundary.

This mountain was officially named in 1917, after a village south of Ypres, Belgium. Canadian troops fought there during the First World War.

St. Julien Mountain (mountain)

82 N/14 - Rostrum Peak
23-33-22-W5
51°50′N 117°01′W
Approximately 125 km north-west of Banff, on the Alberta-B.C. boundary.

This mountain, which is 3091 m in altitude, was named in 1920 after the village 5 km north-east of Ypres. Canadian troops fought there during the First World War.

St. Nicholas Peak (peak)

82 N/9 - Hector Lake
3-31-18-W5
51°38′N 116°29′W

Approximately 85 km north-west of Banff.

There is a striking formation on the side of this peak that A.O. Wheeler thought resembled "Santa Claus," or "St. Nicholas." The name for this 2970 m mountain peak was officially approved 14 January 1909.

St. Piran Mountain (mountain)

82 N/8 - Lake Louise
25-28-17-W5
51°25′N 116°15′W
Approximately 60 km north-west of Banff.

This mountain was named by S.E.S. Allen (see Mount Allen) in 1894 and officially approved 1 March 1904. The 2649 m mountain takes its name from St. Piran, Liggan Bay, Cornwall, England, the birthplace of W.J. Astley, who was the first manager of the Lake Louise Chalet.

Salient Mountain (mountain)

83 E/2 - Resplendent Creek
8-47-5-W6
53°02′N 118°43′W
Approximately 45 km west north-west of Jasper.

The name for this 2810 m mountain was applied in 1923 by A.O. Wheeler. (see Wheeler Flats)

Salt Creek (creek)

82 J/8 - Stimson Creek
35-15-4-W5
50°18′N 114°28′W
Flows east into Pekisko Creek, approximately 40 km south-west of Turner Valley.

S

The name for this creek was officially
approved 12 December 1939. The origin of
the name is unknown.

Salter Creek (creek)
82 J/7 - Mount Head
26-15-5-W5
50°17′N 114°35′W
Flows west into Cataract Creek,
approximately 48 km south-west of Turner
Valley.

The name Salter Creek was officially
approved 8 December 1943. Dr. G.M.
Dawson (see Tombstone Mountain) named
this creek in 1884 after his packer of Scottish
and Indian background who resided on the
Stoney Indian Reserve in Morley, Alberta.

Sam Ridge (ridge)
82 J/9 - Turner Valley
26-18-4-W5
50°33′N 114°27′W
Approximately 17 km south-west of Turner
Valley.

The origin for this name, which was
officially approved 8 December 1943, is
unknown.

Sam Howe Crossing (river crossing)
82 J/9 - Turner Valley
7-19-19-3-W5
50°37′20″N 114°24′15″W
Approximately 10 km west south-west of
Turner Valley.

The name for this crossing was officially
approved 5 May 1983. (see Sam Howe Flats)

Sam Howe Flats (flat)
82 J/9 - Turner Valley
30-19-3-W5
50°37′50″N 114°23′30″W

Approximately 9 km west south-west of
Turner Valley.

This flat was named for Sam Howe, the first
homesteader in the area. In 1888, he took
land on Sheep Creek. The name was
officially approved 5 May 1983.

Samson Lake (lake)
83 C/7 - Job Creek
7-39-20-W5
52°20′25″N 116°51′25″W
Approximately 100 km south-east of Jasper.

(see Leah Lake)

Samson Narrows (narrows)
83 C/12 - Athabasca Falls
52°39′N 117°30′W
Approximately 50 km south-east of Jasper.

Mary Schäffer (see Maligne Lake) named this
feature circa 1907 after Stoney Chief Samson,
who helped her on her journey. (see Samson
Peak)

Samson Peak (peak)
83 C/12 - Athabasca Falls
2-43-25-W5
52°41′N 117°30′W
Approximately 44 km south-east of Jasper.

This 3081 m mountain peak was named by
Mrs. Mary Schäffer in 1911 after a Stoney
Indian Chief, Samson Beaver, who had
drawn a map that enabled her to find
Maligne Lake. When she and her friend
Molly Adams saw Maligne Lake in 1908,
they were the first white women to do so.

Sandbar Creek (creek)
83 L/12 - Nose Creek
5-66-11-W6
54°41′N 119°37′W
Flows south into Nose Creek, approximately
74 km south-west of Grande Prairie.

The precise origin of the name is unknown
but it is likely a descriptive name.

*Sandstone (former locality)
82 J/16 - Priddis
2-21-1-W5
50°45′N 114°02′W
Approximately 20 km south of Calgary.

This Canadian Pacific Railway station was
named in 1907 after the sandstone quarries
in the area. The designation "station" was,
rescinded and changed to a locality in 1970.

Sandstone Creek (creek)
83 F/11 - Dalehurst
19-52-23-W5
53°30′N 117°23′W
Flows north into Athabasca River,
approximately 64 km west of Edson.

There is an abundance of sedimentary rock
composed of compressed sand in the cliffs,
ridges and ledges along this creek. The
descriptive name has been in use since 1875,
and may be found in progress reports of
surveys done at that time.

Sanson Peak (peak)
82 O/4 - Banff
25-12-W5
51°09′N 115°35′W
Approximately 3 km south south-west of Banff.

This feature is 2256 m in altitude and is the
highest peak of Sulphur Mountain. The
name was officially adopted 8 November
1848 after Norman Sanson (1862-1949) who
was the official Meteorologist at Banff for
many years. In 1896 he became the Curator
at the Banff Museum and throughout his life
retained a keen interest in biology.

Saracen Head (mountain)
83 C/10 - George Creek
2-43-21-W5
52°40′N 116°56′W

Approximately 63 km west of Nordegg.

A Saracen was a Muslim nomad of the Syrian and Arabian desert at the time of the Christian Crusades during the Middle Ages. This mountain has a fancied resemblance to a turbaned head, which may have inspired its naming.

Sarbach Creek (creek)

82 N/15 - Mistaya Lake
14-34-20-W5
51°55′N 116°44′W
Flows north-east into Mistaya River, approximately 115 km north-west of Banff.

(see Mount Sarbach)

Sarbach, Mount (mountain)

82 N/15 - Mistaya Lake
3-34-20-W5
51°54′N 116°46′W
Approximately 115 km north-west of Banff.

This 3155 m mountain was named by J.N. Collie in 1897, after Peter Sarbach, the first Swiss guide in Canada. With G.P. Baker and J.N. Collie, he made a first ascent in 1897 on this mountain. Other first ascents included Mount Victoria, Mount Lefroy and Mount Gordon.

Sarcee Butte (butte)

82 O/2 - Jumpingpound Creek
7-24-4-W5
51°01′N 114°32′W
Approximately 25 km west of Calgary.

The locally well-established name for this butte can be traced back to at least 1882. It is named after the Sarcee of Sarsic, an Indian tribe of the Athapascan family. The name was officially approved 12 December 1939.

Sarrail Creek (creek)

82 J/11 - Kananaskis Lakes
14-19-9-W5
50°37′N 115°08′W

Approximately 60 km west south-west of Turner Valley.

(see Mount Sarrail)

Sarrail, Mount (mountain)

82 J/11 - Kananaskis Lakes
3-19-9-W5
50°35′N 115°10′W
Approximately 65 km west south-west of Turner Valley.

This mountain, which is 3170 m in altitude, was named in 1918 after General Maurice Paul Emmanuel Sarrail (1856-1929), a commander of the French 3rd Army in the First World War. After the war, he devoted himself to politics and stood, unsuccessfully, for election as Radical Deputy for Paris. Syria was his next and final appointment.

Saskatchewan Glacier (glacier)

83 C/3 - Columbia Icefield
35-36-23-W5
52°09′N 117°10′W
Approximately 108 km south-east of Jasper.

(see Saskatchewan Mountain)

Saskatchewan Mountain (mountain)

83 C/3 - Columbia Icefield
22-36-22-W5
52°06′N 117°07′W
Approximately 110 km south-east of Jasper.

This mountain, which stands 3289 m in altitude, was named by J.N. Collie (see Collie Creek) in 1898. The mountain is near the headwaters of the North Saskatchewan River, the name of which is taken from the Cree Indian name "kis-is-ska-tche-wan," which means "swift current."

Saskatchewan River Crossing (locality)

82 N/15 - Mistaya Lake
51°58′30″N 116°44′40″W
Approximately 120 km north-west of Banff.

This locality is named for its proximity to the historic Saskatchewan River Crossing, approximately 2 km east on the North Saskatchewan River. It was at this point, where Bear Creek enters the North Saskatchewan, that Mary T.S. Schäffer described one of the best fords on the watercourse for individuals travelling further north. This locality is made up primarily of a commercial development called "The Crossing" which has occupied the location at the junction of Highways 11 and 93 for several years. The development used to occupy a position on the south side of the North Saskatchewan River where the Saskatchewan River Crossing Warden Station is located. The name was made official for use on maps 3 January 1985.

Saskatchewan River Crossing, n.d.

Saskatoon Mountain (mountain)

82 G/9 - Blairmore
21-8-4-W5
49°40′N 114°30′W
Approximately 7 km north-west of Blairmore.

Saskatoon Mountain is 1870 m in altitude. The origin of the name is unknown.

Sassenach, Mount (mountain)

83 E/1 - Snaring
20-49-3-W6
53°14'N 118°24'W
Approximately 46 km north north-west of Jasper.

This is a term that was used by Gaelic inhabitants of Great Britain and Ireland to refer to the Saxons or the English. The connection between this term and the mountain, however, is not clear. The name was officially approved 1 May 1934.

Saturday Night Lake (lake)

83 D/16 - Jasper
23-45-2-W6
52°54'N 118°11'W
Approximately 6 km north-west of Jasper.

The origin of the name is unknown.

Saunders Creek (creek)

83 B/5 - Saunders
19-40-12-W5
52°27'N 115°43'W
Flows south into North Saskatchewan River, approximately 24 km east of Nordegg.

This creek was named after B.J. Saunders, D.L.S., the Commissioner for Ontario on the Ontario-Manitoba Boundary Survey in 1897. He surveyed the 11th Baseline in 1908.

Saurian Mountain (mountain)

83 E/5 - Holmes River
31-51-11-W6
53°27'N 119°36'W
Approximately 119 km north-west of Jasper.

The summit of this 3016 m mountain resembles the back of a prehistoric monster of the lizard species, suborder "sauria." It was officially named in 1925.

Savanna (railway point)

82 G/10 - Crowsnest
11-8-5-W5
49°38'N 114°34'W
Approximately 5 km west of Coleman.

This station was named 20 March 1962 after the nearby creek. (see Savanna Creek)

Savanna Creek (creek)

82 J/1 - Langford Creek
2-14-4-W5
50°09'N 114°27'W
Flows east into Livingstone River, approximately 65 km west north-west of Claresholm.

The origin of the name is unknown.

Sawback Creek (creek)

82 O/5 - Castle Mountain
22-28-13-W5
51°23'N 115°46'W
Flows north into Cascade River, approximately 29 km north north-west of Banff.

(see Sawback Range)

Sawback Lake (lake)

82 O/5 - Castle Mountain
33-27-13-W5
51°21'N 115°46'W
Approximately 23 km north-west of Banff.

(see Sawback Range)

Sawback Range (range)

82 O/5 - Castle Mountain
28-14-W5
51°21'N 115°50'W
Approximately 28 km west north-west of Banff.

The descriptive name was given to this feature by Sir James Hector (see Mount Hector) in 1858. There are vertical beds of

light grey limestone which form the serrated peaks of the range.

Sawtooth Creek (creek)

82 O/14 - Limestone Mountain
19-33-9-W5
51°50'N 115°11'W
Flows south-east into Willson Creek, approximately 75 km north north-east of Banff.

(see Sawtooth Ridge)

Sawtooth Mountain (mountain)

83 C/11 - Southesk Lake
7-41-21-W5
52°30'N 117°01'W
Approximately 83 km south-east of Jasper.

This 2933 m mountain was officially given its presumably descriptive name 5 March 1935.

Sawtooth Ridge (ridge)

82 O/14 - Limestone Mountain
33-9-W5
51°51'N 115°13'W
Approximately 75 km north north-east of Banff.

The descriptive name for this ridge was officially approved 25 November 1941.

Scalp Creek (creek)

82 O/12 - Barrier Mountain
7-32-11-W5
51°44'N 115°43'W
Flows south-east into Red Deer River, approximately 60 km north of Banff.

The origin of the name is unknown.

Scarab Lake (lake)

82 O/4 - Banff
5-25-14-W5
51°06'N 115°55'W

Approximately 26 km south-west of Banff.

Located in the area locally known as the "Egyptian Connection," the outline of this lake resembles a beetle. Since the sacred beetle of ancient Egypt was called the "scarab," the name for this lake is appropriately descriptive.

Scarp Mountain (mountain)

83 D/9 - Amethyst Lakes
22-42-3-W6
52°38'N 118°21'W
Approximately 33 km south-west of Jasper.

A "scarp" is a steep slope or cliff. This steep sloped mountain was given its descriptive name in 1922 by A.O. Wheeler. (see Wheeler Flats)

Scarpe Creek (creek)

82 G/1 - Sage Creek
5-3-W5
49°24'N 114°20'W
Flows north-east into Castle River, off the north face of La Coulotte Ridge, approximately 35 km north-west of Waterton Park.

(see Scarpe Mountain)

Scarpe Mountain (mountain)

82 G/1 - Sage Creek
18-3-3-W5
49°13'N 114°24'W
Approximately 40 km north-west of Waterton Park, on the Alberta-B.C. boundary.

The river which flows through Arras, France is known as the Scarpe. It was on this river, in 1917 and 1918 that Canadian troops fought during the First World War. This 2591 m mountain was named for this battlefield.

Scott Creek (creek)

83 D/8 - Athabasca Pass
32-40-1-W6
52°29'N 118°06'W

Flows north-west into Whirlpool River, approximately 44 km south of Jasper.

(see Mount Scott)

Scott Glacier (glacier)

83 D/8 - Athabasca Pass
40-1-W6
52°26'N 118°05'W
Approximately 49 km south of Jasper.

(see Mount Scott)

Scott Lake (lake)

82 O/2 - Jumpingpound Creek
15-25-6-W5
51°08'N 114°46'W
Approximately 49 km west north-west of Calgary.

The name for this lake was officially approved 2 December 1956 after Leeson's ranching partner, Mr. Scott.

Scott, Mount (mountain)

83 D/8 - Athabasca Pass
15-40-1-W6
52°27'N 118°03'W
Approximately 47 km south of Jasper, on the Alberta-B.C. boundary.

This 3300 m mountain was named after Captain Robert Falcon Scott (1868-1913) of the British Navy. He was the commander of two Antarctic expeditions, one in 1900-1904 and the other in 1910-1912. G.E. Howard named this feature, as well as nearby Scott Creek and Scott Glacier. With a small party of five members, Scott reached the South Pole 18 January 1912, but found a tent there left by Norwegian explorer, Roald Amundsen (1872-1928), who had reached the pole a month before. On the long return from the pole, all the members of the the party died due to lack of food and fuel, and very cold weather.

Scovil Creek (creek)

83 F/5 - Entrance
14-50-27-W5
53°19'N 117°51'W
Flows east into Brule Lake, approximately 18 km west south-west of Hinton.

The name for this creek is taken from the name of a prospector who discovered coal here. He was a representative of the Geological Survey in 1911. No other information is available.

Screwdriver Creek (creek)

82 G/8 - Beaver Mines
23-6-2-W5
49°29'N 114°10'W
Flows north-east into Castle River, approximately 14 km west of Pincher Creek.

No origin information is available for the name of this creek.

Scrimger, Mount (mountain)

82 J/2 - Fording River
15-6-W5
50°15'N 114°46'W
Approximately 60 km south-west of Turner Valley.

Captain Francis A.C. Scrimger, V.C., M.D. (1880-1937), of the 14th Battalion, C.E.F., was a great Canadian hero. On 25 April 1915, when in charge of an advanced dressing station near Ypres, he directed, under heavy fire, the removal of the wounded and personally carried a wounded officer to safety. He displayed a great devotion to the wounded, and it is after him that Mount Scrimger, which stands 2755 m in altitude, was officially named in 1918.

Seabolt Creek (creek)

83 F/5 - Entrance
11-50-26-W5
53°18'N 117°42'W

Flows north-west into Maskuta Creek, approximately 12 km south south-west of Hinton.

This creek and the other features nearby were named after Frank Seabolt, an entrepreneur from Georgia, U.S.A., who formed a partnership with H. King and R.W. Jones. They opened the Hinton Collieries in 1928. Seabolt had been homesteading in the area since 1916, and was one of the first ranchers to build in Hinton.

Seabolt Ridge (ridge)

83 F/5 - Entrance
50-25-W5
53°17′N 117°37′W
Approximately 12 km south of Hinton.

The name for this ridge was officially approved in 1945. (see Seabolt Creek)

Secord, Mount (mountain)

82 G/15 - Tornado Mountain
30-10-5-W5
49°51′20″N 114°39′40″W
Located at the head of North Racehorse Creek, approximately 25 km north of the Crowsnest Pass.

Mount Secord was named in 1981 by the Governments of Alberta and British Columbia, and the Canadian Permanent Committee on Geographical Names to commemorate the contribution made by Richard Secord (1860-1935), a prominent Edmonton pioneer, to the development of western Canada.

Seebe (hamlet)

82 O/3 - Canmore
33-24-8-W5
51°06′N 115°04′W
Approximately 21 km east of Canmore.

This hamlet is located near the Bow River. Its name, Seebe, is a derivation of the Cree Indian word "si-pi," meaning creek.

Seep Creek (creek)

83 E/8 - Rock Lake
16-52-3-W6
53°29′N 118°23′W
Flows south into Wildhay River, approximately 71 km north north-west of Jasper.

The name for this creek was adopted 15 April 1946. The origin of the name is unknown.

Seepage Creek (creek)

83 C/9 - Wapiabi Creek
16-43-17-W5
52°42′N 116°26′W
Flows south-east into Chungo Fork, approximately 34 km north-west of Nordegg.

The origin of the name is unknown.

Seismograph Creek (creek)

82 O/11 - Burnt Timber Creek
16-31-9-W5
51°39′N 115°13′W
Flows north-east into Red Deer River, approximately 60 km north-east of Banff.

The origin of the name is unknown.

*Sentinel (former locality)

82 G/10 - Crowsnest
11-8-5-W5
49°38′N 114°35′W
Approximately 6 km west of Coleman.

This locality was named in 1905, possibly because of its isolated position as a sentinel. The area was originally called "Sentry Siding."

Sentinel Mountain (mountain)

83 C/2 - Cline River
34-36-18-W5
52°08′N 116°29′W
Approximately 110 km west south-west of Rocky Mountain House.

This mountain, which is 2865 m in altitude, is bent into a fold, shaped like an "S." It can be seen from all the valleys looking into the plains, and because of its isolated position, A.P. Coleman (see Mount Coleman) named it Sentinel Mountain.

Sentinel Pass (pass)

82 J/8 - Stimson Creek
15-4-W5
50°15′N 114°29′W
Immediately west of Sentinel Peak, approximately 50 km south-west of Turner Valley.

(see Sentinel Peak)

Sentinel Pass (pass)

82 N/8 - Lake Louise
29-27-16-W5
51°20′N 116°13′W
Approximately 50 km west north-west of Banff.

(see Sentinel Mountain)

Sentinel Peak (peak)

82 J/8 - Stimson Creek
9-15-4-W5
50°15′N 114°29′W
Approximately 50 km south-west of Turner Valley.

This 2340 m peak stands isolated, as a sentinel to the Highwood area. The descriptive name was officially approved 12 December 1939.

Sentry Mountain (mountain)

82 G/10 - Crowsnest
5-8-5-W5
49°36'N 114°38'W
Approximately 10 km west south-west of Coleman.

This 2435 m feature was originally called "Sentinel Mountain" but was officially renamed Sentry Mountain 5 July 1915 to avoid confusion with another "Sentinel Mountain" located nearby.

Serenity Mountain (mountain)

83 D/8 - Athabasca Pass
40-1-W6
52°24'N 118°01'W
Approximately 53 km south of Jasper.

The descriptive name for this 3216 m mountain was approved in 1922.

Seven Mile Creek (creek)

82 O/14 - Limestone Mountain
10-35-10-W5
51°59'N 115°20'W
Flows south-east into Clearwater River, approximately 90 km north of Banff.

The descriptive name for this creek was officially approved 17 December 1941.

Seven Mile Flat (flat)

83 B/3 - Tay River
15-35-10-W5
52°00'N 115°21'W
Approximately 45 km south-west of Caroline.

The name for this flat, which was officially adopted in 1941, is probably descriptive of the length of the flat. The precise origin of the name is unknown.

*denotes rescinded name or former locality.

Seven Sisters Mountain (mountain)

82 G/10 - Crowsnest
11-9-5-W5
49°43'20"N 114°34'40"W
Approximately 11 km north north-west of Coleman.

Seven Sisters Mountain is the well-established local name for this feature, which is located north of Crowsnest Mountain. The name is descriptive of the mountain's seven jagged peaks and was officially adopted 15 February 1978. This 2591 m mountain was known by John Palliser (see Palliser Range) as "The Steeples."

Shadow Lake (lake)

82 O/4 - Banff
25-25-15-W5
51°09'N 115°58'W
Approximately 27 km west south-west of Banff.

The name for this lake, which is a rich blue in colour, is descriptive in that the lake lies in the "shadow" of Mount Ball.

***Shale Banks** (former locality)

83 E/1 - Snaring
22-49-2-W6
53°14'N 118°12'W
Approximately 40 km north of Jasper.

The origin for the name of this locality is unknown.

Shale Creek (creek)

82 G/16 - Maycroft
9-12-4-W5
49°59'N 114°29'W
Flows south into Oldman River, approximately 40 km north of Blairmore.

The origin of the name is unknown.

Shale Pass (pass)

83 E/12 - Pauline Creek
20-54-12-W6
53°36'N 119°44'W

Approximately 50 km south-west of Grande Cache, on the Alberta-B.C. boundary.

(see Big Shale Hill)

Shand Creek (creek)

83 E/15 - Pierre Greys Lakes
22-56-4-W6
53°51'N 118°31'W
Flows north into Lone Teepee Creek, approximately 40 km east of Grande Cache.

This creek was named in 1947 in honour of James Shand-Harvey (1880-1968), a pioneer who arrived at Edmonton in 1905. He was a homesteader, a trapper, and a forest ranger. He surveyed baselines and was a packer for railways and mountaineers. One of Shand-Harvey's first jobs was with the surveying firm of Driscoll and Knight. He led a varied life; he was born in Mauritius, educated at Eton College, and he settled in Canada beyond the Muskeg River in the Hinton area. Much of the information for J.G. MacGregor's book, *Pack Saddles to Tête Jaune Cache* (1962) was furnished by James Shand-Harvey. (see Chetang Ridge)

Shankland Creek (creek)

83 C/8 - Nordegg
29-40-17-W5
52°28'N 116°25'W
Flows north into Wapiabi Creek, approximately 23 km west of Nordegg.

This creek was named in 1932 after William Shankland, who was a forest ranger in the district for over 20 years.

Shanks Creek (creek)

83 C/9 - Wapiabi Creek
13-42-17-W5
52°37'N 116°19'W
Flows north-east into Lookout Creek, approximately 26 km north-west of Nordegg.

(see Shanks Lake)

Shanks Lake (lake)

83 C/8 - Nordegg
24-40-16-W5
52°27'N 116°10'W
Approximately 35 km north-west of
Nordegg.

This lake was named after John Shanks, who
was general manager of Brazeau Colleries
Ltd. from the time it opened in 1914 until
his retirement in 1945. His brother Dave was
mine manager and his son Dave later owned
and operated a garage in Nordegg. Dave
Shanks once ran a sawmill west of Nordegg.
The name was officially approved 30 March
1977.

Shark Lake (lake)

82 J/14 - Spray Lakes Reservoir
10,11-1-22-11-W5
50°52'N 115°26'W
Approximately 100 km south-west of
Calgary.

The Department of Fish and Wildlife
applied to have this lake officially named
after having stocked it in 1985. The lake is
close to Shark Mountain, and takes its name
from its location. (see Mount Shark)

Shark, Mount (mountain)

82 J/14 - Spray Lakes Reservoir
36-21-11-W5
50°50'N 115°25'W
Approximately 90 km west south-west of
Calgary.

This mountain, which is 2786 m in altitude,
was named in 1922 after a destroyer which
was sunk in the Battle of Jutland, during the
First World War.

Sharples Creek (creek)

82 G/16 - Maycroft
31-10-1-W5
49°52'N 114°07'W

Flows west into Callum Creek,
approximately 35 km north-east of
Blairmore.

Charles Sharples was the head manager of
the 20,235 hectare Winder Ranch from
1882-1892. He was the first Justice of the
Peace and Brand Inspector in the district.
This creek may be named after him.

Sheba Creek (creek)

83 F/5 - Entrance
8-51-27-W5
53°23'N 117°56'W
Flows east into West Solomon Creek,
approximately 22 km west of Hinton.

The origin of the name is unknown.

Sheep Creek (creek)

82 O/11 - Burnt Timber Creek
36-30-11-W5
51°36'N 115°25'W
Flows south-east into Panther River,
approximately 50 km north of Banff.

This creek is so named because it is a
favourite haunt of Mountain or Bighorn
Sheep.

Sheep Creek (creek)

83 L/3 - Copton Creek
5-59-7-W6
54°04'N 119°00'W
Flows south-east into Smoky River,
approximately 25 km north north-west of
Grande Cache.

The locally well-established name of this
creek has been in use since 1901. It is
presumably named for the Mountain or
Bighorn Sheep prevalent in the area.

Sheep River Falls (falls)

82 J/10 - Mount Rae
24-19-6-W5
50°37'00"N 114°42'10"W

Approximately 31 km west of Turner Valley.

The Sheep River was so named because
mountain sheep are numerous in this area.
The falls take their name from the river.
Officially approved 20 October 1983, the
name has been locally well-established for
this feature. The falls are also referred to
locally as "Sheep Falls" and "The Falls."

Sheol Mountain (mountain)

82 N/8 - Lake Louise
7-28-16-W5
51°23'N 116°13'W
Approximately 50 km west north-west of
Banff.

This 2776 m mountain was so named in
1894 by S.E.S. Allen (see Mount Allen)
because of the gloomy appearance of the
valley at the base and in contrast to the
nearby Paradise Valley. "Sheol" is the
underworld of the dead in Hebrew.

Sheppard Creek (creek)

82 J/8 - Stimson Creek
8-16-2-W5
50°20'N 114°14'W
Flows east into Stimson Creek,
approximately 35 km south of Turner Valley.

Henry Norman Sheppard was born in
England in 1861. He came to Canada in
1887 and built a home on this creek, which
was formerly known as "Hay Creek." In
1898, Henry N. Sheppard was appointed
Justice of the Peace for the North West
Territories.

Sherman Creek (creek)

83 L/12 - Nose Creek
3-65-13-W6
54°36'N 119°53'W
Flows east into Sulphur Creek, approximately
92 km south-west of Grande Prairie.

The name was suggested by J.V. Butterworth,
a field topographer of the area in 1946, after

Art Sherman, an early forest ranger and trapper. He had a cabin in the area which is now called Sherman Meadows. The other features in the area also take their names from Mr. Sherman.

Sherman Lake (lake)

83 L/12 - Nose Creek
20-64-13-W6
54°33′N 119°56′W
Approximately 110 km south-west of Grande Prairie.

(see Sherman Creek)

Sherman Meadows (meadows)

83 L/5 - Two Lakes
22-61-13-W6
54°18′N 119°52′W
Approximately 60 km north-west of Grande Cache.

(see Sherman Creek)

Shetler Creek (creek)

83 L/5 - Two Lakes
2-63-11-W6
54°25′N 119°33′W
Flows west into Nose Creek, approximately 65 km north north-west of Grande Cache.

(see Pierre Creek)

Shingle Flats (flat)

82 O/12 - Barrier Mountain
51°33′25″N 115°59′50″W
Approximately 50 km north-west of Banff.

There is no origin information available for the name of this flat, which was officially approved 21 January 1985.

Shoe Leather Creek (creek)

83 C/2 - Cline River
25-36-19-W5
52°07′N 116°35′W

Flows east into Entry Creek, approximately 115 km west of Rocky Mountain House.

The name originally suggested for this stream was "Sole Creek" because D.J. Link and other members of the Alpine Club of Canada were hiking along this creek with boot soles which were falling apart. The name Shoe Leather Creek was chosen as an alternative because the name was more picturesque. It was officially approved in 1959.

Short Creek (creek)

83 E/6 - Twintree Lake
34-51-9-W6
53°27′N 119°14′W
Flows east into Smoky River, approximately 100 km north-west of Jasper.

The origin of the name is unknown.

Shoulder Creek (creek)

82 J/15 - Bragg Creek
21-21-7-W5
50°48′N 114°55′W
Flows south-east into Little Elbow River, approximately 55 km west south-west of Calgary.

This creek drains one of the slopes of what is locally known as "Shoulder Mountain," because of its likeness in shape to a shoulder. The stream was officially named 6 October 1949.

Shovel Pass (pass)

83 C/13 - Medicine Lake
15-44-27-W5
52°45′N 117°50′W
Approximately 20 km east of Jasper.

Shovel Pass gets its name from an incident which occurred in 1911. The Otto brothers, pioneer outfitters in the Jasper area, were attempting to transport a boat to Maligne Lake. The trail was rough and the route difficult and at the time, the snow was too

deep to traverse. The brothers fashioned shovels out of timber and scratched a trail through the snowbanks. One of the shovels was left as a reminder of the ingenious haul and was later found by Mrs. Mary T.S. Schäffer, who suggested the name Shovel Pass. (see Big Shovel Pass and Little Shovel Pass)

Shovel Pass, n.d.

Shunda Creek (creek)

83 B/5 - Saunders
22-40-13-W5
52°28′N 115°47′W
Flows south-east into North Saskatchewan River, approximately 65 km west north-west of Rocky Mountain House.

"Shunda" is the Stoney Indian word for "mire." The name, applied in 1912, is descriptive of the ford where the trail crosses the creek.

■ Shunda Lake (lake)

83 C/8 - Nordegg
19-40-15-W5
52°27′N 116°09′W
Approximately 10 km north-west of Nordegg.

(see Shunda Creek)

Shunda Mountain (mountain)

83 C/9 - Wapiabi Creek
41-15-W5
52°32′N 116°08′W
Approximately 8 km north north-west of
Nordegg.

(see Shunda Creek)

Sibbald Creek (creek)

82 O/2 - Jumpingpound Creek
13-24-7-W5
51°02′N 114°51′W
Flows east into Jumpingpound Creek,
approximately 45 km west of Calgary.

Frank Sibbald was the son of Andrew Sibbald
(1833-1933), who settled near Morley in 1875.
This ranching family lived on the creek and
gave it its present name. The name for this
creek was made official 12 December 1939.

Sibbald Flat (flat)

82 O/2 - Jumpingpound Creek
13-24-7-W5
51°02′N 114°51′W
Approximately 45 km west of Calgary.

(see Sibbald Creek)

Sibbald Meadows Pond (reservoir)

82 O/2 - Jumpingpound Creek
20-24-7-W5
51°03′20″N 114°56′55″W
Approximately 50 km west of Calgary.

This reservoir was so named 7 February
1983 because of its proximity to Sibbald
Creek.

Side Creek (creek)

83 B/4 - Cripple Creek
29-37-14-W5
52°12′N 115°58′W
Flows south-east into Cripple Creek,
approximately 31 km south-east of Nordegg.

The origin for the locally well-established
name of this creek is unknown. The
name was officially approved 29 September
1976.

Siffleur Falls (falls)

83 C/1 - Whiterabbit Creek
21-35-17-W5
52°01′30″N 116°22′10″W
Approximately 110 km north-west of Banff.

The name for these falls located on Siffleur
River was approved 22 June 1979. (see
Siffleur Mountain)

Siffleur Mountain (mountain)

82 N/16 - Siffleur River
34-17-W5
51°56′N 116°24′W
Approximately 100 km north-west of Banff.

This mountain, which is 3129 m in altitude,
was named by Sir James Hector (see Mount
Hector) after the "siffleur," also known as
the whistling marmot, one species of
marmot. This small mammal, no bigger
than the common badger of the plains,
inhabits crevices in the rock and possesses a
very shrill whistle. The name was officially
approved 6 February 1912. (see Marmot
Mountain)

■ **Siffleur River** (river)

83 C/1 - Whiterabbit Creek
5-36-17-W5
52°04′30″N 116°24′10″W
Flows north into North Saskatchewan River,
approximately 115 km north-west of Banff.

(see Siffleur Mountain)

Siffleur Wilderness Area (wilderness area)

82 N/15 - Mistaya Lake
1-34-19-W5
Approximately 85 km north-west of Banff.

(see Siffleur Mountain)

Signal Mountain (mountain)

83 C/13 - Medicine Lake
10-45-28-W5
52°52′N 117°58′W
Approximately 9 km east of Jasper.

This 2255 m mountain was given its name in
1916 by M.P. Bridgland (see Mount
Bridgland). This was a site used by forest
rangers to watch for forest fires. A telephone
was located near the summit of this
mountain in order to signal any reports of
fire in the area.

Silverhorn Creek (creek)

82 N/15 - Mistaya Lake
10-33-19-W5
51°49′N 116°36′W
Flows west into Mistaya River,
approximately 100 km north-west of Banff.

(see Silverhorn Mountain)

Silverhorn Mountain (mountain)

82 N/15 - Mistaya Lake
29-32-18-W5
51°46′N 116°30′W
Approximately 90 km north-west of
Banff.

The name for this 2911 m mountain is
descriptive of its snow-covered summit.

Silverton Creek (creek)

82 O/5 - Castle Mountain
26-14-W5
51°16′N 115°55′W
Flows south into Bow River, approximately
26 km west north-west of Banff.

The name for this creek comes from an old
townsite plan which was never registered.
"Silver City" was originally to have been
"Silverton," and this name was adopted for
the creek in 1959 for its historical
significance.

Silvester Creek (creek)

82 J/15 - Bragg Creek
14-22-6-W5
50°52′N 114°43′W
Flows north into Elbow River,
approximately 40 km west of Calgary.

George Silvester (1884-1960) came to Canada
in 1903 from England to homestead near this
creek, which now bears his name. The
creek's name was officially approved 6
October 1949.

Simla, Mount (mountain)

83 E/7 - Blue Creek
3-50-5-W5
53°17′N 118°40′W
Approximately 59 km north-west of Jasper.

The origin of the name of this 2786 m
mountain is unknown. The Rajah and
The Ranee are located nearby. There is a city
and a district by the name of Simla in
mountainous north-western India, which
this feature could resemble, or take its name
from. The name was officially applied to this
feature 1 May 1934.

Simon Creek (creek)

83 D/9 - Amethyst Lakes
34-41-1-W6
52°35′N 118°03′W
Flows east into Whirlpool River,
approximately 33 km south of Jasper.

(see Mount Fraser)

Simon Glacier (glacier)

83 D/9 - Amethyst Lakes
42-3-W6
52°38′N 118°19′W
Approximately 32 km south-west of
Jasper.

(see Mount Fraser)

Simon Peak (peak)

83 D/9 - Amethyst Lakes
35-42-8-W6
52°39′N 118°20′W
Approximately 29 km south-west of Jasper.

Simon Peak is one of the peaks of Mount
Fraser, and it measures 3322 m in altitude.
(see Mount Fraser)

Simpson Pass (pass)

82 O/4 - Banff
24-14-W5
51°05′N 115°50′W
Approximately 22 km south-west of Banff,
on the Alberta-B.C. boundary.

Sir George Simpson (1792-1860) was the
Governor-in-Chief of Rupert's Land for
many years and was also the Governor of
the Hudson's Bay Company in Canada for
over forty years. He crossed over this pass,
which is 2120 m in altitude in 1841, while
on his journey around the world. It is said
that Sir George carved his name and the date
on a tree at the summit, but the tree has
since been cut down.

Sir Douglas, Mount (mountain)

82 J/11 - Kananaskis Lakes
28-20-10-W5
50°43′N 115°21′W
Approximately 75 km west of Turner Valley.

Field Marshal Sir Douglas Haig, K.T.,
G.C.B., Commander-in-Chief of the British
Armies in France, later Earl Haig of
Bemersyde (1861-1928) is the man after
whom this 3589 m mountain was named in
1918.

Sir Harold Mitchell, Mount (mountain)

83 F/3 - Cadomin
9-47-24-W5
53°02′00″N 117°27′45″W
Approximately 92 km south-west of Edson.

The naming request for this 2499 m
mountain was made by Mr. J. Ascott the
Environmental Coordinator for Cardinal
River Coals Ltd. The name was officially
approved 6 November 1983. Sir Harold
Mitchell (1901-1983) was a British
industrialist who was involved in Alberta
coal mining for over 60 years as chairman of
the Luscar group of companies. Sir Harold
resided seasonally at Luscar from the 1920s
through the 1950s and became a well-known
and admired local individual who was
praised for his progressive employee
relations.

Sir James Glacier (glacier)

82 N/15 - Mistaya Lake
33,34-21-W5
51°52′N 116°52′W
Approximately 120 km north-west of Banff.

(see Mount Outram)

Sirdar Mountain (mountain)

83 C/13 - Medicine Lake
35-45-27-W5
52°55′N 117°49′W
Approximately 19 km east of Jasper.

The precise origin of the name of this
mountain, 2804 m in altitude, is not known.
W.P. Hinton (see Hinton) named it in 1916
either because of its "prominence" or after
Field Marshal Lord Kitchener (1850-
1916), who was made Sirdar of the
Egyptian army. "Sirdar" is a term for a
person in command or in a position of
prominence.

Sirius Peak (peak)

83 E/7 - Blue Creek
25-51-5-W6
53°26′N 118°36′W
Approximately 72 km north-west of Jasper.

This 2509 m mountain peak is part of the
Starlight Range and was officially named

1 May 1934. "Sirius" is known as the dog star and is the brightest star in the sky, twenty times brighter than our sun. Its name likely comes from a Greek word meaning "sparkling" or "scorching."

Six Glaciers, Plain of the (plains)

82 N/8 - Lake Louise
28-17-W5
51°24′N 116°16′W
Approximately 55 km west north-west of Banff.

The name for this feature is likely descriptive. No other information is available on its origin.

16¹/₂ Mile Lake (lake)

83 C/12 - Athabasca Falls
52°42′25″N 117°54′30″W
Approximately 27 km south-east of Jasper.

The name is descriptive of this feature's location, 16.5 miles south-east of Jasper townsite.

Skeleton Creek (creek)

82 O/13 - Scalp Creek
25-33-13-W5
51°51′N 115°43′W
Flows north into Forbidden Creek, approximately 78 km north north-west of Banff.

The origin for the name of this creek is unknown.

Skeleton Lake (lake)

82 O/12 - Barrier Mountain
32-30-14-W5
51°37′N 115°56′W
Approximately 55 km north-west of Banff.

The name for this lake was selected because of the presence of an old Indian grave discovered in the vicinity many years ago in which the skeletal remains still rested. The

Indian name for the lake is "huha." The Canadian Permanent Committee on Geographical Names officially approved the name in 1966.

Skene, Mount (mountain)

82 N/15 - Mistaya Lake
26-32-21-W5
51°46′N 116°52′W
Approximately 110 km north-west of Banff.

Named in 1920 by D.B. Dowling (see Dowling Ford), this mountain is 3060 m in altitude. Peter Skene Ogden (1794-1854), the man after whom this mountain was named, was a fur-trader from Quebec who joined the North West Company in 1911 as a Clerk. After 1818, he spent most of his life on the Pacific slope. He became a partner in 1920, and after the union of the Hudson's Bay Company and the North West Company, he was given the rank of chief trader. In 1935, he became a chief factor.

Skoki Lakes (lakes)

82 N/9 - Hector Lake
29-29-15-W5
51°30′N 116°05′W
Approximately 50 km north-west of Banff.

(see Skoki Mountain)

Skoki Mountain (mountain)

82 N/9 - Hector Lake
33-29-15-W5
51°32′N 116°04′W
Approximately 50 km north-west of Banff.

This mountain was named in 1911 by James F. Porter (see Merlin Lake). "Skoki" is the Indian word signifying "marsh" or "swamp."

Skoki Valley (valley)

82 N/9 - Hector Lake
29-15-W5
51°31′N 116°05′W

Approximately 50 km north-west of Banff.

(see Skoki Mountain)

Slacker Creek (creek)

82 J/2 - Fording River
13-5-W5
50°05′N 114°36′W
Approximately 70 km south-west of Turner Valley.

The origin of the name is unknown.

Slate Range (range)

82 N/9 - Hector Lake
29-15-W5
51°31′N 116°03′W
Approximately 50 km north-west of Banff.

The name for this range is taken from the slatey rock of which the mountains in the range are composed.

Sleeping Lake (lake)

82 O/2 - Jumpingpound Creek
33-25-7-W5
51°11′N 114°55′W
Approximately 55 km west north-west of Calgary.

This local Indian name was officially approved 12 December 1939.

Slide Creek (creek)

83 F/4 - Miette
19-47-25-W5
53°04′N 117°38′W
Flows north into Fiddle River, approximately 38 km south-west of Hinton.

(see Slide Mountain)

Slide Mountain (mountain)

83 F/4 - Miette
29-47-25-W5
53°05′N 117°37′W

Approximately 38 km south south-west of Hinton.

Named in 1960 after an extensive rock slide which fell from the western slopes of the mountain. This feature measures 2393 m in altitude. The creek is blocked by the debris and takes its name from the mountain.

Slippery Creek (creek)

83 B/5 - Saunders
20-40-12-W5
52°27′N 115°42′W
Flows south into North Saskatchewan River, approximately 25 km east of Nordegg.

The locally well-established name for this creek was officially approved in 1941 and is likely descriptive.

Smith Creek (creek)

83 C/10 - George Creek
7-42-18-W5
52°36′N 116°35′W
Flows north into Blackstone River, approximately 37 km west north-west of Nordegg.

The name for this creek was given in 1907 by H.L. Seymour of the Dominion Land Survey after one of the men employed staking coal claims in the area.

Smith Creek (creek)

82 O/3 - Canmore
14-20-10-W5
51°02′N 115°17′W
Flows south-west into Bow River, approximately 8 km south-east of Canmore.

The name Smith Creek was officially approved 4 January 1910. The origin of the name is unknown.

Smith Creek (creek)

83 E/16 - Donald Flats
10-57-1-W6
53°54′N 118°04′W

Flows north into Berland River, approximately 72 km east of Grande Cache.

This creek was officially named in 1947. It takes its name from Bill Smith, a well-known trapper, outfitter and guide, as well as former Forest Ranger of the Entrance District.

Smith Lake (lake)

82 O/5 - Castle Mountain
26-14-W5
51°15′N 115°56′W
Approximately 26 km west north-west of Banff.

The name was locally used for many years prior to its official approval in 1959. The lake takes its name from Joe Smith, a man who arrived in Silver City in 1883 and lived there until about 1927.

Smith-Dorrien Creek (creek)

82 J/11 - Kananaskis Lakes
14-20-9-W5
50°41′N 115°09′W
Flows south-east into Lower Kananaskis Lake, approximately 60 km west of Turner Valley.

(see Mount Smith-Dorrien)

Smith-Dorrien, Mount (mountain)

82 J/11 - Kananaskis Lakes
34-20-10-W5
50°44′N 115°17′W
Approximately 70 km west of Turner Valley.

This mountain, which stands 3155 m in altitude, was named in 1918 after General Sir Horace Lockwood Smith-Dorrien (1858-1930), a British soldier involved in the Zulu War, a series of Egyptian-Sudanese campaigns, and the Boer War. He also commanded the 2nd Army, B.E.F. in the First World War. He became Governor of Gibralter in 1918 and retired in 1923.

Smuts Creek (creek)

82 J/14 - Spray Lakes Reservoir
18-22-10-W5
50°52′N 115°21′W
Flows north-west into Spray Lakes Reservoir, approximately 85 km west of Calgary.

(see Mount Smuts)

Smuts, Mount (mountain)

82 J/14 - Spray Lakes Reservoir
25-21-11-W5
50°48′N 115°23′W
Approximately 90 km west south-west of Calgary.

General Jan Christian Smuts (1870-1950) was a South African statesman, soldier and philosopher. He was Prime Minister of South Africa from 1919-1924 and again from 1938-1948. Involved in both World Wars, Smuts' major concern was that of Anglo-Afrikaner unity in South Africa. Smuts held several honours, including many honorary degrees. This 2938 m mountain was named in 1918 after this commander, and the nearby creek also commemorates his name.

Snake Creek (creek)

82 G/16 - Maycroft
31-11-3-W5
49°57′N 114°24′W
Flows west into the Livingstone River, approximately 40 km north of Blairmore.

The Blackfoot Indian name for this creek is "akustsk-siniskway," their word for "a place of many snakes." The locally well-established name is therefore descriptive.

Snake Indian Falls (falls)

83 E/8 - Rock Lake
7-50-2-W6
53°18′N 118°16′W
Approximately 47 km north north-west of Jasper.

(see Snake Indian Mountain)

Snake Indian Mountain (mountain)

83 E/7 - Blue Creek
13-50-7-W6
53°19′N 118°54′W
Approximately 74 km north-west of Jasper.

A small tribe of Indians, the "Snakes," were
massacred by the Assiniboines around 1810.
The "Snakes" lived north of Jasper House.
During a peace feast between the two tribes
where no one was to be armed, the Assiniboine,
having concealed their weapons, fell upon
the Snakes and murdered all of them. The
story of this extermination is noted in Sir
James Hector's journal (see Mount Hector)
and in Moberly's *When Fur Was King* (1929)
(see Moberly Flats). This 2929 m mountain
was named after the "Snakes."

Snake Indian Pass (pass)

83 E/7 - Blue Creek
16-50-7-W6
53°19′N 118°58′W
Approximately 76 km north-west of Jasper.

The name for this pass was officially approved
2 August 1956. (see Snake Indian Mountain)

Snake Indian River (river)

83 F/4 - Miette
35-48-28-W5
53°11′N 117°59′W
Flows north into Athabasca River,
approximately 40 km south-west of Hinton.

(see Snake Indian Mountain)

Snaky Creek (creek)

83 L/5 - Two Lakes
22-62-11-W6
54°22′N 119°34′W
Flows east into Nose Creek, approximately
62 km north north-west of Grande Cache.

The name for this winding creek was
suggested in 1946 by A.M. Floyd, a field
topographer, to describe its shape.

Snaring (locality)

83 E/1 - Snaring
28-47-1-W6
53°04′N 118°05′W
Approximately 23 km north of Jasper.

A tribe of Indians that lived here at one time
dwelled in holes dug in the ground. They
were known as the "Snare Indians" because
they subsisted on animals which they
captured with snares made of green hide. Big
Horn Sheep, deer, and moose were killed in
this manner, according to Sir James Hector.
(see Mount Hector)

Snaring Mountain (mountain)

83 E/1 - Snaring
11-47-3-W6
53°02′N 118°20′W
Approximately 25 km north-west of Jasper.

The name was officially approved 1 May
1934, for this mountain which is 2931 m in
altitude. (see Snaring)

Snaring River (river)

83 E/1 - Snaring
33-46-1-W6
53°00′N 118°05′W
Flows south-east into Athabasca River,
approximately 15 km north of Jasper.

The name of this river was officially
approved 1 May 1934. (see Snaring)

Snow Creek (creek)

82 O/12 - Barrier Mountain
2-30-13-W5
51°32′N 115°43′W
Flows south-east into Panther River,
approximately 40 km north-west of Banff.

This creek is named Snow Creek because
of the length of time the snow remains on
the ground in the area during the spring.
The nearby pass, Snow Creek Summit, takes

its name from this stream, which was
officially named in 1911.

Snow Creek (creek)

83 E/10 - Adams Lookout
26-53-6-W6
53°36′N 118°47′W
Flows south-east into North Berland River,
approximately 38 km south south-east of
Grande Cache.

The origin of the name is unknown.

Snow Creek Summit (pass)

82 O/12 - Barrier Mountain
30-13-W5
51°36′30″N 115°48′40″W
Approximately 45 km north-west of Banff.

(see Snow Creek)

Snow Dome (mountain)

83 C/3 - Columbia Icefield
23-37-24-W5
52°11′N 117°20′W
Approximately 90 km south-east of Jasper,
on the Alberta-B.C. boundary.

The name for this 3456 m snow-covered
dome is descriptive, and was given to this
feature in 1899 by J.N. Collie. (see Collie
Creek)

Snowbird Glacier (glacier)

82 N/15 - Mistaya Lake
8-32-19-W5
51°45′N 116°34′W
Approximately 115 km north-west of Banff.

The Superintendent of Banff recommended
this name after the Snowbird, which
according to the Wildlife Service, actually
exists. The glacier itself lies on a cliff at two
levels and is joined down the centre, thus
giving the effect of a large white bird. This
information was found in the transcripts of
James White, F.R.S.C. (1916).

Snowbowl Basin (basin)

83 C/13 - Medicine Lake
52°45′35″N 117°48′50″W
Approximately 18 km south-east of Jasper.

In the late 1930s, a cabin was built near this basin to accommodate skiers. Earlier, the area had been used to pasture horses believed to have belonged to the Jeffery brothers, who lived in the area. The basin is well-known for the deep snow found there. The name Snowbowl Basin was officially approved for this feature 5 May 1987, but it is locally well-known as "The Snowbowl."

Snowflake Lake (lake)

82 O/12 - Barrier Mountain
25-30-14-W5
51°36′N 115°50′W
Approximately 50 km north-west of Banff.

This lake is located just above Snow Creek Pass where frequent snowstorms occur. Its descriptive name was officially approved in 1966.

Snowshoe Pass (pass)

83 C/12 - Athabasca Falls
52°43′40″N 117°47′25″W
Approximately 25 km south-east of Jasper.

Snowshoe Pass is a locally well-established name. Its precise origin is not known.

Soda Creek (creek)

82 J/2 - Fording River
3-15-13-5-W5
50°04′45″N 114°36′25″W
Flows east into Oldman River, approximately 45 km north of Crowsnest Pass.

This name describes the very soft quality of the water. The name was officially approved 5 May 1983.

Sofa Creek (creek)

82 H/4 - Waterton Lakes
32-1-29-W4
49°05′N 113°51′W
Flows north-west into Lower Waterton Lake, approximately 45 km south south-east of Pincher Creek.

Sofa Creek is derived from Sofa Mountain. The name was officially approved 5 May 1943. (see Sofa Mountain)

Sofa Mountain (mountain)

82 H/4 - Waterton Lakes
1-29-W4
49°03′N 113°47′W
Approximately 50 km south south-east of Pincher Creek.

Named by M.P. Bridgland (see Mount Bridgland), this 2520 m mountain has a peculiar formation on its northern shoulder which extends along the south side of Middle Waterton Lake like a gigantic couch. The name Sofa Mountain was officially approved 15 October 1915.

Sofa Creek Falls (falls)

82 H/4 - Waterton Lakes
1-29-W4
49°03′N 113°49′W
Approximately 50 km south south-east of Pincher Creek.

The name for these falls was officially approved by the Canadian Permanent Committee on Geographical Names 8 June 1971 (see Sofa Mountain)

Solitaire Mountain (mountain)

82 N/15 - Mistaya Lake
13-32-21-W5
51°45′N 116°50′W
Approximately 110 km north-west of Banff, in the Conway Group.

The descriptive name for this 3270 m mountain positioned at the centre of Conway Glacier was applied in 1920.

Solomon Creek (creek)

83 F/5 - Entrance
23-50-27-W5
53°20′N 117°50′W
Flows south-east into Brule Lake, approximately 16 km west south-west of Hinton.

One of the first Iroquois to arrive at Jasper House in 1814 was Dominik Karayinter (later known as Caraconte) who was one of a group of North West Company employees sent from Montreal to Jasper House. Of the group which was sent, he was the only one to reach Jasper House. This creek is named after one of Dominik's descendants, Solomon Caraconte, who lived in this area at one time. The creek was known by some of the natives of the area as "Old House Creek."

Solomon, Mount (mountain)

83 F/5 - Entrance
1-51-27-W5
53°23′N 117°50′W
Approximately 15 km west of Hinton.

Mount Solomon is 1585 m in altitude. (see Solomon Creek)

Sounding Lake (lake)

82 J/11 - Kananaskis Lakes
18-20-8-W5
50°41′35″N 115°06′15″W
Approximately 75 km west south-west of Calgary.

The name for this lake, which was officially approved 8 November 1978, is derived from the sounds that the ice on the lake makes during the spring when warmed by the sun.

South Creek (creek)

83 C/8 - Nordegg
35-39-15-W5
52°24′N 116°03′W
Flows north north-east into North Saskatchewan River, approximately 8 km south of Nordegg.

The name South Creek is the well-established local name for this stream. The origin of the name is unknown but an old trail, known as the south trail, which goes from the North Saskatchewan River Crossing south to Joyce River and the North Ram River, passes its headwaters. The name was officially approved in 1976.

South Berland River (river)

83 E/10 - Adams Lookout
24-54-5-W6
53°40′N 118°37′W
Flows north into Berland River, approximately 40 km south-east of Grande Cache.

(see Berland Range)

South Cabin Creek (creek)

83 E/16 - Donald Flats
26-55-4-W6
53°47′N 118°29′W
Flows north-east into Cabin Creek, approximately 45 km south-east of Grande Cache.

(see Cabin Creek)

South Drinnan Creek (creek)

83 F/4 - Miette
14-48-25-W5
53°09′N 117°34′W
Flows north into Drinnan Creek, approximately 32 km south of Hinton.

(see Drinnan Creek)

South Ghost River (river)

82 O/6 - Lake Minnewanka
35-26-8-W5
51°16′N 115°01′W
Flows north-east into Ghost River, approximately 30 km north-east of Canmore.

(see Ghost River)

South Glasgow Creek (creek)

82 J/15 - Bragg Creek
12-21-7-W5
50°46′N 114°52′W
Flows east into Elbow River, approximately 50 km west south-west of Calgary.

(see Mount Glasgow)

South Hidden Creek (creek)

82 G/15 - Tornado Mountain
11-12-5-W5
49°59′N 114°35′W
Flows north into Hidden Creek, approximately 40 km north of Coleman.

This creek was named for its proximity as a tributary of Hidden Creek, and the name was officially approved in 1960.

South Isaac Creek (creek)

83 C/6 - Sunwapta
2-41-22-W5
52°30′N 117°03′W
Flows north-east into Isaac Creek, approximately 81 km south-east of Jasper.

(see Isaac Creek)

South James River (river)

82 O/14 - Limestone Mountain
15-33-9-W5
51°51′N 115°06′W
Flows north into James River, approximately 80 km north-east of Banff.

The name for this river was officially approved 25 November 1941. (see James Pass)

South Kakwa River (river)

83 L/4 - Kakwa River
11-59-12-W6
54°05′N 119°41′W
Flows north-east into Kakwa River, approximately 42 km north-west of Grande Cache.

"Kakwa" is the Cree Indian word for "porcupine." The name for this river was officially approved 10 September 1959.

South Kananaskis Pass (pass)

82 J/11 - Kananaskis Lakes
19-10-W5
50°38′N 115°17′W
Approximately 91 km west south-west of Calgary, on the Alberta-B.C. boundary.

(see Kananaskis Range)

South Kootenay Pass (pass)

82 G/1 - Sage Creek
12-2-2-W5
49°07′N 114°09′W
Approximately 18 km west north-west of Waterton Park, on the Alberta-B.C. boundary.

This so-called pass is merely a lower elevation of the watershed ridge, and there is no distinct gap. It was shown on Captain Palliser's (see Palliser Range) map as Boundary Pass, but has gained significance and title from the tribe of Indians known as the Kootenays, who formerly inhabited the Waterton Park region and were driven west of the Divide by their enemies. The Kootenay Indians frequented this pass during the years before the arrival of the white man. Although this route was the most difficult to ascend and descend, it was the most direct path to the buffalo lands at the base of

Chief Mountain, just to the east of the Waterton Lakes.

South Lancaster Creek (creek)

83 E/10 - Adams Lookout
5-55-6-W6
53°44′N 118°52′W
Flows north-east into Lancaster Creek, approximately 24 km south-east of Grande Cache.

(see Lancaster Creek)

South Lost Creek (creek)

82 G/8 - Beaver Mines
4-6-4-W5
49°27′N 114°30′W
Approximately 21 km west of Beaver Mines.

(see Lost Creek)

South Maude Lake (lake)

82 J/11 - Kananaskis Lakes
11-20-10-W5
50°40′50″N 115°17′00″W
Approximately 70 km west of Turner Valley.

This small alpine lake was officially named 7 February 1983 for its proximity to Maude Lake. (see Maude Lake)

South Muskeg River (river)

83 E/10 - Adams Lookout
17-54-6-W6
53°40′N 118°50′W
Flows north into Muskeg River, approximately 31 km south-east of Grande Cache.

(see Muskeg River)

South Pass (pass)

83 D/15 - Rainbow
21-46-5-W6
52°59′N 118°40′W
Approximately 41 km west north-west of Jasper.

(see Miette Pass)

South Racehorse Creek (creek)

82 G/15 - Tornado Mountain
17-10-4-W5
49°50′N 114°31′W
Flows north into Racehorse Creek, approximately 20 km north of Coleman.

(see Racehorse Creek)

South Sullivan Creek (creek)

82 J/10 - Mount Rae
9-18-4-W5
50°30′N 114°30′W
Flows north-east into Sullivan Creek, approximately 24 km south-west of Turner Valley.

The name was officially approved 6 September 1951. (see Sullivan Creek)

South Sulphur River (river)

83 E/10 - Adams Lookout
6-53-6-W6
53°34′N 118°52′W
Flows north into Sulphur River, approximately 40 km south south-east of Grande Cache.

(see Sulphur River)

South Todd Creek (creek)

82 G/16 - Maycroft
25-9-3-W5
49°46′N 114°17′W
Flows north into Todd Creek, approximately 20 km north-east of Blairmore.

(see Todd Creek)

South Twin Creek (creek)

82 J/1 - Langford Creek
14-4-W5
50°10′N 114°28′W
Flows south-west into Livingstone River, approximately 65 km north-west of Claresholm.

The origin of the name is unknown. (see North Twin Creek)

South Twin Peak (peak)

83 C/3 - Columbia Icefield
37-25-W5
52°12′15″N 117°26′00″W
Approximately 85 km south-east of Jasper.

(see The Twins)

Southesk Cairn (mountain)

83 C/11 - Southesk Lake
52°43′N 117°08′W
Approximately 50 km east south-east of Jasper.

Southesk Cairn is 2550 m in altitude. (see Southesk River)

Southesk Lake (lake)

83 C/11 - Southesk Lake
52°38′N 117°12′W
Approximately 60 km east south-east of Jasper.

(see Southesk River)

Southesk, Mount (mountain)

83 C/11 - Southesk Lake
52°40′N 117°08′W
Approximately 55 km east south-east of Jasper.

This mountain stands 3125 m in altitude. (see Southesk River)

Southesk Pass (pass)

83 C/11 - Southesk Lake
52°44′N 117°08′W
Approximately 50 km east south-east of Jasper.

(see Southesk River)

Southesk River (river)

83 C/10 - George Creek
22-43-20-W5
52°43'N 116°50'W
Flows north-east into Brazeau River, approximately 90 km east south-east of Jasper.

This river was named in 1907 after James Carnegie, 9th Earl of Southesk (1827-1905). The Earldom of Southesk had been forfeited by the 5th Earl due to his participation in the rebellion of 1745 when the Catholic Young Pretender, "Bonnie Prince Charlie," received some support of the Highland Chiefs, Southesk among them, in his attempt to put his father (King James III) on the throne of the U.K. James III was referred to as the "Old Pretender." In the spring of 1859, the 9th Earl left England for Canada to help his ailing health. A literate person, he read Shakespeare as he hunted and travelled across the northwest. After extensive exploration of the Rockies and foothills area, his party returned to Fort Edmonton 12 October. Here he wrote *Saskatchewan and the Rocky Mountains*, (1875) describing his travels along the Saskatchewan River and in the mountains. (source: "Alberta History," Autumn 1988)

Southfork Lakes (lakes)

82 G/8 - Beaver Mines
29-4-3-W5
49°20'N 114°23'W
Approximately 19 km south-west of Beaver Mines.

The name Southfork Lakes was officially approved 7 August 1958, based on established local usage and on the fact that the mountain nearby is referred to as Southfork Mountain.

Southfork Mountain (mountain)

82 G/8 - Beaver Mines
29-4-3-W5
49°20'N 114°23'W

Approximately 19 km south-west of Beaver Mines.

The precise origin of the name is unknown.

Sparrow's Egg Lake (lake)

82 J/11 - Kananaskis Lakes
6-20-8-W5
50°40'25"N 115°06'20"W
Approximately 60 km west of Turner Valley.

The name for this small permanent lake comes from the Sparrow's Egg Ladyslipper (cypripedium), a rare orchid found on the shores of the lake. The name was officially approved 8 November 1978.

Sparrowhawk, Mount (mountain)

82 J/14 - Spray Lakes Reservoir
12-23-10-W5
50°56'N 115°16'W
Approximately 80 km west of Calgary.

This 3121 m mountain was named in 1922 after a destroyer engaged in the First World War Battle of Jutland. (see Jutland Mountain)

Spectral Creek (creek)

82 O/6 - Lake Minnewanka
31-27-10-W5
51°21'N 115°24'W
Flows south-east into Ghost River, approximately 21 km north north-east of Banff.

(see Spectral Lakes)

Spectral Lakes (lakes)

82 O/6 - Lake Minnewanka
3-28-11-W5
51°22'N 115°28'W
Approximately 21 km north north-east of Banff.

The name for these lakes and the nearby creek was suggested by T.W. Swaddle of

Calgary in 1970. Swaddle proposed that the name Spectral (meaning ghost-like; of ghosts), was concurrent with the "supernatural" idiom long established in the area.

Speers Creek (creek)

82 J/1 - Langford Creek
15-36-12-4-W5
50°02'50"N 114°25'10"W
Approximately 55 km south south-west of Longview.

The name for this creek was officially adopted as "Spears Creek" in 1959. It was named after John "Barbwire" Speers. He acquired the nickname in 1901 while riding night herd for Herb Millar (see Miller Creek) on the Bar U Ranch. One morning, Speers came into camp a little scratched up from the barbed wire that the settlers were beginning to put up around their newly-claimed pastures, and fellow cowhand Joe Lamar jokingly named him "Barbwire Johnny." The spelling was officially changed from "Spears Creek" to Speers Creek 29 July 1986.

Sphinx Creek (creek)

83 F/3 - Cadomin
8-48-24-W5
53°07'N 117°29'W
Flows north-east into Gregg River, approximately 86 km south-west of Edson.

(see Sphinx Mountain)

Sphinx Lake (lake)

82 O/4 - Banff
25-14-W5
51°08'N 115°55'W
Approximately 26 km west south-west of Banff.

See Egypt Lake, Mummy Lake and Scarab Lake, all located in the "Egyptian Connection."

Sphinx Mountain (mountain)

83 F/4 - Miette
22-47-25-W5
53°04′N 117°34′W
Approximately 42 km north-east of Jasper.

This 2460 m mountain was descriptively named 7 July 1960. When viewed from the north and east, this mountain has the appearance of a figure with a lion's body and a human head.

Spider Creek (creek)

83 E/5 - Holmes River
3-52-12-W6
53°27′N 119°40′W
Flows north-east into Jackpine River, approximately 124 km north-west of Jasper.

This creek was named for the number and peculiarity of the spiders seen in this valley.

Spillway Lake (lake)

82 J/11 - Kananaskis Lakes
14-20-9-W5
50°41′50″N 115°08′00″W
Approximately 60 km west of Turner Valley.

The name for this lake is derived from the emergency spillway in the north-west end of Pocaterra Dam. The name was officially applied to this feature 8 November 1978.

Spionkop Ridge (ridge)

82 G/1 - Sage Creek
3-1-W5
49°12′N 114°05′W
Approximately 20 km north-west of Waterton Park.

There was a battle fought in the Boer War (1899-1902) on a ridge of this name near Johannesburg, South Africa and it is after this ridge that the 2444 m ridge was named.

Spirit Island (island)

83 C/12 - Athabasca Falls
26-42-25-W5
52°38′N 117°29′W
Approximately 50 km south south-east of Jasper.

The origin of the name is unknown.

Spotted Wolf Creek (creek)

82 J/11 - Kananaskis Lakes
7-20-8-W5
50°40′30″N 115°05′45″W
Flows north into Pocaterra Creek, approximately 70 km west south-west of Calgary.

"Spotted Wolf" was the Indian name of the Stoney Indian who was George Pocaterra's (see Pocaterra Creek) blood brother and constant companion. His English name was Paul Amos. The name, which was officially approved 12 September 1984, was deemed appropriate, as this stream is a "companion" of Pocaterra Creek.

Sprague, Mount (mountain)

83 E/12 - Pauline Creek
12-55-13-W6
53°44′N 119°47′W
Approximately 45 km west south-west of Grande Cache.

This mountain, which is 2530 m in altitude, was named in 1925 after D.D. Sprague of the Grand Trunk Pacific Railway. Sprague was an engineer who explored in this region in 1906.

Spray Mountains (mountain)

82 J/11 - Kananaskis Lakes
21-10-W5
50°45′N 115°18′W
Approximately 70 km west of Turner Valley.

(see Spray River)

Spray Pass (pass)

82 J/11 - Kananaskis Lakes
2-21-11-W5
50°44′N 115°25′W
Approximately 80 km west of Turner Valley.

(see Spray River)

Spray River (river)

82 O/4 - Banff
25-25-12-W5
51°09′N 115°34′W
Flows north into Bow River, approximately 1 km south of Banff.

The name for this river has been in use since at least 1885 and is descriptive of the "spray" of the falls in the Bow River near its mouth.

Spray Lakes Reservoir (reservoir)

82 J/14 - Spray Lakes Reservoir
23-10-W5
50°54′N 115°20′W
Approximately 80 km west of Calgary.

(see Spray River)

Spreading Creek (creek)

82 N/15 - Mistaya Lake
10-35-18-W5
52°00′N 116°29′W
Flows north into North Saskatchewan River, approximately 110 km north-west of Banff.

This creek served as a landmark to travellers in the mountains near the beginning of this century, and its obvious channels rendered it easily recognizable. The locally well-established name is appropriately descriptive.

Spring-Rice, Mount (mountain)

83 C/3 - Columbia Icefield
35-23-W5
52°01′N 117°14′W
Approximately 110 km south-east of Jasper, on the Alberta-B.C. boundary.

Sir Cecil Arthur Spring-Rice, K.C.M.G. (1839-1918) was a British diplomat and ambassador to the United States. A.O. Wheeler (see Wheeler Flats) named this 3275 m mountain after him in 1920.

Spurling Creek (creek)

82 J/14 - Spray Lakes Reservoir
22-23-10-W5
50°58′N 115°19′W
Flows south-west into Spray Lakes Reservoir, approximately 80 km west of Calgary.

Officially approved in 1976, this name is taken from the species of fish, likened to a smelt, and known as spurling, or spirling, which is found in the creek.

Stairway Peak (peak)

82 N/15 - Mistaya Lake
6-32-19-W5
51°48′N 116°39 W
Approximately 100 km north-west of Banff, on the Alberta-B.C. boundary.

This 2999 m mountain was named by A.O. Wheeler in 1918. The name is descriptive of the formation, which resembles a stairway up the side of the peak.

Stalk Creek (creek)

83 E/9 - Moberly Creek
31-54-3-W6
53°43′N 118°26′W
Flows north-east into Berland River, approximately 96 km north north-west of Jasper.

This creek was named in 1946 by M.E. Nidd, a field topographer in the area. It was near this stream that he watched bear stalk sheep.

Standish Hump (hump)

82 O/4 - Banff
10-24-13-W5
51°04′N 115°47′W

Approximately 20 km south-west of Banff.

"Standish Ridge" was named after Mr. Austin Standish around 1935. Standish was one of the members of a party looking for a site for a lodge — the site eventually chosen would later become known as Sunshine Village. Standish died in 1965. The local name for this 2423 m hump was Standish Hump, and since it is not a ridge, the name was officially corrected 29 July 1986.

Stanley Falls (falls)

83 C/6 - Sunwapta
1-39-24-W5
52°20′N 117°19′W
Approximately 80 km south-east of Jasper.

The origin of the name is unknown.

Star Creek (creek)

82 G/10 - Crowsnest
7-8-4-W5
49°38′N 114°33′W
Flows north-east into Crowsnest River, approximately 2 km west of Coleman.

The name Star Creek is well-established in local usage and was officially applied to this creek in 1962. The exact origin of the name is not known.

Star Creek (creek)

83 E/9 - Moberly Creek
7-54-3-W6
53°39′N 118°26′W
Flows north-west into Moon Creek, approximately 90 km north north-west of Jasper.

The precise origin of the name is unknown.

Star Creek Falls (falls)

82 G/10 - Crowsnest
6-8-4-W5
49°37′35″N 114°32′15″W

Approximately 3 km west south-west of Coleman.

This well-established local name was officially adopted 21 July 1982. The falls are named after the creek on which they are found.

Starlight Range (range)

83 E/7 - Blue Creek
52-5-W6
53°27′N 118°40′W
Approximately 75 km north-west of Jasper.

The name for this range containing Arcturus, Sirius and Vega Peaks was officially adopted 1 May 1934. R.W. Cautley (see Mount Cautley) named the peaks, and it is likely he also named the range.

Station Creek (creek)

82 G/16 - Maycroft
5-11-3-W5
49°52′N 114°23′W
Flows east into Oldman River, approximately 30 km north of Blairmore.

This creek may be named for the Livingstone Ranger Station, which is found close to the point at which Station Creek empties into the Oldman River.

Stearn, Mount (mountain)

83 E/14 - Grande Cache
23-56-9-W6
53°51′N 119°13′W
Approximately 7 km west south-west of Grande Cache.

Mount Stearn is 2013 m in altitude. The origin of the name is unknown.

Steep Creek (creek)

83 L/5 - Two Lakes
9-62-11-W6
54°21′N 119°35′W
Flows east into Nose Creek, approximately 60 km north north-west of Grande Cache.

The descriptive name for this very steep and impassable creek was suggested in 1946 by A.M. Floyd, a field topographer of the area.

Stelfox, Mount (mountain)

82 C/2 - Cline River
31-37-18-W5
52°14′N 116°32′W
Approximately 110 km west south-west of Rocky Mountain House.

This 2134 m mountain was named in 1956 after Henry Stelfox (1883-1974), pioneer of Rocky Mountain House. Mr. Stelfox rendered long and distinguished service to the cause of conservation in the Rocky Mountains and in 1954, his work in this field was recognized by the award of the Julian Crandall Conservation Trophy. Henry Stelfox first came to Alberta in 1906, settling first near Wetaskiwin and then Rocky Mountain House.

Stenton Lake (lake)

82 O/3 - Canmore
35-25-10-W5
51°10′N 115°16′W
Approximately 9 km north-west of Canmore.

James Ernest Stenton (1904-1966) worked as a guide for hunters from 1924-1935 in the area where this lake is located. He later became the National Parks Warden and worked in the Minnewanka area just north of this lake from 1939-1966. Stenton conducted several experiments with trout in an attempt to create a hybrid between lake and brook varieties during his employment with the National Park, eventually producing the hybrid known as "splake." This lake was officially named 5 May 1987, in recognition of J.E. Stenton's contribution to fisheries and the more than 40 years he worked in the area.

Steppe Glacier (glacier)

83 E/3 - Mount Robson
48-7-W6
53°09′N 119°01′W

Approximately 69 km west north-west of Jasper.

This glacier was originally named "Terrace Glacier" by Dominion Land Surveyor A.O. Wheeler (see Wheeler Flats) in 1911, but because that name was already used, the name was changed to "Ermine Glacier" that same year. Some time later, the name Steppe Glacier was officially approved. "Terrace" was chosen because it was descriptive of the formation of the glacier. "Steppe" is a synonym for terrace.

Sterne Creek (creek)

83 E/15 - Pierre Greys Lakes
15-57-7-W6
53°56′N 118°58′W
Flows north into Susa Creek, approximately 12 km east north-east of Grande Cache.

Flight-Lieutenant J.R. Sterne was born in Edmonton in 1922. He enlisted in the Royal Canadian Air Force 10 October 1940, was awarded a Pilot's Flying Badge, and was eventually appointed to a commissioned rank. He was reported missing in action on 16 August 1943. This creek was named in honour of J.R. Sterne, D.F.C.

Stetson Creek (creek)

83 L/5 - Two Lakes
12-62-13-W6
54°21′N 119°49′W
Flows north-west into Torrens River, approximately 68 km north-west of Grande Cache.

This stream runs off Hat Mountain. (see Hat Mountain)

Stevens Creek (creek)

83 B/12 - Harlech
29-43-13-W5
52°44′N 115°51′W
Flows north-east into Nordegg River, approximately 32 km north-east of Nordegg.

The origin of the name is unknown.

Stewart Canyon (canyon)

82 O/5 - Castle Mountain
32-26-11-W5
51°16′N 115°30′W
Approximately 12 km north north-east of Banff.

In 1886, Mr. George A. Stewart (1830-1917) of the Dominion Land Survey was commissioned to survey the area and furnish plans for a proposed townsite. He became the Superintendent of the new park (now Banff National Park) in 1887, a position he held for ten years. This canyon was named after him.

Stewart Creek (creek)

82 O/3 - Canmore
13-14-24-10-W5
51°03′00″N 115°17′45″W
Flows north into Bow River, approximately 6 km south-east of Canmore.

(see Stewart Canyon)

Stewart, Mount (mountain)

83 C/2 - Cline River
33-37-21-W5
52°13′N 116°57′W
Approximately 105 km south-east of Jasper.

Louis B. Stewart, D.L.S., D.T.S. (1861-1937) accompanied A.P. Coleman (see Mount Coleman) on a few expeditions in the Rockies during the late 1800s. He was a Professor of Surveying and Geodesy at the University of Toronto. The name for this 3312 m mountain was made official 5 March 1935.

Still Creek (creek)

83 F/5 - Entrance
30-50-25-W5
53°20′N 117°38′W

Flows north into Cold Creek, approximately 6 km south south-west of Hinton.

The origin of the well-established name for this creek is unknown. The name was approved in 1944.

Stinking Creek (creek)

83 L/4 - Kakwa Falls
27-61-13-W6
54°15′N 119°52′W
Flows north-west into Torrens River, approximately 70 km north-west of Grande Cache.

This creek was once referred to as "South Torrens River;" however, local usage dictated the adoption of the name Stinking Creek 4 December 1958. (see Stinking Springs)

Stinking Springs (springs)

83 L/4 - Kakwa Falls
22-60-13-W6
54°14′N 119°51′W
Approximately 60 km north-west of Grande Cache.

The high concentration of sulphur in the cold water which comes from the springs results in a rotten egg smell and hence, the descriptive name.

Stones Canyon Creek (creek)

82 O/3 - Canmore
30-24-10-W5
51°05′N 115°22′W
Approximately 3 km south-east of Canmore.

The origin of the name is unknown.

Stoney – #142, 143, 144 (Indian reserve)

82 O/2 - Jumpingpound Creek
27-25-6-W5
51°10′N 114°46′W
Immediately west of Calgary.

This reserve was named after the Stoney Indians.

Stoney Squaw Mountain (mountain)

82 O/4 - Banff
11-26-12-W5
51°12′N 115°35′W
Approximately 2 km north north-west of Banff.

This 1868 m mountain was locally named prior to 1884 and the name was officially approved 7 March 1922. The legend connected to this mountain relates that an old Assiniboine ("Stoney" in English) woman took care of her sick husband resting in their lodge at the mountain's foot, by hunting on its top and sloping sides.

Stony Creek (creek)

82 J/7 - Mount Head
16-5-W5
50°23′N 114°38′W
Flows south into Highwood River, approximately 49 km south-west of Turner Valley.

This creek was officially named 8 December 1943. The creek was likely named after the Stoney Indian tribe which occupied the surrounding territory before white men came into the area. The spelling in its official name has been altered.

Stony Creek (creek)

82 O/3 - Canmore
11-24-8-W5
51°02′N 115°01′W
Flows south-west into Lusk Creek, approximately 25 km east south-east of Canmore.

The creek was so named because of its proximity to the Stoney Indian Reserve. Its name, although spelled incorrectly, was officially approved 2 November 1956.

Stony Creek (creek)

82 O/5 - Castle Mountain
34-27-12-W5
51°21′N 115°36′W

Flows south into Cascade River, approximately 18 km north of Banff.

Officially approved 1 September 1960, the origin of this locally well-established name is unknown. It may possibly be an old route used by Stoney Indians.

Storelk Mountain (mountain)

82 J/10 - Mount Rae
24-18-8-W5
50°32′N 114°59′W
Approximately 51 km south-west of Turner Valley.

This well-established local name is derived from its position between Storm Creek and Elk River. The mountain is 2865 m in altitude.

Storm Creek (creek)

82 J/10 - Mount Rae
13-18-7-W5
50°31′N 114°50′W
Flows south-east into Highwood River, approximately 45 km south-west of Turner Valley.

This creek was named in 1884 by Dr. G.M. Dawson (see Tombstone Mountain) because a very heavy rainfall occurred while he was camping on this stream.

Storm Mountain (mountain)

82 J/10 - Mount Rae
8-19-7-W5
50°35′N 114°56′W
Approximately 47 km west of Turner Valley.

Storm Mountain is 3095 m in altitude. (see Storm Creek)

Storm Mountain (mountain)

82 O/4 - Banff
11-26-15-W5
51°13′N 115°59′W

Approximately 30 km west north-west of Banff.

This 3100 m mountain was descriptively named in 1884 by G.M. Dawson (see Tombstone Mountain), after the numerous storm clouds on its summit.

Stornoway, Mount (mountain)

83 E/8 - Rock Lake
36-49-4-W6
53°16′N 118°27′W
Approximately 50 km north north-west of Jasper.

Stornoway is the official residence of the Leader of the Official Opposition in the House of Commons. It was built in 1913 by Asconi Joseph Major. The Majors sold the home to Mrs. Ethel Perley-Robertson, who named it after Stornoway on the Isle of Lewis in the Outer Hebrides, their ancestral home. Princess Juliana of the Netherlands stayed at Stornoway during her visit to Canada during the Second World War. This 2886 m mountain was named 1 May 1934.

Stovepipe Creek (creek)

83 C/9 - Wapiabi Creek
31-41-16-W5
52°35′N 116°18′W
Flows north-west into Lookout Creek, approximately 22 km north-west of Nordegg.

The origin of the name is unknown.

Strachan, Mount (mountain)

82 J/7 - Mount Head
16-6-W5
50°23′N 114°49′W
Approximately 49 km south-west of Turner Valley.

This 2682 m mountain is named after Lieutenant Henry Strachan, V.C. (1889-1917). He came to Canada in 1908 and joined the Canadian Army in 1915. In November 1917, during the First World War, he led a cavalry squadron through the enemy line of machine guns. He then led a charge on an enemy battery, killing seven gunners with his sword.

Strahan, Mount (mountain)

82 N/15 - Mistaya Lake
32-21-W5
51°47′N 116°50′W
Approximately 100 km north-west of Banff.

This 3060 m mountain was named in 1920, after Dr. Aubrey Strahan, the Director of the Geological Survey of Great Britain.

Straight Creek (creek)

82 J/2 - Fording River
13-5-W5
50°06′N 114°39′W
Flows south into Oldman River, approximately 70 km south-west of Turner Valley.

The name appropriately describes the flow pattern of this creek.

Strange, Mount (mountain)

83 E/1 - Snaring
1-49-4-W6
53°12′N 118°27′W
Approximately 44 km north-west of Jasper.

This mountain was one of 12 features named in 1954 for Alberta's 50th Jubilee (see Mount Griesbach). This mountain, which is 2887 m in altitude, was named after Major-General T. Bland Strange (1831-1925). He was a British Army Officer, a Canadian Artillery Officer and Inspector of the Canadian Militia at various times in his military career. Strange served in India in 1857-1858 at many posts. He left the Canadian Service in 1881 and established the Military Colonization Company in the West with a ranch near Calgary.

Strawberry Creek (creek)

82 J/7 - Mount Head
16-5-W5
50°23′N 114°41′W
Flows east into Highwood River, approximately 42 km south-west of Turner Valley.

Strawberries grow in great profusion along this stream. The name was made official 8 December 1943.

Strawberry Ridge (ridge)

82 O/14 - Limestone Mountain
32-9-W5
51°48′N 115°15′W
Approximately 70 km north north-east of Banff.

This ridge was first named by M.P. Bridgland (see Mount Bridgland) and was officially approved 25 November 1941. The precise origin of the name is unknown.

Strubel Lake (lake)

83 B/3 - Tay River
25-37-8-W5
52°12′N 115°00′W
Approximately 20 km south-west of Rocky Mountain House.

Strubel Lake was officially named 12 December 1968. The origin of the name is unknown.

Stuart Knob (peak)

82 O/5 - Castle Mountain
27-14-W5
51°20′N 115°57′W
Approximately 20 km north of Banff.

This 2850 m peak may be named after Benjamin Stuart, son of Helena B. Walcott. (see Helena Ridge).

Stud Creek (creek)

82 O/11 - Burnt Timber Creek
36-30-9-W5
51°37′N 115°09′W

Flows east into Burnt Timber Creek, approximately 55 km north-east of Banff.

This creek was named by M.P. Bridgland (see Mount Bridgland) in 1917. The origin of the name is unknown.

Sturrock Creek (creek)

83 C/9 - Wapiabi Creek
34-41-17-W5
52°34′N 116°23′W
Flows north-west into Wapiabi Creek, approximately 23 km north-west of Nordegg.

Archie Sturrock was the accountant at the Brazeau Mine in 1919. He later became manager of the mine. He often camped on this creek, which now bears his name.

Stutfield Creek (creek)

83 C/6 - Sunwapta
52°16′40″N 117°18′50″W
Flows north-east into Sunwapta River, approximately 84 km south-east of Jasper.

(see Stutfield Peak)

Stutfield Glacier (glacier)

83 C/2 - Cline River
52°14′00″N 117°21′00″W
Approximately 106 km south-east of Jasper.

(see Stutfield Peak)

Stutfield Peak (peak)

83 C/3 - Columbia Icefield
7-38-24-W5
52°15′N 117°25′W
Approximately 85 km south-east of Jasper.

This 3396 m mountain peak was named in 1899 by J.N. Collie (see Collie Creek) after his colleague and joint author of *Climbs and Explorations in the Canadian Rockies* (1903), Hugh E.M. Stutfield. Stutfield was an active member of The Alpine Club (London), and

a member of the party that explored the icefields in 1895. The other features in the area take their names from this peak.

Suicide Creek (creek)

82 G/8 - Beaver Mines
7-5-3-W5
49°22′N 114°23′W
Approximately 17 km south-west of Beaver Mines.

M.P. Bridgland (see Mount Bridgland) named this creek 30 June 1915. No other origin information is available.

Sullivan Creek (creek)

82 J/9 - Turner Valley
20-18-3-W5
50°32′N 114°23′W
Flows north-east into Highwood River, approximately 17 km south south-west of Turner Valley.

This creek is named for John Sullivan and his family, who settled in the area in the late 1880s. The Indian name for the creek is "Makkoye," or "Wolf" Creek. The name Sullivan Creek was officially approved 8 December 1943. (see South Sullivan Creek)

Sullivan, Mount (mountain)

82 N/15 - Mistaya Lake
28-34-21-W5
51°56′N 116°56′W
Approximately 120 km north-west of Banff.

This 2975 m mountain was named for John W. Sullivan, Secretary of the Palliser Expedition. (see Introduction)

Sullivan Peak (peak)

82 N/15 - Mistaya Lake
34-22-W5
51°56′N 116°58′W
Approximately 125 km north-west of Banff.

(see Mount Sullivan)

Sulphur Creek (creek)

83 F/4 - Miette
19-48-26-W5
53°09′N 117°47′W
Flows north into Fiddle River, approximately 35 km south south-west of Hinton.

There are hot sulphur springs located on this creek. The name is likely descriptive of these springs.

Sulphur Creek (creek)

83 L/12 - Nose Creek
3-65-13-W6
54°36′N 119°53′W
Flows north-east into Narraway River, approximately 93 km south-west of Grande Prairie.

(see Stinking Springs)

Sulphur Mountain (mountain)

82 O/4 - Banff
25-12-W5
51°08′N 115°34′W
Approximately 4 km south of Banff.

This 2451 m mountain takes its name from the sulphur hot springs located at its base. (see Samson Peak)

Sulphur Ridge (ridge)

83 L/4 - Kakwa Falls
1-60-13-W6
From 54°06′00″N 119°39′00″W
to 54°13′00″N 119°54′00″W
Approximately 133 km south south-west of Grande Prairie.

There is a spring located at the north-west edge of this ridge. The water in the spring contains a high concentration of sulphur.

Sulphur River (river)

83 E/14 - Grande Cache
19-56-8-W6
53°50′N 119°10′W

Flows north-west into Smoky River, approximately 5 km south-west of Grande Cache.

This river was presumably named for the high content of sulphur in its chemical constitution.

Sulphur Spring (spring)

82 O/3 - Canmore
31-24-10-W5
51°05′N 115°23′W
Approximately 2 km west of Canmore.

(see Sulphur Mountain)

Summit Lake (lake)

82 G/1 - Sage Creek
2-1-1-W5
49°00′N 114°01′W
Approximately 10 km south-west of Waterton Park.

The lake is found at the summit of Kootenay Pass; its name is descriptive.

Summit Lakes (lakes)

83 C/13 - Medicine Lake
52°53′25″N 117°45′35″W
Approximately 25 km east of Jasper.

These two lakes are located at the summit of a pass between the Colin Range and the Queen Elizabeth Range, running from Jacques Lake to Beaver and Medicine lakes. The pass is known locally as "Summit Pass." The descriptive name for these lakes was officially approved 5 May 1987.

Sundance Canyon (canyon)

82 O/4 - Banff
25-12-W5
51°09′N 115°37′W
Approximately 5 km south-west of Banff.

(see Sundance Creek)

Sundance Creek (creek)

82 O/4 - Banff
28-25-12-W5
51°10′N 115°37′W
Flows north into Bow River, approximately 5 km south-west of Banff.

In the early days, Indians met upon the plateau above the falls on this stream and performed their sacred rites during their Sundance fête, according to James White's transcripts. The pass, range and canyon take their name from the creek.

Sundance Pass (pass)

82 O/4 - Banff
25-12-W5
51°06′N 115°33′W
Approximately 10 km south-west of Banff.

(see Sundance Creek)

Sundance Range (range)

82 O/4 - Banff
24-25-12-W5
51°05′N 115°37′W
Approximately 9 km south-west of Banff.

The name for this feature, which was formerly known as "Bourgeau Range," was officially approved 3 July 1958. (see Sundance Creek)

Sundial Mountain (mountain)

83 C/4 - Clemenceau Icefield
3-38-26-W5
52°14′N 117°38′W
Approximately 80 km south-east of Jasper.

The peak of this mountain has a fancied resemblance in shape to a sundial. The descriptive name for this feature, 3182 m in altitude, was officially approved 17 March 1921.

Sunkay Creek (creek)

83 C/8 - Nordegg
30-39-17-W5
52°23′N 116°26′W
Flows south-east into Bighorn River, approximately 26 km west south-west of Nordegg.

The name for this creek was officially approved in 1932. It is an Indian word for "wild horses." Wild horses range along this stream.

Sunset Creek (creek)

83 B/5 - Saunders
18-40-11-W5
52°27′N 115°35′W
Flows south into North Saskatchewan River, approximately 34 km east of Nordegg.

The origin of this locally well-established name is unknown.

Sunset Creek (creek)

83 E/10 - Adams Lookout
29-54-5-W6
53°42′N 118°42′W
Flows east into Berland River, approximately 35 km south-east of Grande Cache.

No information regarding the origin of the name is known.

Sunset Pass (pass)

83 C/2 - Cline River
24-36-21-W5
52°06′N 116°52′W
Approximately 115 km south of Jasper.

The origin of the name is unknown.

Sunset Peak (peak)

83 E/7 - Blue Creek
9-52-7-W6
53°29′N 118°59′W
Approximately 89 km north-west of Jasper.

This 3265 m mountain peak was named in 1929 by R.W. Cautley of the Boundary Commission (see Mount Cautley). The name is likely descriptive.

Sunshine Creek (creek)

82 O/4 - Banff
5-25-13-W5
51°07'N 115°46'W
Flows north into Healy Creek, approximately 16 km south-west of Banff.

The creek was named by A.O. Wheeler (see Wheeler Flats) and was officially approved in 1958. No other origin information is known.

Sunwapta Falls (falls)

83 C/12 - Athabasca Falls
14-41-26-W5
52°32'N 117°39'W
Approximately 50 km south south-east of Jasper.

(see Sunwapta River)

Sunwapta Lake (lake)

83 C/3 - Columbia Icefield
28-37-23-W5
52°13'N 117°14'W
Approximately 100 km south-east of Jasper.

(see Sunwapta River)

Sunwapta Pass (pass)

83 C/3 - Columbia Icefield
25-37-22,23-W5
52°13'N 117°10'W
Approximately 100 km south-east of Jasper.

(see Sunwapta River)

Sunwapta Peak (peak)

83 C/6 - Sunwapta
10-39-23-W5
52°21'N 117°17'W

Approximately 75 km south south-east of Jasper.

This peak is 3265 m in altitude. (see Sunwapta River)

Sunwapta River (river)

83 C/12 - Athabasca Falls
16-41-26-W5
52°32'N 117°41'W
Flows west into Athabasca River, approximately 47 km south south-east of Jasper.

The name "Sunwapta" is the Stoney Indian term meaning "turbulent river." A.P. Coleman (see Mount Coleman) named this river in 1892.

Supply Creek (creek)

83 F/5 - Entrance
10-50-27-W5
53°18'N 117°52'W
Flows east into Brule Lake, approximately 19 km south-west of Hinton.

No information about the origin of the name is known.

Surprise Lake (lake)

83 C/13 - Medicine Lake
52°47'45"N 117°37'25"W
Approximately 23 km east south-east of Jasper.

According to members of the Park Warden Service, this lake was named by former Warden Larry McGuire in the 1950s. The name is likely descriptive of the lake's situation. Both the lake and the valley are completely enclosed with no visible drainage. The name was officially approved 5 May 1987.

Surprise Point (peak)

83 D/9 - Amethyst Lakes
6-43-2-W6
52°40'N 118°16'W

Approximately 26 km south-west of Jasper.

It took the climbing party longer to reach the top of this 2400 m peak than expected. The descriptive name was attached to this feature in 1916 by M.P. Bridgland. (see Mount Bridgland)

Survey Peak (peak)

82 N/15 - Mistaya Lake
36-34-21-W5
51°57'N 116°51'W
Approximately 120 km north-west of Banff.

This 2334 m peak was named in 1898 by J.N. Collie (see Collie Creek). He and H.E.M. Stutfield, an English climbing companion, climbed the peak in order to commence a plane table survey of the area.

Susa Creek (creek)

83 E/15 - Pierre Greys Lakes
23-57-7-W6
53°56'N 118°55'W
Flows east into Muskeg River, approximately 14 km east north-east of Grande Cache.

The origin of the name is unknown.

Swale Creek (creek)

83 C/9 - Wapiabi Creek
36-41-15-W5
52°35'N 116°01'W
Flows north-east into Colt Creek, approximately 12 km north north-east of Nordegg.

No information about the origin of the name is known.

Swan Creek (creek)

83 B/3 - Tay River
24-37-8-W5
52°12'N 115°01'W
Flows north into Prairie Creek, approximately 21 km south-west of Rocky Mountain House.

This well-established local name was officially adopted in 1941. The name was found on a Surveys and Engineering Branch map dated 1940. The name probably predates this map.

Swan Lake (lake)

83 B/3 - Tay River
26-36-9-W5
52°07′N 115°10′W
Approximately 29 km north-west of Caroline.

The origin of the name is unknown.

Swensen, Mount (mountain)

83 E/9 - Moberly Creek
7-53-3-W6
53°33′N 118°26′W
Approximately 80 km north north-west of Jasper.

This 2347 m mountain was named in 1949 after Flying Officer Stanley Powell Swensen of Calgary. He and P/O Gerry are two Albertans listed on the Honour Roll of 47 Canadians who died in the Battle of Britain, during the Second World War.

Swift Creek (creek)

83 E/14 - Grande Cache
18-57-10-W6
53°55′N 119°27′W
Flows east into Sheep Creek, approximately 21 km west north-west of Grande Cache.

The Swift family was the first white family at Hinton. Lewis John Swift was a homesteader who maintained his homestead during the formation of Jasper National Park and throughout the construction of the railway. This creek was likely named after Lewis Swift.

Swoda Creek (creek)

83 E/6 - Twintree Lake
9-50-9-W6
53°18′N 119°15′W

Flows north-west into Smoky River, approximately 93 km north-west of Jasper.

(see Swoda Mountain)

Swoda Mountain (mountain)

83 E/6 - Twintree Lake
34-49-8-W6
53°16′N 119°06′W
Approximately 82 km north-west of Jasper.

Originally known as Pamm Mountain, the name "Swoda" was recommended in 1923, as it is the Stoney Indian name of the nearby Smoky River. The name for the creek, which has its source in glaciers on Swoda Mountain, was changed at the same time. The name "Pamm" was composed of the first letters of the last names of the four members of the Alpine Club of Canada who made the first ascent of this 3003 m mountain in 1913. These were Donald Phillips (see Mount Phillips), A.F. Armistrad, K.D. McClelland and A.H. MacCarthy.

Syncline Brook (brook)

82 G/8 - Beaver Mines
36-4-4-W5
49°21′N 114°24′W
Approximately 20 km south-west of Beaver Mines.

(see Syncline Mountain)

Syncline Hills (hill)

83 E/14 - Grande Cache
27-57-8-W6
53°57′N 119°06′W
Approximately 7 km north of Grande Cache.

(see Syncline Mountain)

Syncline Mountain (mountain)

82 G/8 - Beaver Mines
2-5-4-W5
49°21′N 114°26′W

Approximately 21 km south-west of Beaver Mines.

According to the Alberta and British Columbia Boundary Survey, Part I (1917), Syncline Mountain was named after "a very apparent physical feature." The entire formation of this 2441 m feature comprises a syncline, which is "a fold in which the bed has been forced down in the middle, or up on the sides to form a trough." The rim of this syncline has three peaks, rising a further 152 to 213 metres.

Syncline Ridge (ridge)

83 F/4 - Miette
29-48-27-W5
53°10′N 117°56′W
Approximately 38 km south-west of Hinton.

The name of this ridge was officially approved in 1960. The prominent synclinal fold in the Devonian rocks inspired its descriptive name. (see Syncline Mountain)

Synge, Mount (mountain)

82 N/15 - Mistaya Lake
33-19-W5
51°48′N 116°39′W
Approximately 105 km north-west of Banff.

This 2972 m mountain was officially named 5 November 1918, after Captain Millington Henry Synge, R.E. (1823-1907). This soldier and author prepared a map showing the transcontinental railway route across the Rocky Mountains now followed by the Canadian Pacific Railway.

Table Mountain (mountain)

82 G/8 - Beaver Mines
6-5-2-W5
49°22′N 114°15′W
Approximately 28 km south-west of Pincher Creek.

This 2232 m mountain was named in 1916 due to its striking likeness to a table.

Talbot Lake (lake)

83 F/4 - Miette
35-47-28-W5
53°05′N 117°59′W
Approximately 46 km south-west of Hinton.

(see Mount Talbot)

Talbot, Mount (mountain)

83 E/12 - Pauline Creek
33-53-12-W6
53°37′N 119°43′W
Approximately 49 km south-west of Grande Cache.

The Honourable Peter Talbot (1854-1919) was the "pioneer politician" of Lacombe. He was a Member of the Legislative Assembly of the North West Territories from 1902-1905 and a Senate Member from 1906-1919. This 2373 m mountain was officially named 4 November 1925 after him.

Tallon Peak (peak)

82 G/9 - Blairmore
26-7-3-W5
49°35′N 114°19′W
Approximately 4 km north-east of Bellevue.

There are two possible origins of the name of this mountain. One source states that the name is spelled "Tallon," after L. Tallon, an assistant to W.S. Drewry in surveys of the Rocky Mountains in 1888 and 1892. The second option is suggested by local residents who maintain that it should be spelled "Talon" because the three ridges which radiate from the top resemble an eagle's talon. The name Tallon Peak was made official sometime before 1928.

Talus Lake (lake)

82 J/10 - Mount Rae
7-25-20-8-W5
50°44′N 115°59′W
Approximately 49 km west of Turner Valley.

The lake is situated in a very mountainous area and is surrounded by talus — an accumulation of rock fragments at the base of the steep slopes surrounding it. The descriptive name was officially approved 5 May 1987.

Tangle Ridge (ridge)

83 C/6 - Sunwapta
38-23,24-W5
52°18′N 117°17′W
Approximately 80 km south south-east of Jasper.

This ridge, which is 2598 m in altitude, was named descriptively by Mary T.S. Schäffer because the valley was difficult to travel through.

Tarpeian Rock (rock)

83 C/10 - George Creek
32-41-20-W5
52°34′N 116°50′W
Approximately 50 km west north-west of Nordegg.

This rock was named by R.W. Cautley (see Cautley Creek) because it reminded him of the cliff from which criminals of ancient Rome were killed.

Tatei Ridge (ridge)

83 E/3 - Mount Robson
48-8-W6
53°09′N 119°04′W
Approximately 73 km west north-west of Jasper.

The name of this ridge is the Stoney Indian word for "wind" and was proposed by C.D. Walcott (see Chetang Ridge) of the Alpine Club of Canada in 1912.

Taunton Creek (creek)

83 B/5 - Saunders
23-40-13-W5
52°27′N 115°46′W
Flows south into North Saskatchewan River, approximately 22 km east of Nordegg.

The origin of the name is unknown.

Tawadina Creek (creek)

83 B/5 - Saunders
31-38-11-W5
52°19′N 115°33′W
Flows north-east into Makwa Creek, approximately 40 km south-east of Nordegg.

The name "Tawadina" is an Indian word for "wild goose." It was officially adopted in 1941.

Tay Lake (lake)

83 B/3 - Tay River
5,6-36-8-W5
52°04′N 115°07′W
Approximately 26 km south-west of Caroline.

This locally well-established name was approved in 1978. The origin of the name is unknown.

Tay River (river)

83 B/3 - Tay River
33-35-8-W5
52°03′N 115°05′W

Flows east into Clearwater River, approximately 24 km south-west of Caroline.

This well-established local name was officially adopted in 1941. The name was found on a Surveys and Engineering Branch map dated 1940. The name probably predates this map.

Taylor Creek (creek)

82 N/8 - Lake Louise
15-27-15-W5
51°19′N 116°01′W
Flows east into Bow River, approximately 35 km west north-west of Banff.

This creek is named for its proximity to Taylor Lake. (see Taylor Lake)

Taylor Creek (creek)

83 F/3 - Cadomin
28-47-21-W5
53°05′N 117°01′W
Flows north into Beaverdam Creek, approximately 66 km south-west of Edson.

The origin for the name of this creek is unknown.

Taylor Lake (lake)

82 N/8 - Lake Louise
7-27-15-W5
51°18′N 116°06 W
Approximately 40 km west north-west of Banff.

This lake was named for George Herbert Taylor, a packer for A.O. Wheeler (see Wheeler Flats) on his Dominion Topographic Surveys.

Tecumseh, Mount (mountain)

82 G/10 - Crowsnest
20-8-5-W5
49°40′N 114°38′W
Approximately 11 km west north-west of Coleman.

This mountain, which is 2549 m in altitude, has the name of Tecumseh (Shooting Star), a great Shawnee Indian Chief (1768-1813) who fought alongside Sir Isaac Brock's British troops against American invaders in the War of 1812. Recommended by M.P. Bridgland (see Mount Bridgland), the name Mount Tecumseh finally became official 2 May 1957. Tecumseh was killed in the Battle of Moravian Town, Upper Canada, in 1813.

Teepee Creek (creek)

83 C/8 - Nordegg
27-39-16-W5
52°24′N 116°13′W
Flows east into Haven Creek, approximately 170 km east of Red Deer.

The precise origin of the name is unknown.

Teepee Creek (creek)

83 F/6 - Pedley
23-49-24-W5
53°15′N 117°24′W
Flows east into Gregg River, approximately 74 km south-west of Edson.

The origin for the locally well-established name for this creek, which was officially approved in May 1945, is unknown.

Teepee Pole Creek (creek)

82 O/14 - Limestone Mountain
3-34-8-W5
51°53′N 115°03′W
Flows south-east into James River, approximately 85 km north-east of Banff.

The Stoney Indians, a migratory band, set their teepees up on a flat at the edge of streams such as this. Soon after the 1920s, the tribe stopped coming to the area, but the teepee poles stood for many years as a reminder. The descriptive name for this creek was officially approved 2 December 1941.

Teitge Creek (creek)

83 E/9 - Moberly Creek
34-53-1-W6
53°37′N 118°03′W
Flows north-east into Pinto Creek, approximately 83 km north of Jasper.

This creek was named after a district forest ranger. No other origin information is available.

Tekarra Creek (creek)

83 D/16 - Jasper
10-45-1-W6
52°52′N 118°04′W
Flows north into Athabasca River, approximately 2 km south-east of Jasper.

(see Mount Tekarra)

Tekarra Lake (lake)

83 C/13 - Medicine Lake
52°50′20″N 117°55′10″W
Immediately east of Mount Tekarra, approximately 12 km east of Jasper.

(see Mount Tekarra)

Tekarra, Mount (mountain)

83 C/13 - Medicine Lake
2-45-28-W5
52°50′N 117°57′W
Approximately 10 km east south-east of Jasper.

This prominent mountain was named by Sir James Hector (see Mount Hector) in 1859 after Tekarra, a local Iroquois hunter. He, along with Mr. Moberly (see Moberly Flats), accompanied Hector on a journey toward Athabasca Pass along the Athabasca River in 1859.

Temple (railway point)

82 N/8 - Lake Louise
7-28-15-W5
51°23′N 116°06′W

Approximately 45 km north-west of Banff.

(see Mount Temple)

Temple Lake (lake)
82 N/8 - Lake Louise
51°21′58″N 116°11′00″W
Approximately 50 km west north-west of Banff.

The name for this lake was officially approved 21 January 1985. (see Mount Temple)

Temple, Mount (mountain)
82 N/8 - Lake Louise
32-27-16-W5
51°21′N 116°13′W
Approximately 50 km west north-west of Banff.

This mountain, which is 3543 m in altitude, was named in 1884 by Dr. G.M. Dawson (see Tombstone Mountain), after Sir Richard Temple, President of the Economic Science and Statistics section of the British Association. In the year the mountain was named, Temple was elected leader of the British Association excursion party to the Rockies.

Ten Peaks, Valley of the (peaks)
82 N/8 - Lake Louise
27-16-W5
51°21′N 116°10′W
Approximately 45 km west north-west of Banff.

There are ten mountain peaks which surround this valley formerly called "Desolation Valley." Named in 1895 by S.E.S. Allen (see Mount Allen), each peak was numbered with Stoney Indian numerals, starting from east to west. The names were as follows: Hee jee, No me, Yam ni, Ton sa, Sap ta, Shajk puy, Sagowa, Saknowa, Neptuak, and Wenkchemna. Most of the peaks have been renamed, for example,

Mount Fay, Mount Tuzo, Mount Allen, Deltaform Mountain.

Tent Mountain (mountain)
82 G/10 - Crowsnest
14-7-6-W5
49°33′N 114°42′W
Approximately 17 km south-west of Coleman, forming part of the Alberta-B.C. boundary.

Proposed by M.P. Bridgland (see Mount Bridgland), the name was officially approved 4 April 1911. The shape of this mountain, which is 2197 m in altitude, is likened to that of a tent; hence its name is appropriately descriptive.

Tent Mountain Pass (pass)
82 G/10 - Crowsnest
23-7-6-W5
49°34′N 114°42′W
Approximately 17 km west south-west of Coleman.

(see Tent Mountain)

Terminal Mountain (mountain)
83 D/16 - Jasper
12-44-2-W6
52°46′N 118°09′W
Approximately 12 km south-west of Jasper.

This 2835 m mountain is located at the end of a ridge. Its descriptive name was suggested by M.P. Bridgland (see Mount Bridgland) and was officially approved in 1916.

Terrace Mountain (mountain)
83 C/3 - Columbia Icefield
36-23-W5
52°05′N 117°10′W
Approximately 105 km south-east of Jasper.

This 2917 m mountain was named by Sir James Hector (see Mount Hector) of the Palliser Expedition (see Palliser Range). The

sides of the valley along the stream rise in a distinct terrace formation. The descriptive name was officially approved in 1920.

Terrapin Lake (lake)
82 J/13 - Mount Assiniboine
15-22-12-W5
50°52′N 115°35′W
Approximately 100 km west of Calgary.

(see Terrapin Mountain)

Terrapin Mountain (mountain)
82 J/13 - Mount Assiniboine
22-12-W5
50°53′N 115°35′W
Approximately 105 km west of Calgary, on the Alberta-B.C. boundary.

This 2926 m mountain was officially named in 1918 because of its fancied resemblance to a turtle.

Tershishner Creek (creek)
83 C/8 - Nordegg
34-38-17-W5
52°18′N 116°21′W
Flows south into North Saskatchewan River, approximately 133 km east south-east of Jasper.

Stoney Indians lived nearby and camped on this stream for many years. The term "tershishner" has referred to the creek for a long time, and means "burnt timber."

Tête, Roche (mountain)
83 D/15 - Rainbow
19-45-4-W6
52°53′N 118°34′W
Approximately 33 km west of Jasper.

This mountain, which is 2418 m in altitude, was officially named in 1918. The name was suggested by "Tête Jaune" (Yellow Head), the nickname of a trapper who used to store his furs near Tête Jaune Cache, British Columbia. (see Yellowhead Pass)

Tetley Creek (creek)

82 G/16 - Maycroft
10-10-2-W5
49°49′N 114°11′W
Flows east into Oldman River, approximately 30 km north-east of Blairmore.

The origin of the name is unknown.

Thistle Creek (creek)

83 C/15 - Cardinal River
15-44-20-W5
52°48′N 116°49′W
Flows east into Brazeau River, approximately 83 km east of Jasper.

This name is likely descriptive of the thistles that grow in the area.

Thistle Mountain (mountain)

83 C/14 - Mountain Park
2-44-22-W5
52°46′N 117°04′W
Approximately 69 km east south-east of Jasper.

This mountain, which is 2860 m in altitude, was named after the nearby creek. (see Thistle Creek)

Thompson Creek (creek)

83 F/3 - Cadomin
19-47-21-W5
53°05′N 117°03′W
Flows north into Beaverdam Creek, approximately 70 km south-west of Edson.

The name for this creek was suggested by J.R. Matthews, a local resident. It was officially approved in 1944. No other information is available.

Thompson, Mount (mountain)

82 N/10 - Blaeberry River
20-31-18-W5
51°40′N 116°31′W

Approximately 85 km north-west of Banff.

Mount Thompson is 3084 m in altitude. (see Thompson Pass)

Thompson Pass (pass)

83 C/3 - Columbia Icefield
32-35-23-W5
52°03′N 117°16′W
Approximately 105 km south-east of Jasper, on the Alberta-B.C. boundary.

This pass was named by J.T. Collins after Mr. C.S. Thompson, a member of the Appalachian Mountain Club who came this way at the turn of this century looking for a practical passage across the Continental Divide. Thompson was a very enthusiastic mountaineer, particularly in the Rockies and Selkirks.

Thordarson (railway point)

83 E/15 - Pierre Greys Lakes
32-57-7-W6
53°58′N 119°00′W
Approximately 11 km north-east of Grande Cache.

This railway point was named 8 March 1969. The precise origin is not known; however, it was previously known as Flood and Halvorson. The origin for these names is not known either.

Thoreau Creek (creek)

83 E/9 - Moberly Creek
22-52-4-W6
53°31′N 118°30′W
Flows south-east into Wildhay River, approximately 76 km north north-west of Jasper.

The origin of the locally well-established name for this creek, which was approved in 1946, is unknown.

Thoreau Creek Pass (pass)

83 E/10 - Adams Lookout
53-4-W6
53°32′10″N 118°32′20″W
Approximately 55 km south-east of Grande Cache.

The name for this low mountain pass was officially approved 9 July 1980. The name for this 2200 m pass was taken from its location between Thoreau Creek and Pope Creek. (see Thoreau Creek)

Thorington Tower (mountain)

83 C/5 - Fortress Lake
52°21′30″N 117°30′05″W
Approximately 67 km south-east of Jasper.

Thorington Tower was named for Dr. James Munroe Thorington, an American opthalmologist born in 1874. During the 1920s he began his explorations in the Canadian Rockies, primarily in the present day Banff-Jasper regions. Thorington was involved in a number of pioneer ascents, including North Twin in 1923. He was associated with and an honorary member of numerous alpine, mountaineering and conservation organizations in North America and Europe.

Thornton Creek (creek)

83 C/14 - Mountain Park
4-46-23-W5
52°56′N 117°17′W
Flows north-east into McLeod River, approximately 52 km east north-east of Jasper.

This "small creek which joins the McLeod River about a mile after its start" was named jokingly by one of the new miners after Robert Thornton (see Cheviot Mountain), according to Toni Ross's *Oh! The Coal Branch* (1974).

Thornton, Mount (mountain)

83 E/1 - Snaring
12-48-3-W6
53°08'N 118°19'W
Approximately 33 km north north-west of Jasper.

One of 12 features named in 1954 for Alberta's 50th Jubilee, (see Mount Griesbach), this 2752 m mountain was named after Sir Henry Thornton (1871-1933), President of the Canadian National Railway, 1922-1932. In 1917 his services were commandeered by the British Government, and he became Assistant General-Director of Movements and Railways in France. He was appointed Inspector General of Transportation in 1919 and retained this position until 1922.

Thornton Pass (pass)

83 D/16 - Jasper
52°58'45"N 118°26'45"W
Approximately 26 km north-west of Jasper.

Thornton Pass was named for its proximity to Mount Thornton. (see Mount Thornton)

Three Isle Creek (creek)

82 J/11 - Kananaskis Lakes
30-19-9-W5
50°38'N 115°14'W
Flows east into Upper Kananskis River, approximately 65 km west of Turner Valley.

This creek drains Three Isle Lake.

Three Isle Lake (lake)

82 J/11 - Kananaskis Lakes
26-19-9-W5
50°37'N 115°16'W
Approximately 70 km west of Turner Valley.

No information regarding the origin of the name is known.

Three Lakes Ridge (ridge)

82 G/1 - Sage Creek
3-3-W5
49°14'N 114°24'W
Approximately 40 km north-west of Waterton Park, on the Alberta-B.C. boundary.

The origin of the name is unknown.

Three Sisters, The (mountains)

82 O/3 - Canmore
24-10-W5
51°02'N 115°20'W
Approximately 7 km south of Canmore.

Named by G.M. Dawson (see Tombstone Mountain) in 1886, these three peaks in the same ridge resemble each other. They were previously known as the "Three Nuns" because of a heavy veil of snow found one morning on their north side.

The Three Sisters, ca. 1895

Three Sisters Creek (creek)

82 O/3 - Canmore
21-24-10-W5
51°04'N 115°20'W
Flows north into Bow River, approximately 4 km south-east of Canmore.

The name was officially approved 11 April 1911. (see The Three Sisters)

Threepoint Creek (creek)

82 J/9 - Turner Valley
34-20-2-W5
50°44'N 114°12'W
Flows south-east into Sheep River, approximately 8 km north-east of Turner Valley.

The name Threepoint Creek was officially approved 6 October 1949, because of its proximity to the mountain. (see Threepoint Mountain)

Threepoint Mountain (mountain)

82 J/10 - Mount Rae
19-20-6-W5
50°42'N 114°50'W
Approximately 39 km west of Turner Valley.

The descriptive name for this three-pointed mountain, which is 2595 m in altitude, was approved prior to 1928.

Throne Mountain (mountain)

83 D/9 - Amethyst Lakes
31-42-1-W6
52°40'N 118°08'W
Approximately 24 km south of Jasper.

The summit of this 3120 m mountain resembles the shape of a high-backed chair. The descriptive name was applied by M.P. Bridgland (see Mount Bridgland) in 1916.

Thunder Lake (lake)

83 C/15 - Cardinal River
4-45-19-W5
52°51'N 116°41'W
Approximately 82 km east of Jasper.

Officially approved 24 November 1911. The origin of the name is unknown.

Thunder Mountain (mountain)

82 G/16 - Maycroft
21-10-3-W5
49°50'N 114°21'W

Approximately 25 km north of Blairmore.

Named by M.P. Bridgland (see Mount Bridgland) 30 June 1915, this mountain measures 2352 m in altitude. The origin of the name is unknown.

Thunderbolt Peak (peak)

83 D/9 - Amethyst Lakes
27-42-2-W6
52°39′N 118°13′W
Approximately 27 km south-west of Jasper.

The summit of this 2665 m mountain peak was shattered by lightning. It was officially named in 1916.

Tiecamp Creek (creek)

83 F/5 - Entrance
5-52-24-W5
53°27′N 117°30′W
Flows south-east into Athabasca River, approximately 4 km north-east of Hinton.

The locally well-established name of this stream was officially approved in 1944. No other origin information is known.

Tilted Lake (lake)

82 N/8 - Lake Louise
15-29-15-W5
51°29′N 116°01′W
Approximately 45 km north-west of Banff.

(see Tilted Mountain)

Tilted Mountain (mountain)

82 O/5 - Castle Mountain
15-29-15-W5
51°30′N 116°00′W
Approximately 45 km north-west of Banff.

This 2591 m mountain was named in 1911. The name is descriptive of the rocks composing the mountain.

Timber Creek (creek)

82 J/1 - Langford Creek
35-14-3-W5
50°12′N 114°18′W
Flows north-east into Willow Creek, approximately 55 km north-west of Claresholm.

The origin of the name is unknown.

Timber Creek (creek)

82 O/13 - Scalp Creek
18-34-12-W5
51°56′N 115°41′W
Flows north-west into Clearwater River, approximately 85 km north north-west of Banff.

No information regarding the name is known.

Tip Top Ridge (ridge)

83 E/9 - Moberly Creek
52,53-3,4-W6
53°33′N 118°24′W
Approximately 75 km north north-west of Jasper.

The name was officially changed 19 February 1959. (see Berland Range)

Titkana Peak (peak)

83 E/3 - Mount Robson
23-48-8-W6
53°09′N 119°04′W
Approximately 74 km west north-west of Jasper.

The name of this 2804 m mountain peak is the Stoney Indian word for "bird." It was officially approved in 1912.

*Tod Creek (former locality)

82 G/16 - Maycroft
25-9-3-W5
49°46′N 114°16′W

Approximately 20 km north-east of Blairmore.

This creek was named after William Todd, who settled in the area in 1886. In 1915 it was recognized that the name was incorrectly spelled, but the spelling has not been changed. The post office closed in September 1964.

Todd Creek (creek)

82 G/9 - Blairmore
31-7-1-W5
49°36′N 114°08′W
Flows south into Crowsnest River, approximately 3 km north-east of Lundbreck.

This creek is named after William Todd, who settled in the area in 1886. (see Tod Creek)

Tom Creek (creek)

83 E/16 - Donald Flats
3-56-2-W6
53°49′N 118°12′W
Flows south-east into Big Creek, approximately 63 km east south-east of Grande Cache.

A tree near this stream has the words Tom Creek carved on its trunk. The name was made official in 1947.

Toma Creek (creek)

83 C/14 - Mountain Park
20-45-22-W5
52°53′N 117°10′W
Flows north into Cardinal River, approximately 62 km east of Jasper.

(see Mount Toma)

Toma, Mount (mountain)

83 C/14 - Mountain Park
25-44-23-W5
52°49′N 117°12′W

Approximately 58 km east south-east of Jasper.

Toma was the Iroquois canoe-man who was a member of the Southesk party (see Southesk River) in 1859. The name for this 2760 m mountain was officially approved in 1925.

Tomahawk Mountain (mountain)

82 O/13 - Scalp Creek
20-32-13-W5
51°46'N 115°49'W
Approximately 70 km north north-west of Banff.

The origin of the name of this 2919 m mountain is unknown.

Tombstone Lakes (lakes)

82 J/10 - Mount Rae
13-20-8-W5
50°41'25"N 114°59'25"W
Approximately 50 km west of Turner Valley.

These two small alpine lakes are named because of their proximity to Tombstone Mountain. The name was officially approved 7 February 1983. (see Tombstone Mountain)

Tombstone Mountain (mountain)

82 J/11 - Kananaskis Lakes
11-20-8-W5
50°41'N 115°02'W
Approximately 50 km west of Turner Valley.

This mountain shows a very peculiar collection of pinnacle-like slabs resembling tombstones near its summit. Dr. George Mercer Dawson (1849-1901) was Director of the Geological Survey of Canada from 1875-1901 (see Introduction). He gave this 3035 m mountain its descriptive name in 1884.

Tonquin Hill (hill)

83 D/9 - Amethyst Lakes
26-43-3-W6
52°44'N 118°20'W

Approximately 23 km south-west of Jasper.

This hill was named by E. Deville in 1916 to commemorate the ship "The Tonquin," which carried the Astor Expedition to the mouth of the Columbia River in 1810. The ship, anchored in Clayoquot Sound, Vancouver Island, was attacked by Indians, who slew most of the crew. The survivors blew up the ship, killing all who remained on board. The nearby features take their names from this 2396 m hill.

Tonquin Pass (pass)

83 D/9 - Amethyst Lakes
26-43-3-W6
52°44'N 118°21'W
Approximately 24 km south-west of Jasper.

(see Tonquin Hill)

Tonquin Valley (valley)

83 D/9 - Amethyst Lakes
43-2-W6
52°43'N 118°16'W
Approximately 22 km south-west of Jasper.

(see Tonquin Hill)

Topaz Lake (lake)

83 E/7 - Blue Creek
8-51-6-W6
53°23'N 118°51'W
Approximately 76 km north-west of Jasper.

The name for this lake was adopted 1 May 1934 and is likely descriptive of the colour of the water.

Tornado Mountain (mountain)

82 G/15 - Tornado Mountain
5-12-5-W5
49°58'N 114°39'W
Approximately 40 km north-west of Coleman, on the Alberta-B.C. boundary.

M.P. Bridgland (see Mount Bridgland) recommended approval for the name of this 3236 m mountain in 1915. The name was officially applied 30 June 1915. The name is descriptive name in that the mountain appears to be the storm centre of the locality. Storm clouds and lightning flashes are rumoured to be seen around the summit quite frequently.

Tornado Pass (pass)

82 G/15 - Tornado Mountain
6-12-5-W5
49°58'N 114°40'W
Approximately 40 km north-west of Coleman, immediately west of Tornado Mountain.

(see Tornado Mountain)

Torrens, Mount (mountain)

83 L/5 - Two Lakes
23-61-14-W6
54°17'N 119°59'W
Approximately 75 km west north-west of Grande Cache.

This mountain, which is 2220 m in altitude, was named in 1922, on the suggestion of R.W. Cautley (a Commissioner on the Alberta-British Columbia Boundary Survey, 1918-1924). Sir Robert Richard Torrens (1814-1884) emigrated to Australia in 1840 from Ireland. He was a lawyer, he became the first Premier and Colonial Treasurer of South Australia in 1857, and introduced the well-known Torrens System of Land Titles in 1858.

Torrens Ridge (ridge)

83 L/5 - Two Lakes
19-61-13-W6
54°17'N 119°57'W
Approximately 77 km west north-west of Grande Cache.

(see Mount Torrens)

Torrens River (river)

83 L/5 - Two Lakes
27-62-13-W6
54°23′N 119°52′W
Flows north into Narraway River,
approximately 73 km north-west of Grande
Cache.

(see Mount Torrens)

Tory, Mount (mountain)

83 E/1 - Snaring
15-49-3-W6
53°13′N 118°20′W
Approximately 43 km north north-west of
Jasper.

This 2831 m mountain was one of 12 features
named in 1954 for Alberta's 50th Jubilee (see
Mount Griesbach). It was named after Henry
Marshall Tory (1864-1947). In the 1890s, he
became a professor at McGill University and
helped to establish the University of British
Columbia by founding McGill College in
Vancouver. Tory became the first President
of the University of Alberta in 1908 — a
position he held until 1928, when he left to
become President of the National Research
Council in Ottawa. Dr. Tory was one of the
founders of the Khaki University during the
First World War, was a member of several
Royal Commissions and Enquiries, and was
President and founder of Carleton College,
Ottawa (later Carleton University).

Totem Creek (creek)

82 N/15 - Mistaya Lake
29-33-19-W5
51°52′N 116°40′W
Flows south-west into Mistaya River,
approximately 105 km north-west of Banff.

The origin of the name is unknown.

Totem Tower (peak)

82 N/15 - Mistaya Lake
21-33-19-W5
51°50′N 116°37′W

Approximately 105 km north-west of Banff.

The name for this 3105 m mountain peak
was officially approved by the C.P.C.G.N. in
June 1971, for its proximity to Totem Creek.
(see Totem Creek)

Tower Lake (lake)

82 O/5 - Castle Mountain
27-14-W5
51°18′N 115°55′W
Approximately 28 km north west of Banff.

This lake is not far from the north tower of
Castle Mountain. The name of this lake was
applied because of its proximity to this
tower.

Tower of Babel (mountain)

82 N/8 - Lake Louise
27-16-W5
51°19′N 116°10′W
Approximately 45 km west north-west of
Banff.

This isolated rock monolith east of Moraine
Lake is 2310 m in altitude. W.D. Wilcox (see
Wilcox Pass), forty years a member of the
Alpine Club of Canada, was one of the
pioneers in the Lake Louise district who
wrote about and photographed the Rocky
Mountains. He named the Tower of Babel
for its curious shape and impressive rock
cliff which suggested to him a large tower.
The story of the biblical town of Babel may
be found in Genesis 11:1-9: "All people had
one common language and were able to
settle, build and plan great feats at one time
in history. One of these plans was to build a
tower that would reach the Heavens, so as to
make a name for the people of Babel. This
angered God, and he thwarted their attempt
by confounding their language and scattering
their numbers." Babel, the native name for
Babylon, means the "gate of God." The
application for official approval of this name
was made 25 November 1910.

Towers, The (mountain)

82 J/13 - Mount Assiniboine
22-12-W5
50°53′N 115°36′W
Approximately 105 km west of Calgary, on
the Alberta-B.C. boundary.

The name is descriptive of the numerous
turrets seen on this 2846 m mountain. A.O.
Wheeler (see Wheeler Flats) gave this feature
its name in 1917, and the name was made
official in 1918.

Trail Creek (creek)

82 J/15 - Bragg Creek
22-22-7-W5
50°52′N 114°53′W
Flows north into Prairie Creek,
approximately 50 km west of Calgary.

The name was officially approved 6 October
1949. The origin of the name is unknown.

Trail Creek (creek)

83 F/6 - Pedley
4-52-24-W6
53°28′N 117°28′W
Flows north-west into Athabasca River,
approximately 69 km west south-west of
Edson.

The local name for this creek was made
official 4 May 1944. The origin of the name
is unknown.

Trail Hill (mountain)

82 G/10 - Crowsnest
27-7-6-W5
49°35′N 114°43′W
Approximately 17 km west south-west of
Coleman.

There are actually two peaks to this hill,
which is 1758 m in altitude, making it
appear that there is a trail between the peaks.

Trap Lake (lake)

83 L/12 - Nose Creek
18-65-13-W6
54°37′N 119°58′W
Approximately 95 km south-west of Grande Prairie.

Several traps were observed along the shore of this lake when J.V. Butterworth, a field topographer, and a survey pack-train were in the area circa 1946.

Trapper Creek (creek)

83 C/12 - Athabasca Falls
52°41′25″N 117°36′30″W
Approximately 26 km south-east of Jasper.

This creek was named circa 1930. No other origin information is known.

Trapper Creek (creek)

83 F/3 - Cadomin
18-48-23-W5
53°08′N 117°21′W
Flows north into Mary Gregg Creek, approximately 77 km south-west of Edson.

The origin of the name is unknown.

Trapper Pass (pass)

83 C/12 - Athabasca Falls
52°37′30″N 117°39′30″W
Approximately 35 km south-east of Jasper.

No information regarding the origin of the name is known.

Trapper Peak (peak)

82 N/10 - Blaeberry River
25-31-19-W5
51°41′N 116°35′W
Approximately 90 km north-west of Banff, on the Alberta-B.C. boundary.

This 2984 m mountain peak was named in 1918 after Bill Peyto, a noted guide and trapper of the area. (see Peyto Peak)

Treadmill Ridge (ridge)

83 E/2 - Resplendent Creek
49-7-W6
53°13′N 118°54′W
Approximately 65 km west north-west of Jasper.

The descriptive name for this ridge was applied in 1923 by A.O. Wheeler. (see Wheeler Flats)

Trefoil Lakes (lakes)

83 D/16 - Jasper
22-45-1-W6
52°53′N 118°03′W
Approximately 3 km north-east of Jasper.

The group of three lakes comprising this feature is arranged in a trinity resembling the petals of a clover leaf. The descriptive name was applied in 1914 by H. Matheson, of the Dominion Land Survey.

Trench, The (pass)

83 C/3 - Columbia Icefield
52°08′10″N 117°21′10″W
At the head of the Columbia Glacier, approximately 90 km south-east of Jasper.

The descriptive name of this feature is well-established in local usage.

Trench Creek (creek)

83 E/13 - Dry Canyon
58-13-W6
53°59′N 119°49′W
Flows north-west into Goat Creek, approximately 47 km west north-west of Grande Cache.

This creek is likely named after Ruby Trench Thompson, a pioneer who was one of the first inhabitants of this area at the turn of the century. She was the author of *Pioneers of Athabasca* (Raymond Thompson Company, Washington, 1967), which was an account of life in the wilderness. The name of this creek was officially approved in 1958.

Triad Peak (peak)

83 C/4 - Clemenceau Icefield
30-36-25-W5
52°07′N 117°33′W
Approximately 90 km south-east of Jasper, on the Alberta-B.C. boundary.

The name for this peak, which was officially approved in 1965, is descriptive of its shape.

Trident Lake (lake)

82 N/16 - Siffleur River
6-33-15-W5
51°48′N 116°07′W
Approximately 80 km north-west of Banff.

The position of three adjacent streams gives this lake the appearance of a mythical fish spear, or trident, used by Poseidon or Neptune. The name is therefore appropriately descriptive.

Trident Range (range)

83 D/16 - Jasper
44-2-W6
52°46′N 118°12′W
Approximately 15 km south-west of Jasper.

James Outram (see Mount Outram) suggested the name for this mountain range because of its shape.

Trifid Glacier (glacier)

82 O/5 - Castle Mountain
51°28′50″N 115°57′30″W
Approximately 35 km north-west of Banff.

The descriptive name for this three-lobed glacier was first used by C.D. Walcott in 1922. The locally well-established name was officially approved 23 June 1981.

Tripoli Mountain (mountain)

83 C/14 - Mountain Park
30-45-23-W5
52°54′N 117°20′W

Approximately 50 km east of Jasper.

The locally well-established name for this 2620 m mountain was reported in 1922. The origin of the name is unknown.

Troll Falls (falls)

82 J/14 - Spray Lakes Reservoir
16-11-23-9-W5
50°56′50″N 115°08′40″W
Approximately 78 km west of Calgary.

The falls have been known by the name Troll Falls since about 1972, when Don Gardener, archaeologist and recreational trails consultant, began using the name. The falls are situated in a dark recess in the rocks, and it is possible to crawl in behind the fall. The name was officially approved 12 September 1984, recalling the mythical Scandanavian creatures who lived under bridges, in caves or in other dark dwellings.

Trutch, Mount (mountain)

82 N/10 - Blaeberry River
3-32-21-W5
51°42′N 116°53′W
Approximately 110 km north-west of Banff, on the Alberta-B.C. boundary.

This 3258 m mountain was named in 1920 after Sir Joseph Trutch (1826-1904). He was born in England, went to California in 1849, lured by the Gold Rush, and worked his way north until he arrived in British Columbia in 1859. He became the first Lieutenant Governor of the province in 1871.

Tunnel Mountain (mountain)

82 O/4 - Banff
36-25-12-W5
51°11′N 115°33′W
Approximately 1 km east south-east of Banff.

At one point of the construction on the Canadian Pacific Railway the railway company was prepared to drive a tunnel

through this mountain. The tunnel was never made because another route was found, but the name Tunnel Mountain for this 1692 m feature was retained.

Turbine Canyon (canyon)

82 J/11 - Kananaskis Lakes
12-20-10-W5
50°41′N 115°15′W
Approximately 70 km west of Turner Valley.

The origin of the name is unknown.

Turbulent Creek (creek)

82 J/14 - Spray Lakes Reservoir
14-22-11-W5
50°52′N 115°25′W
Flows south-east into Spray Lakes Reservoir, approximately 90 km west of Calgary.

(see Mount Turbulent)

Turbulent, Mount (mountain)

82 J/14 - Spray Lakes Reservoir
10-23-11-W5
50°56′N 115°28′W
Approximately 90 km west of Calgary.

This mountain, which is 2743 m altitude, was named in 1915 after a destroyer engaged in the Battle of Jutland in the First World War. (see Jutland Mountain)

Turner, Mount (mountain)

82 J/14 - Spray Lakes Reservoir
9-22-11-W5
50°51′N 115°29′W
Approximately 90 km west of Calgary.

This 2813 m mountain was named in 1918 after Lieutenant-General Sir Richard Ernest William Turner, K.C.M.G., K.C.B., D.S.O. (1871-1961). He commanded the Canadian troops who, in April 1915, held their position at Ypres during the first gas attacks of the First World War. He commanded the 2nd Canadian Division in France, 1915-1916.

He then became responsible for the training of all Canadian troops in Britain and later served as Chief of the General staff at Canadian Military Headquarters, London, England. He returned to Canada after the War to serve on the Canadian Pension Tribunal and the Canadian Pension Commission.

Turquoise Lake (lake)

82 N/9 - Hector Lake
18-30-17-W5
51°34′N 116°24′W
Approximately 75 km north-west of Banff.

The name for this lake was applied in 1897 by Mr. G.P. Baker (see Mount Baker) of The Alpine Club (London) to describe the lake's colour.

Turret, The (mountain)

82 J/11 - Kananaskis Lakes
14-1-19-9-W5
50°36′N 115°08′W
Approximately 95 km south south-west of Calgary.

This mountain, which is 2500 m in altitude, has a name descriptive of its appearance. Mrs. H.E. Miller and her husband proposed the name. They own a cottage on the Lower Kananaskis Lake from which they can see The Turret.

Turret Ridge (ridge)

83 E/14 - Grande Cache
10-56-10-W6
53°49′N 119°25′W
Approximately 22 km west of Grande Cache.

The locally well-established name for this ridge was suggested in 1916 by James Outram (see Mount Outram). The descriptive name was made official 1 September 1955.

Turtle Mountain (mountain)

82 G/9 - Blairmore
24-7-4-W5
49°35′N 114°25′W
Approximately 3 km south of Frank.

Turtle Mountain was named in 1880 by
Louis O. Garnett, an early rancher in the
Crowsnest area, due to its striking
resemblance to the shape of a turtle. A rock
slide in 1903 rather destroyed the likeness.

Tuzo, Mount (mountain)

82 N/8 - Lake Louise
8-27-16-W5
51°18′N 116°13′W
Approximately 50 km west north-west of
Banff, on the Alberta-B.C. boundary.

This mountain, which is 3246 m in altitude,
is number seven of the Ten Peaks (see Ten
Peaks, Valley of the) and was named in 1907
in honour of Miss Henrietta L. Tuzo, who
in 1906 was the first to climb it. She was a
charter member of The Alpine Club of
Canada.

Twelve Mile Creek (creek)

83 F/12 - Gregg Lake
10-53-27-W5
53°34′N 117°53′W
Flows north into Wildhay River,
approximately 46 km north-west of Hinton.

This creek is located twelve miles from
Entrance by the old pack trail. The name is
locally well-established.

Twin Cairns (cairn)

82 O/4 - Banff
51°04′30″N 115°48′20″W
Approximately 20 km south-west of Banff.

This feature has two points on its top which
look like cairns. The name was officially
approved 29 July 1986.

Twin Falls (falls)

82 H/4 - Waterton Lakes
1-30-W4
49°01′N 113°52′W
Approximately 50 km south of Pincher
Creek.

The name for this feature was officially
approved by the Canadian Permanent
Committee on Geographical Names 8 June
1971. The origin of the name is unknown.

Twin Lakes (lakes)

82 G/1 - Sage Creek
23-2-2-W5
49°08′N 114°09′W
Approximately 20 km north-west of
Waterton Park.

The exact origin of the name is unknown,
but it is likely descriptive.

Twin Lakes (lakes)

82 O/4 - Banff
11-26-15-W5
51°12′N 115°59′W
Approximately 29 km north north-west of
Banff.

The descriptive name of these two small
lakes was suggested by M.P. Bridgland (see
Mount Bridgland) in 1915.

Twin Lakes (lakes)

83 D/16 - Jasper
35-44-1-W6
52°52′N 118°04′W
Approximately 1 km south-east of Jasper.

The name of these lakes is descriptive.

Twin Peaks (peaks)

82 J/2 - Fording River
8-13-4-W5
50°04′N 114°31′W
Approximately 70 km south-west of Turner
Valley.

H.M. Stutfield (see Stutfield Peak) and J.N.
Collie (see Collie Creek) gave this peak its
descriptive name.

Twins, The (peaks)

83 C/3 - Columbia Icefield
25-37-25-W5
52°12′50″N 117°26′00″W
Approximately 85 km south-east of Jasper.

This double-headed mountain was named by
H.M. Stutfield (see Stutfield Peak) and J.N.
Collie (see Collie Creek) in 1898. The entire
massif is called The Twins, but its two peaks
are named North Twin (3684 m) and South
Twin (3559 m). The decision to name the
peaks separately was approved 28 February
1980.

Twins Tower (peak)

83 C/3 - Columbia Icefield
52°13′45″N 117°26′55″W
Approximately 80 km south-east of Jasper.

This peak is so named because of its
proximity to the Twins (see the Twins). The
name was officially approved 7 November
1984.

Twintree Creek (creek)

83 E/6 - Twintree Lake
3-52-9-W6
53°27′N 119°14′W
Flows north-west into Smoky River,
approximately 102 km west north-west of
Jasper.

(see Twintree Mountain)

Twintree Lake (lake)

83 E/6 - Twintree Lake
10-51-8-W6
53°23′N 119°06′W
Approximately 87 km west north-west of
Jasper.

(see Twintree Mountain)

Twintree Mountain (mountain)

83 E/6 - Twintree Lake
10-51-8-W6
53°23′N 119°09′W
Approximately 92 km west north-west of Jasper.

At the time of the mountain's naming in 1918, the lake which it dominates contained two small islets. Each islet had a lone spruce tree on it, hence the name of the lake. The 2544 m mountain was named because of the lake, and the creek takes its name from the mountain.

Two Lakes (lakes)

83 L/5 - Two Lakes
29-62-12-W6
54°22′N 119°45′W
Approximately 70 km north-west of Grande Cache.

The locally well-established name for these lakes has been in use for at least 30 years. The individual lakes are sometimes referred to as "Two Lakes North" and "Two Lakes South."

Two Cabin Creek (creek)

83 E/14 - Grande Cache
5-57-8-W6
53°53′N 119°09′W
Flows west into Smoky River, approximately 2 km west north-west of Grande Cache.

The origin of the name is unknown.

Two Dam Creek (creek)

83 B/3 - Tay River
37-10-W5
52°12′N 115°21′W
Flows west into Prairie Creek, approximately 35 km south-west of Rocky Mountain House.

This name, which was officially adopted in 1941, was given to the creek because of the remains of two very prominent beaver dams located on the creek. The name was suggested by Mr. H.P. Brownlee, a statistician for the Department of Trade and Industry, and the name is locally well-established.

Two Jack Lake (lake)

82 O/4 - Banff
26-11-W5
51°14′N 115°30′W
Approximately 7 km north-east of Banff.

The lake was named for John "Captain Jack" Standley (1865-1946) and John "Jack" Watters (1878-1950), both of whom were closely associated with the area. The name was officially approved in 1959.

Two O'Clock Creek (creek)

83 C/1 - Whiterabbit Creek
5-36-17-W5
52°04′N 116°24′W
Approximately 105 km west south-west of Rocky Mountain House.

The Canadian Permanent Committee on Geographical Names approved the name for this creek 24 February 1977. It was so named because meltwater from the mountain snow and ice increased its flow so much by the afternoon that crossing it became dangerous. Travellers usually arranged to make their crossing before two o'clock in the afternoon.

Two Pines (hill)

82 J/15 - Bragg Creek
13-23-5-W5
50°58′N 114°34′W
Approximately 30 km west of Calgary.

The name was officially approved 6 October 1949. No other origin information is known.

Two Valley Creek (creek)

83 C/13 - Medicine Lake
19-45-27-W5
52°54′N 117°55′W
Flows south into Maligne River, approximately 12 km east of Jasper.

The origin of the name is unknown.

Tyrrell Creek (creek)

82 O/12 - Barrier Mountain
22-31-13-W5
51°40′N 115°46′W
Flows south-east into Red Deer River, approximately 55 km north north-west of Banff.

(see Mount Tyrrell)

Tyrrell, Mount (mountain)

82 O/12 - Barrier Mountain
31-31-13-W5
51°42′N 115°50′W
Approximately 60 km north north-west of Banff.

This 2755 m mountain was named after Joseph Burr Tyrrell (1858-1957). An assistant to Dr. G.M. Dawson, (see Tombstone Mountain) in surveys of the Rocky Mountains in 1883, Tyrrell was a member of the Geological Survey (1880-1897) and president of the Champlain Society (1927-1932). He published many reports and papers on his explorations as well as on geological, mining, and historical subjects.

Tyrwhitt, Mount (mountain)

82 J/11 - Kananaskis Lakes
3-19-8-W5
50°35′N 115°01′W
Approximately 55 km west south-west of Turner Valley, on the Alberta-B.C. boundary.

Rear Admiral Sir Reginald Y. Tyrwhitt, Commander of the Horwich Force, commanded British destroyer flotillas during the First World War. This 2874 m mountain was officially named in 1918.

Unnecessary Mountain (mountain)

82 G/9 - Blairmore
16-9-2-W5
49°46′15″N 114°13′10″W
Approximately 34 km north-west of Pincher Creek, between Todd and Wildcat Creeks.

During the Depression, the Lynch-Stauntons had a hired man who thought it unnecessary to farm to the top of the ridge. The feature is inconveniently located in the middle of a field. The name was officially approved 21 July 1982.

Unwin, Mount (mountain)

83 C/12 - Athabasca Falls
16-42-25-W5
52°36′N 117°32′W
Approximately 47 km south-east of Jasper.

This 3268 m mountain was named by Mary Schäffer (see Maligne Lake) after her second guide, Sidney J. Unwin. Unwin, who joined The Alpine Club of Canada in 1908, was the first to climb this mountain. He was killed in action during the First World War in 1917.

Upper Bertha Falls (falls)

82 H/4 - Waterton Lakes
1-30-W4
49°02′N 113°56′W
Approximately 50 km south of Pincher Creek.

This name was officially approved by the Canadian Permanent Committee on Geographical Names 8 June 1971. The origin of the name is unknown.

Upper Colefair Lake (lake)

83 C/12 - Athabasca Falls
52°44′00″N 117°51′40″W
Approximately 21 km south-east of Jasper.

The origin of this locally well-established name is unknown.

Upper Kananaskis Falls (falls)

82 J/11 - Kananaskis Lakes
21-19-9-W5
50°37′35″N 115°11′15″W
Approximately 65 km west of Turner Valley.

These falls may be found on the Upper Kananaskis River and were officially named 8 November 1978. (see Kananaskis Range)

Upper Kananaskis Lake (lake)

82 J/11 - Kananaskis Lakes
22-19-9-W5
50°37′N 115°09′W
Approximately 60 km west of Turner Valley.

(see Kananaskis Range and Hogue Island)

Upper Kananaskis River (river)

82 J/11 - Kananaskis Lakes
21-19-9-W5
50°37′N 115°10′W
Flows south-east into Upper Kananaskis Lake, approximately 65 km west of Turner Valley.

(see Kananaskis Range)

Upper Longview Lake (lake)

83 C/7 - Job Creek
22-39-21-W5
52°22′30″N 116°55′40″W
Approximately 95 km south-east of Jasper.

(see Lower Longview Lake)

Upper Spray Falls (falls)

82 O/3 - Canmore
7-24-9-W5
51°02′N 115°15′W
Approximately 10 km south-east of Canmore.

(see Spray River)

Upper Waterton Lake (lake)

82 H/4 - Waterton Lakes
14-1-30-W4
49°02′N 113°54′W
Approximately 50 km south of Pincher Creek.

The name is used for specific identification, as each lake is a clearly defined entity in its own right. The name was officially approved 7 September 1969. (see Waterton Lakes)

Upright Mountain (mountain)

83 E/2 - Resplendent Creek
31-48-6-W6
53°11′N 118°51′W
Approximately 63 km west north-west of Jasper.

This mountain was so named in 1911 because the strata of which it is composed has been upheaved to an almost vertical position.

Upright Pass (pass)

83 E/2 - Resplendent Creek
11-48-6-W6
53°08′N 118°47′W
Approximately 54 km west north-west of Jasper.

This pass is 1972 m in altitude. (see Upright Mountain)

Utopia Creek (creek)

83 F/4 - Miette
34-47-26-W5
53°06′N 117°43′W
Flows north-east into Fiddle River, approximately 37 km south south-west of Hinton.

This stream is part of the east watershed of Utopia Mountain. Its name was made official in 1960. (see Utopia Mountain)

Utopia Mountain (mountain)

83 F/4 - Miette
32-47-26-W5
53°06′N 117°46′W
Approximately 39 km south south-west of
Hinton.

This 2602 m mountain was the surveyors'
refuge from flies. It was named in 1916 by
M.P. Bridgland. (see Mount Bridgland)

Valad Peak (peak)

83 C/11 - Southesk Lake
14-41-24-W5
52°32′N 117°21′W
Approximately 63 km south-east of Jasper.

This 3150 m mountain peak was named by
Howard Palmer in 1923 after his guide of
mixed blood background who accompanied
H.A.F. McLeod to Maligne Lake in 1875.

Valley Creek (creek)

83 L/12 - Nose Creek
14-64-12-W6
54°33′N 119°41′W
Flows north-east into Nose Creek,
approximately 86 km south south-west of
Grande Prairie.

The name for this creek which runs from
the Valley of Two Lakes was proposed in
1946 by J. Haan, a forest ranger in the area.

Valley Head Mountain (mountain)

83 C/10 - George Creek
11-41-21-W5
52°31′N 116°56′W
Approximately 57 km west of Nordegg.

The name for this mountain was officially
approved 5 March 1935. It is descriptive of
its location; the mountain is at the head of
the Brazeau River Valley.

Vam Creek (creek)

82 O/11 - Burnt Timber Creek
3-31-8-W5
51°38′N 115°03′W
Flows north into Red Deer River,
approximately 60 km north-east of Banff.

The name Vam Creek was taken from field
notes of 1959-1960, from the Department of
Highways Survey Branch. No information
about the origin was given.

Vavasour, Mount (mountain)

82 J/14 - Spray Lakes Reservoir
10-21-11-W5
50°46′N 115°27′W
Approximately 95 km west south-west of
Calgary.

This 2835 m mountain was named in 1918
after Lieutenant M. Vavasour, R.E., who
died in 1866. (see Mount Warre)

Vega Peak (peak)

83 E/7 - Blue Creek
25-51-6-W6
53°26′N 118°45′W
Approximately 75 km north-west of Jasper.

This 2491 m mountain peak is part of the
Starlight Range. It was named by R.W.
Cautley (see Mount Cautley) after the star,
"Vega," which is the fourth brightest star in
the night sky.

Verdant Creek (creek)

83 D/9 - Amethyst Lakes
7-43-1-W6
52°41′N 118°07′W
Flows north-west into Astoria River,
approximately 22 km south south-west of
Jasper.

The name of this creek was officially
approved 24 November 1927. It is likely
descriptive, as "verdant" means green-
coloured in French.

Verdant Pass (pass)

83 D/9 - Amethyst Lakes
28-42-1-W6
52°39′N 118°05′W
Approximately 26 km south of Jasper.

(see Verdant Creek)

Vermilion Lakes (lake)

82 O/4 - Banff
34-25-12-W5
51°11′N 115°36′W
Approximately 2 km west of Banff.

The name for this feature is a translation of
the Cree Indian word "wiyaman." There are
ferrunginus (i.e., reddish-brown colour or
containing iron) beds in the vinicity of the
lakes. Vermilion is the same colour as these
beds. Another Indian name applied to the
feature is "kianiskkotiki," Indian for "chain
of lakes joining each other."

Vermilion Pass (pass)

82 N/1 - Mount Goodsir
26-15-W5
51°14′N 116°03′W
Approximately 35 km west north-west of
Banff, on the Alberta-B.C. boundary.

(see Vermilion Lakes)

Vermilion Range (range)

82 O/5 - Castle Mountain
28-13-W5
51°23′N 115°42′W
Approximately 8 km north-west of Banff.

(see Vermilion Lakes)

Veronique Creek (creek)

83 E/15 - Pierre Greys Lakes
18-57-5-W6
53°56′N 118°43′W
Flows south into Muskeg River, approximately
25 km east north-east of Grande Cache.

The name was officially adopted in 1947. The origin of the name is unknown.

Vertex Peak (peak)

83 D/16 - Jasper
34-43-2-W6
52°45′N 118°12′W
Approximately 16 km south-west of Jasper.

This 2957 m mountain peak has a very sharp triangular summit. Its descriptive name was applied to this feature by M.P. Bridgland (see Mount Bridgland) in 1916.

Vetch Creek (creek)

83 B/3 - Tay River
28-29-37-W5
52°13′N 115°06′W
Approximately 21 km south-west of Rocky Mountain House.

The name Vetch Creek was officially applied to this creek in 1941 and was suggested by Mr. H.P. Brownlee, a statistician for the Department of Trade and Industry. The creek enters Prairie Creek near Vetchland settlement. Vetch is a plant of the pea family and is used for forage. Wild vetch grows in abundance in this area.

Vicary Creek (creek)

82 G/16 - Maycroft
23-10-4-W5
49°50′N 114°25′W
Flows north into Racehorse Creek, approximately 25 km north of Blairmore.

The origin of the name is unknown.

Victor Lake (lake)

83 E/14 - Grande Cache
35-56-8-W6
53°53′N 119°05′W
Approximately 2 km east of Grande Cache.

There is some suggestion that the lake may be named after Victor Gay, a member of a local survey party, but the precise origin of the name is unknown.

Victoria Glacier (glacier)

82 N/8 - Lake Louise
28-17-W5
51°23′N 116°16′W
Approximately 55 km west north-west of Banff.

(see Mount Victoria)

Victoria, Mount (mountain)

82 N/8 - Lake Louise
3-28-17-W5
51°23′N 116°17′W
Approximately 60 km west north-west of Banff, on the Alberta-B.C. boundary.

This mountain, which is 3464 m in altitude, was named in 1897 by J.J. McArthur, a Dominion Land Surveyor who introduced the system of photography for the survey of the Rocky Mountains in 1887. The mountain was named to commemorate Queen Victoria (1819-1910). It was previously known as Mount Green.

Victoria Peak (peak)

82 G/8 - Beaver Mines
13-4-2-W5
49°18′N 114°08′W
Approximately 25 km south-west of Pincher Creek.

This mountain peak was named by J.J. McArthur of the Dominion Land Survey in 1915 after Queen Victoria (1819-1901). (see Mount Victoria)

Victoria Ridge (ridge)

82 G/1 - Sage Creek
3-2-W5
49°14′N 114°10′W
Approximately 27 km north-west of Waterton Park.

(see Mount Victoria)

Victoria Cross Range (range)

83 E/1 - Snaring
47-3-W6
53°02′N 118°21′W
Approximately 11 km north-west of Jasper.

The peaks in this range are named after winners of the Victoria Cross since the First World War. This range includes Mounts Pattison, Kinross, McKean, Kerr and Zengel.

Villeneuve Creek (creek)

83 F/4 - Miette
25-48-27-W5
53°10′N 117°48′W
Flows north into Fiddle River, approximately 34 km south-west of Hinton.

Frank Villeneuve was a prospector in the area who, with Alfred Lamoreau, found coal on the lower slope of Roche Miette and on the north bank of the Athabasca River. No other information is available.

Vimy Peak (peak)

82 H/4 - Waterton Lakes
1-29-W4
49°02′N 113°51′W
Approximately 50 km south south-east of Pincher Creek.

The name Vimy commemorates the struggle for Vimy Ridge by Canadian troops in April 1917. The battle was without doubt the greatest Canadian victory of the First World War. The three-day battle saw 7,707 casualties from Canada but ended in victory 9 April 1917. From Vimy Peak, which has an altitude of 2385 m, one of the best viewpoints of the Waterton Lakes and the central portion of Waterton Lakes National Park may be had.

Vimy Ridge (ridge)

82 H/4 - Waterton Lakes
1-29-W4
49°01′N 113°50′W

Approximately 52 km south south-east of
Pincher Creek.

(see Vimy Peak)

Vine Creek (creek)

83 E/1 - Snaring
21-47-1-W6
53°04′N 118°05′W
Flows south-east into Athabasca River,
approximately 22 km north of Jasper.

The local name for this creek was established
due to the abundance of red bear berries
along its shore. No other information is
available.

Virl Lake (lake)

83 D/16 - Jasper
16-45-2-W6
52°52′N 118°14′W
Approximately 10 km west of Jasper.

The origin for the name of this lake,
suggested in 1914 by H. Matheson, D.L.S., is
unknown.

Vista Creek (creek)

83 D/16 - Jasper
19-44-2-W6
52°45′N 118°19′W
Flows north into Meadow Creek,
approximately 20 km south-west of
Jasper.

(see Vista Peak)

Vista Lake (lake)

82 N/1 - Mount Goodsir
22-26-15-W5
51°14′N 116°01′W
Approximately 35 km west north-west of
Banff.

The precise origin of the name is unknown.

Vista Pass (pass)

83 D/16 - Jasper
43-3-W6
52°45′N 118°24′W
Approximately 24 km south-west of Jasper.

(see Vista Peak)

Vista Peak (peak)

83 D/16 - Jasper
5-44-3-W6
52°46′N 118°24′W
Approximately 25 km south-west of Banff.

The name for this 2795 m mountain peak
describes the long narrow view seen from
this point and has been in local usage since
at least 1917. The surrounding features take
their names from this peak.

Vogel Creek (creek)

83 E/16 - Donald Flats
53°46′N 118°25′W
Flows south-east into Cabin Creek,
approximately 54 km south-east of Grande
Cache.

Fred Vogel had a trap line on this creek
which was officially named after him in
1947.

Volcano Creek (creek)

82 J/10 - Mount Rae
1-21-6-W5
50°45′N 114°43′W
Flows north into Threepoint Creek,
approximately 26 km west of Turner Valley.

(see Volcano Ridge)

Volcano Ridge (ridge)

82 J/10 - Mount Rae
20-6-W5
50°44′N 114°42′W
Approximately 29 km west of Turner Valley.

The names Volcano Ridge and Volcano
Creek were officially approved 6 September
1951. The name is descriptive of the ridge's
volcanic formation.

Vulture Glacier (glacier)

82 N/9 - Hector Lake
28-30-18-W5
51°36′00″N 116°48′20″W
Approximately 20 km south-west of Banff.

The origin of the name is unknown.

Wa-Wa Ridge (mountain ridge)

82 0/4 - Banff
51°05′30″N 115°47′20″W
Approximately 18 km south-west of Banff.

The name Wa-Wa Ridge is a well-established local name for this feature, located at Sunshine Ski Resort. It is used by the Banff Warden Service, climbers, skiers and local residents. Though the name is well-established in usage, its origin is not known.

Wabasso Lake (lakes)

83 D/16 - Jasper
44-1-W6
52°49′N 118°01′W
Approximately 7 km south-east of Jasper.

This small chain of lakes was formed by the damming up of numerous streams by beavers. "Wabasso" is an Indian word for "rabbit." No other origin information is available for the name of these lakes.

Waitabit Peak (peak)

82 N/10 - Blaeberry River
34-31-21-W5
51°42′N 116°54′W
Approximately 110 km north-west of Banff, on the Alberta-B.C. boundary.

There is a Waitabit Creek in British Columbia, after which this peak is likely named. Its precise origin is unknown but may have to do with waiting a bit, perhaps for the sky to clear.

Waite Valley (valley)

82 J/9 - Turner Valley
8-20-4-W5
50°42′N 114°25′W
Approximately 8 km west of Turner Valley.

No origin information about the name is known.

Waldie Creek (creek)

82 J/8 - Stimson Creek
25-17-4-W5
50°28′N 114°25′W
Flows north-east into Highwood River, approximately 25 km south-west of Turner Valley.

There is no definite date as to when E.F. Waldy, a tall, cheerful Englishman arrived in the area to settle on this creek, but it was circa 1894. The Indian name for this creek is "ke-to-ke," for "Prairie Chicken Creek." The creek was officially named Waldie Creek 12 December 1939.

Wales Glacier (glacier)

83 C/4 - Clemenceau Icefield
52°10′00″N 117°37′30″W
Approximately 80 km south-east of Jasper, on the Alberta-B.C. boundary.

Prior to 20 October 1983, which was the official approval date for the name of this feature, the name for this glacier was approved only in British Columbia, the province which shares this glacier with Alberta. (see Wales Peak)

Wales Peak (peak)

83 C/4 - Clemenceau Icefield
52°11′05″N 117°39′15″W
Approximately 80 km south-east of Jasper.

The name for this 3109 m mountain peak was originally suggested by American climber A.J. Ostheimer III in 1927. Its precise origin is unknown; however, the name has been extensively used by local alpinists and residents since the early 1950s. This peak is located at the head of the Wales Glacier and was officially named 5 May 1987.

Walker, Mount (mountain)

82 N/10 - Blaeberry River
4-32-21-W5
51°43′N 116°55′W

Approximately 110 km north-west of Banff.

This mountain, which is 3303 m in altitude, was named in August 1987 after Horace Walker, past President of The Alpine Club (London). The name was officially approved 3 January 1911.

Wall of Jericho (mountain)

82 N/8 - Lake Louise
19-29-15-W5
51°30′N 116°06′W
Approximately 55 km north-west of Banff.

This long, thin, low mountain, 2910 m in altitude, appears to be tumbling to pieces very rapidly. The appearance of this feature inspired James F. Porter (see Merlin Lake) to christen it the Wall of Jericho. The story of the battle in which these famous walls fell may be found in the Bible, in the Book of Joshua, Chapter 6, verses 20-21.

Walter Peak (peak)

82 N/14 - Rostrum Peak
34-22-W5
51°57′09″N 117°06′06″W
Approximately 135 km north-west of Banff, on the Alberta-B.C. boundary.

(see Christian Peak)

Walton Creek (creek)

83 E/10 - Adams Lookout
17-55-7-W6
53°45′N 119°00′W
Flows west into Sulphur River, approximately 17 km south south-east of Grande Cache.

The name for this creek was made official in 1947. The origin of the name is unknown.

Wampum Peak (peak)

82 N/16 - Siffleur River
34-15-W5
51°54′N 116°02′W
Approximately 85 km north north-west of Banff.

"Wampum" are beads made from shells and string which were once used as currency and decoration by the Indians. This 2850 m mountain peak may have strata resembling these beads.

Wampus Creek (creek)

83 F/3 - Cadomin
23-48-23-W5
53°10′N 117°17′W
Flows north-east into McLeod River, approximately 73 km south-west of Edson.

The origin of the name is unknown.

Wanyandie Creek (creek)

83 L/2 - Bolton Creek
5-59-6-W6
54°04′N 118°52′W
Flows west into Smoky River, approximately 25 km north-east of Grande Cache.

This creek may be named after Vincent Wanyandi (the spelling of this name may have been corrupted) of Iroquois ancestry who was born circa 1850 near Jasper House. He resided all his life in this area.

Wapiabi Creek (creek)

83 C/9 - Wapiabi Creek
14-42-17-W5
52°37′N 116°20′W
Flows north into Blackstone River, approximately 24 km north-west of Nordegg.

This creek was officially named in 1910 and bears the Stoney Indian word for a grave. There is a grave on its banks.

Wapiabi Gap (gap)

83 C/8 - Nordegg
30-40-17-W5
52°29′N 116°25′W
Approximately 123 km east south-east of Jasper.

(see Wapiabi Creek)

Wapiti Mountain (mountain)

82 O/12 - Barrier Mountain
11-32-13-W5
51°43′N 115°45′W
Approximately 65 km north north-west of Banff.

The Wapiti River is derived from glaciers in this area north of Bare Range, and it is from the river that this mountain is named. "Wapiti" is the Indian word for "elk." Plenty of elk fed around the river at the time of its naming.

Wapta Icefield (icefield)

82 N/10 - Blaeberry River
31-19-W5
51°38′N 116°33′W
Approximately 85 km north-west of Banff.

"Wapta" is the Stoney Indian word for "river." This feature was named by Jean Habel (see Mount Habel) in 1897.

Waputik Glacier (glacier)

82 N/9 - Hector Lake
6-30-17-W5
51°33′N 116°23′W
Approximately 70 km north-west of Banff.

(see Waputik Range)

Waputik Icefield (icefield)

82 N/9 - Hector Lake
6-30-17-W5
51°32′N 116°24′W
Approximately 75 km north-west of Banff.

(see Waputik Range)

Waputik Mountains (range)

82 N/9 - Hector Lake
27-29-17-W5
51°35′N 116°25′W
Approximately 75 km north-west of Banff.

(see Waputik Range)

Waputik Peak (peak)

82 N/9 - Hector Lake
30-17-W5
51°30′N 116°19′W
Approximately 65 km north-west of Banff.

Waputik Peak is 2736 m in altitude. (see Waputik Range)

Waputik Range (range)

82 N/8 - Lake Louise
29-17-W5
51°32′N 116°20′W
Approximately 70 km north-west of Banff.

"Waputik" is the Stoney Indian word for "white goat." The range was named by Dr. G.M. Dawson (see Tombstone Mountain) in 1884. At that time, this was a favourite haunt of Rocky Mountain Goats. The Mountain Goat is not a true goat, but an antelope closely related to the European chamois. It has a heavy coat of long white hair, which provides protection against the winds of the mountains where it makes its home.

Ward, Mount (mountain)

82 G/15 - Tornado Mountain
20-9-5-W5
49°45′N 114°39′W

Approximately 15 km north-west of Coleman, on the Alberta-B.C. boundary.

Captain A.C. Ward, a member of the Royal Engineers, was the Secretary for the British Boundary Commission studying the area from Lake of the Woods to the Rockies, in the years 1872-1874 (see Cameron Lake). This 2530 m mountain was named after Captain Ward and the name was officially approved 16 March 1917.

Warden Creek (creek)

83 F/3 - Cadomin
6-49-24-W5
53°12′N 117°30′W
Flows east into Gregg River, approximately 83 km south-west of Edson.

Named in 1925, this creek is crossed by a trail once used by Jasper Park wardens. The Park Warden's cabin was situated at the mouth of the stream.

Warden Rock (peak)

82 O/12 - Barrier Mountain
29-31-12-W5
51°42′N 115°40′W
Approximately 60 km north of Banff.

The descriptive name for this 2696 m mountain, which resembles a warden on guard, is locally well-established.

Ware Creek (creek)

82 J/9 - Turner Valley
23-20-4-W5
50°43′N 114°28′W
Flows east into Threepoint Creek, approximately 16 km west north-west of Turner Valley.

The name for this creek was officially approved 8 December 1943. (see John Ware Ridge)

Ware, Mount (mountain)

82 J/10 - Mount Rae
3-20-6-W5
50°40′N 114°45′W
Approximately 16 km north-west of Turner Valley.

The name of this 2118 m mountain was officially approved 6 September 1951. (see John Ware Ridge)

Warre Creek (creek)

82 J/14 - Spray Lakes Reservoir
21-21-11-W5
50°48′N 115°26′W
Flows north-east into Spray River, approximately 90 km west south-west of Calgary.

(see Mount Warre)

Warre, Mount (mountain)

82 J/14 - Spray Lake Reservoir
16-21-11-W5
50°47′N 115°27′W
Approximately 95 km west south-west of Calgary.

Lieutenant Sir Henry James Warre, K.C.B. (1819-1898), in 1845-1846 crossed the Rockies with Lieutenant M. Vavasour (see Mount Vavasour) on a reconnaissance of the Oregon Territory to ascertain what military action should be taken should there be another war with the United States. Warre served in the Crimean War and became Commander-in-Chief at Bombay. He is the author of *Sketches in North America and the Oregon Territory* (1848). This mountain, which is 2743 m in altitude, was named in 1918 after Lieutenant Warre.

Warren Creek (creek)

83 C/11 - Southesk Lake
From 52°35′58″N 117°44′10″W
to 52°10′00″N 117°48′00″W
Approximately 59 km south-east of Jasper.

Named for its proximity to Mount Warren, the name of this creek is well-established in local usage. (see Mount Warren)

Warren, Mount (mountain)

83 C/11 - Southesk Lake
27-41-24-W5
52°34′N 117°22′W
Approximately 60 km south-east of Jasper.

Mary T.S. Schäffer named this 3140 m mountain after her head guide. She "called [it] for Warren, through whose grit and determination we ever found the lake (Maligne Lake) and through whose energies we plied the lake with a raft." Mount Warren was named in 1911.

Warrior Mountain (mountain)

82 J/11 - Kananaskis Lakes
31-18-9-W5
50°34′N 115°14′W
Approximately 70 km west south-west of Turner Valley, on the Alberta-B.C. boundary.

Officially approved in 1918, this 2926 m mountain was named after a British man-of-war cruiser engaged in the Battle of Jutland, during the First World War. (see Jutland Mountain)

Warspite Cascade (falls)

82 J/11 - Kananaskis Lakes
18-20-9-W5
50°41′40″N 115°13′30″W
Approximately 65 km west of Turner Valley.

This feature comprises a series of small waterfalls on Warspite Creek at the point where the creek descends from the hanging valley of the north-east cirque of Mount Warspite. The water flows over a series of ledges. The name Warspite Cascade was made official 8 November 1978. (see Mount Warspite)

Warspite Creek (creek)

82 J/11 - Kananaskis Lakes
20-20-9-W5
50°42′30″N 115°12′15″W
Flows east into Smith-Dorrien Creek,
approximately 65 km west of Turner
Valley.

(see Mount Warspite)

Warspite, Mount (mountain)

82 J/11 - Kananaskis Lakes
8-20-9-W5
50°41′N 115°13′W
Approximately 65 km west of Turner
Valley.

This mountain, which is 2819 m in altitude,
was named in 1922 after a cruiser engaged in
the Battle of Jutland (see Jutland Mountain).
H.M.S. Warspite, nicknamed "The Old
Lady," was built in 1912 and served in both
World Wars.

Warwick Creek (creek)

83 C/5 - Fortress Lake
20-38-25-W5
52°17′N 117°32′W
Flows north-east into Athabasca River,
approximately 75 km south-east of Jasper.

(see Warwick Mountain)

Warwick Mountain (mountain)

83 C/4 - Clemenceau Icefield
6-38-25-W5
52°14′N 117°35′W
Approximately 80 km south-east of Jasper.

Due to the feature's castellated appearance
and its proximity to Mount King Edward, a
name taken from Warwick Castle,
Warwickshire, England was adopted. The
mountain was named by A.O. Wheeler and
measures 2906 m in altitude.

Washout Creek (creek)

83 B/4 - Cripple Creek
13-35-13-W5
52°00′N 115°43′W
Flows south into Clearwater River,
approximately 58 km south-east of Nordegg.

There is no origin information available for
the locally well-established name for this
creek.

Washy Creek (creek)

83 E/14 - Grande Cache
17-57-7-W6
53°55′N 119°00′W
Flows south-east into Peavine Lake,
approximately 9 km east north-east of
Grande Cache.

This creek was named after "Washy Joe," a
Stoney Indian who lived in the Smoky River
country. His real name was "Sagapatura,"
meaning "long hair." He was invited to pose
as a mixed blood native to earn scrip, by
Donald Macdonald, who renamed him Joe
Atkins. The problems with the name made
it necessary to eventually transform it to
"Washy Joe."

Wasootch Creek (creek)

82 J/14 - Spray Lakes Reservoir
19-23-8-W5
50°59′N 115°06′W
Flows north-west into Kananaskis River,
approximately 65 km west of Calgary.

The name for this creek dates back to 1863.
The Stoney Indian word for "hail" is "wa-
sootch." This stream was named by Captain
Palliser. (see Palliser Range)

Wastach Pass (pass)

82 N/8 - Lake Louise
30-27-16-W5
51°20′N 116°15′W
Approximately 55 km west north-west of
Banff.

This 2544 m pass was named by S.E.S. Allen
(see Mount Allen). "Wastach" is the Stoney
Indian word for "beautiful."

Watchman Creek (creek)

83 C/3 - Columbia Icefield
35-23-W5
52°03′N 117°12′W
Flows east into Castleguard River,
approximately 110 km south-east of Jasper.

(see Watchman Peak)

Watchman Lake (lake)

83 C/3 - Columbia Icefield
35-23-W5
52°03′N 117°14′W
Approximately 110 km south-east of Jasper.

(see Watchman Peak)

Watchman Peak (peak)

83 C/3 - Columbia Icefield
35-23-W5
52°03′N 117°14′W
Approximately 110 km south-east of Jasper,
on the Alberta-B.C. boundary.

This 3009 m mountain was named by James
Outram (see Mount Outram) because of its
towering, isolated appearance. The other
features nearby take their name from the
mountain.

Watchtower, The (mountain)

83 C/13 - Medicine Lake
33-44-27-W5
52°49′N 117°50′W
Approximately 17 km east south-east of
Jasper.

The descriptive name for this mountain,
whose peak stands like a tower, was given in
1916 by M.P. Bridgland. (see Mount
Bridgland)

Watchtower Creek (creek)

83 C/13 - Medicine Lake
52°51′50″N 117°47′15″W
Flows north into Medicine Lake,
approximately 17 km east of Jasper.

The creek was named circa 1940 by the
Jeffery brothers (see Jeffery Creek) and was
officially approved 5 May 1987. (see The
Watchtower)

Waterfall Peaks (peaks)

83 C/6 - Sunwapta
6-40-23-W5
52°25′N 117°18′W
Approximately 73 km south-east of Jasper.

The first use of the name Waterfall Peaks is
found in connection with an ascent made by
Lillian Gest and Polly Prescott in 1938. A
cabin located to the north-east of the peaks
(2640 m, 2920 m, 2880 m and 2960 m from
north-west to south-east peak) was known as
Waterfall Cabin; the Wardens' entries begin
calling it so in 1928. The name for the four
peaks was made official 14 August 1975.

Waterfalls Creek (creek)

83 C/2 - Cline River
5-37-19-W5
52°09′N 116°41′W
Flows north-east into Cline River,
approximately 125 km south-east of Jasper.

This 13 km long creek contains many (five
to nine) waterfalls along its course. The
major falls on the creek drains the final lake,
dropping over 275 m to the valley floor, of
which an estimated 185 m is a vertical
cascade worn into the cliff face. The
descriptive name for this creek was officially
approved in 1975.

Waterfowl Lakes (lake)

82 N/15 - Mistaya Lake
21-33-19-W5
51°50′N 116°37′W
Approximately 110 km north-west of Banff.

The descriptive name refers to the large
number of ducks seen on the lake at the
time of its naming in 1898.

Waterton Lakes (lakes)

82 H/4 - Waterton Lakes
1-29-W4
49°03′N 113°54′W
Approximately 35 km south of Pincher
Creek.

This chain of lakes lying in the main valley
of Waterton Lakes National Park consists of
four bodies of water connected by a river:
Maskinonge Lake, and the Middle, Upper
and Lower Waterton Lakes. The chain of
lakes was named by Thomas Blakiston (see
Mount Blakiston) after Charles Waterton
(1782-1865), a famous English naturalist. He
was widely known for his research into the
sources of Indian poisons and ornithological
work. He expended a large part of his
fortune in creating a sanctuary for birds. (see
Dardanelles and Bosporous)

Waterton Lakes, 1920

Waterton Lakes (national park)

82 H/4 - Waterton Lakes
1-30-W4
49°05′N 113°55′W
Approximately 35 km south of Pincher
Creek.

This park was named for the lakes. (see
Waterton Lakes)

Waterton Park (hamlet)

82 H/4 - Waterton Lakes
1-30-W4
49°03′N 113°55′W
Approximately 50 km south of Pincher
Creek.

(see Waterton Lakes)

Watridge Creek (creek)

82 J/14 - Spray Lakes Reservoir
13-22-11-W5
50°52′N 115°24′W
Flows north-east into Spray Lakes Reservoir,
approximately 85 km west of Calgary.

The origin of the name is unknown.

Watridge Lake (lake)

82 J/14 - Spray Lakes Reservoir
22-11-W5
50°51′N 115°26′W
Approximately 90 km west of Calgary.

No origin information for the name is
known.

Watson Creek (creek)

83 F/3 - Cadomin
24-47-23-W5
53°05′N 117°15′W
Flows north-east into McLeod River,
approximately 77 km south-west of Edson.

This creek was officially named 6 May 1923.
The origin of the name is unknown.

Wawa Creek (creek)

83 C/9 - Wapiabi Creek
24-43-15-W5
52°43′N 116°02′W

Flows east into Nordegg River, approximately 27 km north of Nordegg.

The origin of the name is unknown.

Weary Creek Gap (gap)

82 J/7 - Mount Head
35-16-7-W5
50°24′N 114°51′W
Approximately 50 km south-west of Turner Valley.

The surveyors of the Alberta-British Columbia Boundary Commission passed through this area in 1915 and named this gap Weary Creek Gap because it led down to the locally known "Weary Creek."

Wedge, The (mountain)

82 J/14 - Spray Lakes Reservoir
22-9-W5
50°51′N 115°07′W
Approximately 70 km west of Calgary.

The name for this 2652 m mountain describes its summit.

Wedge Mountain (mountain)

82 G/10 - Crowsnest
30-8-4-W5
49°39′N 114°31′W
Approximately 4 km north north-west of Coleman.

This mountain is approximately 1870 m in altitude. The name is likly descriptive of its appearance.

Wedge Pond (pond)

82 J/14 - Spray Lakes Reservoir
22-9-W5
50°52′N 115°09′W
Approximately 70 km west of Calgary.

This feature takes its name from the nearby mountain (see The Wedge). The name was officially approved in 1977.

Weed, Mount (mountain)

82 N/10 - Mistaya Lake
32-32-18-W5
51°48′N 116°32′W
Approximately 92 km north-west of Banff.

This 3080 m mountain was named after G.M. Weed of the Appalachian Mountain Club of Boston, Mass. He made a number of first ascents in the Canadian Rockies, including Mount Murchison, Mount Neptuak, Mount Balfour, Howse Peak, Mount Freshfield and Mount Forbes.

Weeping Wall (cliff)

83 C/3 - Columbia Icefield
52°09′30″N 117°00′25″W
Approximately 100 km south-east of Jasper.

Water from melting snow falls more than 100 m over the face of this cliff. The descriptive name for the feature was officially approved 28 February 1980.

Weiss, Mount (mountain)

83 C/6 - Sunwapta
25-39-25-W5
52°23′N 117°28′W
Approximately 67 km south south-east of Jasper.

J.A. (Joe) Weiss came to Canada from Switzerland in 1921. After spending some time in British Columbia, he moved to Jasper (in the winter of 1925-1926). He lived there for the next 45 years, exploring, acting as a guide, and photographing the mountains. He made several first ascents of mountains in the area. During his years in Jasper Park, Mr. Weiss kept a photo record, and his pictures are internationally known. This 3090 m mountain peak was officially named after Mr. Weiss 14 August 1975.

Wenkchemna Glacier (glacier)

82 N/8 - Lake Louise
27-16-W5
51°19′N 116°13′W

Approximately 50 km west north-west of Banff.

(see Wenkchemna Peak)

Wenkchemna Pass (pass)

82 N/8 - Lake Louise
27-17-W5
51°19′N 116°16′W
Approximately 55 km west north-west of Banff, on the Alberta-B.C. boundary.

(see Wenkchemna Peak)

Wenkchemna Peak (peak)

82 N/8 - Lake Louise
24-27-17-W5
51°20′N 116°16′W
Approximately 55 km west north-west of Banff, on the Alberta-B.C. boundary.

This 3170 m peak is the tenth peak in the Valley of the Ten Peaks. "Wenkchemna" is the Stoney Indian numeral for "ten." It was named by S.E.S. Allen (see Mount Allen) in 1894. (see Ten Peaks, Valley of the)

Wenkchemna Peaks (peaks)

82 N/8 - Lake Louise
27-17-W5
51°19′N 116°16′W
Approximately 55 km west north-west of Banff.

(see Wenkchemna Peak)

West Glacier (glacier)

82 N/15 - Mistaya Lake
19-33-21,22-W5
51°51′N 116°58′W
Approximately 120 km north-west of Banff.

(see East Glacier)

West Castle Mountain (mountain)

82 G/8 - Beaver Mines
11-4-3-W5
49°17′N 114°19′W
Approximately 21 km south south-west of Beaver Mines.

(see Castle Peak)

West Castle River (river)

82 G/8 - Beaver Mines
16-5-3-W5
49°24′N 114°21′W
Flows north into Castle River, approximately 30 km west south-west of Pincher Creek.

(see Castle River)

West Elk Pass (pass)

82 J/11 - Kananaskis Lakes
5-19-8-W5
50°35′N 115°05′W
Approximately 60 km west south-west of Turner Valley, on the Alberta-B.C. boundary.

(see Elk Pass)

West Muskeg River (river)

83 E/10 - Adams Lookout
8-54-6-W6
53°39′N 118°52′W
Flows east into Muskeg River, approximately 31 km south south-east of Grande Cache.

(see Muskeg River)

West Solomon Creek (creek)

83 F/5 - Entrance
16-51-27-W5
53°23′N 117°55′W
Flows north-east into Solomon Creek, approximately 20 km west of Hinton.

(see Solomon Creek)

West Sulphur River (river)

83 E/10 - Adams Lookout
30-52-6-W6
53°32′N 118°52′W
Flows north-east into South Sulphur River, approximately 44 km south south-east of Grande Cache.

(see Sulphur River)

West Wind Creek (creek)

82 O/3 - Canmore
6-24-9-W5
51°01′N 115°15′W
Flows north-east into Wind Creek, approximately 11 km south-east of Canmore.

(see Wind Mountain)

Westover Lake (lake)

82 O/2 - Jumpingpound Creek
11-25-6-W5
51°07′N 114°44′W
Approximately 40 km west north-west of Calgary.

This lake was officially named 12 December 1939 after a local rancher.

Westrup Creek (creek)

82 J/1 - Langford Creek
29-13-2-W5
50°07′N 114°14′W
Flows north-east into Langford Creek, approximately 45 km west north-west of Claresholm.

Harry and Dick Westropp arrived from Ireland in the 1890s and worked for various ranchers in the area. Each set up his own homestead in 1896, but by 1924, both had returned to Ireland. A misspelling of their name resulted in official approval of the name Westrup Creek for the feature on 3 March 1960.

Whaleback Ridge (ridge)

82 G/16 - Maycroft
11-2-W5
49°55′N 114°12′W
Approximately 35 km north-east of Blairmore.

This mountain was named by James Outram (see Mount Outram) and Mr. Whymper in 1916, due to its resemblance to the shape of the back of a whale.

Wheeler Flats (flats)

82 O/4 - Banff
51°05′30″N 115°46′30″W
Approximately 17 km south-west of Banff.

Wheeler Flats was named for A.O. Wheeler (1860-1945), a Dominion Land Surveyor who worked on the Alberta-British Columbia Boundary Survey, as the representative for British Columbia. He was author of *The Selkirk Mountains* (1912) and *The Selkirk Range* (1905). The Alpine Club of Canada frequently refers to Wheeler Flats in its publications. The name was officially adopted 20 October 1983. (see Introduction)

Whirlpool Pass (pass)

83 D/8 - Athabasca Pass
33-40-2-W6
52°29′N 118°14′W
Approximately 44 km south south-east of Jasper.

(see Whirlpool River)

Whirlpool River (river)

83 C/12 - Athabasca Falls
26-43-28-W5
52°44′N 117°57′W
Flows north into Athabasca River, approximately 18 km south south-east of Jasper.

This river was named in 1859 by Sir James Hector (see Mount Hector) after the numerous eddies in the stream.

Whisker Creek (creek)

83 C/7 - Job Creek
33-40-2-W5
52°28'N 116°49'W
Flows west into Job Creek, approximately 95 km east south-east of Jasper.

The name Whisker Creek was given to this feature circa 1920 by Ray and William Mustard, sons of Harvey M. Mustard, who operated a guiding and outfitting business in Mountain Park prior to 1920. Ray Mustard, a retired forest ranger noted that he, his father and his brother were the first outfitters in the area and that they named the stream Whisker Creek because Harvey, who chewed tobacco, washed his moustache in the creek. (personal interview, June 1976)

Whisker Lakes (lakes)

83 C/7 - Job Creek
16-40-19-W5
52°26'N 116°41'W
Approximately 105 km east south-east of Jasper.

These four small lakes are located at the headwaters of Whisker Creek. The name Whisker Lakes was officially approved 15 July 1977. (see Whisker Creek)

Whiskey Creek (creek)

82 O/4 - Banff
26-12-W5
51°11'N 115°34'W
Flows south-west into Forty Mile Creek, approximately 1 km north-west of Banff.

The name for this creek was officially adopted 3 October 1957. The exact origin of the name is unknown.

Whiskey Creek (creek)

82 J/16 - Priddis
7-22-3-W5
50°51'N 114°24'W
Flows east into Fish Creek, approximately

20 km west south-west of Calgary.

The name for this stream was officially approved 6 October 1949 and is said to recall the activities of "Moonshine Wilson," a still owner and operator in the area.

Whiskey Jack Crossing (crossing)

82 O/14 - Limestone Mountain
33-32-9-W5
51°47'N 115°13'W
Approximately 75 km north north-east of Banff.

The name for this crossing is taken from the Canada jay, or whiskey-jack.

Whiskeyjack Creek (creek)

83 F/5 - Entrance
12-51-25-W5
53°23'N 117°33'W
Flows north-west into Hardisty Creek approximately 3 km south of Hinton.

The name Whiskeyjack Creek was suggested by the Forestry Training Branch because of the prevalence of these birds in the valley. It was officially adopted 14 April 1965. The name is from the Cree word "weskuchanis," which translates as "little blacksmith."

Whisky Ridge (ridge)

82 J/9 - Turner Valley
31-18-3-W5
50°34'N 114°25'W
Approximately 15 km south-west of Turner Valley.

The name for this ridge was officially approved 8 December 1943. Local legend has it that the ridge was so named because a man named Shorty McLaughlin operated a still in these hills.

Whistler Mountain (mountain)

82 G/8 - Beaver Mines
35-4-3-W5
49°20'N 114°17'W
Approximately 15 km south south-west of Beaver Mines.

The precise origin of this name is unknown, but it could be after the siffleur, or whistling marmot, which is found in this area. (see Siffleur Mountain)

Whistlers, The (mountain)

83 D/16 - Jasper
30-44-1-W6
52°49'N 118°08'W
Approximately 6 km south-west of Jasper.

This green, wooded peak, which is 2464 m in altitude, derives its name from the colonies of whistling marmots found among the bare rocks near its top. Dr. E. Deville, one-time Surveyor General, applied the name in 1916. (see Siffleur Mountain)

Whistlers Creek (creek)

83 D/16 - Jasper
34-44-1-W6
52°50'N 118°03'W
Flows east into Athabasca River, approximately 5 km south of Jasper.

(see The Whistlers)

Whistlers Pass (pass)

83 D/16 - Jasper
22-44-2-W6
52°48'N 118°11'W
Approximately 11 km south-west of Jasper.

This pass is 2377 m in altitude. (see The Whistlers)

Whistling Valley (valley)

82 O/4 - Banff
25-14-W5
51°06'N 115°56'W

Approximately 26 km west south-west of Banff.

The sound of the wind which gathers in this valley resembles that of the siffleur, or whistling marmot. (see Siffleur Mountain)

White Creek (creek)

83 G/16 - Maycroft
7-12-4-W5
49°59′N 114°24′W
Flows west into Livingstone River, approximately 40 km north of Blairmore.

The origin of the name is unknown.

White Creek (creek)

83 F/6 - Pedley
4-51-22-W5
53°23′N 117°11′W
Flows north-west into McLeod River, approximately 54 km south-west of Edson.

The name was officially approved 21 December 1944, but its exact origin is unknown.

White, Mount (mountain)

82 O/12 - Barrier Mountain
7-31-13-W5
51°38′N 115°48′W
Approximately 55 km north north-west of Banff.

Dr. G.M. Dawson (see Tombstone Mountain) named this 2755 m mountain in 1884 after James White (1863-1928), a geographer who was an assistant to Dawson in surveys of the southern Rockies. He joined the Geographical Survey of Canada in 1884 and in 1899 was appointed Chief Geographer of the Department of the Interior. He served on the Commission of Conservation in various capacities and later became chairman of the Advisory Board on Wildlife Protection.

White Man Creek (creek)

82 J/14 - Spray Lakes Reservoir
33-21-11-W5
50°50′N 115°27′W
Flows north into Currie Creek, approximately 95 km west south-west of Calgary.

(see White Man Pass)

White Man Mountain (mountain)

82 J/14 - Spray Lakes Reservoir
5-21-11-W5
50°46′N 115°29′W
Approximately 95 km west south-west of Calgary, on the Alberta-B.C. boundary.

This mountain is 2977 m in altitude. (see White Man Pass)

White Man Pass (pass)

82 J/14 - Spray Lakes Reservoir
21-11-W5
50°46′N 115°29′W
Approximately 95 km west south-west of Calgary, on the Alberta-B.C. boundary.

The name of this 2168 m pass is a translation of an Indian name, "tshakooap-te-ha-wapta," which probably refers to Father Jean de Smet's (see Mount de Smet) crossing of the Rockies by it in 1845. It is called "White Man's Pass" in Captain Palliser's Report, circa 1865. (see Palliser Range)

White Pyramid (peak)

82 N/15 - Mistaya Lake
13-33-20-W5
51°50′N 116°42′W
Approximately 105 km north-west of Banff.

The name for this 3210 m mountain peak is likely descriptive.

Whiteaves, Mount (mountain)

82 N/10 - Blaeberry River
8-32-20-W5
51°43′N 116°48′W

Approximately 105 km north-west of Banff, on the Alberta-B.C. boundary.

This 3150 m mountain was named in 1920 after Joseph Frederick Whiteaves, LL.D., F.R.S.C. (1835-1909). This British-born paleontologist emigrated to Canada in 1862 and became curator of the museum of the Montreal Natural History Society for 12 years. In 1876, he was appointed to the staff of the Geological Survey of Canada and was one of the original Fellows of the Royal Society of Canada.

Whitecap Mountain (mountain)

83 E/1 - Snaring
32-47-2-W6
53°06′N 118°15′W
Approximately 21 km north north-west of Jasper.

This 2865 m mountain is snow-capped. Its descriptive name was applied in 1916 by Bridgland. (see Mount Bridgland)

Whitecrow Mountain (mountain)

83 D/9 - Amethyst Lakes
30-41-2-W6
52°34′N 118°15′W
Approximately 37 km south-west of Jasper.

A number of Clark's Crow (commonly known as white crow) were seen on this 2831 m mountain while some members of The Alpine Club of Canada were on it. A.O. Wheeler (see Wheeler Flats) applied the name in 1922.

Whitegoat Creek (creek)

83 C/2 - Cline River
33-37-18-W5
52°13′N 116°30′W
Flows south-east into Abraham Lake, approximately 130 km north-west of Banff.

This creek was named in 1975. (see Whitegoat Peaks)

Whitegoat Creek (creek)

83 C/7 -Job Creek
7-38-18-W5
52°15′N 116°33′W
Flows east into North Saskatchewan River,
approximately 105 km west south-west of
Rocky Mountain House.

This creek was formerly known as Cline
Creek. (see Whitegoat Peaks)

Whitegoat Lakes (lake)

83 C/1 - Whiterabbit Creek
22-37-18-W5
52°11′N 116°29′W
Approximately 130 km north-west of Banff.

(see Whitegoat Peaks)

Whitegoat Peaks (peak)

83 C/2 - Cline River
20-36-19-W5
52°06′N 116°41′W
Approximately 125 km south-east of Jasper.

The locally well-established name for this
feature was officially approved in 1959.
There were five goats seen near the campsite
of D.J. Link, a surveyor in the area who
suggested the name.

Whitehorn (mountain)

82 N/8 - Lake Louise
10-29-16-W5
51°28′N 116°09′W
Approximately 55 km north-west of Banff.

The locally well-established name of this
2637 m feature is descriptive and was
officially approved in 1959.

Whitehorse Creek (creek)

83 C/13 - Medicine Lake
24-46-25-W5
52°50′N 117°30′W

Flows east into McLeod River,
approximately 51 km east north-east of
Jasper.

The origin of the name is unknown.

Whitemans Pond (reservoir)

82 O/3 - Canmore
25-24-11-W5
51°04′10″N 115°24′45″W
Approximately 4 km south-west of
Canmore.

This locally well-established name was
officially approved 7 February 1983. No
other origin information is known.

Whiterabbit Creek (creek)

82 N/16 - Siffleur River
9-35-16-W5
52°00′N 116°13′W
Flows north-east into North Saskatchewan
River, approximately 100 km north north-
west of Banff.

The name of this creek is a translation of a
Stoney Indian name.

Whiteshield Mountain (mountain)

83 E/6 - Twintree Lake
11-50-10-W6
53°18′N 119°22′W
Approximately 97 km west north-west of
Jasper.

The descriptive name for this 2684 m
mountain derives from the snow and ice
formation on the east side of the feature.
The name was approved in 1924.

Whitney Creek (creek)

82 G/8 - Beaver Mines
12-5-2-W5
49°22′N 114°09′W
Flows north into Mill Creek, approximately
18 km south-west of Pincher Creek.

There is no origin information available for
this locally well-established name.

Whyte, Mount (mountain)

82 N/8 - Lake Louise
24-28-17-W5
51°25′N 116°16′W
Approximately 55 km north-west of Banff.

Sir William Methuen Whyte (1843-1914) was
a native of Scotland and came to Canada in
1863. He worked for the Grand Trunk
Railway for twenty years, occupying several
different positions during that time. He
moved to the Canadian Pacific Railway in
1884 and held several official titles during
the building of the Canadian Pacific Railway
to the Pacific Coast. He eventually became
Vice-President of the Railway in 1910. This
mountain, which is 2983 m in altitude, was
named for Sir William Whyte.

Wickman Creek (creek)

83 F/2 - Foothills
35-48-19-W5
53°11′20″N 116°40′00″W
Approximately 46 km south-east of Edson.

The name for this creek was officially
approved 9 July 1980. The origin of the
name is unknown.

Wigmore Creek (creek)

82 O/12 - Barrier Mountain
2-30-13-W5
51°32′N 115°44′W
Flows north into Panther River,
approximately 40 km north-west of Banff.

This creek was named after Sam Wigmore, a
prospector who had a copper claim on the
creek.

Wigmore Lake (lake)

82 O/5 - Castle Mountain
20-29-12-W5
51°29′N 115°42′W

Approximately 36 km north north-west of Banff.

Officially approved by the Canadian Permanent Committee on Geographical Names 17 August 1964, this lake takes its name from the nearby creek. (see Wigmore Creek)

Wigwam Creek (creek)

82 O/11 - Burnt Timber Creek
18-31-9-W5
51°38′N 115°16′W
Flows north into Red Deer River, approximately 55 km north north-east of Banff.

This creek was officially given this name in 1918 because an Indian camp was located nearby.

Wigwam Creek (creek)

83 F/6 - Pedley
28-49-24-W5
53°15′N 117°27′W
Flows east into Teepee Creek, approximately 76 km south-west of Edson.

The Geographic Board of Canada approved the name for this branch of Teepee Creek in 1927. The name was originally applied to this creek after an Indian camp.

Wilcox Lake (lake)

83 C/3 - Columbia Icefield
52°14′55″N 117°11′50″W
Approximately 93 km south-east of Jasper.

Wilcox Lake is so named for its proximity to Wilcox Pass and Mount Wilcox. The name was officially approved 7 November 1984. (see Wilcox Pass)

Wilcox, Mount (mountain)

83 C/3 - Columbia Icefield
38-23-W5
52°15′N 117°14′W
Approximately 90 km south-east of Jasper.

This mountain, which was named in 1899, is 2884 m in altitude. (see Wilcox Pass)

Wilcox Pass (pass)

83 C/3 - Columbia Icefield
9-38-23-W5
52°15′N 117°14′W
Approximately 90 km south-east of Jasper.

Walter Dwight Wilcox, the author of *The Rockies of Canada* (1906) was probably the first white man to traverse the pass. He was also one of the pioneer tourist-explorers and mountaineers in the Canadian Rockies. This 2347 m mountain pass was named by J.N. Collie (see Collie Creek) in 1899 after W.D. Wilcox.

Wilcox Peak (peak)

83 C/3 - Columbia Icefield
8-38-23-W5
52°15′N 117°15′W
Approximately 90 km south-east of Jasper.

(see Wilcox Pass)

Wild Hay (station)

83 F/12 - Gregg Lake
1-53-27-W5
53°34′N 117°51′W
Approximately 77 km north of Jasper.

This station located on the Canadian National Railway was likely named after the Wildhay River, in turn, descriptively named from the wild hay grown in meadows around where it flows.

Wildcat Creek (creek)

82 G/9 - Blairmore
11-9-2-W5
49°43′N 114°10′W
Flows east into Todd Creek, approximately 15 km north of Lundbreck.

The origin of the name is unknown.

Wildcat Creek (creek)

83 F/12 - Gregg Lake
16-55-27-W5
53°44′N 117°59′W
Flows north into Hightower Creek, approximately 46 km north-west of Hinton.

No origin information is known.

Wildflower Creek (creek)

82 N/8 - Lake Louise
2-29-15-W5
51°27′N 116°01′W
Flows west into Baker Creek, approximately 43 km north-west of Banff.

The origin is not known for the locally well-established name, originally proposed by C.D. Walcott. (see Introduction)

Wildflower Lake (lake)

82 O/5 - Castle Mountain
28-14-W5
51°25′N 115°59′W
Approximately 40 km north-west of Banff.

The descriptive name for this lake is appropriate since the surrounding area is noted for its colourful flora.

Wildhorse Creek (creek)

82 O/11 - Burnt Timber Creek
15-31-10-W5
51°39′N 115°21′W
Flows south into Red Deer River, approximately 55 km north north-east of Banff.

Reports of wild horses at the head of creek inspired its naming in 1919. The name was made official 12 December 1939.

Wildhorse Lakes (lakes)

83 F/5 - Entrance
31-49-26-W5
53°16′N 117°48′W

Approximately 18 km south-west of Hinton.

The origin of the name is unknown.

Wildman Creek (creek)

82 O/2 - Jumpingpound Creek
32-25-6-W5
51°10′N 114°48′W
Flows north into Chiniki Creek,
approximately 45 km west north-west of
Calgary.

D.A. Nichols suggested the name for this
creek in 1929 after Daniel Wildman, the
local Indian interpreter.

Wileman Creek (creek)

82 J/7 - Mount Head
25-17-5-W5
50°28′N 114°34′W
Flows north into Flat Creek, approximately
31 km south-west of Turner Valley.

Mr. Harry J. Wileman was born in England
in 1901. He came to Alberta in 1921 and
became a forest ranger five years later. He
served in the Slave Lake district for a
number of years. He came to the Highwood
Forest Ranger Station in 1935 and served
faithfully until a fatal accident on 19 August
1943. He was a strong, active man, well-
known and highly respected among his
peers. The creek may be named after this
forest ranger. The name was officially
approved 8 December 1943.

Wilkinson Creek (creek)

82 J/7 - Mount Head
26-15-5-W5
50°17′N 114°35′W
Flows north into Cataract Creek, approximately
48 km south-west of Turner Valley.

This creek, which was officially named
8 December 1943, might be named after an
early homesteader, Philip Wilkinson, who
came to Canada from London, England in
1901.

William A. Switzer Park (provincial park)

83 F/5 - Entrance
8-52-26-W5
53°30′N 117°48′W
Approximately 13 km west north-west of
Hinton.

This park, which was formerly known as
Entrance Provincial Park, was renamed in
1974 after William A. Switzer (1921-1969).
He was a three-term Mayor of Hinton, prior
to entering provincial politics, where he was
Liberal Member of the Legislative Assembly
for Hinton, 1965-1969.

William Booth, Mount (mountain)

83 C/1 - Whiterabbit Creek
14-36-37-W5
52°05′N 116°19′W
Approximately 115 km north-west of Banff.

This 2728 m mountain was named in 1965
after Mr. William Booth (1829-1912), the
founder of the Salvation Army, on the
centenary of its founding.

Williams, Mount (mountain)

82 J/11 - Kananaskis Lakes
20-10-W5
50°43′N 115°22′W
Approximately 75 km west of Turner Valley.

This 2730 m mountain was officially named
in 1918 after Major-General Victor W.
Williams, an officer of the Canadian
Expeditionary Force who was taken prisoner
at Zillebeke, Flanders, Belgium June 1916.

Willingdon, Mount (mountain)

82 N/16 - Siffleur River
19-32-16-W5
51°45′N 116°15′W
Approximately 80 km north-west of Banff.

This 3061 m mountain was named after
Freeman Freeman-Thomas, First Marquis of
Willingdon (1866-1941). This British

statesman and colonial administrator was the
Governor General of Canada from
1926-1931. He then served in India, where he
was appointed to the post of Viceroy and
then served as Governor General until 1936,
when he returned to Britain.

Willmore Wilderness Park (park)

83 E - Mount Robson
8-52-8-W6
53°45′N 119°30′W

This 4568 square kilometre area is a
wilderness park of mountain peaks, glaciers,
alpine meadows and river valleys. It was
established in 1959 and was named in 1965
after the Honourable Norman A. Willmore
(1909-1965). Mr. Willmore came to Canada
from North Dakota in 1915 and began his
public life in 1942 as a town council member
in Edson. After being elected as a Social
Credit Member of the Legislative Assembly
for Edson in 1944, 1948 and 1952, he was
appointed Minister of Industries and Labour,
holding the position until 2 August 1955,
when he became Minister of Lands and
Forests, a position he held until his death.

Willoughby Ridge (ridge)

82 G/9 - Blairmore
7-4-W5
49°34′N 114°29′W
Approximately 8 km south-east of Blairmore.

This 1686 m ridge is part of the Crowsnest
volcanics. The upthrust of volcanic rocks
through the limestone ranges which formed
this ridge is unique in western Canada. The
ridge itself is named after the Willoughby
family, who were very early settlers in the area
and were in charge of a boarding house near
Sulphur Spring at the base of Turtle Mountain.

Willow Creek (creek)

83 E/8 - Rock Lake
2-51-3-W6
53°22′N 118°20′W

Flows south into Snake Indian River, approximately 58 km north north-west of Jasper.

The origin of the name is unknown.

Willow Valley (valley)

82 G/16 - Maycroft
9-10-3-W5
49°46′00″N 114°17′30″W
Approximately 15 km north-east of Blairmore.

The origin of the name for this valley is locally well-established and is descriptive of the flora of the valley. The Willow Valley School District is at the junction of Todd and South Todd Creek.

Willson Creek (creek)

82 O/14 - Limestone Mountain
10-33-9-W5
51°49′N 115°11′W
Flows east into James River, approximately 75 km north north-east of Banff.

This creek was named in 1908 after a veteran of the Riel Rebellion and the Boer War, Colonel Willson (1850-1927). During the First World War, he served as commander of the "D" Company, 4th Battalion, Canadian Expeditionary Force, throughout operations at Messines, Kemmel Hill and Ypres.

Wilson Creek (creek)

83 C/1 - Whiterabbit Creek
36-17-W5
52°06′N 116°25′W
Flows west north-west into Abraham Lake, approximately 122 km north north-west of Banff.

This creek flows through the ranch of Thomas Wilson who was one of the first white men to visit the Upper Kootenay Plains.

Wilson Icefield (icefield)

83 C/2 - Cline River
22-35-20-W5
52°01′05″N 116°50′00″W
Approximately 125 km south-east of Jasper.

The name for this icefield, formerly Wilson Glacier, is derived from Mount Wilson, on which the feature is located. The change was made official 14 April 1980. (see Mount Wilson)

Wilson, Mount (mountain)

83 C/2 - Cline River
20-35-20-W5
52°01′N 116°47′W
Approximately 130 km south-east of Jasper.

This mountain, which is 3260 in altitude, was named by J.N. Collie (see Collie Creek) in 1898. It commemorates Tom Edmunds Wilson (1859-1933), a well-known guide of Banff, originally from Ontario. During his life, he had several careers, including volunteer militia, a member of the North West Mounted Police (22 September 1880-16 May 1881), a packer for the C.P.R. in 1881, and eventually settled as a trapper and a home-steader north-east of Morley circa 1886. An assignment with the Dominion Topographic Survey between 1889-1893 left him with his claim to fame. (see E.J. Hart, *Diamond Hitch*, 1979 p.1-16).

Wilson Range (range)

82 H/4 - Waterton Lakes
1-29-W4
49°00′N 113°52′W
Approximately 55 km south south-east of Pincher Creek.

Thomas Blakiston of the Palliser Expedition (see Palliser Range) had originally given the name "Wilson" to a single peak. Only later did the name apply to the entire range. Lieutenant Charles William Wilson, R.E., (1836-1903), was the Secretary to the British Boundary Commission of 1858-1862, which

determined the international boundary from the Pacific to the Rockies (see Cameron Lake). Wilson kept a daily journal in which he commented on the survey and on social life in the frontier towns of that era. This journal was edited and introduced by G.F.G. Stanley and published in 1970 under the title, *Mapping the Frontier*. The Wilson Range is listed in the *Canadian Gazette* of 8 January 1916.

Winchester Creek (creek)

82 O/11 - Burnt Timber Creek
51°35′35″N 115°29′50″W
Flows north north-east into Panther River, approximately 60 km north north-east of Banff.

The well-known local name for this creek was in use for over thirty years before its official approval 29 July 1986. Apparently, the name originated after a Winchester rifle was found (allegedly by local Indians) beside the creek. The rifle was very old, and the butt was completely rotten.

Wind Creek (creek)

82 O/3 - Canmore
24-9-W5
51°03′N 115°15′W
Flows north-west into Pigeon Creek, approximately 11 km south-east of Canmore.

The name of this creek was officially approved 5 August 1948. (see Wind Mountain)

Wind Mountain (mountain)

82 J/14 - Spray Lakes Reservoir
18-23-9-W5
50°57′15″N 115°14′45″W
Approximately 75 km west of Calgary.

This mountain was originally named either "Windy" or Wind Mountain by Eugène Bourgeau, a member of the Palliser Expedition (1857-1860) (see Mount Bourgeau) in 1858 because of the clouds gathering and curling around the summit. It was officially approved 17 June 1983.

Wind Ridge (ridge)

82 O/3 - Canmore
2-24-10-W5
51°01′N 115°18′W
Approximately 9 km south south-east of Canmore.

(see Wind Mountain)

Windfall Creek (creek)

82 O/14 - Limestone Mountain
28-32-9-W5
51°46′N 115°14′W
Flows north into James River, approximately 70 km north north-east of Banff.

The origin of the name is unknown.

Window Mountain Lake (lake)

82 G/15 - Tornado Mountain
29-9-5-W5
49°45′40″N 114°38′20″W
Located immediately north of Mount Ward, approximately 15 km north-west of Coleman.

This small alpine lake north of the Crowsnest Pass was first stocked with Rainbow Trout in July of 1957. The name Window Mountain Lake is a locally well-established one and was officially approved 9 May 1978.

Windsor Mountain (mountain)

82 G/8 - Beaver Mines
9-4-2-W5
49°17′N 114°13′W
Approximately 30 km south-west of Pincher Creek.

In 1858, Thomas Blakiston (see Mount Blakiston) named this feature Castle Mountain, and it is still locally known by

some residents as such. However, the name was changed to Windsor Mountain in 1915 to avoid confusion with Castle Mountain north of Castle Railway station. There is a slight likeness in shape to Windsor Castle when the mountain is seen against the western sky. Because of this resemblance, its official name is now Windsor Mountain.

Windsor Ridge (ridge)

82 G/1 - Sage Creek
9-4-2-W5
49°17′N 114°13′W
Approximately 28 km south-west of Pincher Creek.

(see Windsor Mountain)

Windy Peak (peak)

82 J/1 - Langford Creek
13-14-4-W5
50°10′N 114°25′W
Approximately 60 km north-west of Claresholm.

The locally well-established name of this 2237 m peak is likely descriptive.

Windy Point (point)

83 C/8 - Nordegg
8-38-17-W5
52°15′N 116°23′W
Approximately 50 km south-east of Jasper.

The origin of the name is unknown.

Windy Ridge (ridge)

82 J/13 - Mount Assiniboine
16-23-12-W5
50°57′40″N 115°37′00″W
Approximately 105 km west of Calgary, on the Alberta-B.C. boundary.

The name "Windy Pass" was originally submitted in 1979 by D.F. Pearson, B.C. Representative on the Canadian Permanent Committee on Geographical Names, but was

rejected on the grounds that the feature is not a pass. The name Windy Ridge was submitted 1 February 1983 by Mr. Pearson based on information from Assistant District Manager for Mount Assiniboine Provincial Park that the name was applied to the northern slope of Og Mountain. The name was officially approved 29 January 1986.

Winston Churchill Range (range)

83 C/6 - Sunwapta
38,39,40-23,24,25-W5
52°21′N 117°28′W
Approximately 65 km south-east of Jasper.

Located at the south end of Jasper National Park, this picturesque 200 square mile mountain range was named in 1965 after Sir Winston Leonard Spencer Churchill, K.G., P.C., O.M., C.H. (1874-1965) to honour the character and distinguished services in the preservation of democratic ideals. This English stateman and writer was a correspondent for the *Morning Post* during the Boer War, and wrote about his several adventures while in South Africa. He returned to England to be elected to Parliament, during which he made a dramatic move by switching parties. He continued to become a forceful speaker, and maintained that the only way to keep peace was to prepare for war. During the First World War, Churchill was sent to France. He devoted much attention to the control and development of the air force and to both military and civilian airplane experiment and development. He returned to the Conservative party in 1924. Churchill was awarded the Nobel Prize for Literature in 1953. Besides many foreign decorations and honorary degrees, Churchill was the recipient of several medals, prizes and other tributes including the naming of this mountain range. He was knighted in 1953.

Winter Creek (creek)

83 F/12 - Gregg Lake
20-52-26-W5
53°30′N 117°48′W

Flows east into Jarvis Creek, approximately 18 km north-west of Hinton.

This was where the pack saddle horses belonging to the Forest Service were wintered. The name of the creek is well-established in local usage.

Wintering Creek (creek)

82 G/16 - Maycroft
22-10-4-W5
49°49′N 114°28′W
Flows south-east into Racehorse Creek, approximately 25 km north of Blairmore.

The origin of the name is unknown.

Wintour, Mount (mountain)

82 J/11 - Kananaskis Lakes
17-20-8-W5
50°42′N 115°04′W
Approximately 55 km west of Turner Valley.

Captain C. Wintour and his 4th Destroyer Flotilla bore the brunt of a fierce naval battle at Jutland, during the First World War. He was killed in the fight, and his name is commemorated on this 2700 m mountain.

Wolf Creek (creek)

82 J/10 - Mount Rae
21-19-4-W5
50°37′N 114°30′W
Flows north-east into Sheep River, approximately 16 km south-west of Turner Valley.

The name for this creek was officially approved 6 September 1951, but its origin is unknown.

Wolf Creek (creek)

83 L/6 - Chicken Creek
29-63-8-W6
54°29′N 119°10′W
Flows north-east into Cutbank River, approximately 68 km north of Grande Cache.

This creek was likely named after the wolf, the coarse furred, carnivorous, dog-like mammal of genus canis.

Wolf Lake (lake)

83 F/1 - Raven Creek
12-49-15-W5
53°13′N 116°03′W
Approximately 47 km south-east of Edson.

The origin of the name is unknown.

Wolverine Creek (creek)

83 E/14 - Grande Cache
25-55-10-W6
53°46′N 119°21′W
Flows north-west into Smoky River, approximately 19 km west south-west of Grande Cache.

This creek is likely named after the wolverine, a mammal related to the weasel.

Wolverine Mountain (mountain)

83 E/6 - Twintree Lake
14-50-9-W6
53°19′N 119°12′W
Approximately 89 km west north-west of Jasper.

Named in 1925, the precise origin for the name of this 2777 m mountain is unknown.

Wonder Pass (pass)

82 J/13 - Mount Assiniboine
22-12-W5
50°54′N 115°35′W
Approximately 105 km west of Calgary, on the Alberta-B.C. boundary.

(see Wonder Peak)

Wonder Peak (peak)

82 J/13 - Mount Assiniboine
22-12-W5
50°53′N 115°34′W

Approximately 100 km west of Calgary, on the Alberta-B.C. boundary.

The name for this 2852 m peak is descriptive and was officially approved in 1913.

Woolley Creek (creek)

83 C/6 - Sunwapta
27-38-24-W5
51°18′N 117°20′W
Flows east into Sunwapta River, approximately 82 km south south-east of Jasper.

(see Mount Woolley)

Woolley, Mount (mountain)

83 C/6 - Sunwapta
30-38-24-W5
51°18′N 117°25′W
Approximately 77 km south south-east of Jasper.

This 3460 m mountain was named by J.N. Collie (see Collie Creek), in 1898, after Hermann Woolley, of Caucasion and Alpine Mountaineering fame. He was a fellow climber of Collie, who accompanied him on many climbing expeditions.

Woolley Shoulder (ridge)

83 C/6 - Sunwapta
30-38-24-W5
51°17′N 117°24′W
Approximately 78 km south south-east of Jasper.

(see Mount Woolley)

Worthington, Mount (mountain)

82 J/11 - Kananaskis Lakes
22-19-10-W5
50°37′N 115°18′W
Approximately 70 km west of Turner Valley, on the Alberta-B.C. boundary.

This 2972 m mountain was named after Lieutenant-Colonel Don Worthington, who was killed in action in 1944 while commanding the 7th Battalion, B.C. Regiment. The name was officially approved 3 May 1956.

Wroe Creek (creek)

83 F/12 - Gregg Lake
8-55-26-W5
53°45′N 117°50′W
Flows north-east into Pinto Creek, approximately 41 km north-west of Hinton.

The name was officially approved in 1945. The origin of the name for this creek is unknown.

Wynd (locality)

83 D/16 - Jasper
7-45-1-W6
52°52′N 118°08′W
Approximately 3 km west of Jasper.

The precise origin of the name of this locality on the Canadian National Railway line is not known.

Yara Creek (creek)

82 O/11 - Burnt Timber Creek
17-31-9-W5
51°38′N 115°15′W
Flows south-east into Red Deer River, approximately 55 km north-east of Banff.

No origin information for the name is known.

Yellowhead Mountain (mountain)

83 D/15 - Rainbow
45-4-W6
52°52′N 118°35′W
Approximately 33 km west of Jasper.

(see Yellowhead Pass)

Yellowhead Pass (pass)

83 D/16 - Jasper
23-45-4-W6
52°53′N 118°28′W

Approximately 26 km west of Jasper.

The Yellowhead Pass, which is 1131 m in altitude, is the lowest crossing of the Continental Divide on the North American continent. It was first discovered by Europeans around 1826. This pass was for many years the route through which fur traders proceeded to the New Caledonian Department of the Hudson's Bay Company. The pass is thought to be named after one of three people: Francois Décoigne, Pierre Hatsination or Pierre Bostonais. Francois Décoigne worked for both the North West Company and the Hudson's Bay Company from the 1790s to the 1810s. This trader, who was reputed to be blonde, was manager in 1814 of Jasper House, on Brule Lake near to this pass. Other sources, including Hudson's Bay Company documents such as letters by Colin Robertson, use the nickname "Tête Jaune (Yellowhead)" interchangeably with the name Pierre Hatsination (or Hathawiton) to describe an Iroquois with some white blood. The white blood gave his hair a light tinge, thus gaining him the nickname "Tête Jaune." He was hired, but not indentured, to the Hudson's Bay Company as a guide and trapper and escorted trading expeditions into the Yellowhead Pass area and New Caledonia (a large area around present-day Prince George, B.C.). *The Canadian Encyclopedia* gives the source of the nickname as "a blond Iroquois trapper, Pierre Bosstonais, who hunted and trapped in the area." As the biographical data (with the exception of the last names) of Bostonais and Hatsinaton is identical, perhaps they were the same person.

York Creek (creek)

82 G/9 - Blairmore
3-8-4-W5
49°37′N 114°28′W
Flows north into Crowsnest River, approximately 40 km west north-west of Pincher Creek.

This creek is said to be named after York, the cathedral city in England.

Younghusband Ridge (ridge)

83 C/4 - Clemenceau Icefield
52°14′10″N 117°48′50″W
Approximately 75 km south south-east of Jasper, on the Alberta-B.C. boundary.

The name of this ridge was first suggested in 1927 by A.J. Ostheimer III, an American alpinist, to honour Lieutenant Colonel Sir Francis Younghusband, a member of an unsuccessful Everest expedition. It is not known whether Younghusband climbed in the Rockies, but the name was considered appropriate and was officially approved 5 May 1987.

Zebra Mountain (mountain)

83 E/9 - Moberly Creek
53-3-W6
53°33′N 118°24′W
Approximately 79 km north north-west of Jasper.

The origin of the name is unknown.

Zenda Creek (creek)

83 E/10 - Adams Lookout
7-53-6-W6
53°34′N 118°52′W
Flows west into Sulphur River, approximately 40 km south south-east of Grande Cache.

(see Anthony Creek)

Zengel, Mount (mountain)

83 D/16 - Jasper
14-46-2-W6
52°58′N 118°11′W
Approximately 12 km north of Jasper.

This 2560 m mountain was named in 1951. Sergeant Raphael Louis Zengel, V.C., M.M., won the British Empire's highest military decoration while serving with the 45th Battalion in France. He was born in Fairbutt, Minnesota, U.S.A., in 1894 and

later moved to Rocky Mountain House,
where he enlisted in the Canadian Infantry
in 1915. On 27 September 1918, Sergeant
Zengel won the Victoria Cross for attacking
a German machine gun emplacement. In
March of that same year, he won the
Military Medal for taking command of his
platoon after the platoon officer and platoon
sergeant had been wounded.

Zephyr Creek (creek)

82 J/7 - Mount Head
16-5-W5
50°23'N 114°35'W
Flows north into Highwood River,
approximately 38 km south-west of Turner
Valley.

Officially approved 8 December 1943, the
name for this creek may be taken from
Zephyr, the god of the West Wind in Greek
mythology.

Zigadenus Lake (lake)

82 N/9 - Hector Lake
20-29-15-W5
51°30'N 116°05'W
Approximately 50 km north-west of Banff.

The name for this lake was suggested by
James F. Porter, an American who made an
extensive exploration of the area in 1911 (see
Merlin Lake). There was an abundance of
the white camas flower, genus Zygadenus
elegans (the spelling has been corrupted in
the official name), around its shore, which
gave the name for this lake its source.

BIBLIOGRAPHY

Alberta Historical Review, Volume 11, number 1. Calgary, Alberta: Historical Society of Alberta, Winter 1963.

Alberta Wilderness Association. *The Willmore Wilderness Park*. Alberta Wilderness Association, 1973.

Allen, Samuel E.S. *Explorations Among the Watershed Rockies of Canada*. s.n., [s.1.: 1894?].

Anderson, Anne. *Plains Cree Dictionary in the "y" Dialect*. Edmonton, Alberta: Anne Anderson, 1975.

Anderson, Frank W. "Frontier Guide to Mystic Jasper and the Yellowhead Pass." (Frontier book; number 30). Frontiers Unlimited; Calgary, Alberta: Frank W. Anderson, 1973.

Appleby, Edna (Hill). *Canmore: The Story of an Era*. Canmore, Alberta: E. Appleby, 1975.

Babcock, Doug. "Crowsnest Pass — Preliminary Survey of Historical Themes." Unpublished manuscript on file with Historic Sites Service, Edmonton, Alberta: In-house report, June, 1977.

Baird, David M. *Jasper National Park, Behind the Mountains and Glaciers*. The Geological Survey of Canada, Department of Mines and Technical Surveys, Ottawa, Ontario: Queen's Printer and Controller of Stationery, 1963.

Barber, Richard. *The Life and Campaigns of the Black Prince*. London, England: Folio Society, 1979.

Bearberry Wapitana Society of Bearberry. *Recollections of the Homestead Trails: History of Bearberry and Sunberry Valleys*. Bearberry, Alberta: Bearberry Wapitana Society of Bearberry, 1978.

Berton, Pierre. *The National Dream: The Great Railway, 1871-1881*. Toronto, Ontario: McClelland and Stewart Ltd., 1970.

Blairmore Lions Club. *The Story of Blairmore Alberta 1911-1961*. Blairmore, Alberta: The Blairmore Lions Club, 1961.

Boles, Glen W. with Robert Kruszyna and William L. Putnam. *The Rocky Mountain of Canada South*. New York: The American Alpine Club, Inc., 1979.

Bond, Courtney C.J. *Surveyors of Canada, 1867-1967*. Ottawa, Ontario: C.I.S., 1966.

Brewster, F.O. "Pat." *Banff, The Rockies, and the West: Tales of the Early Days*. Banff, Alberta: Dorothy Cranstone, for the estate of F.O. Brewster, 1984.

_____. *Weathered Wood: Ancedotes and History of the Banff-Sunshine Area*. Banff, Alberta: Banff Crag and Canyon, 1977.

Bruce, George L. *Sea Battles of the 20th Century*. London, England: Hamlyn, 1975.

Burles, Mary-Jo and Marjorie Haugen. *Cowley: Sixty Years a Village, 1906-1966*. Cowley, Alberta: Mary-Jo Burles, 1966.

Burpee, Lawrence Johnstone. *Jungling in Jasper*. Ottawa, Ontario: L.J. Burpee, 1929.

_____. *Sandford Fleming: Empire Builder*. Toronto, Ontario: The Oxford University Press, 1915.

Canada, Geographic Board of. *Place Names of Alberta*. Ottawa, Ontario: Geographic Board of Canada, 1928.

Canadian Encyclopedia, The: Volume I. Edmonton, Alberta: Hurtig Publishers Ltd., 1985.

Cardston and District Historical Society. *Chief Mountain Country: A History of Cardston and District*. Cardston, Alberta: Cardston and District Historical Society, 1978-1987.

Cautley, R.W. and A.O. Wheeler. *Report of the Commission Appointed to Delimit the Boundary Between the Provinces of Alberta and British Columbia*. Parts I-III, Ottawa, Ontario: Office of the Surveyor General, 1917-1925.

Cayley Women's Institute. *Under the Chinook Arch: A History of Cayley and Surrounding Area*. Cayley, Alberta: Cayley Women's Institute, 1967.

Chamber's Encyclopedia. Oxford; New York: Pergamon Press, 1967.

Cheadle, W.B. and Viscount W.F. Milton. *The North West Passage by Land*. London, England: Cassell, Pelter and Galpin, 1965.

Civil Service Bulletin. Volume XXX, number 11: December, 1950.

Claresholm History Book Club. *Where the Wheatlands Meet the Range*. Claresholm, Alberta: Claresholm History Book Club, 1974.

Cochrane and Area Historical Society. *Big Hill Country*. Cochrane, Alberta: Cochrane and Area Historical Society, 1977.

Coleman, A.P. *The Canadian Rockies: New and Old Trails*. Toronto, Ontario: H. Frowde, 1912.

Collie, J.N. and Hugh E.M. Stutfield. *Climbs and Explorations in the Canadian Rockies.* London, England: Longmans, Green and Company Publishers, 1903.

Collier's Encyclopedia. London, England: P.F. Collier, 1982.

Cormack, R.G.H. *Wild Flowers of Alberta.* Edmonton, Alberta: Department of Industry and Development, 1967.

Costello, John and Terry Hughes. *Jutland, 1916.* London, England: Weidenfeld & Nicolson, 1976.

Cousins, William J. *A History of the Crow's Nest Pass.* Lethbridge, Alberta: The Historic Trails Society of Alberta, 1981.

Cowan, Ian McTaggert and J.A. Munro. *Canadian Alpine Journal.* (various volumes), 1946(?).

Crowsnest Pass Historical Society. *Crowsnest and its People.* Coleman, Alberta: Crowsnest Pass Historical Society, 1979.

Cuncliff, Barry. *Rome and Her Empire.* New York: McGraw-Hill Book Company Limited, 1978.

Daffern, Tony and Gillean Daffern. *Kananaskis Country: A Guide to Hiking and Skiing Trails.* Calgary, Alberta: Rocky Mountain Books, 1979.

Dawson, George M. *Report on the Region in the Vicinity of the Bow and Belly Rivers, North West Territory.* By G.M. Dawson; assisted by R.G. McConnell. Montreal, Quebec: Dawson Brothers, 1884.

Dempsey, Hugh. (ed.). *Alberta History.* Calgary, Alberta: Historical Society of Alberta, Autumn, 1988.

_____. "Indian Names for Alberta Communities." Occasional Paper number 4. Calgary, Alberta: Glenbow Alberta Institute, 1969.

Denton, G.H. "Glaciers of the Rocky Mountains" in *Mountain Glaciers of the Northern Hemisphere.* W.O. Field (ed.). Hanover, N.H.: Technical Information Analysis Center, Cold Regions Research and Engineering Laboratory, 1975.

Department of the Interior. *A Description of and Guide to Jasper Park.* Ottawa, Ontario: Department of the Interior, 1917.

De Smet, (Father) P.J. *Oregon Missions and Travel Over the Rocky Mountains in 1845-46.* New York: Edward Dunigan, 1847.

De Winton and District Historical Committee. *Sodbusting to Subdivision.* De Winton, Alberta: De Winton and District Historical Committee, 1978.

Dictionary of Canadian Biography: Volume 9, 1861-1870. Toronto, Ontario: The University of Toronto Press, 1976.

East Longview Historical Society. *Tales and Trails, 1900-1972.* Longview, Alberta: compiled and edited by the East Longview Historical Society, 1973.

Edwards, Ralph. *Trails to the Charmed Land.* Saskatoon, Saskatchewan: H.R. Larson, 1950.

Encyclopedia Britannica. Chicago, Ill.: Encyclopedia Britannica, 1972.

Encyclopedia Canadiana. Toronto, Ontario: Grolier of Canada, 1977.

Encyclopedia of Canadian Biography, Volume 1. Montreal, Quebec: Canadian Press Syndicate, Canada, 1904.

Eskrick, Muriel. *Road to Ya Ha Tinda: A Story of Pioneers.* Sundre, Alberta: Muriel Eskrick, first printed in 1960; second edition, 1969.

Finch, David A. "Turner Valley Oilfield Development 1914-1945." Calgary, Alberta: University of Calgary, October 1980.

Fleming, Sandford. "Confidential Memorandum on the Canadian Pacific Railway by Sandford Fleming, Engineer in Chief." Ottawa, Ontario: 1874.

_____. "Progress Report on the Canadian Pacific Railway Exploratory Survey." Ottawa, Ontario: 1872.

_____. "Report on Surveys and Prelimiary Operations on the Canadian Pacific Railway up to January 1877." Ottawa, Ontario: 1877.

_____. *Canadian Pacific Railroad: Report of Progress on the Exploration and Surveys up to January, 1874.* Ottawa, Ontario: MacLean, Roger, 1874.

Foothills Historical Society. *Chaps and Chinooks: A History West of Calgary.* Calgary, Alberta: Foothills Historical Society, 1976.

Fort Macleod Gazette. "A Trip to the Rockies." August 29, 1984.

Franchère, Gabriel. *Narrative of a Voyage to the North West Coast of America.* New York: Redfield, 1854.

Fraser, Esther. *The Canadian Rockies: Early Travels and Explorations.* Edmonton, Alberta: M.G. Hurtig, Ltd., 1969.

Frontiers Unlimited. *Frontier Guide to Waterton — Land of Leisure.* Calgary, Alberta: Frontiers Unlimited, 1968.

Frontiers Unlimited. *The Romantic Crowsnest Pass.* (Frontier book; number 5) Calgary, Alberta: Frontiers Unlimited, 1963.

Fryer, Harold. *Ghost Towns of Alberta.* Langley, British Columbia: Stagecoach Publishing Co., 1976.

Funk and Wagnalls New Encyclopedia. New York: Funk and Wagnalls Inc., 1972.

Gazetteer of Canada: Alberta. Ottawa, Ontario: Minister of Supply and Services Canada, Canadian Government Publishing Centre, 1988.

Gest, Lillian. *History of Mount Assiniboine in the Canadian Rockies.* Banff, Alberta: Banff Crag and Canyon, 1979.

Glover, Richard (ed.). *David Thompson's Narrative, 1784-1812.* Toronto, Ontario: Champlain Society, 1962.

Grant, (Rev.) George M. *Ocean to Ocean: Sandford Fleming's Expedition Through Canada in 1872. Being a Diary Kept During a Journey from the Atlantic to the Pacific with the Expedition of the Engineer-in-Chief of the Canadian Pacific and Intercolonial Railways.* Toronto, Ontario: James Campbell and Son, 1973.

Granum History Committee. *Leavings by Trail, Granum by Rail.* Granum, Alberta: Granum History Committee, ca. 1976.

Haig, Bruce. *In the Footsteps of Thomas Blakiston — 1858.* Lethbridge, Alberta: Bruce Haig; assisted by "Trek" students, Hamilton Junior High School, 1980.

Hart, E.J. *Diamond Hitch.* Banff, Alberta: Summerthought Ltd., 1979.

_____. *The Brewster Story: From Pack Trail to Tour Bus.* Banff, Alberta: Brewster Transport Company, 1981.

Hart, Hazel. *History of Hinton.* Hinton, Alberta: H. Hart, 1980.

Helenius, Alice Cummings. *It's Good to Remember.* Alice Cummings Helenius; drawing by Tom Dison; foreword by Dan Brockman. Puyallup, Washington: Valley Press, 1977.

High River Pioneers and Oldtimers Association. *Leaves From the Medicine Tree: A History of the Area Influenced by the Tree and Biographies of Pioneers Who Came Under Its Spell Prior to 1900.* High River, Alberta: High River Pioneers and Oldtimers Association, 1960.

Holmgren, Eric J. *Over 2000 Place Names of Alberta.* 3rd ed. Saskatoon, Saskatchewan: Western Producer Prairie Books, 1976.

Holtz, Ann and Mark Rasmussen. "History of the Jasper Park and Entrance Area." Unpublished manuscript on file with Historic Sites Service, Edmonton, Alberta: In-house report for Dr. J. Lunn, November 1979.

Hough, Richard. *The Great War at Sea — 1914-1918.* Oxford: Oxford University Press, 1983.

Ingersoll, Ernest. *Canadian Guide Book, Part II.* Charles G.D. Roberts (ed.). New York: D. Appleton, 1899.

Ings, Fred W. *Before the Fences (Tales From the Midway Ranch).* Jim Davis (ed.). Calgary, Alberta: McAra Printing Limited, 1980.

Irving, Commander John. *The Smoke Screen of Jutland.* New York: D.McKay Co., 1967.

Jordan, Mabel E. *The McDougall Memorial United Church, Morley, Alberta: The Church of the Mountain Stoneys, a souvenir booklet.* Calgary, Alberta(?): Mabel Jordan, 1957?

Kane, Paul. *Wanderings of an Artist Among the Indians of North America.* London, England: Longmans, Brown, Green, Longmans and Roberts, 1859.

Kemp, Herbert Douglas. *The Department of the Interior in the West, 1873-1883.* Edmonton, Alberta: University of Alberta, 1950.

Kruszyna, Robert with William L. Putnam. *The Rocky Mountains of Canada North.* New York: The American Alpine Club, Inc., 1985.

Lee, Sidney & Leslie Stephen, (ed.). *Dictionary of National Biography.* London, England: 1909.

Liddell, Kenneth E. *Roamin' Empire of Southern Alberta.* Calgary, Alberta: Frontiers Unlimited, 1963(?).

_____. *This is Alberta.* Toronto, Ontario: Ryerson Press, [c.1952].

Luxton, Eleanor G. *Banff, Canada's First National Park: A History and a Memory of Rocky Mountains Park.* Banff, Alberta: Summerthought Ltd., 1975.

Lynch-Staunton, C. *A History of the Early Days of Pincher Creek.* Lethbridge, Alberta: Women's Institute of Alberta, 1942.

MacEwan, Grant. *Calgary Cavalcade: From Fort to Fortune.* Edmonton, Alberta: Institute of Applied Art, 1958.

MacGregor, James Grierson. *Edmonton, A History.* Edmonton, Alberta: Hurtig Publishers Ltd., 1967.

_____. *The Land of Twelve Foot Davis: A History of the Peace River Country.* Edmonton, Alberta: Institute of Applied Art, 1952.

_____. *Pack Saddles to Tête Jaune Cache.* Edmonton, Alberta: McClelland and Stewart Ltd., 1962.

_____. *Peter Fidler — Canada's Forgotten Surveyor, 1769-1822.* Toronto, Ontario: McClelland and Stewart Ltd., 1966.

_____. *Vision of an Ordered Land: The Story of the Dominion Land Survey.* Saskatoon, Saskatchewan: Western Producer Prairie Books, 1981.

MacLeod Gazette. December 15, 1899.

Macmillan Dictionary of Canadian Biography, The. 3rd ed. W. Stewart Wallace (ed.). Toronto, Ontario: The Macmillan Company of Canada Ltd., 1963.

Mardon, Ernest G. *Community Names of Alberta.* Lethbridge, Alberta: University of Lethbridge, 1972.

"Marl Lake Trail." Pamphlet published by Alberta Recreation, Parks and Wildlife, Parks Division, 1984(?)

McDougall United Church. *The Pioneers: McDougall Church, Our First 100 Years 1871-1971.* Edmonton, Alberta: s.n. 1971(?).

McNaughton, Margaret. *Overland to Cariboo.* Toronto, Ontario: William Briggs, 1896; Toronto, Ontario: J.J. Douglas Ltd., 1973.

Middleton, Drew. *Crossroads of Modern Warfare.* Garden City, New Jersey: Doubleday, 1983.

Millarville History Society. *Foothills Echoes.* Millarville, Alberta: Millarville History Society, 1979.

Millarville, Kew, Priddis and Bragg Creek Historical Society. *Our Foothills.* Millarville,

Alberta: Recorded by the Millarville, Kew, Priddis and Bragg Creek Historical Society, 1975.

Mills, David. *The Place Names of Lancashire.* London, England: B.T. Batsford Ltd., 1976.

Mitchell, B.W. *Trail Life in the Canadian Rockies.* Toronto, Ontario: MacMillan Company, 1924.

Moberly, Henry John. *When Fur Was King.* London; Toronto: Dent, 1929.

Morgan, Henry James (ed.). *Canadian Men and Women of the Time.* Toronto, Ontario: William Briggs, 1912.

Moss, E.H. *Flora of Alberta.* Toronto, Ontario: University of Toronto Press, 1959.

Muller, Robert A. *Physical Geography Today: A Portrait of a Planet.* 2nd ed. New York: C.R.M. Random House, 1978.

Nanton and District Historical Society. *Mosquito Park Roundup: Nanton Parkland.* Nanton, Alberta: Nanton and District Historical Society, 1975.

Nicles, Trudy. "History of the Native Peoples of the Grande Cache Area." Grande Cache, Alberta: T. Nicles, April 1974.

Nordegg, Martin. *The Possibilities of Canada are Truly Great: Memoirs, 1906-1924.* Edited and with an introduction by T.D. Regehr. Toronto, Ontario: MacMillan of Canada, c1971.

Okotoks and District Historical Society. *A Century of Memories.* Okotoks, Alberta: Okotoks and District Historical Society, 1983.

Ommanney, Simon C.L. *Perennial Snow and Ice Features.* Ottawa, Ontario: Inland Waters Directorate, Water Resources Branch, 1976.

Outram, (Sir) James. *In the Heart of the Canadian Rockies.* New York: MacMillan, 1905.

Patterson Raymond M. *Far Pastures.* Sidney, British Columbia: Gray's Publishing, 1963.

Patton, Brian and Bart Robinson. *The Canadian Rockies Trail Guide.* Banff, Alberta: Summerthought, 1978.

_____. *Parkways of the Canadian Rockies.* Banff, Alberta: Summerthought, 1982.

Pincher Creek Historical Society. *Prairie Grass to Mountain Pass.* Pincher Creek, Alberta: Pincher Creek Historical Society, 1974.

Robbins, Keith. *The First World War.* Oxford, England: Oxford University Press, 1984.

Roberts, Charles G.D. and Arthur L. Tunnell (eds.). *A Standard Dictionary of Canadian Biography.* Toronto, Ontario: Trans Canada Press, 1934.

Rocky Mountain House Jubilee Committee. *Stories of the Pioneers of the West Country, 1912-1962.* Rocky Mountain House, Alberta: Rocky Mountain House Jubilee Committee, 1962.

Rocky Mountain House Reunion Historical Society. *The Days Before Yesterday.* Rocky Mountain House, Alberta: Rocky Mountain House Reunion Historical Society, 1977.

Ross, Alexander. *Fur Hunters of the Far West,* Volume II. London, England: Smith, Elder and Company, 1855.

Ross, Toni. *Oh! The Coal Branch.* Edmonton, Alberta: Mrs. Toni Ross, 1974.

Sandford Fleming, Pioneer Engineer. Marcia Shortreed (ed.). Inaugural lectures, Sandford Fleming Foundation, delivered at the University of Waterloo, March 1977. Sandford Educational Press.

Schäffer, Mary T.S. *Hunter of Peace: Mary T.S. Schäffer's Old Indian Trails of the Canadian Rockies.* E.J. Hart (ed.). Banff, Alberta: The Whyte Foundation, 1983.

Sessional Papers. 25B Volume XLI number 10, 1906-07. (Surveyor's Field Notes).

Sheep River Historical Society. *In the Light of the Flares: History of Turner Valley Oilfields.* Sheep River, Alberta: Sheep River Historical Society, 1979.

Sheppard, Bert. *Just About Nothing.* High River, Alberta: B. Sheppard, n.d.

_____. *Spitzee Days.* Calgary, Alberta: High River Historical Community; John D. McAra, 1971.

Southesk, Earl of. *Saskatchewan and the Rocky Mountains*. Edmonton, Alberta: M.G. Hurtig Ltd., 1969.

Spry, Irene (ed.). *The Palliser Papers*. Toronto, Ontario: The Champlain Society, 1968.

Stavely Historical Book Society. *The Butte Stands Guard: Stavely and District*. Stavely, Alberta: Stavely Historical Book Society, 1976.

Stelfox, H. *Rambling Thoughts of a Wandering Fellow, 1903-68*. Edmonton, Alberta: I.D.B. Printing, 1972.

Stovel, H.C. *50 Switzerlands in One: Banff the Beautiful; Canada's National Park: Where to Go and What to See in and Around Banff*. Banff, Alberta: Banff Board of Trade; Crag and Canyon, 1915(?).

Sykes, Egerton. *Everyman's Dictionary of Non-Classical Mythology*. London, England: J.M. Dent and Sons Ltd., 1952.

Thomas, Harriet Hartley. *From Barnacle to Banff: A Story of the Rising of the Rockies From the Depth of the Ocean to the Height of a World Famous Resort*. Revised edition. Calgary(?), Alberta: H.H. Thomas, 1973.

Thompson, Don W. *Men and Meridians: The History of Surveying and Mapping in Canada, Volume 2, 1867-1917*. Ottawa, Ontario: Queen's Printer, 1967.

Thompson, Ruby Trench. *Pioneers of Athabasca*. Aldewood Manor: Raymond Thompson Company, 1967.

Thorington, James M. *A Climber's Guide to the Rocky Mountains of Canada. 3rd edition*. New York: American Alpine Club, Inc., 1966.

_____. *The Glittering Mountains of Canada*. Philadelphia: John W. Lea, 1925.

Wallace, Stewart M. (ed.). *The MacMillan Dictionary of Canadian Biography*. Toronto, Ontario: MacMillan, 1963.

Webster's New Geographical Dictionary. Springfield, Mass.: Merriam-Webster Inc., 1972.

White, James. *Place Names in the Rocky Mountains Between the 40th Parallel and the Athabasca River*. Ottawa, Ontario: (Proceedings and Transactions/Royal Society of Canada; 3rd series, V.10), 1916.

Who was Who and Why, Volume XI (1917-18). Toronto, Ontario: International Press Limited, 1918(?)

Who was Who in America, Volume I (1897-1942). Chicago, Ill.: The A.N. Marquis Company, 1943.

Wilcox, Walter D. *The Rockies of Canada*. 3rd edition. New York: T.P. Putnam and Sons, 1909.

_____. "The Valley of the Hidden Lakes." Reprinted from the Bulletin of the Geographical Society of Philadelphia, Volume XX, numbers 3 and 4 (April and July), 1913-14.

Williams, M.B. *The Banff-Jasper Highway*. Ottawa, Ontario: Department of the Interior, 1948.

_____. *Jasper Trails*. Ottawa, Ontario: Department of the Interior, circa 1920.

_____. *Through the Heart of the Rockies and Selkirks*. 3rd edition. Ottawa, Ontario: Department of the Interior, 1924.

_____. *Waterton Lakes National Park*. Ottawa, Ontario: Department of the Interior, circa 1927; Reprinted 1982.

Wilson, T.E. *Trail Blazer of the Canadian Rockies*. Hugh Dempsey (ed.). Calgary, Alberta: Glenbow Alberta Institute, 1972.

PHOTOGRAPHS

*Mount Balfour
and Waputik Icefield*

Athabasca Falls

Barrier Lake

The Cave

Cirrus Mountain

Coral Creek Canyon and Coral Creek

Corona Ridge

Crescent Falls

Crowsnest Mountain

Mount Ernest Ross

Frank

Flathead Range

Island Lake

Kootenay Plains

Longview Hill

Lyons Creek

Lundbreck Falls

McConnel Falls

Mount Outram

Mount Michener

Shunda Lake

Siffleur River